Advances in Energy Harvesting Methods

Niell Elvin • Alper Erturk
Editors

Advances in Energy Harvesting Methods

Springer

Editors
Niell Elvin
The City College of New York
New York, NY
USA

Alper Erturk
Georgia Institute of Technology
Atlanta, GA
USA

The Chapter, "Stretching the Capabilities of Energy Harvesting: Electroactive Polymers based on Dielectric Elastomers," by Roy D. Kornbluh, Ron Pelrine, Harsha Prahlad, Annjoe Wong-Foy, Brian McCoy, Susan Kim, Joseph Eckerle and Tom Low was adapted from "Dielectric elastomers: Stretching the capabilities of energy harvesting," by Roy D. Kornbluh, Ron Pelrine, Harsha Prahlad, Annjoe Wong-Foy, Brian McCoy, Susan Kim, Joseph Eckerle and Tom Low, in MRS Bulletin, Volume 37 (March 2012), pp. 246–253. Reprinted with the permission of Cambridge University Press.

ISBN 978-1-4614-5704-6 ISBN 978-1-4614-5705-3 (eBook)
DOI 10.1007/978-1-4614-5705-3
Springer New York Heidelberg Dordrecht London

Library of Congress Control Number: 2012952283

Printed on acid-free paper

Springer is part of Springer Science+Business Media (www.springer.com)

Preface

For more than a decade, the research field of mechanical energy harvesting for low-power electricity generation has received growing interest by academia and industry. The ultimate goal in energy harvesting research is to enable energy-autonomous small electronic devices that can scavenge ambient mechanical energy and convert it to electrical power. The potential applications of energy harvesting technologies span over many different industries and areas of applications including: wireless sensor networks employed to monitor civil infrastructure systems, unmanned aerial vehicles, battery-free medical sensors implanted in the human body, and long-term sensors used for animal tracking.

By potentially eliminating the need to replace or periodically recharge the batteries in autonomous electrical devices, the research in energy harvesting has the potential to not only allow unprecedented monitoring of engineered and natural systems over almost arbitrarily long periods of time but also to achieve this goal both economically and sustainably.

Research into energy harvesting started with the fundamental and conceptual efforts for converting simple harmonic vibrations into electricity. However in the last couple of years, research efforts have focused on converting other forms of mechanical energy, such as impulse-type kinetic energy from human gait, random ambient vibrations, surface strain energy of civil engineering structures, wind and water flow energy, and acoustic energy of air-borne and structure-borne waves. Alternative materials and transduction mechanisms have also been investigated for mechanical-to-electrical energy conversion in addition to the developments in MEMS power harvester architectures and fabrication methods. Novel electrical architectures have allowed for improved mechanical to electrical conversion, and developments in ultralow-power integrated circuits have focused on being able to achieve greater electronic functionality with less electrical power.

Our aim with this volume is to bring together research advances in energy harvesting by focusing on different transduction mechanisms and forms of mechanical excitation. The title *Advances in Energy Harvesting Methods* therefore refers not only to the conversion (i.e., transduction) methods but also to the physical nature of the excitation and the system-level energy conversion problem (which is often a multi-physics problem), such as the harvesting of random vibrations through base excitation of a piezoelectric cantilever or that of airflow energy harvesting through aeroelastic vibrations based on various fluid–structure interaction mechanisms.

It is worth adding that this book is essentially focused on the harvesting of kinetic or strain energy induced within the harvester in a variety of multi-physics problems (through direct vibrations, flow excitation, sound waves, etc.) and excludes other contemporaneous energy scavenging methods, such as solar and thermoelectric energy harvesting. Likewise, the focus of the book is specifically placed on low-power energy harvesting. In this regard, for instance, the harvesting of flow energy discussed herein is not an alternative to replace large windmills and wind turbines, but it could rather be a component powering the wireless structural health monitoring sensors of such large systems to reduce their inspection and maintenance costs.

The state-of-the-art research efforts predominantly covered in this book include examples of energy scavenging using piezoelectric transduction, electromagnetic induction, electrostatic transduction, as well as electroactive polymer harvesting. Each one of these alternative methods of mechanical energy harvesting has its own advantages (and disadvantages) depending on the specific application that is being considered; for example, in many cases geometric scale of the application as well as the form and characteristics of the mechanical energy input could dictate the best transduction mechanism. To further elaborate on this point, a specific example is the case of piezoelectric transduction. In the last couple of years, most research in energy harvesting has focused on piezoelectric transduction due to its high-power density and ease of application. However, electrostatic energy conversion has peculiar advantages for MEMS fabrication and implementation while a magnet-coil arrangement that exploits electromagnetic induction can outperform piezoelectric transduction for low-frequency and large-displacement oscillations.

Although early efforts on vibration-based energy harvesting focused on simple tuned oscillators for exploiting the resonance phenomenon under harmonic excitation, researchers have recently directed their attention toward more sophisticated and commonly encountered forms of excitation such as ambient random vibrations. As could be expected, these more sophisticated forms of excitation (e.g., varying frequency or broadband excitation) require more advanced energy harvester configurations for effective power generation. Recent research in vibrational energy harvesting has thus focused on nonlinear scavenging systems with potentially improved performance under broadband (or wideband) excitation in comparison with their linear resonant counterparts. This book also devotes a number of chapters to recent efforts on broadband energy harvesting methods by exploiting nonlinear dynamic phenomena.

First we would like to thank the authors for their high-quality contributions. We would also like to thank Michael Luby and Merry Stuber of Springer for their interest and enthusiasm throughout the editing process of this volume. We sincerely hope you enjoy reading this collection of leading advances in energy harvesting.

New York, NY, USA Niell Elvin
Atlanta, GA, USA Alper Erturk

Contents

Part I
Introduction

Chapter 1
Introduction and Methods of Mechanical Energy Harvesting

Niell Elvin and Alper Erturk

Abstract The harvesting of various forms of mechanical energy, ranging from kinetic and surface strain energy to flow-induced aeroelastic and hydroelastic vibrations, has been investigated extensively over the last decade. The goal of this book is to cover the state-of-the-art research advances in energy harvesting with a focus on different transduction mechanisms and forms of mechanical excitation. The following chapters include various examples of energy scavenging using piezoelectric transduction, electromagnetic induction, electrostatic transduction, as well as electroactive polymer harvesting. The aim of this first chapter is to provide a brief introduction to the literature and fundamentals of energy harvesting methods discussed through this volume along with an outline of the present book.

1.1 Introduction

The goal in energy harvesting is to enable self-powered electronic devices by scavenging ambient energy for various wireless electronic applications ranging from structural health monitoring to medical implants [1–4]. The energy conversion methods that have been used for transforming mechanical (mostly vibrational and kinetic) energy into electrical energy are the piezoelectric [5–7], electromagnetic [8–11], electrostatic [12–15], and magnetostrictive [16] or magnetoelectric [17] composite-based conversion methods as well as the use of electroactive polymers,

---

N. Elvin (✉)
Department of Mechanical Engineering, The City College of New York, 160 Convent Avenue, Steinman Hall ST – 228, New York, NY10031, USA
e-mail: elvin@me.ccny.cuny.edu

A. Erturk
G. W. Woodruff School of Mechanical Engineering, Georgia Institute of Technology, 771 Ferst Dr. (Love Building) Room 126, Atlanta, GA30332, USA
e-mail: alper.erturk@me.gatech.edu

N. Elvin and A. Erturk (eds.), *Advances in Energy Harvesting Methods*,
DOI 10.1007/978-1-4614-5705-3_1,

such as dielectric elastomers [18, 19] and ionic polymer-metal composites [20–22]. Low-power electricity generation from kinetic energy and ambient vibrations can be argued to be a relatively well-established field with several review articles having already appeared in the literature [23–27].

It has been argued in several review articles [24, 25, 27] that piezoelectric energy harvesting remains the most widely researched harvesting method due to its ease of application, high voltage output without requiring post-processing for voltage multiplication or bias input, high-power density, as well as relatively mature thin-film and thick-film manufacturing methods [28, 29] that can be used for fabricating devices at different geometric scales. However, other techniques such as electromagnetic induction and electrostatic transduction methods also have specific advantages. For instance, although it remains a challenge to develop and fabricate effective MEMS electromagnetic energy harvesters due to the poor transduction properties of planar magnets and the limited number of induction loops for such small-scale devices [30, 31], electromagnetic induction is very convenient for harvesting kinetic energy with large deflections as long as the geometric scale allows for sufficiently strong electromechanical coupling with a proper coil-magnet arrangement. Likewise, in spite of its bias voltage requirement as a particular downside, electrostatic energy harvesters are very conveniently implemented MEMS fabrication techniques [14, 32]. Consequently, while focusing on various forms of excitation and the multi-physics aspects of power scavenging, throughout this book we intend to cover the advances in energy harvesting methods with examples involving various transduction mechanisms.

The examples of state-of-the-art energy harvesters discussed for different excitation forms (such as deterministic and stochastic) and different physical ambient conditions (such as flow excitation, human gait, and vibrations) in this text predominantly consider piezoelectric transduction, electromagnetic induction, electrostatic transduction, and electroactive polymer techniques. Therefore the following section provides a brief introduction to each of these energy conversion methods along with useful references from the respective literature. As previously mentioned, several review articles [23–27] are available for the reader's reference on the efforts of energy harvesting using these mechanical energy harvesting methods. This chapter ends with an outline of the book along with the motivation for each section.

1.2 Methods and Materials for Energy Conversion

1.2.1 Piezoelectric Transduction

Piezoelectricity is a two-way coupling between the mechanical and the electrical behaviors of materials belonging to certain classes, particularly ceramics and crystals. Piezoelectric materials, such as the ceramic PZT (lead zirconate titanate), the single crystal PMN-PT (lead magnesium niobate-lead titanate), or the semi-crystalline polymer PVDF (polyvinylidene difluoride), exhibit the so-called direct and converse piezoelectric effects. In the simplest terms, these materials produce electric polarization when strained mechanically (direct effect), and they become strained mechanically when subjected to electric polarization (converse effect). The concept of energy harvesting leverages primarily the direct piezoelectric effect to convert mechanical and structural vibrations (i.e., kinetic energy) into electricity. It is useful to note that the converse effect is still exhibited by the material during energy harvesting and manifests itself in the form of shunt damping in strongly coupled energy harvesters [4, 33].

Figure 1.1 shows various piezoelectric energy harvester configurations. If the poling axis and the mechanical strain axis are perpendicular to one another, the device uses the 31-mode of piezoelectricity (where the 3- and 1- axes are the poling and strain directions respectively). Examples are the bimorph cantilever[1] under base excitation (Fig. 1.1a) or a thickness-poled piezoelectric plate mounted on the surface of a large structure (such as a bridge or a pipe) to harvest strain fluctuations of the host structure (Fig. 1.1b). On the other hand, if the poling and strain axes are coincident, the piezoelectric device is said to be operating in the 33-mode. Typical examples are piezoelectric stacks (made of several thickness-poled piezoelectric layers) as depicted in Fig. 1.1c and monolithic cylindrical or cuboid piezoelectrics such as the thickness-poled annulus shown in Fig. 1.1d. Cantilevered energy harvesters employing interdigitated electrodes [34] also have their strain and electric field directions coincident and exploit the 33-mode in bending. In all cases, the electrode leads are connected to an electrical circuit, which could be as simple as an electrical resistor (which is often used for estimating AC power levels), or more complex circuits such as an AC–DC converter followed by a DC–DC regulator to charge a storage component [35, 36]. Typically, high voltage levels (on the order of volts to tens of volts) are extracted directly from the material itself without needing preconditioning (such as using a voltage multiplication circuit) and piezoelectric transduction does not require a bias voltage. The high voltage output is associated with a low current, which might occasionally be an issue if the leakage current level of the storage component (a battery or a capacitor) or preconditioning circuit is larger than the level of the generated current.

The cantilever configuration shown in Fig. 1.1a has been of particular interest in the energy harvesting community due to its excellent energy harvesting capability when designed to exploit resonance under harmonic excitation. The electrical power generation performance of resonant energy harvesters depend on the level of system's electromechanical coupling[2] (strong or weak) and mechanical damping (high-

---

[1]A bimorph is a configuration that uses an elastic substructure sandwiched between two thickness-poled piezoelectric layers; a unimorph (not discussed here) is composed of a single thickness-poled piezoelectric layer attached to an elastic substructure.

[2]Electromechanical coupling of a piezoelectric energy harvester depends not only on the amount of the piezoelectric material used but also on the structural design of the harvester (such as the location of the piezoelectric material on the cantilever and the way it is bonded to its substrate).

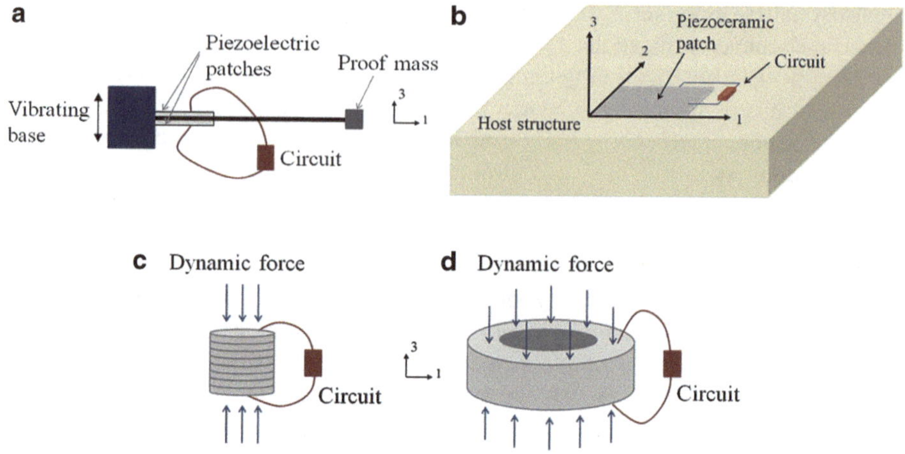

Fig. 1.1 Typical piezoelectric energy harvester configurations: (**a**) bimorph piezoelectric cantilever under base excitation and (**b**) piezoelectric patch harvesting surface strain energy of a large structure using the 31-mode; (**c**) multi-layer piezoelectric stack and (**d**) monolithic annulus under compressive loading using the 33-mode (3-direction is the poling direction in all cases)

or low mechanical quality factor) [4, 6, 37]. Strong electromechanical coupling and light mechanical damping (low-quality factor) are preferred for resonant energy harvesting. In contrast the mechanical quality factor of the harvester has little or no effect on the harvesting performance of a patch attached to a large structure (Fig. 1.1b) or to a compressed stack (Fig. 1.1c) at low frequencies (which are typical of off-resonant conditions that would be found in many applications such as for piezoelectric laminates embedded and strained in a shoe for power scavenging during walking).

1.2.2 Electromagnetic Induction

The conversion of kinetic or vibrational energy into electricity using electromagnetic induction exploits the well-known Faraday's law. That is electricity generation in electromagnetic (inductive) energy harvesting is due to the relative motion between a conductor (such as a coil) and a magnetic field (created by a magnet). Typically, a mechanical oscillator is designed to have a magnet-coil arrangement, which moves relative to each other in response to mechanical excitation. The configuration shown in Fig. 1.2 depicts an electromagnetic energy harvester design that consists of a cantilever combined with a magnet-coil arrangement. The permanent magnet rigidly attached at the tip of an elastic cantilever as an inertial mass (or proof mass) oscillates inside the fixed coil in response to base excitation [10, 11]. The alternative arrangement is to have a moving coil oscillating relative to a fixed

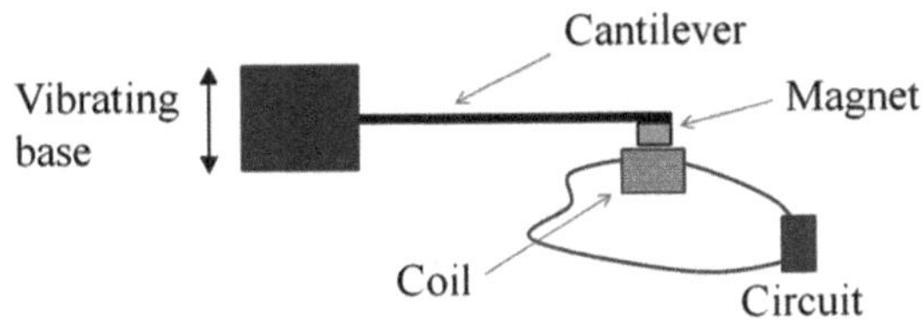

Fig. 1.2 Schematic of a typical electromagnetic energy harvester subjected to base excitation. The permanent magnet that serves as the proof mass of an elastic cantilever oscillates inside a coil due to the ambient base excitation; this relative motion between magnet and coil induced a current that is delivered to the electrical circuit

magnet [10]. The amount of electrical power output depends on the strength of the magnetic field, number of turns of the coil, and the relative velocity between the coil and the magnet. Under resonant operation, the power output is significantly affected by the quality factor of the mechanical oscillator and on the internal resistance of the generator coil.

As for piezoelectric transduction, electromagnetic induction does not require the device to have an initial bias voltage. Similarly, the oscillatory electrical response needs to be rectified and converted to a DC signal in order to charge a storage component. In contrast to piezoelectric transduction, electromagnetic energy harvesting results in low voltage and high current outputs (associated with a much lower optimal circuit resistance as compared to piezoelectric energy harvesters). Consequently, a voltage multiplier circuit is often required to reach the required voltage level of typical off-the-shelf storage components.

1.2.3 Electrostatic Transduction

In electrostatic (capacitive) energy harvesting, ambient mechanical vibrations are used to move the charged capacitor plates (or electrode fingers) of a variable capacitor against the electrostatic forces between the electrodes. These electrodes of the capacitor are separated by air, vacuum, or an insulating dielectric material. Typically a dielectric material is used to both increase the harvested energy and to prevent the capacitor faces from touching under the applied mechanical load [12–14]. The two most common approaches in this method of energy harvesting are charge-constrained and voltage-constrained mechanisms as detailed in [12, 23]. Unlike piezoelectric and electromagnetic energy harvesting methods, electrostatic energy harvesting requires a DC voltage supplied by a battery to oppositely charge the capacitor plates or electrode fingers (i.e. the so-called bias voltage). The oscillatory motion induced by the ambient vibration results in a cyclic variation of the capacitance of the device from a maximum to a minimum value, and the energy transfer per cycle is highly dependent on the ratio of this maximum to minimum capacitance [1].

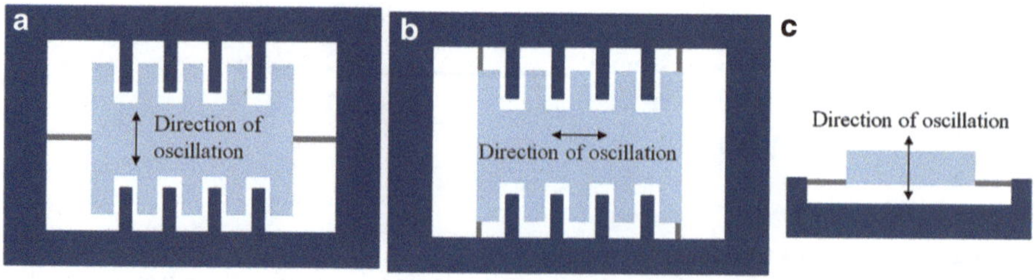

Fig. 1.3 Schematics of electrostatic energy harvesting configurations using in-plane and out-of-plane vibrations: (**a**) in-plane overlap varying; (**b**) in-plane gap closing; and (**c**) out-of-plane gap closing

The basic classification of electrostatic energy harvesters that exploit dynamic capacitance variation in response to mechanical excitation is given in Fig. 1.3 [1, 23]. Two of the configurations (Fig. 1.3b) in this figure employ in-plane vibrations of the electrode fingers while the third approach (Fig. 1.3c) uses the out-of-plane vibrations of the one of the capacitor plate relative to the other. The capacitance variation in Fig. 1.3b is due to the changing overlap area of the fingers and the changing gap between the fingers, respectively, while the gap closing between relatively large electrode plates causes the capacitance variation in Fig. 1.3c. All three configurations are capable of producing roughly the same power output [1]. The in-plane overlap and out-of-plane gap closing mechanisms require high spring deflections to create maximal power output. Particularly in the in-plane overlap mechanism, large deflections may create relative rotation of the oscillating component due to a lack of symmetry in the excitation, thus leading to the touching and possible shorting of the electrode fingers. Roundy et al. [1] suggest that the most favorable architecture is the in-plane gap closing configuration due to its smaller displacement requirement for the same power output, and hence more stable behavior as compared to the in-plane overlap varying mechanism. Furthermore the in-plane gap closing method has a higher power density as compared to the out-of-plane gap closing mechanism. Nevertheless both overlap varying and gap closing mechanisms have been widely researched especially in the MEMS energy harvesting literature [12, 13, 15].

1.2.4 Electroactive Polymers

The two types of electroactive polymers (EAPs) that have been studied for energy harvesting are dielectric elastomers (DEs) [18, 19] and ionic polymer-metal composites (IPMCs) [20–22]. According to the commonly used classification, DEs are *electronic* EAPs whereas IPMCs are *ionic* EAPs. Both electronic and ionic EAPs exhibit coupling between the mechanical stress (or strain) and electrical potential (or charge). The electromechanical coupling in electronic EAPs is due to

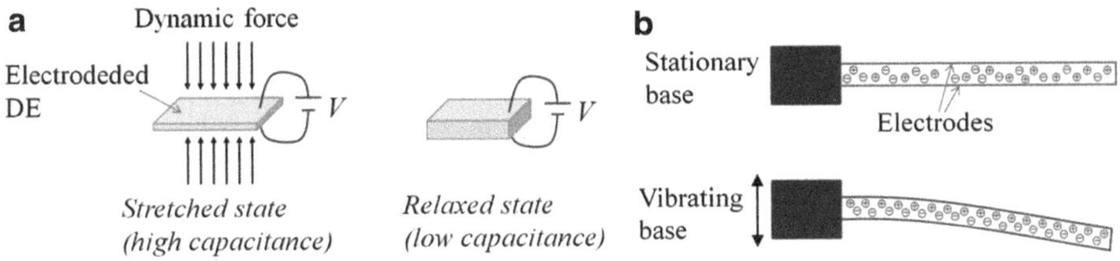

Fig. 1.4 Schematics of basic energy conversion mechanisms in DEs and IPMCs: (**a**) stretched and relaxed states of a DE in the presence of a bias voltage applied to the surface electrodes and (**b**) migration of free cations due to a charge concentration gradient in an IPMC cantilever resulting from bending deformation

polarization-based or electrostatic mechanisms while ionic EAPs exhibit electromechanical coupling due to diffusion or conduction of charged species in the polymer network [38].

A typical DE-based energy harvester is a soft polymer material, such as natural rubber, bracketed between two conductive electrodes (Fig. 1.4a). The mechanism of power generation using dielectric elastomers is analogous to that of electrostatic (capacitive) energy harvesting discussed in the previous section. DEs are elastically deformable insulators that respond to applied mechanical loading and can convert the mechanical work of the resulting deformation into electricity [19]. As in the case of electrostatic energy harvesting, a DC voltage input is required to oppositely charge the electrodes, and a cycle of energy harvesting involves a change of device capacitance from a maximum to a minimum value. A complete energy harvesting cycle consists of the following: (1) charging stage (to a high-charge value) by means of a battery associated with stretching of the DE and therefore increased capacitance, (2) switching to the open-circuit conditions and thickening of the DE at the fixed high-charge state, (3) switching to the output storage device and further thickening associated with loss of tension, hence capacitance reduction until the low-charge state, and (4) switching back to the open-circuit condition at the low-charge state that is associated with increased tension and reduced thickness [18]. In addition to the electromechanical properties of the polymer, various modes of failure (namely electrical breakdown, electromechanical instability, loss of tension, and rupture by overstretching) determine the limits of maximum energy conversion using DEs [18].

IPMCs are composed of ionic polymers, such as Nafion or Flemion, coated by conductive electrodes, which are typically made of gold or platinum. IPMCs exhibit a form of two-way coupling that is analogous to electromechanical coupling in piezoelectric materials. However, the mechanism of electromechanical coupling in IPMCs is based on the motion of ionic species upon application of an electric field or mechanical deformation. An IPMC-based energy harvester relies on the formation of a charge concentration gradient resulting from bending of the ionic polymer (Fig. 1.4b). The free cations within the IPMC migrate from the high-density to low-density region, resulting in an accumulation of charges at the electrode region and a potential difference across the electrodes [21, 39, 40]. Therefore, under dynamic bending of an IPMC cantilever in response to base excitation, an AC output can be generated analogous to piezoelectric energy harvesting. However, power density in IPMC-based energy harvesting is substantially lower (especially as compared to ceramic-based piezoelectrics) with typical power outputs on the order of nanowatts [41, 42].

1.3 Outline of the Book

As detailed in the next paragraphs of this introductory chapter, the major sections of this book are organized to include multiple chapters on the following subjects that cover the use of various transduction mechanisms for:

- Broadband energy harvesting and nonlinear dynamics
- Nonharmonic and spectral excitation
- Fluidic energy harvesting
- Advances in electronics
- Materials development and MEMS fabrication

Broadband energy harvesting and nonlinear dynamics: Research efforts toward enabling broadband (or wideband) energy harvesters are motivated by the limitations of well-studied resonant energy harvesters. Resonant energy harvesters (such as cantilevered oscillators) exploit the linear resonance phenomenon under harmonic excitation. A small mismatch between the excitation frequency and the mechanical resonant frequency of the harvester (due to manufacturing tolerances, ambient temperature fluctuations, or varying excitation frequency, among other reasons) results in drastically reduced power output, especially for lightly damped (i.e., high quality factor) harvesters. Exploiting nonlinear dynamic effects has been investigated as a promising way of enabling broadband energy harvesting. The goal is to enhance the frequency bandwidth by introducing nonlinearities through various mechanisms such as magnetoelastic buckling [43–45], purely elastic buckling [46, 47], stoppers [48, 49], and bilinear stiffness [50]. Following an extensive review article on the subject of broadband energy harvesting methods, this section of the book first covers broadband MEMS electrostatic energy harvesting using nonlinear springs and then bistable energy harvesters employing piezoelectric and electromagnetic transduction.

Nonharmonic and spectral excitation: As mentioned in the previous paragraph, the ambient mechanical excitation is often more sophisticated than a simple harmonic input. Therefore, in addition to enabling harvesters that can outperform the conventional designs under nonharmonic excitation, more advanced tools are required to model and analyze the electromechanical response to nonharmonic and arbitrary spectral forms of mechanical energy. Impulse-type excitation has broad frequency content and is most commonly associated with energy harvesting from human gait. Harvesting energy using shoe inserts strained during walking was previously studied in detail [51]. An alternative form covered in this section of the book employs a shoe-mounted cantilever that is exposed to base excitation associated with walking. The resulting excitation form resembles periodic impulse excitation of the energy harvester. Another way of harvesting gait energy disused here is achieved by focusing on the knee joint and employing the rotational excitation of piezoelectric bimorphs through plucking, that is essentially a mechanical frequency-up conversion technique. This section also investigates random excitation of energy harvesters with a special focus on stiffness nonlinearities, in particular bistable energy harvesters studied under harmonic excitation in the previous section.

Fluidic energy harvesting: The motivation in fluidic energy harvesting is to generate low-power electricity in fluidic media (e.g. air, water, etc.) by implementing an appropriate design and transduction mechanism. In-air and underwater flow energy harvesting through aeroelastic [52–54] and hydroelastic [55, 56] vibrations have been extensively researched by several groups. The goal in this research field is to create simple and efficient alternatives to small-scale windmills and wind turbines [57, 58]. The section starts with an investigation of energy harvesting from IPMCs in aquatic media. Two subsequent chapters then focus on harvesting wind energy through bluff body-based and airfoil-based energy harvesters. Finally, acoustic energy harvesting using sonic crystals is discussed as an alternative to well-investigated Helmholtz resonators [59].

Advances in electronics: In most cases, the mechanical energy harvesters discussed in this work require conversion of the alternating output (i.e. AC) to a stable DC signal in order to charge a storage component. Typically, AC–DC conversion is followed by DC–DC regulation to achieve the maximum power transfer to the electrical load [35, 36]. Performance enhancement in weakly coupled piezoelectric energy harvesters using switching circuits has been investigated in great detail in the existing literature [60], and there is currently significant research interest focused on the broad field of energy harvesting circuits. With this in mind, this section starts with an instructive review article on power conditioning circuits used in piezoelectric, electromagnetic, and electrostatic energy harvesting. After this fundamental review chapter, two subsequent chapters focus on novel integrated circuits used in MEMS piezoelectric and electrostatic energy harvesters.

Materials development and MEMS fabrication: Alternative materials and transduction methods are of interest in state-of-the-art energy harvesting research. Although not discussed in this book, the development of lead-free piezoelectric materials [61] with sufficiently strong electromechanical coupling is a present challenge to enable environment-friendly piezoelectric energy harvesters. Likewise, it is essential to enhance the magnetic properties in microscale devices to improve the feasibility of MEMS electromagnetic energy harvesting [30, 31]. We point our attention in this last section first to the relatively less studied yet promising dielectric elastomer-based energy harvesting. After that the focus is placed on materials and devices for MEMS piezoelectric energy harvesting, which is followed by a chapter on performance enhancement in energy harvesting using high permeability magnetic materials.

References

1. Roundy S, Wright PK, Rabaey JM (2004) Energy scavenging for wireless sensor networks: with special focus on vibrations. Kluwer Academic, Boston, MA
2. Priya S, Inman DJ (2009) Energy harvesting technologies. Springer, New York
3. Beeby S, White N (2010) Energy harvesting for autonomous systems. Artech House Publishers, Boston, MA
4. Erturk A, Inman DJ (2011) Piezoelectric energy harvesting. Wiley, Chichester
5. Roundy S, Wright PK (2004) A piezoelectric vibration based generator for wireless electronics. Smart Mater Struct 13:1131–1142
6. Dutoit NE, Wardle BL (2006) Performance of microfabricated piezoelectric vibration energy harvesters. Integr Ferroelectr 83:13–32
7. Erturk A, Inman DJ (2009) An experimentally validated bimorph cantilever model for piezoelectric energy harvesting from base excitations. Smart Mater Struct 18:025009
8. Williams CB, Yates RB (1996) Analysis of a micro-electric generator for microsystems. Sensors Actuators A Phys 52:8–11
9. Amirtharajah R, Chandrakasan AP (1998) Self-powered signal processing using SPIE vibration-based power generation. IEEE J Solid State Circuits 33:687–695
10. Glynne-Jones P, Tudor MJ, Beeby SP, White NM (2004) An electromagnetic, vibration-powered generator for intelligent sensor systems. Sens Actuators A Phys 110:344–349
11. Elvin NG, Elvin AA (2011) An experimentally validated electromagnetic energy harvester. J Sound Vib 330:2314–2324
12. Meninger S, Mur-Miranda JO, Amirtharajah R, Chandrakasan A, Lang JH (2001) Vibration-to-electric energy conversion. IEEE Trans VLSI Syst 9:64–76
13. Roundy S, Wright PK, Rabaey J (2003) A study of low level vibrations as a power source for wireless sensor nodes. Comput Commun 26:1131–1144
14. Mitcheson PD, Miao P, Stark BH, Yeatman E, Holmes A, Green T (2004) MEMS electrostatic micropower generator for low frequency operation. Sens Actuators A Phys 115:523–529
15. Tvedt LGW, Nguyen DS, Halvorsen E (2010) Nonlinear behavior of an electrostatic energy harvester under wide-and narrowband excitation. J Microelectromech Syst 19:305–316
16. Wang L, Yuan F (2008) Vibration energy harvesting by magnetostrictive material. Smart Mater Struct 17:045009
17. Dong S, Zhai J, Li J, Viehland D, Priya S (2008) Multimodal system for harvesting magnetic and mechanical energy. Appl Phys Lett 93:103511
18. Koh SJA, Zhao X, Suo Z (2009) Maximal energy that can be converted by a dielectric elastomer generator. Appl Phys Lett 94:262902
19. Kornbluh RD, Pelrine R, Prahlad H, Wong-Foy A, Mccoy B, Kim S, Eckerle J, Low T (2011) From boots to buoys: promises and challenges of dielectric elastomer energy harvesting. Proceedings of SPIE 7976, 797605
20. Tiwari R, Kim KJ, Kim SM (2008) Ionic polymer-metal composite as energy harvesters. Smart Mater Struct 4:549–563
21. Brufau-Penella J, Puig-Vidal M, Giannone P, Graziani S, Strazzeri S (2008) Characterization of the harvesting capabilities of an ionic polymer metal composite device. Smart Mater Struct 17:015009
22. Aureli M, Prince C, Porfiri M, Peterson SD (2010) Energy harvesting from base excitation of ionic polymer metal composites in fluid environments. Smart Mater Struct 19:015003
23. Beeby SP, Tudor MJ, White NM (2006) Energy harvesting vibration sources for microsystems applications. Meas Sci Technol 17:R175–R195
24. Anton SR, Sodano HA (2007) A review of power harvesting using piezoelectric materials (2003–2006). Smart Mater Struct 16:R1–R21
25. Priya S (2007) Advances in energy harvesting using low profile piezoelectric transducers. J Electroceram 19:167–184

26. Hudak NS, Amatucci GG (2008) Small-scale energy harvesting through thermoelectric, vibration, and radiofrequency power conversion. J Appl Phys 103:101301
27. Cook-Chennault KA, Thambi N, Sastry AM (2008) Powering MEMS portable devices – a review of non-regenerative and regenerative power supply systems with special emphasis on piezoelectric energy harvesting systems. Smart Mater Struct 17:043001
28. Choi WJ, Jeon Y, Jeong JH, Sood R, Kim SG (2006) Energy harvesting MEMS device based on thin film piezoelectric cantilevers. J Electroceram 17:543–548
29. Fang HB, Liu JQ, Xu ZY, Dong L, Wang L, Chen D, Cai BC, Liu Y (2006) Fabrication and performance of MEMS-based piezoelectric power generator for vibration energy harvesting. Microelectron J 37:1280–1284
30. Kulah H, Najafi, K (2004) An electromagnetic micro power generator for low-frequency environmental vibrations. IEEE 237–240
31. Koukharenko E, Beeby S, Tudor M, White N, O'Donnell T, Saha C, Kulkarni S, Roy S (2006) Microelectromechanical systems vibration powered electromagnetic generator for wireless sensor applications. Microsyst Technol 12:1071–1077
32. Torres EO, Rincon-Mora GA (2006) Electrostatic energy harvester and li-ion charger circuit for micro-scale applications. IEEE 65–69
33. Lesieutre GA, Ottman GK, Hofmann HF (2004) Damping as a result of piezoelectric energy harvesting. J Sound Vib 269:991–1001
34. Jeon Y, Sood R, Jeong J, Kim SG (2005) MEMS power generator with transverse mode thin film PZT. Sens Actuators A Physical 122:16–22
35. Ottman GK, Hofmann HF, Bhatt AC, Lesieutre GA (2002) Adaptive piezoelectric energy harvesting circuit for wireless remote power supply. IEEE Trans Power Electron 17:669–676
36. Ottman GK, Hofmann HF, Lesieutre GA (2003) Optimized piezoelectric energy harvesting circuit using step-down converter in discontinuous conduction mode. IEEE Trans Power Electron 18:696–703
37. Shu Y, Lien I (2006) Analysis of power output for piezoelectric energy harvesting systems. Smart Mater Struct 15:1499
38. Leo DJ (2007) Engineering analysis of smart material systems. Wiley, Hoboken, NJ
39. Sadeghipour K, Salomon R, Neogi S (1992) Development of a novel electrochemically active membrane and 'smart' material based vibration sensor/damper. Smart Mater Struct 1:172
40. Shahinpoor M, Bar-Cohen Y, Simpson J, Smith J (1998) Ionic polymer-metal composites (IPMCs) as biomimetic sensors, actuators and artificial muscles-a review. Smart Mater Struct 7:R15
41. Farinholt KM, Pedrazas NA, Schluneker DM, Burt DW, Farrar CR (2009) An energy harvesting comparison of piezoelectric and ionically conductive polymers. J Intell Mater Syst Struct 20:633–642
42. Giacomello A, Porfiri M (2011) Underwater energy harvesting from a heavy flag hosting ionic polymer metal composites. J Appl Phys 109:084903
43. Cottone F, Vocca H, Gammaitoni L (2009) Nonlinear energy harvesting. Phys Rev Lett 102:080601
44. Erturk A, Hoffmann J, Inman DJ (2009) A piezomagnetoelastic structure for broadband vibration energy harvesting. Appl Phys Lett 94:254102
45. Stanton SC, McGehee CC, Mann BP (2010) Nonlinear dynamics for broadband energy harvesting: investigation of a bistable piezoelectric inertial generator. Physica D 239:640–653
46. Arrieta AF, Hagedorn P, Erturk A, Inman DJ (2010) A piezoelectric bistable plate for nonlinear broadband energy harvesting. Appl Phys Lett 97:104102
47. Masana R, Daqaq MF (2011) Relative performance of a vibratory energy harvester in mono- and bi-stable potentials. J Sound Vib 330:6036–6052
48. Soliman M, Abdel-Rahman E, El-Saadany E, Mansour R (2008) A wideband vibration-based energy harvester. J Micromech Microeng 18:115021
49. Blystad LCJ, Halvorsen E (2011) A piezoelectric energy harvester with a mechanical end stop on one side. Microsyst Technol 17:505–511

50. Sousa VC, Anicezio MD, De Marqui C, Erturk A (2011) Enhanced aeroelastic energy harvesting by exploiting combined nonlinearities: theory and experiment. Smart Mater Struct 20:094007
51. Shenck NS, Paradiso JA (2001) Energy scavenging with shoe-mounted piezoelectrics. IEEE Micro 21:30–42
52. Akaydin HD, Elvin N, Andreopoulos Y (2010) Wake of a cylinder: a paradigm for energy harvesting with piezoelectric materials. Exp Fluids 49:291–304
53. Erturk A, Vieira WGR, De Marqui C, Inman DJ (2010) On the energy harvesting potential of piezoaeroelastic systems. Appl Phys Lett 96:184103
54. Bryant M, Garcia E (2011) Modeling and testing of a novel aeroelastic flutter energy harvester. J Vib Acoust 133:011010
55. Allen JJ, Smits AJ (2001) Energy harvesting eel. J Fluids Struct 15:629–640
56. Pobering S, Ebermeyer S, Schwesinger N (2009) Generation of electrical energy using short piezoelectric cantilevers in flowing media. Proceedings of SPIE 7288, 728807
57. Myers R, Vickers M, Kim H, Priya S (2007) Small scale windmill. Appl Phys Lett 90:054106-054106-3
58. Xu F, Yuan F, Hu J, Qiu Y (2010) Design of a miniature wind turbine for powering wireless sensors. Proceedings of SPIE 7647, 764741
59. Liu F, Phipps A, Horowitz S, Ngo K, Cattafesta L, Nishida T, Sheplak M (2008) Acoustic energy harvesting using an electromechanical Helmholtz resonator. J Acoust Soc Am 123:1983
60. Guyomar D, Badel A, Lefeuvre E, Richard C (2005) Toward energy harvesting using active materials and conversion improvement by nonlinear processing. IEEE Trans Ultrason Ferroelectr Freq Control 52:584–595
61. Shrout TR, Zhang SJ (2007) Lead-free piezoelectric ceramics: alternatives for PZT? J Electroceram 19:113–126

Part II
Broadband Energy Harvesting and Nonlinear Dynamics

Chapter 2
Broadband Vibration Energy Harvesting Techniques

Lihua Tang, Yaowen Yang, and Chee Kiong Soh

Abstract The continuous reduction in power consumption of wireless sensing electronics has led to immense research interests in vibration energy harvesting techniques for self-powered devices. Currently, most vibration-based energy harvesters are designed as linear resonators that only work efficiently with limited bandwidth near their resonant frequencies. Unfortunately, in the vast majority of practical scenarios, ambient vibrations are frequency-varying or totally random with energy distributed over a wide frequency range. Hence, increasing the bandwidth of vibration energy harvesters has become one of the most critical issues before these harvesters can be widely deployed in practice. This chapter reviews the advances made in the past few years on this issue. The broadband vibration energy harvesting techniques, covering resonant frequency tuning, multimodal energy harvesting, and nonlinear energy harvesting configurations are summarized in detail with regard to their merits and applicability in different circumstances.

2.1 Introduction

Portable devices and wireless sensors are conventionally powered by chemical batteries. The use of batteries not only leads to their costly replacement especially for sensors at inaccessible locations, but also causes pollution to the environment. Besides, batteries also place limitation on the miniaturization of micro- or nano-electromechanical systems. With the advances in integrated circuits, the size and power consumption of current electronics have dramatically decreased. For example, a wireless sensor now can be powered at less than 100 μW. Hence, in the past few years, ambient energy harvesting as power supplies for small-scale

L. Tang • Y. Yang (✉) • C.K. Soh
School of Civil and Environmental Engineering, Nanyang Technological University,
50 Nanyang Avenue, Singapore, Singapore 639798
e-mail: tanglh@ntu.edu.sg; cywyang@ntu.edu.sg; csohck@ntu.edu.sg

N. Elvin and A. Erturk (eds.), *Advances in Energy Harvesting Methods*,
DOI 10.1007/978-1-4614-5705-3_2,

electronics has evoked great research interest from various disciplines, including material science, mechanical, civil, and electrical engineering.

Different energy sources existing in the environment around a system, such as sunlight, wind, and mechanical vibration, can be the options for energy harvesting. Among them, pervasive vibration sources are suitable for small-scale power generation of low-power electronics and thus have attracted more research attention. Current solutions for vibration-to-electricity transduction are mostly accomplished via electrostatic [1, 2], electromagnetic [2, 3], or piezoelectric methods [4, 5]. Various models, including analytical models [2, 6], finite element models ([3, 7]) and equivalent circuit models [8, 9], have been established to investigate the energy harvesting capability of each method. No matter which principle was exploited, most of the previous research work focused on designing a linear vibration resonator, in which the maximum system performance is achieved at its resonant frequency. If the excitation frequency slightly shifts, the performance of the harvester can dramatically decrease. Since the majority of practical vibration sources are present in frequency-varying or random patterns, how to broaden the bandwidth of vibration energy harvesters becomes one of the most challenging issues before their practical deployment.

This chapter presents a review of recent advances in broadband vibration energy harvesting. The state-of-the-art techniques in this field, covering resonant frequency tuning, multimodal energy harvesting, and nonlinear energy harvesting configurations, are summarized in detail with regard to their merits and applicability in different circumstances.

2.2 Resonant Frequency Tuning Techniques

When the excitation frequency is known *a priori*, the geometry and dimensions of a conventional linear harvester can be carefully selected to match its resonant frequency with the excitation frequency. However, when the excitation frequency is unknown or varies in different operational conditions, the harvester with pre-tuned resonant frequency is unable to achieve optimal power output. Hence, in practice a conventional linear harvester is expected to incorporate a resonance tuning mechanism to increase its functionality. According to Roundy and Zhang [10], the resonance can be tuned "*actively*" or "*passively*". The active mode requires continuous power input for resonance tuning. While in the passive mode, intermittent power is input for tuning and no power is required when frequency matching is completed, that is until the excitation frequency varies again.

Resonance tuning methods can be categorized into mechanical, magnetic, and piezoelectric methods. Furthermore, the tuning process can be implemented manually or in a self-tuning way. Manual tuning is very difficult to implement during operation. A fine self-tuning implementation is expected not only to cover the targeted frequency range but also to be capable of self-detecting the frequency

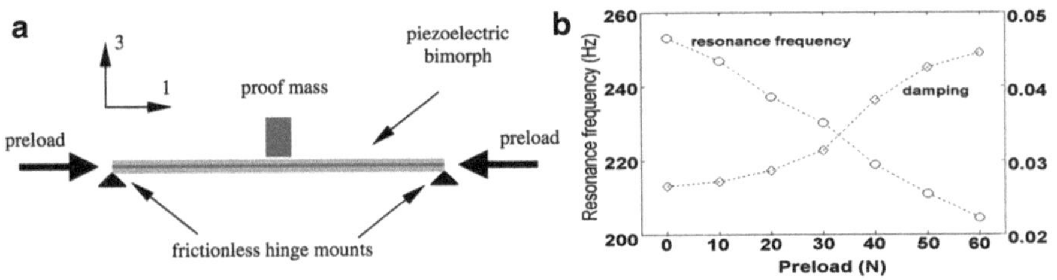

Fig. 2.1 (**a**) Schematic of a simply supported bimorph energy harvester and (**b**) resonance frequency and damping ratio versus tuning preload ([11], copyright: IOP Publishing)

change, automatic control, and of being self-powered consuming as little (harvested energy as possible).

2.2.1 Mechanical Methods

2.2.1.1 Manual Tuning

From elementary of vibration theory, the resonance of a system can be tuned by changing the stiffness or mass. Usually, it is more practical to change the stiffness rather than the mass of the system. Leland and Wright [11], Eichhorn et al. [12], and Hu et al. [13] proposed to apply axial preload to alter the stiffness in their energy harvesting devices, thus tuning the resonant frequencies. In Leland and Wright's work, an axial compressive load was applied on a simply supported bimorph energy harvester (Fig. 2.1a). In their experimental test on the prototype with a 7.1 g proof mass, it was determined that before the bimorph failure, a compressive axial preload can reduce its resonant frequency by up to 24%. Over the frequency range of 200–250 Hz, this protype achieved a power output of 300–400 μW under a 1g excitation acceleration. The power output was relatively flat over this range and even decreased at low frequencies, which could be explained by the increased damping ratio due to the applied preload, as shown in Fig. 2.1b. Besides, the design presented was intended to operate in "passive" mode, where the preload was manually tuned. However, the energy required for the tuning procedure was not addressed. Furthermore, the resonant frequency could only be tuned unidirectionally since only the compressive preload was considered.

Eichhorn et al. [12] presented a cantilever tunable energy harvester by applying prestress at its free end. Figure 2.2a shows the generator and the schematic of the entire setup. The arms connected the tip of the beam and two wings. A revolution of the screw generated compression on the spring, which applied the force on the arms. This force was then applied to the free end of the cantilever through the wings. Below the fracture limit, a resonance shift from 380 to 292 Hz was achieved by applying preload from 7 N to 22.75 N, as shown in Fig. 2.2b. The quality factor

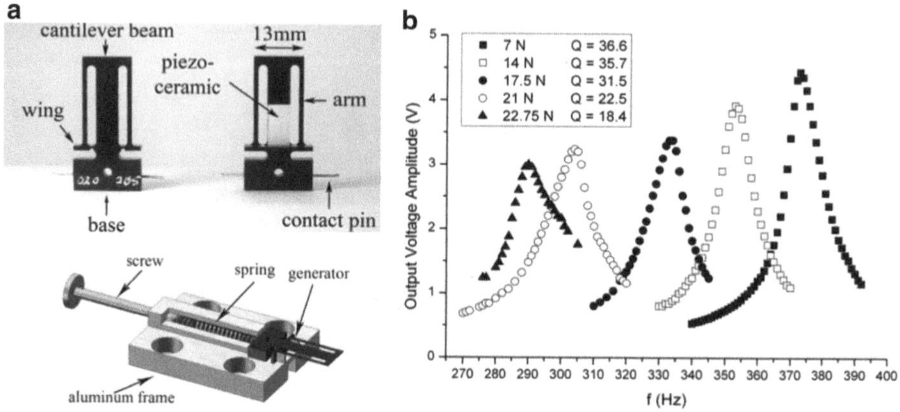

Fig. 2.2 (**a**) Schematic of tunable generator and entire setup and (**b**) resonance curves with various prestresses [12]

was reduced, which means damping arose with increased compression, similar to the finding by Leland and Wright [11].

Analytically, Hu et al. [13] derived the governing equations of a cantilever piezoelectric bimorph with an axial preload and investigated its feasibility and resonance characteristics. The resonance can be adjusted either higher or lower with a tensile or compressive load, respectively. In their model, it was reported that a tensile load of 50 N increased the resonance from 129.3 to 169.4 Hz while the same compressive load decreased the resonance from 129.3 to 58.1 Hz.

Instead of considering the bending mode, some researchers have investigated a tunable resonator working in extensional mode, termed XMR [14, 15]. The XMR presented by Morris et al. [14] was formed by suspending a seismic mass with two piezoelectric membranes (PVDF). Pretensioning two rectangular membranes (with dimensions of $2\,l \times w \times h$ and Young's modulus E) by a rigid link with length of $2u_{\mathrm{p}}$ and deflecting the link by Δu, as shown in Fig. 2.3a, the force–deflection characteristics of the rigid link were found to be

$$F = \frac{Ewh}{l^3}\left(6u_{\mathrm{p}}^2\Delta u + 2\Delta u^3\right). \tag{2.1}$$

For sufficiently small deflection, the natural frequency can be approximated as

$$f_{\mathrm{N}} = \frac{1}{2\pi}\sqrt{\frac{k}{m}} = \frac{1}{2\pi}\sqrt{\frac{\frac{\mathrm{d}F}{\mathrm{d}(\Delta u)}}{m}} \approx u_{\mathrm{p}}\frac{1}{2\pi}\sqrt{\frac{6Ewh}{ml^3}}. \tag{2.2}$$

Hence, the resonant frequency can be tuned by adjusting the link length that symmetrically pretensions both piezoelectric sheets. Similar force–deflection relationships and natural frequency expressions can be found for other rigidly

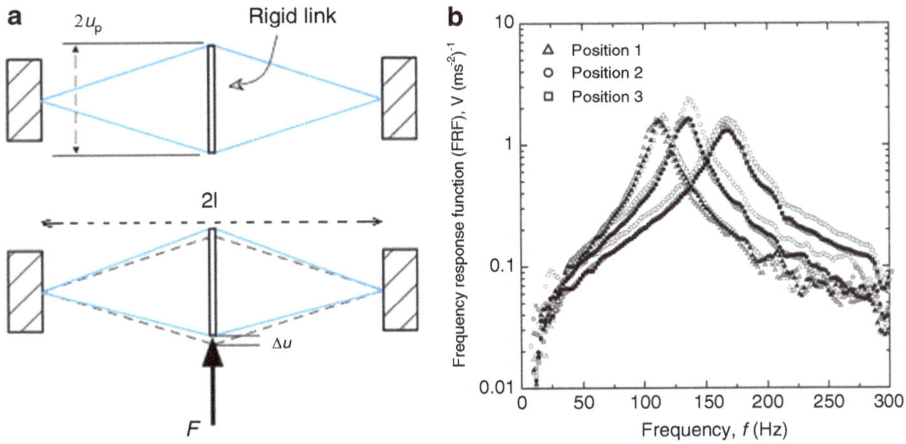

Fig. 2.3 (**a**) Schematic of XMR with two pretensioned membranes by a rigid link and (**b**) frequency responses for three adjustment positions ([14], copyright: IOP Publishing)

coupled and transversely loaded membrane. For the fabricated XMR prototype with a circular configuration, the frequency response functions were obtained by tuning the preloading screw at three random adjustment positions, as shown in Fig. 2.3b. For the developed prototype, it was found that a resonance shift between 80 and 235 Hz can be easily achieved with a change of pretension displacement of around 1.25 mm. Morris et al. [14] claimed that this was not the upper limit of their XMR, which would be constrained by the mechanical failure of the device. However, the capability of self-tuning or sequential tuning during operation of the XMR was not investigated.

A similar investigation was pursued by Loverich et al. [16], in which the resonance can be tuned by adjusting the pre-deflection of the circular plate, as shown in Fig. 2.4. The resonant frequency could be experimentally varied between 56 and 62 Hz by adjusting the boundary location by approximately 0.5 mm. Furthermore, they also made use of nonlinearity of the pre-deflected plate. Similar force–deflection characteristics were obtained as Eq. (2.1). It was found that the stiffness was nearly linear and the system had a high quality factor Q for low vibration amplitudes, while the resonance frequency shifted and Q reduced for high vibration amplitudes. This feature of nonlinear stiffness also provided an auto-protection mechanism, which is important when mechanical robustness is required for high vibration levels.

Rather than applying the axial or in-plane preload, adjusting the gravity center of the tip mass is another idea to adjust the resonance of a cantilever. Wu et al. [17] presented such a device in which the proof mass consisted of a fixed part and a movable part, as shown in Fig. 2.5a. The gravity center of the whole proof mass can be adjusted by driving the movable screw. The mass of the fixed part was much lighter than that of the movable part such that the adjustable distance of the gravity

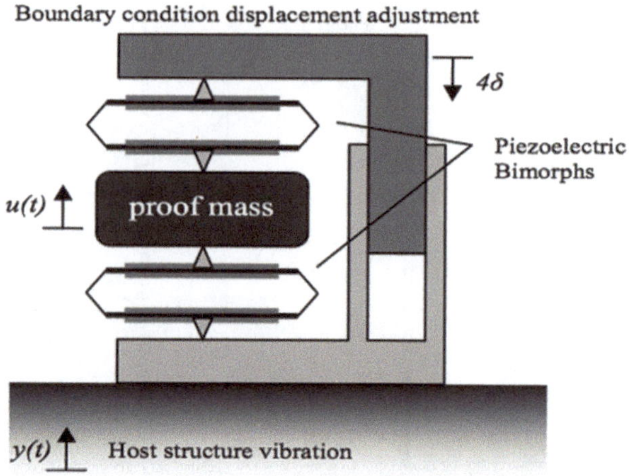

Fig. 2.4 Energy harvester configuration with adjustable boundary condition for inducing large deformation in bimorph plates ([16], copyright: SPIE)

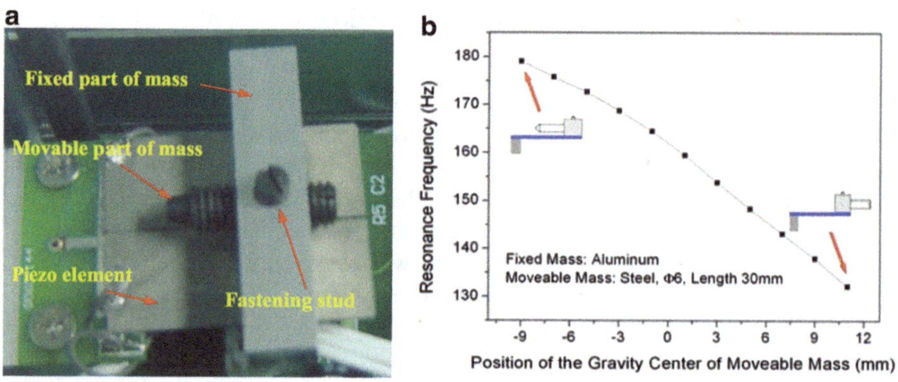

Fig. 2.5 (**a**) Piezoelectric harvester with moveable mass and (**b**) its resonant frequency versus position of gravity center of moveable mass [17]

center of the proof mass and in turn the frequency tunability can be increased. In their prototype, the adjustable resonant frequency range could cover 130–180 Hz by tuning the gravity center of the tip mass up to 21 mm, as illustrated in Fig. 2.5b.

2.2.1.2 Self-Tuning

In Sect. 2.2.1.1, the resonance tuning of the reported devices were implemented manually (usually using screws), which is not favorable for real-time application during operation. To address this problem, some researchers [18, 19] developed novel passive self-tuning harvesters.

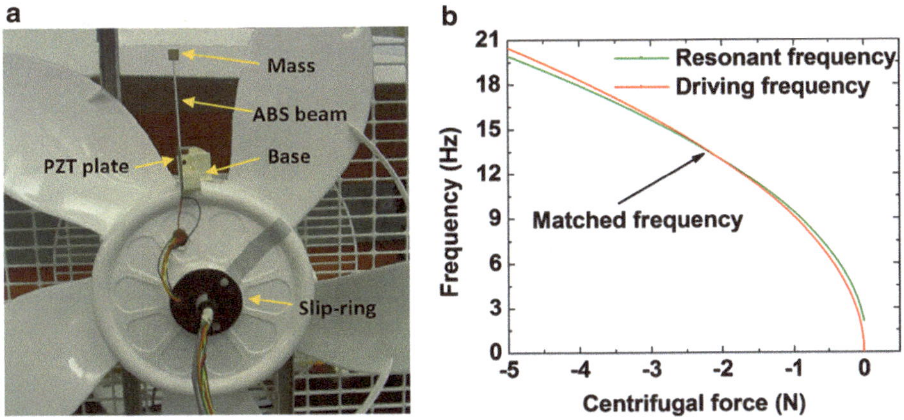

Fig. 2.6 (**a**) Experiment setup for harvesting energy from rotational vibration and (**b**) predicted driving frequency and resonant frequency versus centrifugal force ([18], copyright: American Institute of Physics)

Jo et al. [19] presented a design that was composed of a cantilever couple with different lengths. This cantilever couple was movable laterally and had two operational phases. The horizontal inertia forces exerted to two equal proof masses change with the excitation frequency and become maximum when the excitation frequency matches the resonant frequency. The difference between the horizontal inertia forces is the key to switch the harvester between the two phases. This harvester is self-tunable and no power is required in the tuning procedure. Each cantilever has two resonant peaks as the excitation frequency changes. Although the resonant frequency only switches between two phases and thus can not cover a continuous frequency range, such device is still significantly more efficient than a conventional cantilever harvester without a self-tuning mechanism.

Different from previous research on harvesting energy from translational base excitation, Gu and Livermore [18] focused on rotational motion. A passive self-tuning piezoelectric energy harvester was designed in which centrifugal force was exploited to adjust the stiffness and thus its resonant frequency. The harvester consisted of a radially oriented cantilever beam mounted on a rotational body, as shown in Fig. 2.6a. Since the centrifugal force was related to both driving frequency and resonant frequency of the harvester, the harvester could be designed such that the resonant frequency was exactly equal to the driving frequency at 13.2 Hz. In addition, within a range of 6.2 Hz–16.2 Hz, the two frequencies matched well, as shown in Fig. 2.6b. Thus, the harvester always worked at or near its resonance. In their experiment, the self-tuning harvester could achieve a much wider bandwidth of 8.2 Hz as compared to 0.61 Hz of the untuned harvester. However, this device is only applicable for rotational motion.

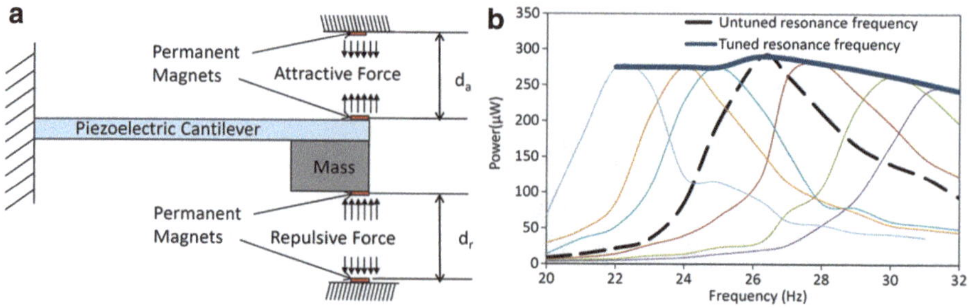

Fig. 2.7 (**a**) Schematic of resonance tunable harvester and (**b**) power output versus tuned resonant frequency in experiment ([20], copyright: IOP Publishing)

2.2.2 *Magnetic Methods*

2.2.2.1 Manual Tuning

Applying magnetic force to alter the effective stiffness of a harvester is another option for resonance tuning. Challa et al. [20] proposed a tunable cantilever harvester in which two magnets were fixed at the free end of the beam, while the other two magnets were fixed at the top and bottom of the enclosure of the device, as shown in Fig. 2.7a. All magnets were vertically aligned so that attractive and repulsive magnetic forces could be generated on each side of the beam. By tuning the distance between the magnets using a spring-screw mechanism, the prototype with a volume of 50 cm^3 was tunable over the range of 22–32 Hz with a power output of 240–280 μW operating at an acceleration of 0.8 m/s^2. Power output was undermined as the damping increased during the tuning procedure, as shown in Fig. 2.7b. Given the maximum tuning distance of 3 cm, the required energy was estimated to be 85 mJ and it would take 320 s for each tuning procedure. Thus such device can only work where the excitation frequency changes infrequently.

Reissman et al. [21] demonstrated a tuning technique using variable attractive magnetic force, as shown in Fig. 2.8a. The resonance of the piezoelectric energy harvester could be tuned bidirectionally by adjusting a magnetic slider. This is a much simplified design as compared to the design of Challa et al. [20]. The effective stiffness of the piezoelectric beam was dependent on the structural component K_m, the electromechanical component K_e that varied with external resistive loading R_l, and the magnetic stiffness $K_{magnetic}$ that varied with the relative distance D between the two magnets, i.e.,

$$K_{eff} = (K_m + K_e(R_l)) + K_{magnetic}(D). \tag{2.3}$$

By tuning the vertical relative distance D_y of the two magnets, the resonance could be tuned bidirectionally. For a specific D_x, the maximum frequency achieved was 99.38 Hz at $D_y = 0$, and the lowest frequency was 88 Hz at $D_y = 1.5$ cm, as

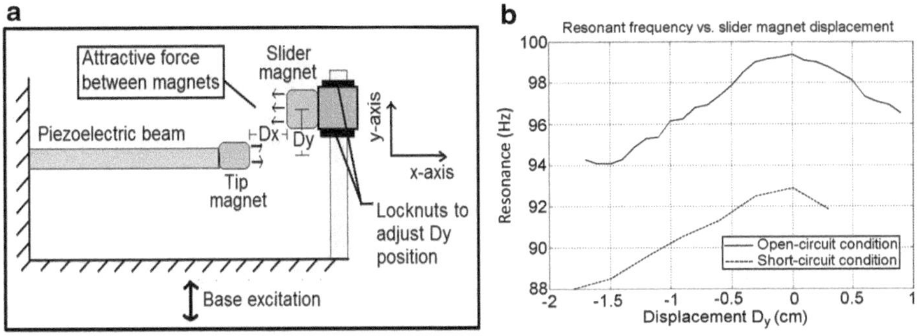

Fig. 2.8 (**a**) Schematic of proposed resonance-tunable harvester and (**b**) open- and short-circuit resonant frequencies with variable D_y ([21], copyright: SPIE)

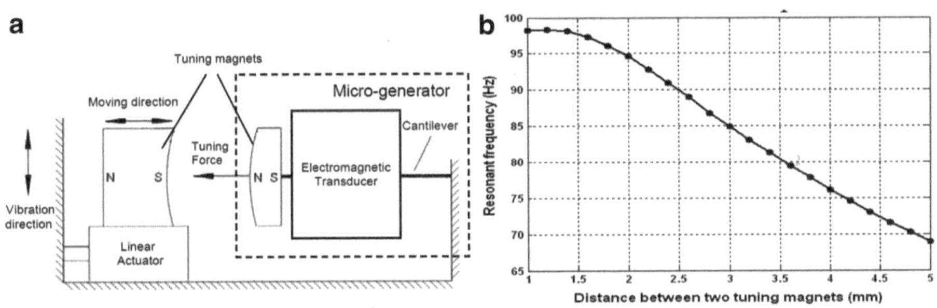

Fig. 2.9 (**a**) Schematic of tuning mechanism and (**b**) resonant frequency versus distance between two magnets [22]

shown in Fig. 2.8b. Hence, the total bandwidth of the harvester was 11.38 Hz, including the resonant frequency shift from short-circuit to open-circuit condition due to the piezoelectric coupling.

In the aforementioned two designs, no "smart" controller for resonance tuning process was implemented.

2.2.2.2 Self-Tuning

Zhu et al. [22] proposed a similar setup as Reissman et al. [21], but they further implemented an automatic controller for resonance tuning. A schematic of the tuning mechanism is shown in Fig. 2.9a. The microcontroller woke up periodically, detected the output voltage of the generator and gave instructions to drive a linear actuator to adjust the distance *D* between the two magnets. In their experimental test, the resonant frequency was tuned from 67.6 to 98 Hz when *D* was changed from 5 to 1.2 mm, but it could not be further increased when *D* was smaller than 1.2 mm, as shown in Fig. 2.9b. At a constant acceleration of 0.588 m/s^2, the power output of 61.6–156.6 μW over the tuning range could be achieved. They found that the damping of the micro-generator was not affected by the tuning mechanism over most of the tuning range. However, the damping was increased and the output power was less than expected if the tuning force became larger than the inertial force caused by vibration. Additionally, the energy consumed for the tuning procedure in their design was 2.04 mJ mm^{-1}. They claimed that the linear actuator and microcontroller would be ultimately powered by the generator itself to form a closed-loop tuning system. However, experimentally, the tuning system was still powered by a separate power supply for preliminary evaluation. Another drawback was that the control system detected the resonance by comparing the output voltage with a predefined threshold. Thus, such a system could suffer from inefficient detection of the frequency change direction and from mistaken triggering if there was certain change in the excitation amplitude.

Following the work of Zhu et al. [22], Ayala-Garcia et al. [23] presented an improved tunable kinetic energy harvester based on thc same tuning mechanism. The phase difference between the harvester and the base was measured in the closed-loop control, which was used to indicate the direction to tune the magnets. A tuning range of 64.1–77.86 Hz (i.e., bandwidth of 13.76 Hz) was achieved by varying the distance between magnets from 5 to 3 mm. However, under the excitation of 0.588 m/s^2, this device required more than 2 h to accumulate enough energy in the supercapacitor of 0.55 F for one tuning process. Challa et al. [24] also improved their previous design [20] by implementing an automatic control system for the tunable harvester. In this improved version, the output power of 736 μW–1 mW was achieved over the tuning range of 13–22 Hz. However, the energy of 3.2–3.9 J was consumed during the tuning process, which required 72–88 min to recover for the next tuning.

Although the above magnetic tuning harvesters implemented automatic control systems, they were not self-powered. Hence, strictly speaking, they have not achieved complete "self-tuning." Besides, the required energy for one tuning procedure is a huge burden in these devices, thus they are only suitable for the vibration scenarios where small and infrequent frequency changes occur. The magnets and control systems also increase the complexity of system design and integration.

2.2.3 *Piezoelectric Methods*

A piezoelectric transducer used as an actuator can alter the stiffness of a system. In fact, the stiffness of the piezoelectric material itself can be varied with various shunt electrical load. Hence, piezoelectric transducers provide another option for resonance tuning. It should be emphasized that the notion "*piezoelectric methods*" refers to the methods for resonance tuning using piezoelectric transducers. The energy generation method could be electrostatic, electromagnetic, or piezoelectric conversion.

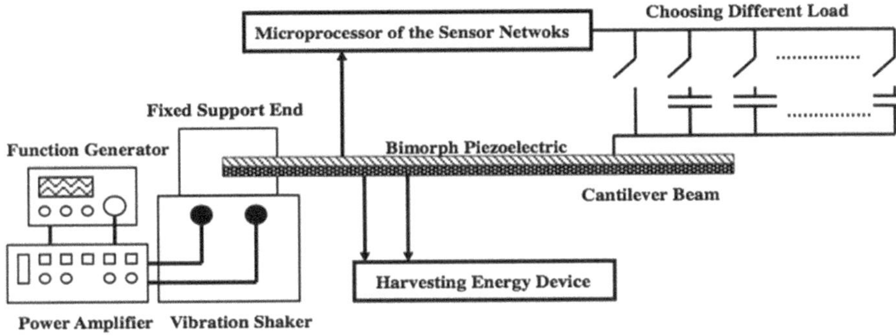

Fig. 2.10 Experiment setup of the tunable energy harvesting system ([25], copyright: SPIE)

Wu et al. [25] presented a piezoelectric bimorph generator in which the upper piezoelectric layer connected to various capacitive loads was used for tuning purpose; the lower layer was used for energy harvesting to charge a supercapacitor, as shown in Fig. 2.10. The tunable bandwidth of the generator was 3 Hz from 91.5 to 94.5 Hz, which was much narrower than achieved by the other aforementioned designs. In the two demo tests, the device was excited under a chirp and random vibration from 80 to 115 Hz, an average harvested power of 1.53 mW and 1.95 mW were generated, respectively, when the real-time tuning system was turned on. These results corresponded to an increase of 13.4% and 27.4% respectively as compared to the output when the tuning system was turned off. A microcontroller was utilized to sample the external frequency and adjusted the capacitive load to match the external vibration frequency in real time, in other words, the device was tuned actively. The continuous power required by the microcontroller system was on μW level.

Peters et al. [26] proposed another novel tunable resonator whose mechanical stiffness and hence the resonance could be adjusted through two piezoelectric actuators. The free actuator swung around the axis of rotation with a deflection angle α, as shown in Fig. 2.11a. By applying a voltage on the actuators, both ends of the actuators were deflected by $\Delta y(V_{op})$, as shown in Fig. 2.11b. Such deformation caused an additional hinge moment and thus a stiffer structure. One of their fabricated resonators achieved a large tuning of over 30% from an initial frequency of 78 Hz, using a tuning voltage of only ±5 V, as shown in Fig. 2.11c. A discrete control circuit, which exploited the phase characteristic of the resonator, was implemented to actively control the resonance tuning. However, the power consumption of around 150 mW was supplied externally, which significantly outweighed the harvested power (1.4 mW). Thus, the development of a low-power CMOS integration control circuit was recommended for practical closed-loop automatic tuning.

Roundy and Zhang [10] investigated the feasibility of active tuning mechanism. Via an analytical study, they demonstrated that an "active" tuning actuator never resulted in a net increase in power output for the actuator shown in Fig. 2.10. The fabricated piezoelectric generator, with an active tuning actuator is shown

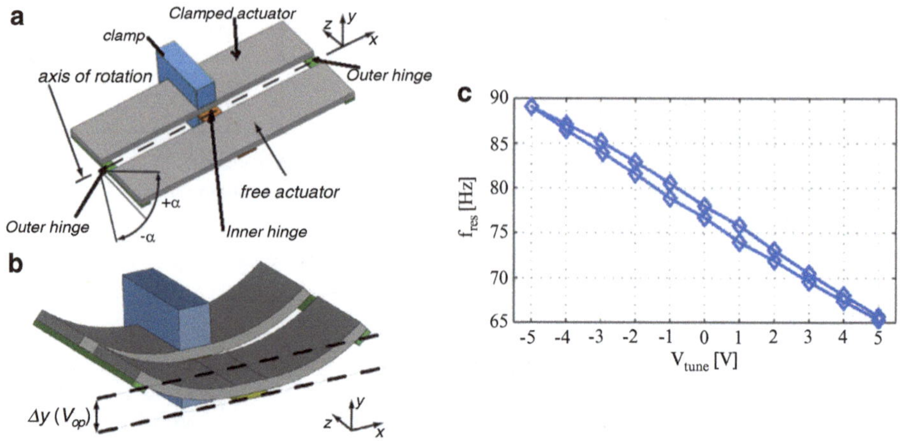

Fig. 2.11 (**a**) Tunable resonator with one clamped and one free actuator; (**b**) Both ends of the actuators are deflected by $\Delta y(V_{\mathrm{op}})$ with applied tuning voltage; (**c**) Measured resonance frequency versus applied tuning voltage ([26], copyright: IOP Publishing)

Fig. 2.12 Schematic of a piezoelectric bender, in which the surface electrode is etched to a scavenging and a tuning part ([10], copyright: SPIE)

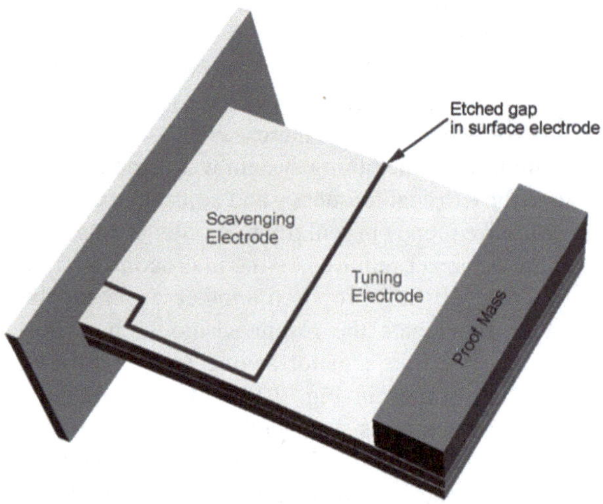

in Fig. 2.12. The electrode was etched to create a scavenging and a tuning part. Through three experimental test cases, it was found that the change in power output (82 μW) as a result of tuning was significantly smaller than the power needed to continuously drive the actuator (440 μW), which verified the conclusion of their analytical study. They suggested that "passive" tuning mechanism was worth more attention.

Lallart et al. [27] proposed a low-cost self-tuning scheme in which self-detection of frequency change and self-actuation were implemented. The schematic of the system is shown in Fig. 2.13. One layer of the piezoelectric bimorph was used

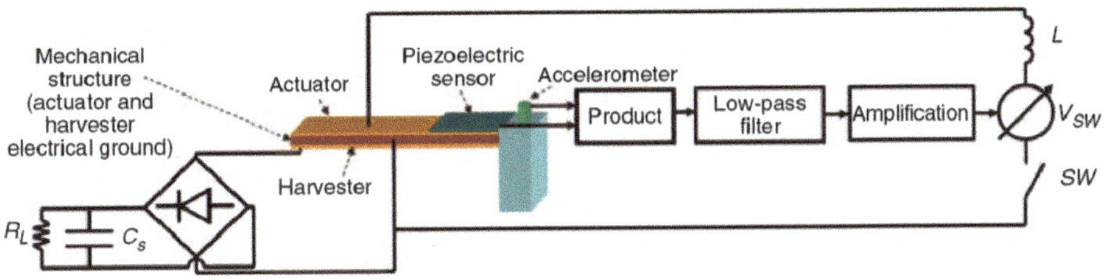

Fig. 2.13 Schematic of self-tuning system ([27], copyright: SAGE Publications)

as harvester and another layer as actuator to tune the stiffness of the structure via an external switching voltage source. An additional piezoelectric sensor and an accelerometer recorded the beam deflection and base acceleration. The self-detection of frequency change was based on the average product of these two signals, which gave the phase information and instructed the closed-loop control to apply the actuation voltage V_S. The most critical part of the required power for tuning in this device was the power for actuation P_{act}. V_S and P_{act} were given by

$$V_S = \pm\beta\frac{\alpha}{C_0}\langle V_{base} \times V_{cant}\rangle \tag{2.4a}$$

$$P_{act} = \frac{\omega}{2\pi}\frac{1+\gamma}{1-\gamma}\frac{\alpha_{act}^2}{(C_0)_{act}}\beta_0^2\cos(\varphi)^2 u_M^2, \tag{2.4b}$$

where β and β_0 are the user-defined tuning coefficients and φ is the phase between the beam deflection signal V_{cant} and base acceleration signal V_{base}. (Other terms can be found in Lallart et al. [27]). The power needed for actuation is therefore dependent on φ, and can be higher than the harvested power when the excitation frequency is far away from the resonance (Fig. 2.14). The actuation power is zero when the harvester approaches the resonance ($\varphi = \pi/2$). However, frequency detection and information processing modules of the system worked in real time from a continuous external power supply. Thus, this tuning system worked in "active" mode. The proposed system was estimated to achieve a positive net power output and to increase the bandwidth by a factor of 2 (from 4.1 to 8.1 Hz) near the resonance of 112 Hz, as shown in Fig. 2.14. This result is different from the conclusion by Roundy and Zhang [10], in which they could wrongly derive the net power by using the maximum power rather than the average power for actuation [28].

Instead of "active" tuning, Wischke et al. [29] reported a design of a tunable electromagnetic harvester in which the resonance was adjusted in a "semi-passive" way. Figure 2.15a shows the schematic of the design. The maximum tunable frequency range covered 56 Hz between 267 to 323 Hz by applying the voltage −100 V +260 V to the piezoelectric bimorph actuator. This was equivalent to 18% of the basic open-circuit resonant frequency of 299 Hz. More than 50 μW with optimal resistive loading were continuously achieved across the tunable frequency

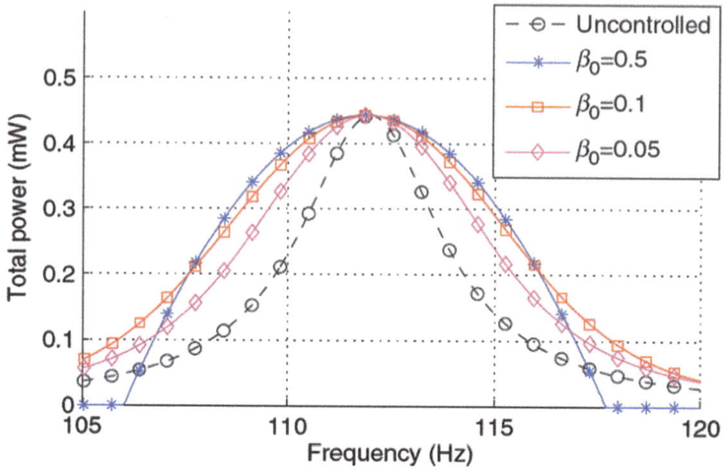

Fig. 2.14 Estimation of net power by proposed technique ([27], copyright: SAGE Publications)

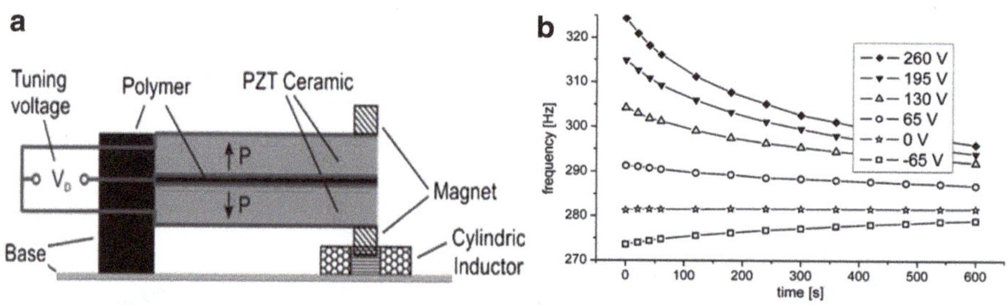

Fig. 2.15 (**a**) Schematic of the device and (**b**) time response of the harvester's operating frequency after the control voltage was disconnected ([29], copyright: IOP Publishing)

range. However, once the control voltage was disconnected, the frequency drifted away from the initial adjusted value due to leakage of the piezoceramic, as shown in Fig. 2.15b. This drift was more intense for high control voltages (>130 V). The charge had to be refreshed periodically to maintain the desired resonant frequency. Hence, the tuning mechanism was defined as "semi-passive" since it is different from the "passive" principle, in which the charge on the piezoceramic and accordingly the adjusted frequency would remain constant after disconnecting the control voltage. In order to reduce the frequency drifting and the energy consumed, the tuning range was suggested to be limited to 25 Hz by applying a voltage of −65 V +130 V, which was still feasible for sensor nodes. To further reduce the energy required for tuning, the shorter electrode of 10 mm length was used, which could achieve 80% of the tuning range, i.e., 20 Hz. Hence, given the power output of 50 μW, 20% circuit efficiency and 200 μJ required for tuning, the resonant frequency of the harvester could be tuned across 20 Hz every 20 s.

2.2.4 Comments on Resonant Frequency Tuning Techniques

Table 2.1 compares the reported resonance tuning methods with regard to tunability (frequency change Δf/average frequency f_{ave}), tuning load, tuning energy required, and whether the harvester is self-tunable.

- *Mechanical methods*. From Table 2.1, generally, mechanical tuning can achieve the largest tunability. However, most of the tunable designs using mechanical method required manual adjustment of the system parameters, such as the preload, pre-deflection, or gravity center of the tip mass. Tuning screws were widely used in these adjusting procedures, which makes it difficult to implement automatically during operation. The mechanical work required for tuning was not addressed in the literature reviewed. Only a few self-tuning mechanical methods [18, 19] enabled the harvesters to be self-adaptive to the vibration environment by exploiting the frequency-dependent inertia force. These devices were capable of automatic tuning during operation without external power input. However, they were applicable for specific conditions. For example, the device by Gu and Livermore [18] only worked for rotational vibration, and the design by Jo et al. [19] only had two working phases (similar to the 2DOF harvesters discussed in Sect. 2.3).
- *Magnetic methods*. Using magnets for resonance tuning can achieve moderate tunability. Automatic control and tuning can be implemented to adjust the distance between the magnets by using linear actuators. Thus, automatic tuning can be achieved during operation. However, the control and tuning systems of reported devices were still powered externally, which means that they were not completely "self-tuning." Moreover, the required energy for tuning was a huge burden in these devices compared to the harvested energy. Thus they are only suitable for the scenarios where small and infrequent frequency changes occur. The use of magnets and control systems also increase the complexity of system design and integration.
- *Piezoelectric methods*. As shown in Table 2.1, piezoelectric methods provide the smallest tunability as compared to the mechanical and magnetic methods. However, since the piezoelectric transducer itself functions as both the controller and tuning component, it is convenient to implement automatic tuning by applying voltage to the transducer or switching the shunt electrical load. In some reported designs [10, 26], the power required for active tuning significantly outweighed the harvested power. However, Wu et al. [25] reported that the required tuning power was only in μW level such that net power increase could be obtained. The reason for this difference is that the concept in Wu et al. [25] was piezoelectric shunt damping where power was only required to continuously switch the shunt electrical load, rather than to apply the voltage to the actuator. The latter usually consumes much more power [27]. Besides, when voltage is applied to the actuator, the leakage of piezoceramic increases the power consumption. Although the shunt damping concept requires small power,

Table 2.1 Summary of reported resonant frequency tuning methods

Author	Methods	Tuning range	Tunability ($\Delta f/f_{ave}$)	Load (force, distance, voltage)	Energy or power for tuning	Manually tuning	Self-tuning: Self-detection	Self-tuning: Automatic control	Self-tuning: Self-powered
Leland and Wright [11]	Mechanical (passive)	200–250 Hz (7.1 g tip mass)	22.22%	Up to 65 N	–	√	–	–	–
Eichhorn et al. [12]	Mechanical (passive)	292–380 Hz	26.19%	Up to 22.75 N	–	√	–	–	–
Hu et al. [13]	Mechanical (passive)	58.1–169.4 Hz	97.85%	−50–50 N	–	√	–	–	–
Morris et al. [14]	Mechanical (passive)	80–235 Hz (can be wider)	≥98.41%	≈1.25 mm	–	√	–	–	–
Loverich et al. [16]	Mechanical (passive)	56–62 Hz	10.17%	0.5 mm	–	√	–	–	–
Wu et al. [17]	Mechanical (passive)	130–180 Hz	32.26%	21 mm	–	√	–	–	–
Jo et al. [19]	Mechanical (passive)	Switch (24 Hz ⟺ 32 Hz)	–	–	–	–	√ (Self-adaptive, no need of external power)		
Gu and Livermore [18]	Mechanical (passive)	6.2–16.2 Hz	89.29%	–	–	–	√ (Self-adaptive, no need of external power)		
Challa et al. [20]	Magnetic (passive)	22–32 Hz	37.04%	3 cm	85 mJ	√	–	–	–

Reissman et al. [21]	Magnetic (passive)	88–99.38 Hz	12.15%	1.5 cm	–	√	–	–	–
Challa et al. [24]	Magnetic (passive)	13–22 Hz	51.43%	10 mm	3.2 J–3.9 J	–	√	√	×
Zhu et al. [22]	Magnetic (passive)	67.6–98 Hz	36.71%	3.8 mm	2.04 mJ mm^{-1}	–	√	√	×
Ayala-Garcia et al. [23]	Magnetic (passive)	64.1 –77.86 Hz	19.39%	2 mm	0.191 J	–	√	√	×
Wu et al. [25]	Piezoelectric (active)	91.5–94.5 Hz	3.23%	–	μW level(controller)	–	√	√	Not mentioned
Peters et al. [26]	Piezoelectric (active)	66–89 Hz(actuator PL140)	29.68%	±5 V	150 mW (discrete controller)	–	√	√	×
Roundy and Zhang [10]	Piezoelectric (active)	64.5–67 Hz	3.80%	5 V	440 μW	–	√	√	×
Lallart et al. [27]	Piezoelectric (active)	8.1 Hz(around resonance of 112 Hz)	7.23%	V_S[Eq. (2.4a)]	P_{act}[Eq. (2.4b)]	–	√	√	×
Wischke et al. [29]	Piezoelectric (semi-passive)	20 Hz (10 mm long electrode)	≈6.7%	−65–+130 V	200 μJ	–	√	√	×

Note: – not applicable, √ implemented, × not implemented

it provides the lowest tunability, as compared to other piezoelectric methods (Table 2.1).

- *Active tuning versus passive tuning*. Active tuning is usually implemented by piezoelectric tuning methods. Generally, it requires more power input than passive tuning, and the tuning power may outweigh the harvested power. However, a net power increase is still possible in active tuning mode if resonance tuning is only required in a very limited range [25, 27]. Passive tuning requires less power input to periodically detect and change the frequency, which is suitable when the excitation frequency varies infrequently. However, if the harvested power can sustain the continuous power required for tuning, an active tuning harvester can work for the excitation with constantly changing frequency or under random excitation, such as the case studied by Wu et al. [25].

2.3 Multimodal Energy Harvesting

In practice, energy harvesters are multiple degree-of-freedom (DOF) systems or distributed parameter systems. Thus one of the vibrational modes of the harvester can be excited when the driving frequency approaches the natural frequency associated with the particular mode. If multiple vibration modes of the harvester are utilized, useful power can be harvested i.e. a wider bandwidth can be covered for efficient energy harvesting. Here, such techniques are termed "multimodal energy harvesting."

Some researchers have reported on theoretical investigations of exploiting the translation and rotation modes of a rigid body [30] or multiple translation modes of lumped parameter systems [31, 32] for multimodal energy harvesting. However, in practice, multimodal energy harvesters are usually implemented by exploiting multiple bending modes of a continuous beam or by an array of cantilevers.

2.3.1 Exploiting Multiple Bending Modes of a Continuous Beam

Roundy et al. [33] first proposed the idea of multiple-DOF system incorporating multiple proof masses attached on a clamped–clamped beam to achieve wider bandwidth. One implementation of this idea was the multifrequency electromagnetic harvester developed by Yang et al. [34]. Other than this work, most of the reported studies in the literature exploited a multimodal harvester with a cantilever beam configuration, in which the first two bending modes were used (in other words, a 2DOF vibration energy harvester).

Tadesse et al. [35] presented a cantilever harvester integrated as part of a hybrid energy harvesting device. The harvester consisted of a cantilever beam with bonded piezoelectric plates and a permanent magnet attached at the tip, which

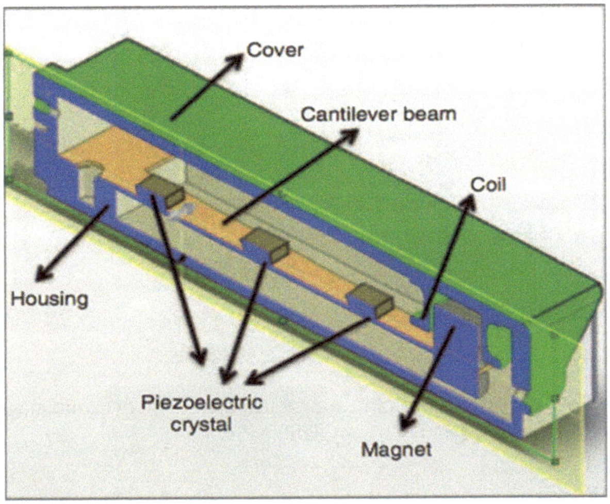

Fig. 2.16 Schematic of the multimodal energy harvesting device ([35], copyright: SAGE Publications)

oscillated within a stationary coil fixed to the housing, as shown in Fig. 2.16. The electromagnetic scheme generated high output power for the first mode, while the piezoelectric scheme was efficient for the second mode. However, the first resonance and the second resonance of such device were far away from each other (20 Hz and 300 Hz). Such discrete effective bandwidth may only be helpful when the vibration source has a rather wide frequency spectrum. The increased size may be another drawback since the permanent magnet is usually difficult to scale down. Besides, a drastic difference of matching loads for electromagnetic and piezoelectric harvesting presents a difficulty in interface circuit design to combine the power outputs from the two schemes.

Ou et al. [36] theoretically modeled a two-mass cantilever beam for broadband energy harvesting. Although two useful modes were obtained, similar to Tadesse et al. [35], they were quite far apart at 26 Hz and 174 Hz, respectively. Arafa et al. [37] presented a similar 2DOF cantilever piezoelectric energy harvester in which one proof mass functioned as a dynamic magnifier (Fig. 2.17a). For the prototype they fabricated, the power output with the dynamic magnifier reached 230 $\mu W/m/s^2$, increasing the power of a conventional harvester without magnifier by a factor of 13.12. Besides, it was observed (see Fig. 2.17b) that the two modes in Fig. 2.17b that the two modes were much closer as compared to those in Ou et al. [36] and Tadesse et al. [35]. However, the magnifier with a spring beam length of 70 mm and a magnifier mass of 11.2 g significantly increases the volume and weight of the original harvester composed of a 52 mm piezoelectric bimorph and a proof mass of 2.06 g.

Erturk et al. [38] exploited an L-shaped cantilever piezoelectric structure for multimodal energy harvesting, as shown in Fig. 2.18a. With proper parameter selection, the second natural frequency was approximately double the first, as shown in Fig. 2.18b. However, how to avoid mode-shape-dependent voltage cancelation was a critical issue. For the three piezoelectric segments combined in series,

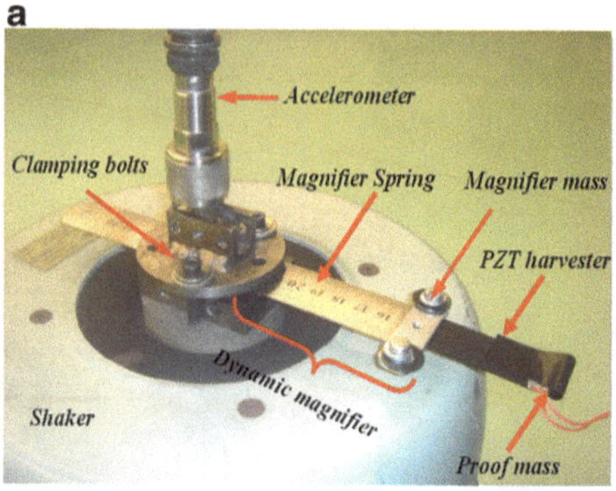

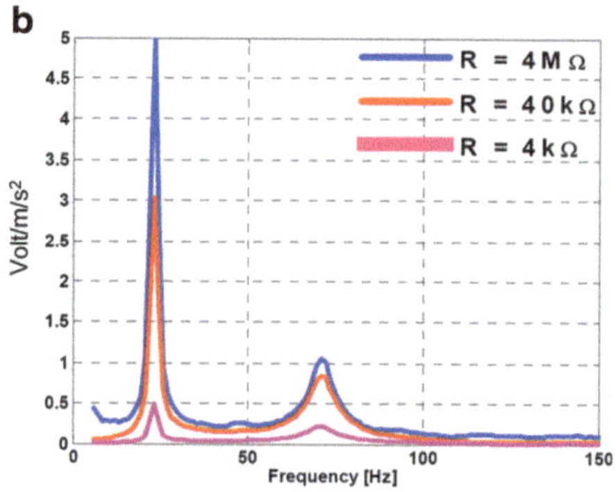

Fig. 2.17 (**a**) Piezoelectric energy harvester with dynamic magnifier and (**b**) its voltage responses of first two modes ([37], copyright: SPIE)

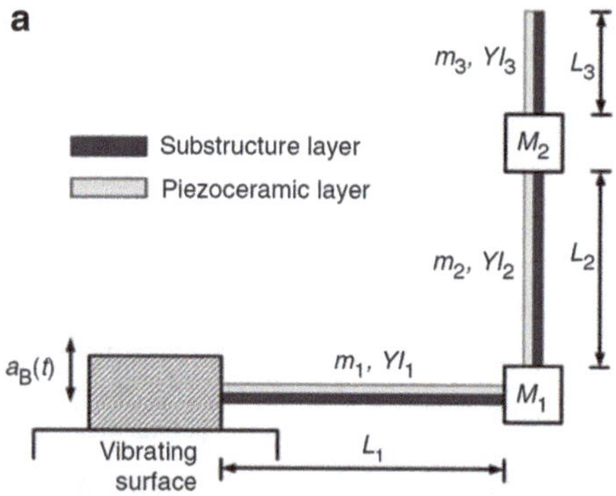

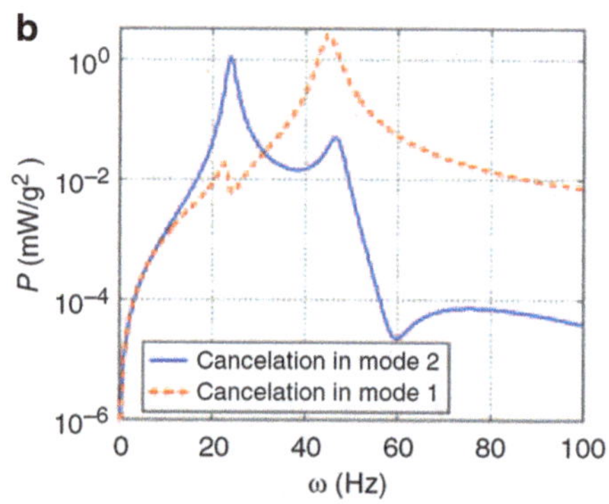

Fig. 2.18 (**a**) Schematic of L-shaped piezoelectric energy harvester. (**b**) Power frequency response function for 50 kΩ load resistance ([38], copyright: SAGE Publications)

cancelation occurred for the second mode. Changing the leads from the first piezoelectric segment in a reverse manner could avoid the cancelation of the second mode but this caused the cancelation for the first mode instead. Thus a more sophisticated interface circuit is required to adaptively change the electrode leads or to deliver the energy separately to avoid voltage cancelation.

Berdy et al. [39] reported a wide-band vibration energy harvester composed of a cantilevered symmetric meandering bimorph and a distributed proof mass. The concept of this design is shown in Fig. 2.19. The fabricated prototype successfully achieved two closely spaced resonant modes at 33 Hz and 43.3 Hz with measured RMS output powers of 107.3 μW and 74.9 μW, respectively, at a peak acceleration of $0.2 \times g$. In a wide bandwidth of 32.3–45 Hz, the output power remained above 25 μW. Another advantage of this device was that the sensing electronics and circuit board could be used as the distributed proof mass thus achieving a compact system.

Wu et al. [40] developed a novel compact 2DOF energy harvester, as shown in Fig. 2.20a. This device was fabricated from the conventional SDOF harvester by

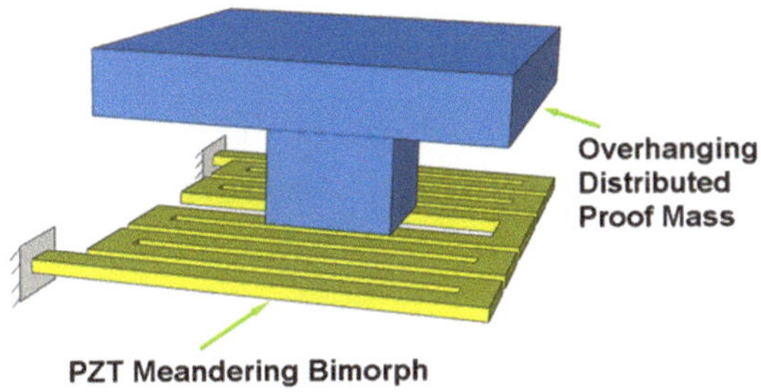

Fig. 2.19 Concept of meandering energy harvester with distributed proof mass

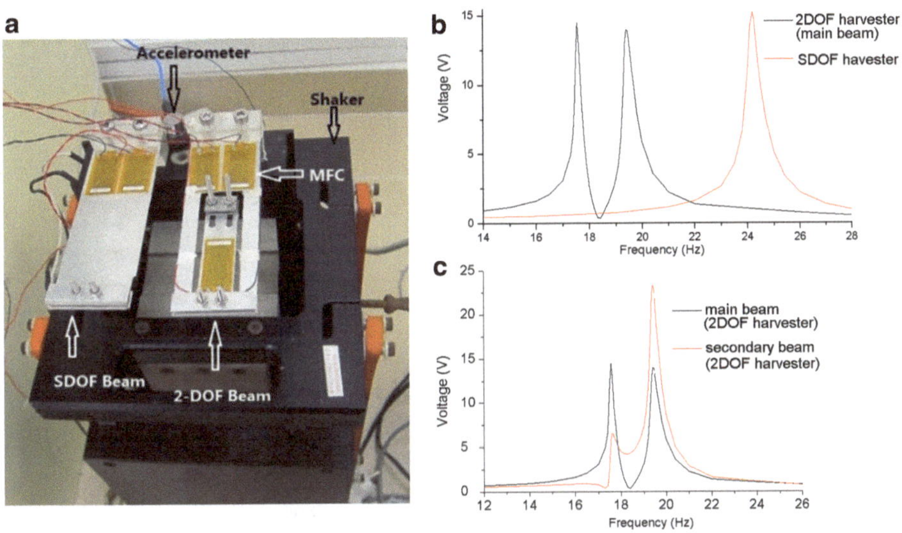

Fig. 2.20 (**a**) Conventional SDOF (proof mass $M_1 = 7.2$ g) and proposed 2DOF harvesters (proof mass $M_1 = 7.2$ g on main beam and $M_2 = 11.2$ g on secondary beam) installed on seismic shaker. (**b**) Comparison of open-circuit voltages from conventional SDOF and proposed 2DOF harvester. (**c**) Comparison of open-circuit voltages from main beam and secondary beam of proposed 2DOF harvester [40]

cutting out a secondary beam inside the main beam. Compared to the conventional SDOF harvester, this device was able to generate two close effective peaks in voltage response with properly selected parameters, as shown in Fig. 2.20b. Thus, multimodal energy harvesting was achieved with only a slight increase of the system volume. Besides, significant voltage output could be obtained from the secondary beam (Fig. 2.20c), which was not utilized due to the low strain level in the conventional SDOF configuration. Thus, this device efficiently utilized the material of the cantilever beam. Moreover, as compared to previously reported 2DOF harvester designs, it was more compact and could have two resonant frequencies much closer to each other.

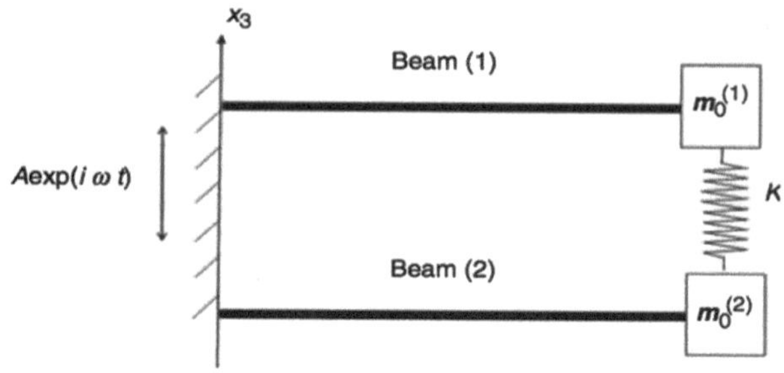

Fig. 2.21 Schematic of two beams with two end masses elastically connected ([41], copyright: SAGE Publications)

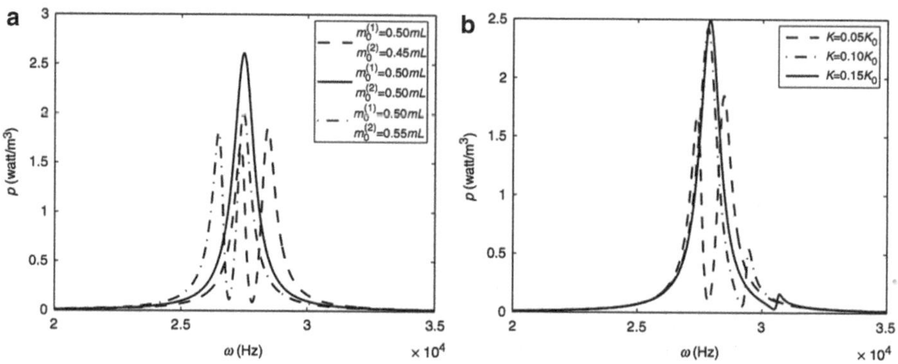

Fig. 2.22 Power density versus frequency for (**a**) different end mass pairs with a fixed spring stiffness and (**b**) different spring stiffness with a fixed mass pair ($m_0^{(1)} \neq m_0^{(2)}$) ([41], copyright: SAGE Publications)

2.3.2 Cantilever Array Configuration

Different from the discrete bandwidth corresponding to the multiple modes of a single beam, multiple cantilevers or cantilever arrays integrated in one energy harvesting device can easily achieve continuous wide bandwidth if the geometric parameters of the harvester are appropriately selected. Similar to the configurations in Sect. 2.3.1, sophisticated interface circuits are required to avoid charge cancelation due to the phase difference between the cantilevers in array configurations.

Yang and Yang [41] suggested using connected or coupled bimorph cantilever beams for energy harvesting, whose resonant frequencies could be tuned to be very close to each other. Figure 2.21 shows the schematic of the design, and Fig. 2.22 shows the theoretical prediction of power output versus frequency. Similar to Wu et al. [40], two close modes and thus wider bandwidth could be achieved as compared to a single-beam harvester. The amplitude and location of the resonances were found to be sensitive to the end spring and end masses.

Kim et al. [42] developed a 2DOF harvester composed of two piezoelectric cantilevers connected by a common proof mass, as shown in Fig. 2.23a. Although

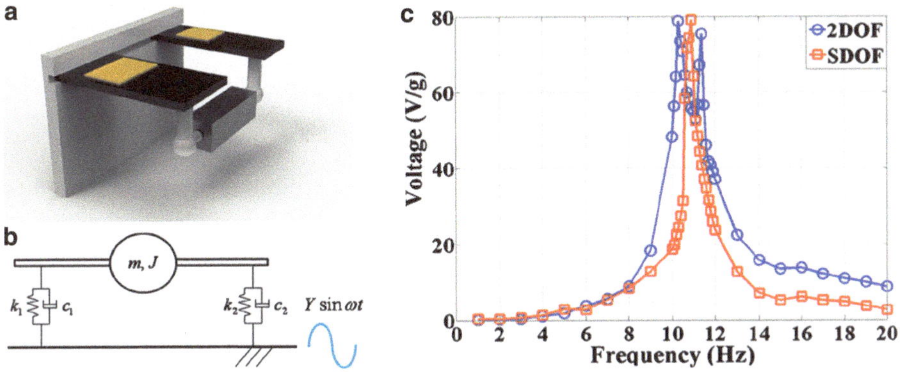

Fig. 2.23 (**a**) Schematic view of proposed device; (**b**) simplified mechanical model; (**c**) Frequency response comparison between proposed 2DOF and SDOF harvesters ([42], copyright: American Institute of Physics)

this design is categorized as a cantilever array configuration in this chapter, it should be emphasized that the underlying principle is to exploit the translational and rotational DOFs of the rigid mass (Fig. 2.23b). The two modes could be designed to be very close to each other and 280% increase in bandwidth at a voltage level of 55 V/g was achieved from a single piezoelectric cantilever in the 2DOF harvester, as compared to a conventional SDOF device in their experiment (Fig. 2.23c).

Other than these previous two designs of two coupled cantilevers, most of the research attempts were made to develop multimodal devices with more cantilevers to tailor and cover desired bandwidth for specific applications [43–47]. Different from Yang and Yang [41] and Kim et al. [42], these cantilevers were usually quasi-uncoupled. Each cantilever was regarded as one substructure of the harvester and thus the first mode of each cantilever was one of the vibration modes of the harvester.

Shahruz [43] designed an energy harvester that consists of piezoelectric cantilevers of various lengths and tip masses attached to a common base (Fig. 2.24a). It was capable of resonating at various frequencies by properly selecting the length and tip mass of each beam and thus provided voltage response over a wide frequency range (Fig. 2.24b). Such combination of cantilevers into a single device created a so-called "mechanical band-pass filter."

Xue et al. [44] presented a similar design of a broadband energy harvester using multiple piezoelectric bimorphs (PB) with different thickness of piezoelectric layers. They found that the bandwidth of their PB array configuration could be tailored by choosing an appropriate connection pattern (mixed series and parallel connections). Connecting multiple bimorphs in series could broaden the bandwidth. Comparing the single bimorph harvester and a 10-bimorph array configuration, their numerical results showed that not only the power magnitude of the energy harvesting system was increased but also the bandwidth (output power >10 μW) was widened from (97,103)Hz to (87,115)Hz. Furthermore, the bandwidth could be shifted to the dominant frequency range by changing the number of bimorphs in

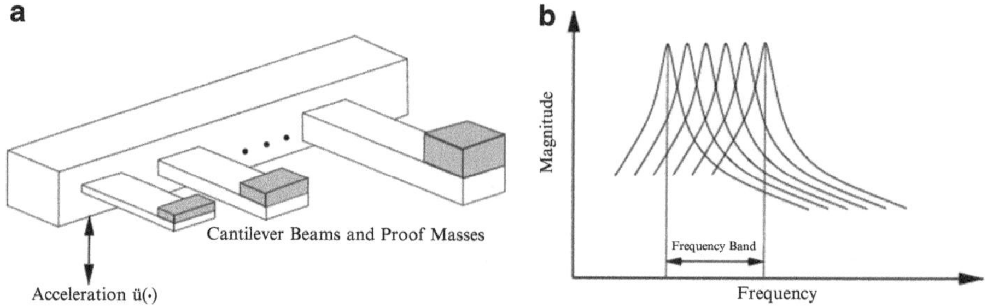

Fig. 2.24 (**a**) Mechanical band-pass filter and (**b**) its transfer function ([43], copyright: Elsevier)

parallel. This shift was due to the change in the electrical boundary condition when increasing or decreasing bimorphs in parallel.

Ferrari et al. [45] developed another multifrequency converter and investigated its feasibility and efficiency for powering a wireless sensor. This device consisted of three piezoelectric bimorph cantilevers with the same dimensions of 15 mm × 1.5 mm × 0.6 mm but with different tip masses ($m_1 = 1.4$ g, $m_2 = 0.7$ g, $m_3 = 0.6$ g). When excited by mechanical vibrations, the device charged the storage capacitor and regularly delivered the energy to the wireless sensor and measurement transmission module. Under resonant excitation, i.e., at either f_1, f_2, or f_3, the corresponding single cantilever in the array could alone trigger the transmission, but a single cantilever could not do so at off-resonance frequency f_4. Conversely, with the complete converter array, the converted energy was high enough to trigger the transmission for all the tested frequency, including f_4. Besides, the shorter switching time (two measure-and-transmit operations) was obtained using the converter array rather than a single cantilever. It was claimed that the wider bandwidth and improved performance were worth the modest increase in size of the proposed array device.

Broadband energy harvesters with cantilever array were also implemented compatibly with current standard MEMS fabrication techniques [46, 47]. Liu et al. [46] implemented such a MEMS-based broadband cantilever array harvester, as shown in Fig. 2.25a. In their experimental test, a phase difference in voltage output from each cantilever was observed, which impaired the voltage output of this three cantilever device (Fig. 2.25b). Thus, the DC voltage across the capacitor after rectification was only 2.51 V, and the maximum DC power output was about 3.15 μW. To address this problem, separate rectifier for each cantilever was required, which increased the total DC voltage to 3.93 V and the maximum DC power output to 3.98 μW. With the wider bandwidth 226–234 Hz and the improved output, such a device was claimed to be promising in applications of ultra-low-power wireless sensor networks. However, the more complicated rectification circuit may cause significant energy loss in these MEMS-scale devices especially for low-level or *off-resonance* excitations. Low-voltage-drop rectification techniques using "active diode" may alleviate this problem in such cases [48].

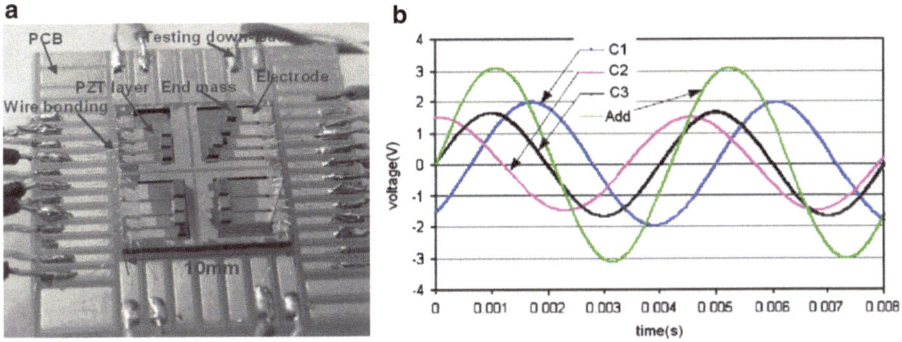

Fig. 2.25 (**a**) Schematic of generator array prototype and (**b**) AC output of three cantilevers in an array and their direct serial connection ([46], copyright: Elsevier)

Sari et al. [47] implemented a micro broadband energy harvester using electromagnetic induction. The developed device generated power via the relative motion between a magnet and coils fabricated on 35 serially connected cantilevers with different lengths. It was reported that 0.4 μW continuous power with 10 mV voltage was generated, covering a wide external vibration frequency range of 4.2–5 kHz. The test was carried out at an acceleration level of $50 \times g$, which was much higher than the $0.5 \times g$ in the test of Liu et al. [46]. The cantilever size had a very similar scale but the power output from the device by Sari et al. [47] was much less than that from the device by Liu et al. [46], which indicated that the piezoelectric conversion was more favorable for vibration energy harvesting on the MEMS scale. Furthermore, the voltage level of 10 mV from the harvester by Sari et al. [47] was more challenging for AC–DC rectification and energy storage.

2.3.3 Comments on Multimodal Energy Harvesting

Multimodal energy harvesting can be implemented by exploiting multiple bending modes of a continuous beam or by exploiting a cantilever array integrated in one device where the first mode of each cantilever is one of the vibration modes of the device. Compared with the resonance tuning techniques, multimodal energy harvesting does not require tuning and hence is much easier to implement. The concerns for multimodal energy harvesting include:

- *Bandwidth.* The multiple bending modes of a continuous beam are usually far away from one another and thus the effective bandwidth is discrete. Some novel structures like L-shaped beams [38], cut-out beams [40], and cantilevered meandering beams [39] can be considered to achieve close and effective resonant peaks. However, in general, only the first two modes can contribute to effective multimodal energy harvesting. By using cantilever arrays, the targeted bandwidth can be covered continuously by proper selection of the system parameters (see Fig. 2.24b).

- *Power density*. Multimodal energy harvesting increases the bandwidth but is however accompanied by an increased volume or weight of the device. Thus the overall power density (power/volume or power/weight) may be sacrificed. For example, in the cantilever array configuration, only one cantilever or a subset of the array is active and effective for energy generating while the other cantilevers are at an *off-resonance* status. Hence, with the known dominant spectrum of the ambient vibration, the harvester should be carefully designed with a proper number and dimensions of the cantilevers such that the device can cover the targeted bandwidth with the least sacrifice of power density.
- *Complex interface circuit*. Multimodal energy harvesting requires more complex interface circuit than that for a single-mode harvester. A critical electrical issue is to avoid mode shape dependent voltage cancelation in a continuous beam or the cancelation due to the phase difference between cantilevers in array configurations. More sophisticated interface circuits are required to adaptively change the electrode leads or to deliver the energy separately (i.e., each piezoelectric segment in a continuous beam or each cantilever in a cantilever array configuration is connected to a separate load or rectifier). An interface circuit is also required to address the drastic difference in matching load for different energy transduction mechanisms in the hybrid energy harvesting scheme based on a continuous beam [35].

2.4 Nonlinear Energy Harvesting Configurations

In Sect. 2.2 we presented several resonance tuning techniques using magnets [20–22]. Actually these magnets introduce not only a change in the linear stiffness but also a change in the nonlinear stiffness. The nonlinear behavior becomes apparent when the harvester experiences oscillation with significant amplitude. Such nonlinearity also benefits wideband energy harvesting.

As reported in the available literature, nonlinearities in energy harvesters are considered from two perspectives, i.e., nonlinear stiffness [49–57] and nonlinear piezoelectric coupling [58, 59]. Compared to the nonlinear piezoelectric coupling, which results from the manufacturing process of piezoelectric materials, the nonlinear stiffness of a harvester is relatively easier to achieve and control. This section reviews recent advances in designing broadband energy harvesters with nonlinear stiffness.

The dynamics of a general oscillator can be described as

$$\ddot{x} = -\frac{\mathrm{d}U(x)}{\mathrm{d}x} - \gamma\dot{x} + f(t), \tag{2.5}$$

where x represents the oscillator position; γ represents the viscous damping; $f(t)$ is the ambient vibration force; and $U(x)$ is the potential function. If an electromagnetic

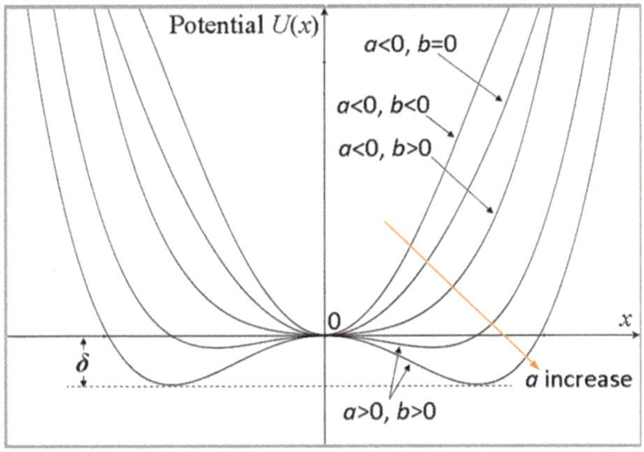

Fig. 2.26 Potential function for different Duffing oscillators

generator is considered, γ includes the viscous damping caused by electromagnetic coupling. Details on this kind of electrical viscous damping can be found in El-Hami et al. [3] or Mann and Sims [50]. For a piezoelectric generator, the damping caused by piezoelectricity cannot be modeled as a viscous damper [60] and Eq. (2.5) should be modified by adding a coupling term as

$$\ddot{x} = -\frac{\mathrm{d}U(x)}{\mathrm{d}x} - \gamma\dot{x} + \kappa V + f(t), \tag{2.6}$$

where κ represents the electromechanical coupling coefficient and V is the voltage on the electrical load. The circuit equations for the piezoelectric and the electromagnetic harvesters are quite different due to differences in their internal impedances. They are not given here but they can be readily found in the literature related to electromagnetic and piezoelectric transductions, such as El-Hami et al. [3] and Erturk et al. [53].

Duffing-type nonlinear oscillator

For a Duffing-type oscillator, the potential energy function $U(x)$ can be considered in a quadratic form as [61, 62],

$$U(x) = -\frac{1}{2}ax^2 + \frac{1}{4}bx^4. \tag{2.7}$$

Thus the Duffing-type oscillator has the cubic nonlinear spring force as

$$F(x) = -ax + bx^3. \tag{2.8}$$

The potential function $U(x)$ for different Duffing oscillators is shown in Fig. 2.26. $U(x)$ is symmetric and bistable for $a > 0$ and $b > 0$ and monostable for $a \leq 0$. In the bistable case, two minima at $x_m = \pm\sqrt{a/b}$ are separated by a barrier δ at $x = 0$.

Piecewise-linear oscillator

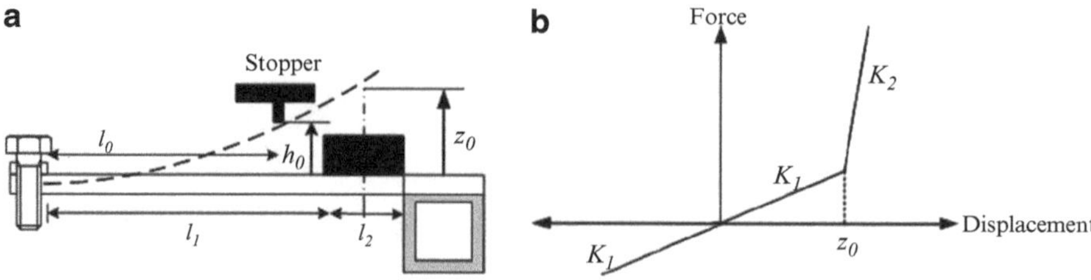

Fig. 2.27 (**a**) Typical mechanical stopper configuration in vibration energy harvester and (**b**) its piecewise-linear stiffness ([63], copyright: IOP Publishing)

Other than the Duffing-type oscillator, some researchers also attempted to exploit piecewise-linear stiffness to increase the bandwidth of vibration energy harvesters. Using mechanical stoppers is one common way to introduce the piecewise-linear stiffness [56, 63–65]. A typical setup of a vibration energy harvester with a mechanical stopper and its nonlinear stiffness are illustrated in Fig. 2.27.

This section reviews both Duffing-type nonlinear harvesters and harvesters with mechanical stoppers. Their benefits on improving the performance of vibration energy harvester are discussed in the following parts.

2.4.1 Monostable Nonlinear Configuration

Substituting Eq. (2.7) into Eq. (2.5) gives the forced Duffing's equation [49, 50, 66],

$$\ddot{x} + \gamma\dot{x} - ax + bx^3 = f(t). \tag{2.9}$$

For $a \leq 0$, it can be used to describe a monostable system. $b > 0$ determines a hardening response, while $b < 0$ a softening response.

Ramlan et al. [49] investigated the hardening mechanism of the nonlinear monostable harvester. Their numerical and analytical studies showed that ideally, the maximum amount of power harvested by a system with a hardening stiffness was the same as the maximum power harvested by a linear system, regardless of the degree of nonlinearity. However, this might occur at a different frequency depending on the degree of nonlinearity, as shown in Fig. 2.28. Such a device has a larger bandwidth over which the significant power can be harvested due to the shift in the resonant frequency.

Mann and Sims [50] presented a design for electromagnetic energy harvesting from nonlinear oscillations due to magnetic levitation. Figure 2.29a shows the schematic of the system where two outer magnets are oriented to repel the center magnet, thus suspending it with a nonlinear restoring force. The derived governing equation has the same form as Eq. (2.9). Figure 2.29b,c shows the experimental velocity response and theoretical predictions under low and high harmonic base

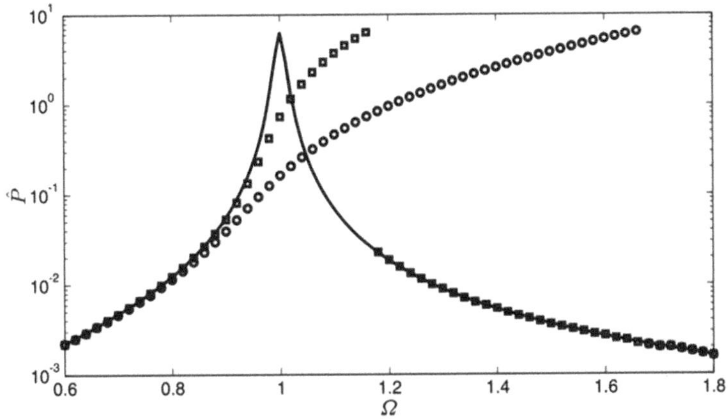

Fig. 2.28 Numerical solution for nondimensional power harvested with damping ratio $\zeta = 0.01$ and excitation amplitude $Y = 0.5$: Linear system (*solid line*), hardening system with nonlinearity $b = 0.001$ (*open square*) and $b = 0.01$ (*open circle*) [b is the coefficient of the nonlinear term in Eq. (2.9)] ([49], copyright: Springer Science+Business Media)

excitation levels, respectively. At low excitation level, the frequency response (Fig. 2.29b) was similar to the response of a linear system. However, at high excitation level, the response curve was bent to the right (Fig. 2.29c). Thus, relatively large amplitudes persisted over a much wider frequency range. Both experiment and theoretical analysis captured the jump phenomena near the primary resonance and the multiple periodic attractors, as shown in Fig. 2.29c. However, such a hardening device only broadened the frequency response in one direction (the peak response shifts to the right).

Stanton et al. [51] proposed another monostable device for energy harvesting using the piezoelectric effect. The device consisted of a piezoelectric beam with a magnetic end mass interacting with the fields of oppositely poled stationary magnets, as shown in Fig. 2.30. The system was modeled by an electromechanically coupled Duffing's equation similar to Eq. (2.9), except that the piezoelectric coupling term κV should be added as in Eq. (2.6). By tuning the nonlinear magnetic interactions around the end mass (i.e., tuning the distance d), both hardening and softening responses may occur, as shown in Fig. 2.31, which allows the frequency response to be broadened bidirectionally. In the experimental validation, a linearly decreasing frequency sweep was performed for the softening case. Different from Ramlan et al. [49], it was shown that not only a wider bandwidth but also a better performance could be obtained by the monostable configuration, as compared to the linear configuration (with stationary magnets removed), as shown in Fig. 2.32. This might be due to the change of damping due to the magnets used in the experiment [20, 22], while a constant damping was used in the analysis by Ramlan et al. [49].

Previous monostable designs have a larger bandwidth due to the shift in the resonant frequency. This nonlinear advantage if the high-energy attractor regime

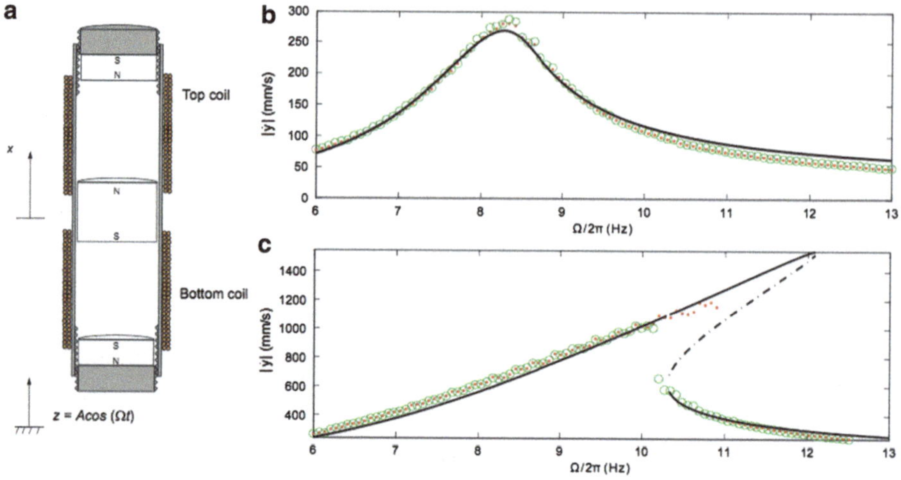

Fig. 2.29 (**a**) Schematic of the magnetic levitation system; experimental velocity response; and theoretical predictions from forward (*red dots*) and reverse (*green circles*) frequency sweep under two excitation levels: (**b**) 2.1 m/s^2 and (**c**) 8.4 m/s^2. Theoretical predictions include stable solutions (*solid line*) and unstable solutions (*dashed line*) ([50], copyright: Elsevier)

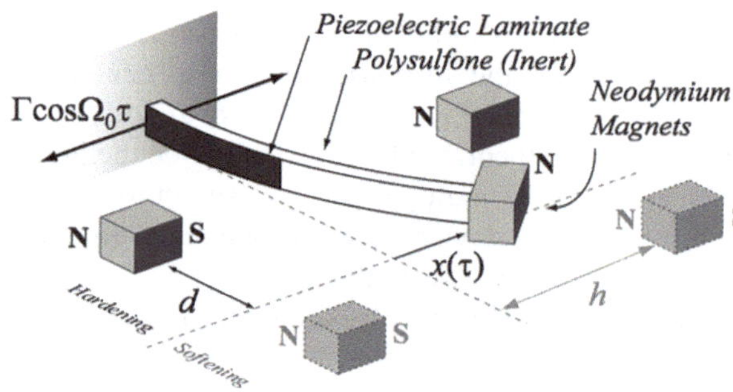

Fig. 2.30 Schematic of proposed nonlinear energy harvester ([51], copyright: American Institute of Physics)

is realized [51]. A linearly decreasing or increasing frequency sweep can capture the high-energy attractor and hence improve the bandwidth for the softening and hardening cases. Unfortunately, such conditions cannot be guaranteed in practice. Certain means of mechanical or electrical disturbance or perturbation is required once the nonlinear devices enter low-energy orbits; otherwise little power can be harvested. Previous reported studies did not address the required momentary perturbation if the harvester is in the low-energy branch and the requisite actuation energy. Furthermore, under a White Gaussian excitation, Daqaq [67] demonstrated that the hardening-type nonlinearity failed to provide any enhancement of output power over typical linear harvesters. Under colored Gaussian excitations, the expected output power even decreased with a hardening-type nonlinearity. This suggested that the monostable configuration may be only applicable for frequency sweep excitations.

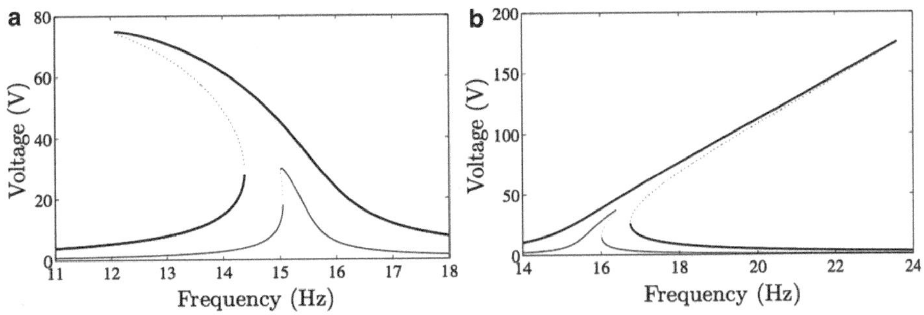

Fig. 2.31 Predicted response amplitudes of output voltage for (**a**) $d = 5$ mm and (**b**) $d = -2$ mm, corresponding to softening and hardening cases, respectively. *Solid lines* correspond to stable solutions while the dotted line to unstable solutions. The *lighter line* and *darker line* correspond to low- and high excitation levels, respectively ([51], copyright: American Institute of Physics)

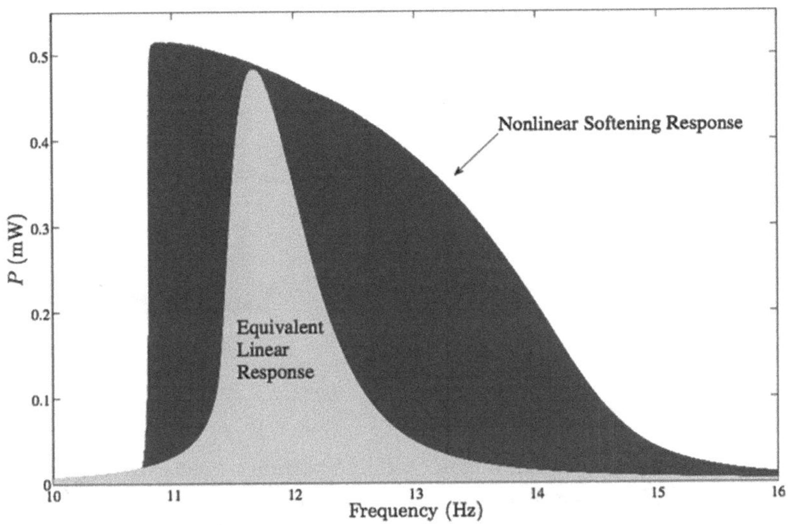

Fig. 2.32 Comparison of energy harvesting performances of nonlinear and linear configurations under the same excitation amplitude of $0.3 \times g$ ([51], copyright: American Institute of Physics)

2.4.2 Bistable Nonlinear Configuration

For $a > 0$, Eq. (2.9) can be used to describe a bistable nonlinear system. In this section, we discuss in detail how to exploit the properties of the nonlinearity of a bistable system to improve energy harvesting performance over a wide range of ambient vibration frequencies, subjected to either periodic forcing or stochastic forcing.

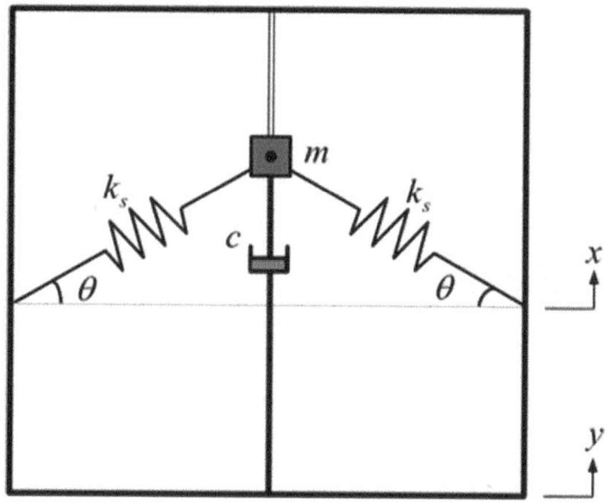

Fig. 2.33 Arrangement of mass-spring-damper generator for the *snap-through* mechanism ([49], copyright: Springer Science+Business Media)

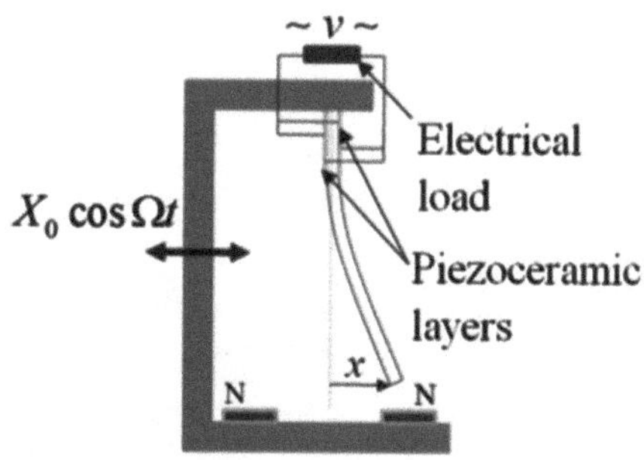

Fig. 2.34 The piezomagnetoelastic generator ([53], copyright: American Institute of Physics)

2.4.2.1 Periodic Forcing

A periodically forced oscillator can undergo various types of large-amplitude oscillations, including chaotic oscillation, large-amplitude periodic oscillation, and large-amplitude quasiperiodic oscillation. The behavior depends on the design of the device, the frequency, and amplitude of the forcing and the damping [66]. One physically realizable energy harvester with bistable nonlinear stiffness was proposed by Ramlan et al. [49], utilizing a so called "*snap-through*" mechanism. The setup consisted of two linear oblique springs connected to a mass and a damper, as shown in Fig. 2.33, yielding a nonlinear restoring force in the x direction. This mechanism has the effect of steepening the displacement response of the mass as a function of time, resulting in a higher velocity for a given input excitation. Numerical results revealed that this mechanism could provide much better performance than the linear mechanism when the excitation frequency was much less than the natural frequency.

Bistable nonlinear stiffness can also be created by using magnets. Erturk et al. [53] and Erturk and Inman [68] pursued such method in designing a broadband piezomagnetoelastic generator. The device consisted of a ferromagnetic cantilever beam with two piezoceramic layers attached at the root for energy generation and with two permanent magnets near the free end, as illustrated in Fig. 2.34. For an

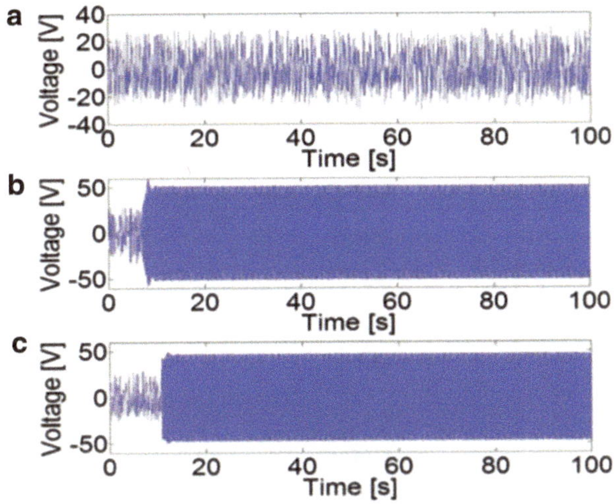

Fig. 2.35 Experimental voltage histories: (**a**) Chaotic strange attractor motion (excitation: $0.5 \times g$ at 8 Hz); (**b**) Large-amplitude periodic motion due to the excitation amplitude (excitation: $0.8 \times g$ at 8 Hz); (**c**) Large-amplitude periodic motion due to a disturbance at $t = 11\ s$ (excitation: $0.5 \times g$ at 8 Hz) ([53], copyright: American Institute of Physics)

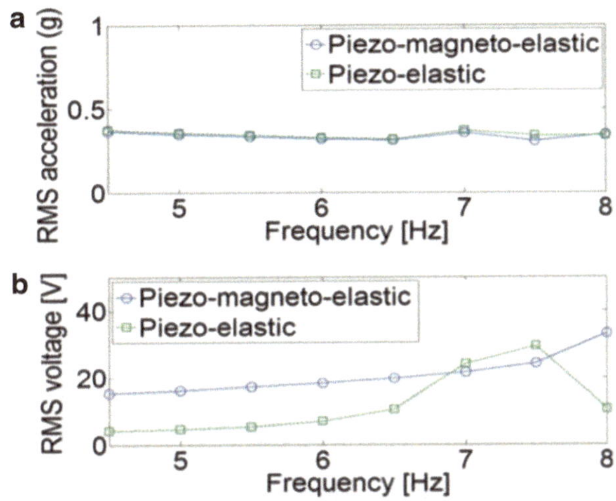

Fig. 2.36 (**a**) Root-mean-square (RMS) acceleration input at different frequencies (average value: $0.35 \times g$); (**b**) Open-circuit RMS voltage output over a wide frequency range ([53], copyright: American Institute of Physics)

initial deflection at one of the stable equilibriums, the voltage response could be chaotic strange attractor motion or large-amplitude periodic motion (limit cycle oscillation), under small or large harmonic excitations, as shown in Fig. 2.35a,b. The large-amplitude periodic motion could also be obtained under small excitation level by applying a disturbance or equivalently an initial velocity condition, as shown in Fig. 2.35c. Thus a large-amplitude response could be obtained at *off-resonance* frequencies as well as broadband performance, with a clear advantage over the linear piezoelastic configuration (with two magnets removed), as shown in Fig. 2.36. However, for small excitation amplitude, actuation energy is required to perturb the beam and hence drive the system into high-energy orbits, which was not addressed.

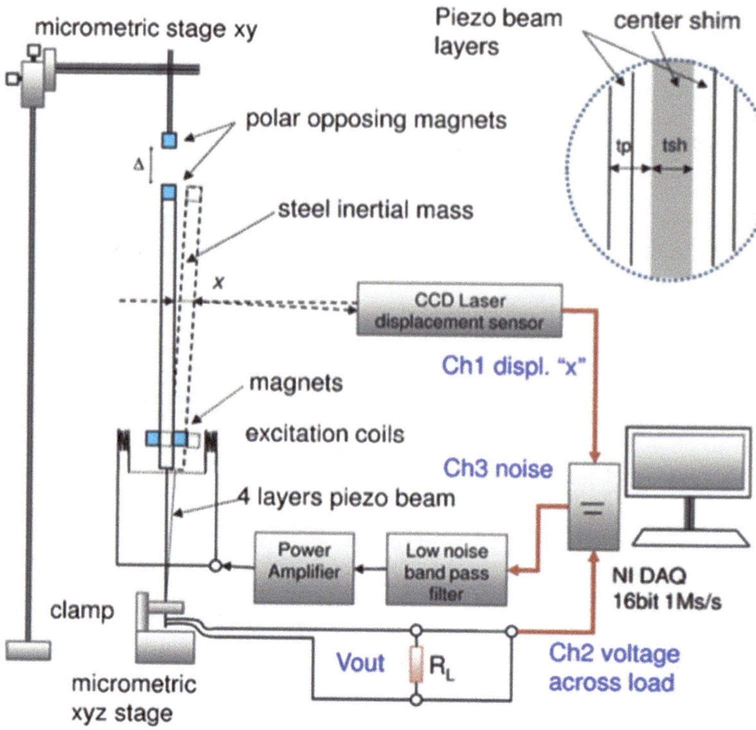

Fig. 2.37 Schematic of the experimental apparatus ([61], copyright: American Physical Society)

2.4.2.2 Stochastic Forcing

For a bistable system, stochastic forcing can also induce transitions between the stable equilibria of the system, and thus causing large-amplitude oscillations. Cottone et al. [61] realized a piezoelectric inverted pendulum by using the bistable mechanism (polar opposing magnets with a small separation distance Δ). Figure 2.37 shows the schematic of their experimental apparatus. The random vibration made the pendulum swing with small oscillations around each equilibrium or with large excursions from one equilibrium position to another. However, for extremely small Δ, the pronounced potential energy barrier confined the pendulum swing within one potential well. For specific Δ and noise level, the deflection of the pendulum x_{RMS} reached a maximum and hence the maximum power could exceed 4–6 times the power obtained when the magnets were far away (quasi-linear), as shown in Fig. 2.38.

Ferrari et al. [69], Lin and Alphenaar [70], Andò et al. [71], and Stanton et al. [52] extended this idea to study the energy harvesting performance of bistable cantilevers with repulsive magnets under wide-spectrum vibrations. From these studies, the critical issue for the broadband energy harvesting involves how to enable

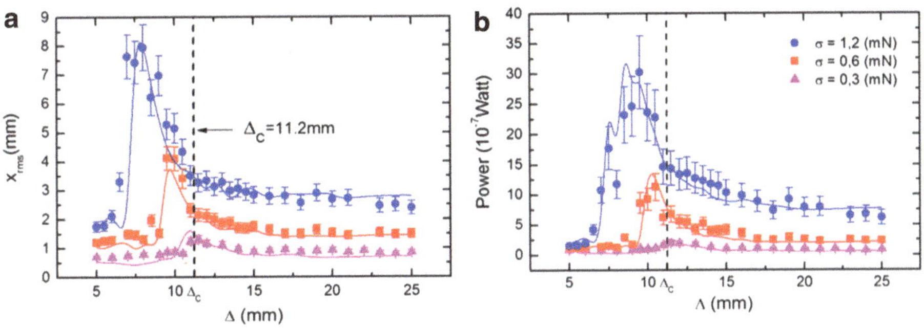

Fig. 2.38 (**a**) Position x_{RMS} and (**b**) power versus Δ for three different noise levels ([61], copyright: American Physical Society)

the harvester to readily transit between the two stable states, which is dependent on the excitation amplitude, frequency, and the extent of nonlinearity. For the bistable pendulum and a more general bistable dynamic system, Cottone et al. [61] concluded that (1) the raising of the response x_{RMS} is mainly due to the growth of the separation between the two minima of the potential function and (2) the drop of x_{RMS} is mainly due to the decrease in the jump probability caused by the increase of the potential barrier height δ (Fig. 2.26).

In order to increase the probability of transition between the potential wells and thus to further enhance the performance of a bistable system, some researchers [72, 73] have proposed to exploit the phenomenon of stochastic resonance. Stochastic resonance can occur if the dynamics of the system is forced such that the potential barrier oscillates, and this forcing matches with the mean time between transitions—i.e. the inverse Kramer's rate [74]. For a beam clamped at both ends, the SDOF bistable model is shown in Fig. 2.39. This is similar to the *snap-through* setup by Ramlan et al. [49], except that the distance A–A' between boundaries can be modulated at frequency ω and hence the potential barrier is modulated. Thus the parametrically forced dynamics of the system is defined as [72]

$$\ddot{\xi} + c\dot{\xi} - \mu\left(1 - \eta\cos\left(\omega t\right)\right)\xi + \xi^3 = Q(t), \tag{2.10}$$

where ξ is the nondimensional coordinate; c is the damping coefficient; μ is a measure of the compressive load acting on the beam; η and ω are the magnitude and frequency of forcing for modulation, respectively; and $Q(t)$ is the external noise. With this model, McInnes et al. [72] demonstrated that the properly tuned system in stochastic resonance by forcing (i.e., the forcing matched with inverse Kramer's rate) apparently experienced larger amplitude vibrations than those of the unforced mechanism, which was confined in a single potential well, as shown in Fig. 2.40. Thus, significantly more energy could be obtained. However, if the system was untuned, the net energy generated by the forcing mechanism could be less than the unforced mechanism, due to the energy consumed to force the boundaries to oscillate.

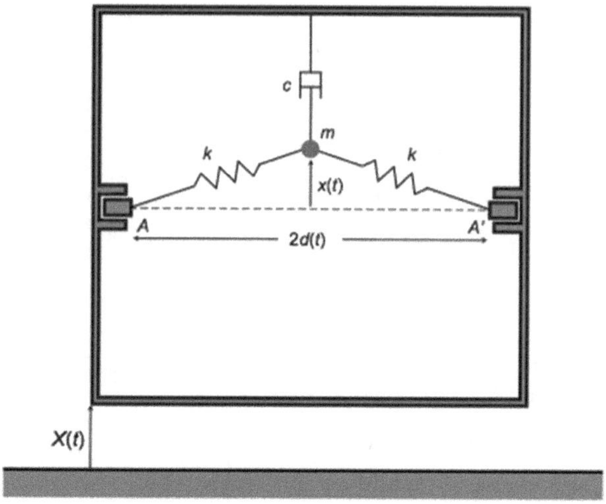

Fig. 2.39 One degree-of-freedom beam model, in which the distance A–A' can be modulated at frequency ω ([72], copyright: Elsevier)

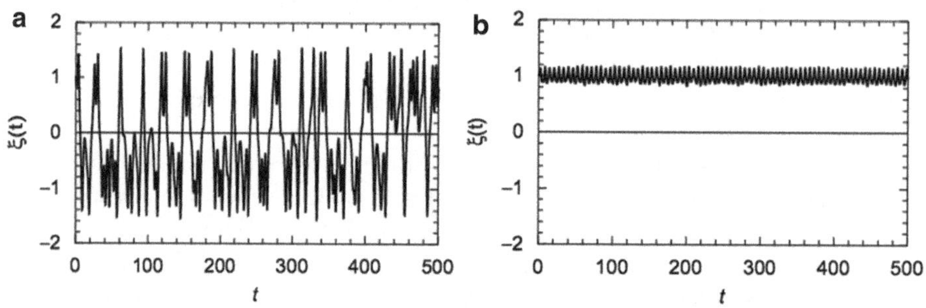

Fig. 2.40 Tuned system in stochastic resonance with $\omega = 1.2$: (**a**) response with forcing $\eta = 0.7$ and (**b**) response without forcing $\eta = 0$ ([72], copyright: Elsevier)

2.4.3 Configuration with Mechanical Stoppers

Piecewise-linear stiffness is another type of nonlinearity which can be introduced by mechanical stoppers. Soliman et al. [63] presented a micro-electromagnetic harvester incorporating such a mechanism, as shown in Fig. 2.27a. They investigated the benefit of such an architecture using mechanical stoppers via analytical, numerical, and experimental studies. They found that the new architecture increased the bandwidth of the harvester during a frequency upsweep, while maintaining the same bandwidth in a downsweep, as shown in Fig. 2.41. Similar to the Duffing-type hardening configuration, jump phenomenon and multiple solutions were observed during the frequency upsweep. They further investigated the benefit of stopper when the vibration frequency randomly changed in a range of 13.8 Hz centered around the natural frequency. In their numerical simulation, the harvester with one-sided stopper collected energy at a lower power level but for a larger fraction of time (due to a larger bandwidth), resulting in 30% more overall collected energy (Fig. 2.42).

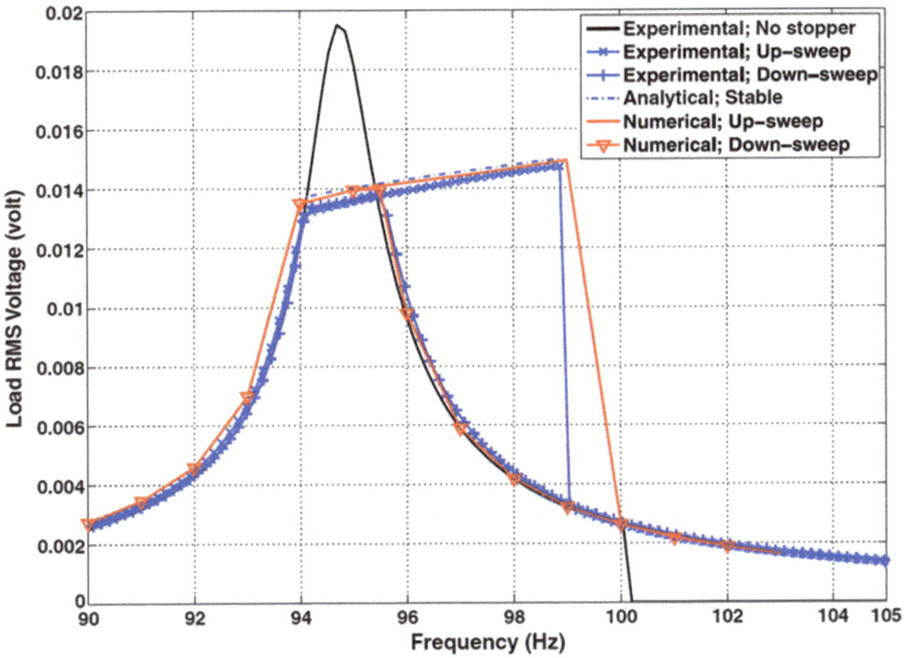

Fig. 2.41 Analytical, numerical, and experimental frequency responses of RMS load voltage from harvester with and without stopper ([63], copyright: IOP Publishing)

Blystad and Halvorsen [65] reported an experimental study on a piezoelectric harvester with a one-sided mechanical stopper under broadband vibrations. This device had a similar trend under sinusoidal sweep vibrations as in Soliman et al. [63]. Under colored noise vibrations, although wider bandwidth was achieved when the stopper became effective as the excitation level increased, the power output was smaller than the harvester without stopper (Fig. 2.43).

With increased bandwidth but lowered power level, the advantage of the harvester with stoppers is questionable. Soliman et al. [56] presented an optimization procedure for a harvester with stopper. They found that the performance of such a device is dominated by two factors: the stiffness ratio (k_2/k_1, refer to Fig. 2.27b) and the velocity of the beam at the impact point. These factors are controlled by the stopper height h_0 and the offset distance l_0 of the stopper from the cantilever support. Thus, in an environment with a known vibration probability density function (PDF), l_0 and h_0 should be tuned to tailor the upsweep bandwidth to better fit the given PDF, while h_0 should be set as high as possible to minimize contact damping and maximize energy collection.

Blystad et al. [64] presented circuit simulations to further investigate the effects of different two-sided stopper models and various interface circuits on the piezoelectric energy harvesting performance. Under harmonic excitations, they found that the output power was nearly unaffected by the stopper model used (elastic, critically damped, and completely inelastic stopper models). However, the

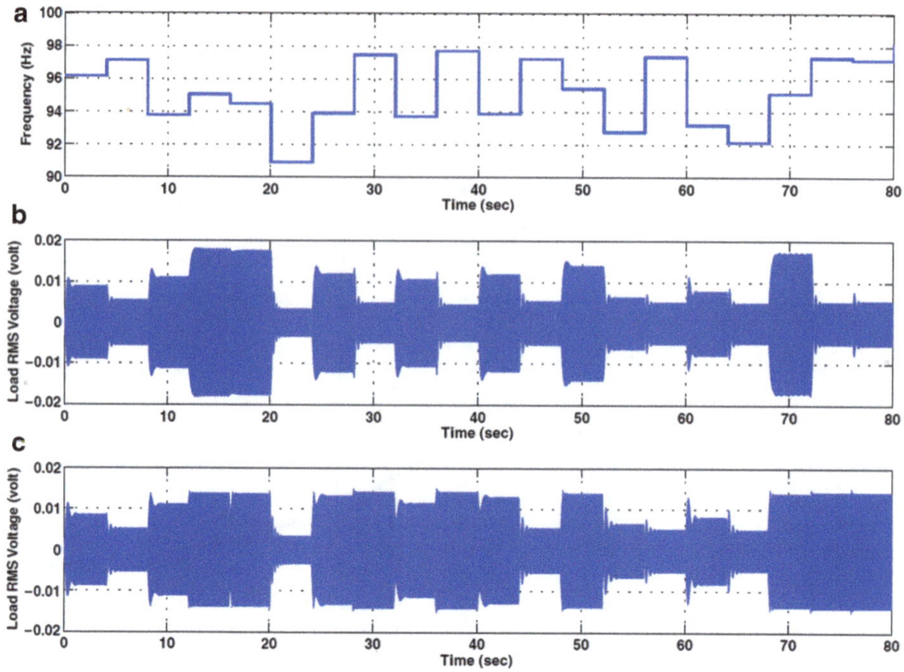

Fig. 2.42 (**a**) Time history of base excitation frequency and (**b**) RMS load voltage of no-stopper and (**c**) one-sided stopper harvesters by numerical simulation ([63], copyright: IOP Publishing)

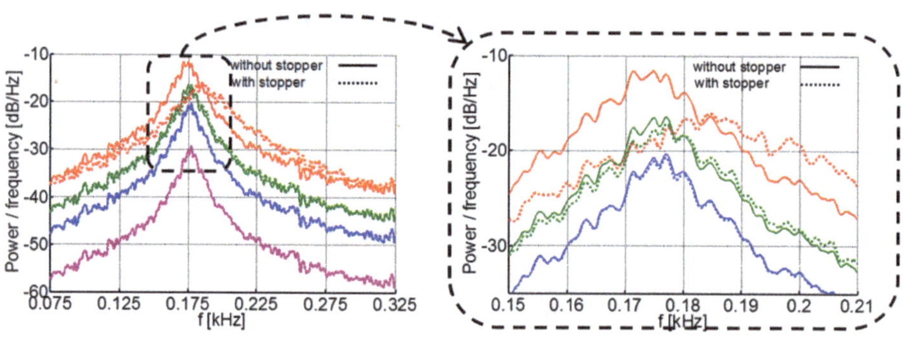

Fig. 2.43 Power spectral density of energy harvester output for increasing spectral density of excitation signal S_a. Without end stop (*solid line*) and with end stop (*dashed line*). $S_a = 0.087$, 0.82, 2.3, $8.0 \times 10^{-4} g^2 \mathrm{Hz}^{-1}$ starting from the lowest curve ([65], copyright: IOP Publishing)

stopper implementation did affect the jump phenomenon and thus the bandwidth during frequency upsweep. As to the different interface circuits, at low-excitation level (stopper not in effect), the sophisticated interfaces SECE and SSHI did not significantly enhance the performance. This is because the system they modeled was not weakly coupled. Similar result can be found in Tang and Yang [75]. At high excitation level, the SECE and SSHI interfaces were found to lead to much better performance, especially for SSHI. This is because when the stopper becomes effective, the dynamics of the harvester is controlled by the stopper and less affected by piezoelectric coupling. Thus the system can be regarded as weakly coupled, for which case SECE and SSHI have been proved to be capable of enhancing system performance [75, 76]. Under random excitations, Blystad et al. [64] found that the damping in the stopper model significantly affected the output power, which is different from the harmonic excitation case. Furthermore, they found that SECE gave significant larger output power than the standard interface at large random excitation levels (i.e. frequent impacts with the stopper). Moreover, less power for large excitation level was observed as compared to the no-stopper configuration. This was consistent with the findings in Soliman et al. [56, 63] and in their experimental work [65] mentioned before.

2.4.4 Comments on Nonlinear Energy Harvesting Configuration

This section concentrates on exploiting the nonlinearity of a system for broadband energy harvesting, with a focus on nonlinear stiffness. The nonlinear energy harvester can be a Duffing-type oscillator with cubic nonlinear stiffness typically introduced by using magnets. It can also be a piecewise-linear oscillator with nonlinearity caused by a mechanical stopper.

- *Monostable nonlinear configuration.* In both hardening and softening monostable configurations, the resonance curve can be bent to the right or left when the nonlinearity is engaged. When the nonlinearity is sufficiently strong, a broad bandwidth energy harvesting could be achieved. The advantage imparted by the nonlinearity depends on the implementation of high-energy attractor. A linearly decreasing or increasing frequency sweep for the softening or hardening case respectively can capture the high-energy attractor motion and hence acquire a large-amplitude response over a wide bandwidth. However, such characteristics limit its practical application, i.e., the monostable energy harvester can only work in the condition that a slow and proper frequency sweep excitation exists. Besides, since multi-value and jump phenomenon near resonance occur with increased nonlinearity (Figs. 2.29c and 2.31), a mechanism should be implemented to perturb and drive the system into high-energy orbits in case the system vibrates in a low-energy branch. Otherwise the harvester provides much lower output power.
- *Bistable nonlinear configuration.* For a bistable system, large-amplitude oscillation can occur under both periodic and stochastic forcing. Under high-level periodic forcing or low-level forcing with proper perturbation, the bistable harvester can be driven into high-energy orbits. Hence, it can outperform the linear device over a wide bandwidth. Under stochastic forcing, the bistable

system also shows significant performance improvement when the system parameters are properly selected, such as the distance between magnets (Δ in Cottone et al. [61]). The performance of the bistable harvester can be further improved by exploiting the stochastic resonance, in which the boundary should be properly forced to periodically change the potential barrier height and hence the probability of the large-amplitude transition between the two stable states. However, actively achieving this by using actuators in such methods require external energy input and are quite difficult to implement.

- *Configuration with mechanical stoppers.* Under harmonic excitation, incorporating mechanical stoppers increases the bandwidth of the harvester during a frequency upsweep, while maintaining the same bandwidth in a downsweep. With the consideration of the lowered power level, the harvester with stopper during a downsweep definitely provides worse performance as compared to the harvester without stopper. Under random excitation, the performance of the harvester with stopper is controlled by the stopper height h_0 and the offset distance l_0 of the stopper from the cantilever support. Thus, in an environment with a known vibration probability density function (PDF), l_0 and h_0 should be tuned to tailor the upsweep bandwidth to better fit the given PDF and h_0 should be set as high as possible to minimize the contact damping and to maximize energy collection. Thus, it is possible to have better performance with a stopper in harmonic upsweep and random scenarios if the enlarged bandwidth can be tuned properly to have more significant influence on the overall harvested energy than the influence by the lowered power level. However, harvesters with mechanical stoppers may suffer from noise, fatigue, and mechanical wear.

2.5 Conclusions

The fundamental drawback of linear resonating harvesters, i.e., the narrow bandwidth, limits their application in practical scenarios where the ambient vibration source is frequency-variant or random. This chapter summarized recent advances in broadband energy harvesting techniques, including resonance tuning techniques, multimodal energy harvesting, and nonlinear techniques. Obviously, there are some other broadband techniques that cannot be categorized into the three groups described in this chapter, for example, the frequency up-conversion technique by magnetic excitation [77] and the optimal inductor technique of Renno et al. [78] (the optimal inductance level may not be practical and synthetic inductors are required).

Thus there appears to be no "one-fits all" broadband energy harvesting solution. Each technique reviewed in this chapter is only preferable in specific conditions. A suitable technique for broadband vibration energy harvesting should be selected according to the type of excitation (periodic or stochastic), the variation of frequency (infrequent or frequent), the excitation level and the targeted frequency range, etc. The merits, weakness, and applicability of current techniques are summarized in Table 2.2, it provides some guidance for developing vibration-based energy

Table 2.2 Merits, weakness, and applicability of various broadband energy harvesting techniques

Methods		Merits, weakness, and applicability
Resonant frequency tuning	Active tuning	• Usually implemented by piezoelectric methods • Limited tunability • Piezoceramic leakage aggravates tuning power consumption • Net power increase is achievable in limited range • Easy to implement automatic and continuous tuning during operation • Applicable when excitation frequency varies frequently or for random excitation in a limited range, given affordable tuning power consumption
	Passive tuning	• Mostly by mechanical and magnetic methods • Relatively large tunability • For manual tuning or using controller for self-tuning: (i) Large power consumption for actuation (ii) Difficult to implement self-powered automatic control system for tuning (iii) Applicable when excitation frequency varies infrequently (iv) Complexity in system design and integration, especially for magnetic methods • Some exceptions in mechanical methods using inertia force: (i) Totally passive (no need of external tuning power) (ii) Self-adaptive to excitation frequency (iii) Continuous tuning during operation (iv) Applicable for specific conditions (e.g., rotational motion)
Multimodal energy harvesting		• Much easier to implement than resonance tuning techniques • Exploits multiple bending modes of a continuous beam or a cantilever array configuration • For a continuous beam, usually only first two modes contribute to effective energy harvesting • For a cantilever array configuration, continuous and larger bandwidth can be achieved but with significant sacrifice of power density • Should be designed with proper parameters to cover the targeted frequency range with least sacrifice of power density • Requires complex interface circuit to avoid voltage cancelation
Nonlinear techniques	Monostable	• Applicable for excitations with slow and proper frequency sweep • Requires certain perturbation when the harvester enters low-energy orbits
	Bistable	• Applicable for high-level periodic excitation • Applicable for low-level periodic excitation but with certain perturbation mechanism to drive the harvester into high-energy orbits • Applicable for random excitation with properly designed potential barrier at given excitation level • Further performance enhancement achievable by using stochastic resonance, which requires proper actuation and extra power
	Mechanical stopper	• Increases bandwidth by frequency upsweep • Increases bandwidth under random excitation with properly selected stopper parameters • Lowered power level • Modest performance improvement with enlarged bandwidth when tuned to fit a given vibration spectrum and thus to have a more significant influence on overall harvested energy • Suffers from noise, fatigue, and mechanical wear

harvesters. It is envisioned that, with further improvement of these broadband techniques, the concept of vibration energy harvesting will approach practical deployment in industrial applications as well as in our daily life.

References

1. Roundy S, Wright PK, Rabaey J (2003) A study of low level vibrations as a power source for wireless sensor nodes. Comput Commun 26:1131–1144
2. Mitcheson PD, Green TC, Yeatman EM, Holmes AS (2004) Architectures for vibration-driven micropower generators. J Microelectromech Syst 13:429–440
3. El-Hami M, Glynne-Jones P, White NM, Beeby S, James E, Brown AD, Ross JN (2001) Design and fabrication of a new vibration-based electromechanical power generator. Sens Actuators A 92:335–342
4. Anton SR, Sodano HA (2007) A review of power harvesting using piezoelectric materials (2003–2006). Smart Mater Struct 16:R1–R21
5. Yang YW, Tang LH, Li HY (2009) Vibration energy harvesting using macro-fiber composites. Smart Mater Struct 18:115025
6. Erturk A, Inman DJ (2008) A distributed parameter electromechanical model for cantilevered piezoelectric energy harvesters. J Vib Acoust 130:041002
7. De Marqui C Jr, Erturk A, Inman DJ (2009) An electromechanical finite element model for piezoelectric energy harvester plates. J Sound Vib 327:9–25
8. Yang YW, Tang LH (2009) Equivalent circuit modeling of piezoelectric energy harvesters. J Intell Mater Syst Struct 20:2223–2235
9. Elvin NG, Elvin AA (2009) A general equivalent circuit model for piezoelectric generators. J Intell Mater Syst Struct 20:3–9
10. Roundy S, Zhang Y (2005) Toward self-tuning adaptive vibration based micro-generators. Proc SPIE 5649:373–384
11. Leland ES, Wright PK (2006) Resonance tuning of piezoelectric vibration energy scavenging generators using compressive axial preload. Smart Mater Struct 15:1413–1420
12. Eichhorn C, Goldschmidtboeing F, Woias P (2008) A frequency tunable piezoelectric energy converter based on a cantilever beam. In: Proceedings of PowerMEMS, pp 309–312
13. Hu Y, Xue H, Hu H (2007) A piezoelectric power harvester with adjustable frequency through axial preloads. Smart Mater Struct 16:1961–1966
14. Morris DJ, Youngsman JM, Anderson MJ, Bahr DF (2008) A resonant frequency tunable, extensional mode piezoelectric vibration harvesting mechanism. Smart Mater Struct 17:065021
15. Youngsman JM, Luedeman T, Morris DJ, Andersonb MJ (2010) A model for an extensional mode resonator used as a frequency-adjustable vibration energy harvester. J Sound Vib 329:277–288
16. Loverich J, Geiger R, Frank J (2008) Stiffness nonlinearity as a means for resonance frequency tuning and enhancing mechanical robustness of vibration power harvesters. Proc SPIE 6928:692805
17. Wu X, Lin J, Kato S, Zhang K, Ren T, Liu L (2008) A frequency adjustable vibration energy harvester. In: Proceedings of PowerMEMS, pp 245–248
18. Gu L, Livermore C (2010) Passive self-tuning energy harvester for extracting energy from rotational motion. Appl Phys Lett 97:081904
19. Jo SE, Kim MS, Kim YJ (2011) Passive-self-tunable vibrational energy harvester. In: Proceedings of 16th international solid-state sensors, actuators and microsystems conference (TRANSDUCERS), pp 691–694
20. Challa VR, Prasad MG, Shi Y, Fisher FT (2008) A vibration energy harvesting device with bidirectional resonance frequency tunability. Smart Mater Struct 17:015035

21. Reissman T, Wolff EM, Garcia E (2009) Piezoelectric resonance shifting using tunable nonlinear stiffness. Proc SPIE 7288:72880G
22. Zhu D, Roberts S, Tudor J, Beeby S (2008) Closed loop frequency tunning of A vibration-based microgenerator. In: Proceedings of PowerMEMS, pp 229–232
23. Ayala-Garcia IN, Zhu D, Tudor MJ, Beeby SP (2010) A tunable kinetic energy harvester with dynamic over range protection. Smart Mater Struct 19:115005
24. Challa VR, Prasad MG, Fisher FT (2011) Towards an autonomous self-tuning vibration energy harvesting device for wireless sensor network applications. Smart Mater Struct 20:025004
25. Wu W, Chen Y, Lee B, He J, Peng Y (2006) Tunable resonant frequency power harvesting devices. Proc SPIE 6169:61690A
26. Peters C, Maurath D, Schock W, Mezger F, Manoli Y (2009) A closed-loop wide-range tunable mechanical resonator for energy harvesting systems. J Micromech Microeng 19:094004
27. Lallart M, Anton SR, Inman DJ (2010) Frequency self-tuning scheme for broadband vibration energy harvesting. J Intell Mater Syst Struct 21:897–906
28. Zhu D, Tudor J, Beeby S (2010) Strategies for increasing the operating frequency range of vibration energy harvesters: a review. Meas Sci Technol 21:022001
29. Wischke M, Masur M, Goldschmidtboeing F, Woias P (2010) Electromagnetic vibration harvester with piezoelectrically tunable resonance frequency. J Micromech Microeng 20:035025
30. Jang S-J, Rustighi E, Brennan MJ, Lee YP, Jung H-J (2011) Design of a 2DOF vibrational energy harvesting device. J Intell Mater Syst Struct 22:443–448
31. Aldraihem O, Baz A (2011) Energy harvester with a dynamic magnifier. J Intell Mater Syst Struct 22:521–530
32. Tang X, Zuo L (2011) Enhanced vibration energy harvesting using dual-mass systems. J Sound Vib 330:5199–5209
33. Roundy S, Leland ES, Baker J, Carleton E, Reilly E, Lai E, Otis B, Rabaey JM, Wright PK, Sundararajan V (2005) Improving power output for vibration-based energy scavengers. IEEE Pervasive Comput 4:28–36
34. Yang B, Lee C, Xiang W, Xie J, He JH, Krishna Kotlanka R, Low SP, Feng H (2009) Electromagnetic energy harvesting from vibrations of multiple frequencies. J Micromech Microeng 19:035001
35. Tadesse Y, Zhang S, Priya S (2009) Multimodal energy harvesting system: piezoelectric and electromagnetic. J Intell Mater Syst Struct 20:625–632
36. Ou Q, Chen X, Gutschmidt S, Wood A, Leigh N (2010) A two-mass cantilever beam model for vibration energy harvesting applications. In: Proceedings of 6th annual IEEE conference on automation science and engineering (CASE), pp 301–306
37. Arafa M, Akl W, Aladwani A, Aldrarihem O, Baz A (2011) Experimental implementation of a cantilevered piezoelectric energy harvester with a dynamic magnifier. Proc SPIE 7977:79770Q
38. Erturk A, Renno JM, Inman DJ (2009) Modeling of piezoelectric energy harvesting from an L-shaped beam-mass structure with an application to UAVs. J Intell Mater Syst Struct 20: 529–544
39. Berdy DF, Jung B, Rhoads JF, Peroulis D (2011) Increased-bandwidth, meandering vibration energy harvester. In: Proceedings of 16th international solid-state sensors, actuators and microsystems conference (TRANSDUCERS), pp 2638–2641
40. Wu H, Tang LH, Yang YW, Soh CK (2011) A novel 2-DOF piezoelectric energy harvester. 22nd international conference on adaptive structures and technologies (ICAST), Corfu, Greece, 10–12 October, paper no. 077
41. Yang Z, Yang J (2009) Connected vibrating piezoelectric bimorph beams as a wide-band piezoelectric power harvester. J Intell Mater Syst Struct 20:569–574
42. Kim I-H, Jung H-J, Lee BM, Jang S-J (2011) Broadband energy-harvesting using a two degree-of-freedom vibrating body. Appl Phys Lett 98:214102
43. Shahruz SM (2006) Design of mechanical band-pass filters for energy scavenging. J Sound Vib 292:987–998
44. Xue H, Hu Y, Wang Q (2008) Broadband piezoelectric energy harvesting devices using multiple bimorphs with different operating frequencies. IEEE Trans Ultrason Ferroelectr Freq Control 55:2104–2108

45. Ferrari M, Ferrari V, Guizzetti M, Marioli D, Taroni A (2008) Piezoelectric multifrequency energy converter for power harvesting in autonomous microsystems. Sens Actuators A 142:329–335
46. Liu J, Fang H, Xu Z, Mao X, Shen X, Chen D, Liao H, Cai B (2008) A MEMS-based piezoelectric power generator array for vibration energy harvesting. Microelectron J 39: 802–806
47. Sari I, Balkan T, Kulah H (2008) An electromagnetic micro power generator for wideband environmental vibrations. Sens Actuatators A 145–146:405–413
48. Cheng S, Jin Y, Rao Y, Arnold DP (2009) A bridge voltage doubler AC/DC converter for low-voltage energy harvesting applications. In: Proceedings of PowerMEMS, pp 25–28
49. Ramlan R, Brennan MJ, Mace BR, Kovacic I (2010) Potential benefits of a non-linear stiffness in an energy harvesting device. Nonlinear Dyn 59:545–558
50. Mann BP, Sims ND (2009) Energy harvesting from the nonlinear oscillations of magnetic levitation. J Sound Vib 319:515–530
51. Stanton SC, McGehee CC, Mann BP (2009) Reversible hysteresis for broadband magnetopiezoelastic energy harvesting. Appl Phys Lett 95:174103
52. Stanton SC, McGehee CC, Mann BP (2010) Nonlinear dynamics for broadband energy harvesting: investigation of a bistable piezoelectric inertial generator. Physica D 239:640–653
53. Erturk A, Hoffmann J, Inman DJ (2009) A piezomagnetoelastic structure for broadband vibration energy harvesting. Appl Phys Lett 94:254102
54. Marinkovic B, Koser H (2009) Smart sand—a wide bandwidth vibration energy harvesting platform. Appl Phys Lett 94:103505
55. Hajati A, Kim S-G (2011) Ultra-wide bandwidth piezoelectric energy harvesting. Appl Phys Lett 99:083105
56. Soliman MSM, Abdel-Rahman EM, El-Saadany EF, Mansour RR (2009) A design procedure for wideband micropower generators. J Microelectromech Syst 18:1288–1299
57. Lin J, Lee B, Alphenaar B (2010) The magnetic coupling of a piezoelectric cantilever for enhanced energy harvesting efficiency. Smart Mater Struct 19:045012
58. Triplett A, Quinn DD (2009) The effect of non-linear piezoelectric coupling on vibration-based energy harvesting. J Intell Mater Syst Struct 20:1959–1967
59. Stanton SC, Erturk A, Mann BP, Inman DJ (2010) Nonlinear piezoelectricity in electroelastic energy harvesters: modeling and experimental identification. J Appl Phys 108:074903
60. Erturk A, Inman DJ (2008) Issues in mathematical modeling of piezoelectric energy harvesters. Smart Mater Struct 17:065016
61. Cottone F, Vocca H, Gammaitoni L (2009) Nonlinear energy harvesting. Phys Rev Lett 102:080601
62. Gammaitoni L, Neri I, Vocca H (2009) Nonlinear oscillators for vibration energy harvesting. Appl Phys Lett 94:164102
63. Soliman MSM, Abdel-Rahman EM, El-Saadany EF, Mansour RR (2008) A wideband vibration-based energy harvester. J Micromech Microeng 18:115021
64. Blystad L-CJ, Halvorsen E, Husa S (2010) Piezoelectric MEMS energy harvesting systems driven by harmonic and random vibrations. IEEE Trans Ultrason Ferroelectr Freq Control 57:908–919
65. Blystad L-CJ, Halvorsen E (2011) An energy harvester driven by colored noise. Smart Mater Struct 20:025011
66. Moehlis J, DeMartini BE, Rogers JL, Turner KL (2009) Exploiting nonlinearity to provide broadband energy harvesting. In: Proceedings of ASME dynamic systems and control conference, DSCC2009-2542
67. Daqaq MF (2010) Response of uni-modal Duffing-type harvesters to random forced excitations. J Sound Vib 329:3621–3631
68. Erturk A, Inman DJ (2011) Broadband piezoelectric power generation on high-energy orbits of the bistable Duffing oscillator with electromechanical coupling. J Sound Vib 330:2339–2353
69. Ferrari M, Ferrari V, Guizzetti M, Andò B, Baglio S, Trigona C (2010) Improved energy harvesting from wideband vibrations by nonlinear piezoelectric converters. Sens Actuators A 162:425–431

70. Lin J, Alphenaar B (2010) Enhancement of energy harvested from a random vibration source by magnetic coupling of a piezoelectric cantilever. J Intell Mater Syst Struct 21:1337–1341
71. Andò B, Baglio S, Trigona C, Dumas N, Latorre L, Nouet P (2010) Nonlinear mechanism in MEMS devices for energy harvesting applications. J Micromech Microeng 20:125020
72. McInnes CR, Gorman DG, Cartmell MP (2008) Enhanced vibrational energy harvesting using nonlinear stochastic resonance. J Sound Vib 318:655–662
73. Formosa F, Büssing T, Badel A, Marteau S (2009) Energy harvesting device with enlarged frequency bandwidth based on stochastic resonance. In: Proceedings of PowerMEMS, pp 229–232
74. Wellens T, Shatokhin V, Buchleitner A (2004) Stochastic resonance. Rep Prog Phys 67:45–105
75. Tang LH, Yang YW (2011) Analysis of synchronized charge extraction for piezoelectric energy harvesting. Smart Mater Struct 20:085022
76. Shu YC, Lien IC, Wu WJ (2007) An improved analysis of the SSHI interface in piezoelectric energy harvesting. Smart Mater Struct 16:2253–2264
77. Wickenheiser AM, Garcia E (2010) Broadband vibration-based energy harvesting improvement through frequency up-conversion by magnetic excitation. Smart Mater Struct 19:065020
78. Renno JM, Daqaq MF, Inman DJ (2009) On the optimal energy harvesting from a vibration source. J Sound Vib 320:386–405

Chapter 3
MEMS Electrostatic Energy Harvesters with Nonlinear Springs

Einar Halvorsen and Son Duy Nguyen

Abstract Design of nonlinear proof mass suspensions is one among several possible strategies that can be adopted to enlarge the operating frequency range of energy harvesters. Emphasizing continuous mode operation, this chapter gives a brief overview of the working principles of electrostatic energy harvesters. We argue that nonlinear springs are particularly well suited to make nonlinear suspensions for MEMS electrostatic energy harvesters. We then discuss from a theoretical point of view how nonlinear springs will modify the vibration spectrum of the devices and what can be expected from them in terms of performance. Different nonlinear spring designs are presented together with recent experimental results on characterization of micromachined devices. With frequency sweeps or white-noise vibration, nonlinear devices have shown dramatic increases in bandwidth compared to their linear counterparts. Experiments with band-limited noise show that the use of nonlinear springs is a viable method to increase the harvester tolerance towards variations in vibration bandwidth and center frequency.

3.1 Introduction

A major weakness of traditional resonant vibration energy harvesters is their lack of capability to give appreciable output power when the vibration is different from a harmonic vibration at, or at least very close to, the resonant frequency. The use of nonlinear stiffness as a means to enlarge the bandwidth of energy harvesters was first

E. Halvorsen (✉) • S.D. Nguyen
Department of Micro and Nano Systems Technology, Faculty of Technology and Maritime Sciences, Vestfold University College, P.O. Box 2243, NO-3103 Tønsberg, Norway
e-mail: Einar.Halvorsen@hive.no; Duy.S.Nguyen@hive.no

N. Elvin and A. Erturk (eds.), *Advances in Energy Harvesting Methods*,
DOI 10.1007/978-1-4614-5705-3_3,

Table 3.1 List of reported methods to create nonlinearities of proof mass suspension

Source	Approach	Nonlinearity	Device scale	Transducer
Soliman et al. [4]	End-stops (for beam)	Hardening	Mesoscale	Electromagnetic
Blystad and Halvorsen [5]	End-stops (for proof mass)	Hardening	Mesoscale	Piezoelectric
Matsumoto et al. [6]	End-stops (for proof mass)	Hardening	MEMS	Electrostatic
Liu et al. [7]	End-stops (for proof mass)	Hardening	MEMS	Piezoelectric
Marinkovic and Koser [8]	Stretching at large deflections	Hardening	MEMS	Piezoelectric
Marzencki et al. [9]	Stretching at large deflections	Hardening	MEMS	Piezoelectric
Tvedt et al. [10]	Stretching at large deflections	Hardening	MEMS	Electrostatic
Burrow and Clare [1,2]	Magnetic forces	Bistable	Mesoscale	Electromagnetic
Stanton et al. [11]	Magnetic forces	Hardening	Mesoscale	Piezoelectric
Cottone et al. [12]	Magnetic forces	Bistable	Mesoscale	Piezoelectric
Erturk et al. [13]	Magnetic forces	Bistable	Mesoscale	Piezoelectric
Barton et al. [14]	Magnetic forces	Bistable	Mesoscale	Piezoelectric
Stanton et al. [11]	Magnetic forces	Softening	Mesoscale	Piezoelectric
Nguyen et al. [15]	Beam geometry	Softening	MEMS	Electrostatic
Nguyen et al. [16]	Beam geometry	Softening	MEMS	Electrostatic

studied by Burrow and Clare [1,2] and later followed up by a number of other works considering either of the three transduction mechanisms as shown by the examples in Table 3.1. Designing a nonlinear stiffness is one among several strategies that can be followed in enlarging the bandwidth of energy harvesters [3].

As indicated in the table, there are several ways that a nonlinear stiffness can be obtained. Here we focus on electrostatic energy harvesters where the nonlinear stiffness arises from the way the spring suspensions of the proof mass are designed. This approach is particularly well suited for micromachined electrostatic energy harvesters with in-plane motion. It hinges on the great freedom in choice of spring geometry that is offered by the technology.

We give a short introduction to MEMS electrostatic energy harvesters and discuss the mathematical modeling necessary to understand the devices and making assessments on the usefulness of the nonlinear spring approach. Then we consider nonlinear spring suspensions from an phenomenological point of view with emphasis on the particularities arising from mere geometric effects without any reliance on prestress or external forces. Motivated by this analysis, we present two methods to obtain the desired nonlinearities and summarize the most important findings of our recent characterization of microfabricated devices.

3.2 Vibration Waveforms and Spectra

Real world vibration spectra are seldom monochromatic but have instead a rich spectral content. One example is given in Fig. 3.1a below, which shows the estimated spectral density S_a of an acceleration time series measured in a car tire when driving at 50 km/h [17, 18]. This data set is for vibrations tangential to the tire circumference. It shows strong spectral peaks at the revolution frequency of the wheel and at a great number of its higher harmonics. Above about 100 Hz, the spectrum smears into a band that extends up to about 1 kHz. When driving speed is changing, the spectral peaks will shift much compared to their widths, while the smeared out band will only distort somewhat. One possible energy harvesting strategy is therefore to target the continuous band of the spectrum rather than the peaks. This demands an energy harvester that can respond effectively to a wideband vibration spectrum.

The spectrum shown in Fig. 3.1b was measured on the fan belt cage of a Heating, Ventilation and Air Conditioning (HVAC) system [19]. It has substantial weight below 200 Hz and two strong peaks at low frequencies. The peaks are reported to vary substantially in strength between measurements taken some days apart. Hence there can be variability in the spectral content that one would like an energy harvester to display a certain level of tolerance against.

In designing and characterizing energy harvesters, it is quite possible to use measured vibration waveforms in simulations [18, 20] and tests. In the absence of standard, agreed-upon measured benchmark waveforms, a model waveform with a few parameters has the advantage of being simple to define precisely and communicate to others.

For narrowband vibrations, it is customary to use a harmonic waveform with a certain acceleration amplitude A and angular frequency $\omega = 2\pi f$. In practice, a swept frequency (chirp) or swept amplitude is used. If the sweep rate is low enough, this approach ensures conditions close to sinusoidal steady state, but some artifacts can occur [4, 10, 21].

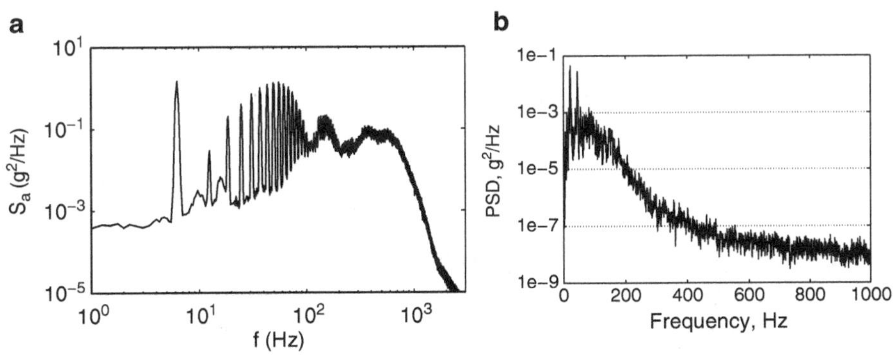

Fig. 3.1 Vibration spectrum examples. (**a**) Spectrum measured on the inner liner of a car tire while driving at 50 km/h. From [18]. (**b**) Vibration spectrum measured on the fan belt cage of a HVAC system. From [19]

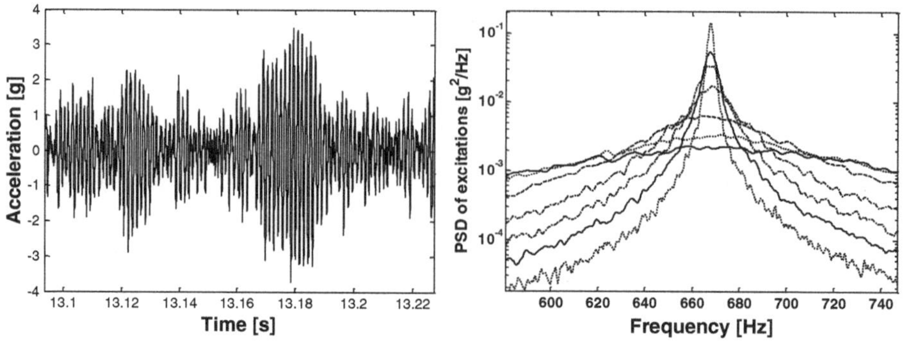

Fig. 3.2 Colored noise acceleration in time domain ($f_c = 668\,\text{Hz}$ and $\Delta = 50\,\text{Hz}$) and in frequency domain ($f_c = 668\,\text{Hz}$, fixed RMS acceleration and varying the bandwidth from 2 Hz to 150 Hz)

For energy harvesters driven by broadband vibrations, we have advocated the use of Gaussian white noise as a model signal [22, 23]. It has the advantage of being characterized by only one parameter, the spectral density S_a. We choose to normalize it so that the acceleration autocorrelation function is given by

$$K_{aa}(t - t') = \langle a(t)a(t') \rangle = S_a \delta(t - t'), \tag{3.1}$$

where $\langle \cdots \rangle$ denotes the statistical expectation, that is, it is a two-sided spectral density. This model signal allows application of the great body of knowledge on stochastic processes [24, 25] in theoretical treatments as well as in numerical treatments [26]. For experiments and use in conventional simulation tools, one can introduce a cutoff frequency by filtering white noise with a sufficiently high cutoff frequency not to influence the harvester behavior [12, 22, 23, 27, 28]. The noise is then strictly speaking "colored," but if the cutoff is sufficiently high, the low-pass or band-pass nature of the device makes the response independent of the actual value of the cutoff and the noise can be treated as white.

To model situations where the bandwidth and center frequency of the vibrations may vary, one can use white noise filtered by a band-pass filter. For a second-order filter, the spectrum is then given by

$$S_a(f) = \frac{1}{\pi} \frac{\Delta A^2 f^2}{(f_c^2 - f^2)^2 + \Delta^2 f^2}, \tag{3.2}$$

where A is the RMS acceleration, f_c is the center frequency of the colored noise acceleration, and Δ is the full bandwidth (at 3 dB). A is kept fixed when comparing the average output power under different bandwidths and center frequencies of the vibrations. Examples are shown in Fig. 3.2.

In the following sections all these model vibration waveforms will eventually be considered.

3.3 Electrostatic Harvesters

We give a brief introduction to electrostatic energy harvesters and the mathematical modeling that is most relevant for the nonlinear spring devices.

3.3.1 Device Principles

Electrostatic energy harvesters are often based on a capacitance varying with inertial mass movement as in electrostatic accelerometers [29]. Many varieties are possible; some are shown in Fig. 3.3. They can be classified based on the motion of the proof mass relative to the substrate and whether it is the capacitor gap or overlap that is modulated by the proof mass motion [30]. Hence, varieties (a), (e), and (f) are conventionally called *out-of-plane gap closing*, *in-plane overlap varying*, and *in-plane gap closing*, respectively. By the same logic, varieties (b), (c), and (d) should be referred to as in-plane overlap varying transducers. Other possibilities are modulation of both overlaps and gaps [31] or multi-axis devices [32].

The electrical work $\mathrm{d}W$ done on a capacitor when adding a charge $\mathrm{d}Q$ to it is $\mathrm{d}W = V\mathrm{d}Q$. When the device is made to traverse a closed contour c in the $V-Q$ plane, the total energy converted per cycle is $W = \oint_c V\mathrm{d}Q$ given by the area enclosed by the contour. Some conversion techniques rely on electronic switching circuitry to shape the contour, for example, the two canonical conversion cycles referred to as *charge constrained* and *voltage constrained*, respectively [33, 34]. Other more complicated cycles are also possible [35].

An alternative way to operate an electrostatic harvester is to build a bias into the device. This makes the electrostatic energy harvester more like a piezoelectric

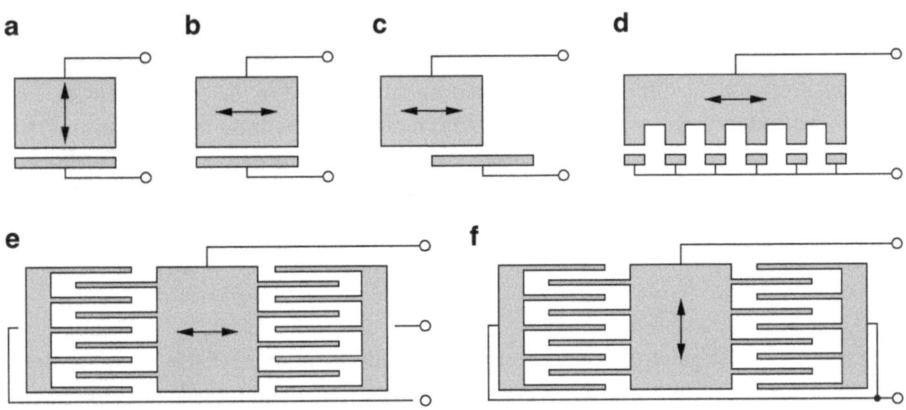

Fig. 3.3 Electrostatic energy harvester concepts based on the variation of a capacitance due to motion of a proof mass. Proof mass motion is indicated by double arrows. Suspensions are not shown

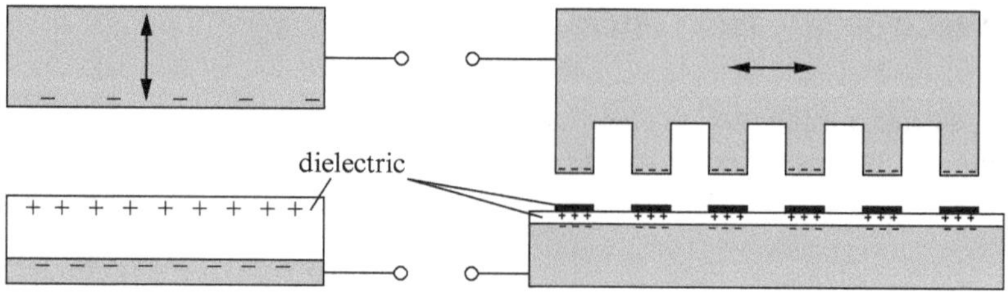

Fig. 3.4 Electrostatic energy harvesters biased by electret

one in being able to also deliver power to a passive load, in which case, it is said to operate in continuous mode [34]. Possible internal biasing sources are the following: material work function differences [36], floating electrodes charged by tunneling [37], betavoltaic sources [38], and *electrets* [39]. Electrets are electrically polarized materials, that is, dielectrics that have a trapped net charge distribution inside [40]. The use of electrets is by far the most common method of providing an internal bias to electrostatic harvesters. The dielectric could, for example, be a polymer [41,42] or a SiO_2/Si_3N_4 structure [43,44].

Two examples are shown in Fig. 3.4. The leftmost example is simply an out-of-plane gap-closing variety with charges embedded in the dielectric. The rightmost example has a striped electret pattern on an otherwise homogeneous substrate. The electrodes on the proof mass are patterned to match the patterned electret. Recent advances in fabricating electrets in vertical gaps show that they may also be used to bias devices such as those in Fig. 3.3e,f [45].

3.3.2 *Mathematical Modeling*

Since the electric field and flux density are linearly related to the charge distribution in any material configuration made from linear dielectrics and perfect conductors, the total electrostatic energy must be quadratically related to the charge density. For a two-electrode system with overall charge neutrality, the amount of charge Q on one of the electrodes specifies the charge distribution completely if the distribution of fixed charges is known. If one electrode is movable and its displacement given by the coordinate x, we can write the electrostatic energy as

$$E_{\mathrm{e}}(x, Q) = E_{\mathrm{e}}^{0}(x) + V_0(x)Q + \frac{1}{2C(x)}Q^2, \tag{3.3}$$

which is the most general quadratic form possible. It is valid if the structure does not deform with variation of Q when x is held constant. The transducer force is

$$F_{\mathrm{e}} = \frac{\partial E_{\mathrm{e}}}{\partial x} = E_{\mathrm{e}}^{0\prime}(x) + V_0'(x)Q + \frac{1}{2}\left(\frac{1}{C(x)}\right)' Q^2, \tag{3.4}$$

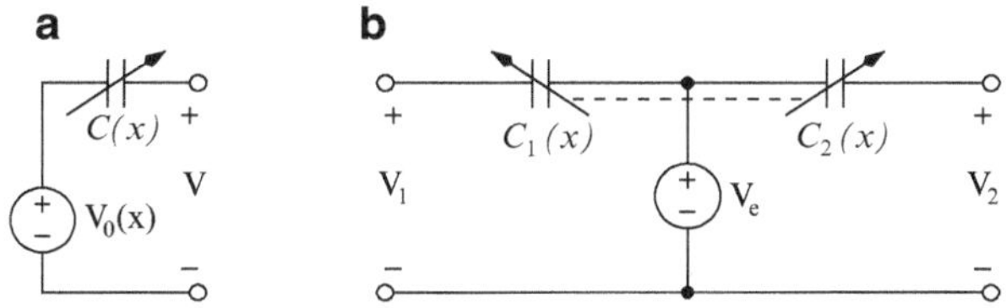

Fig. 3.5 Circuit diagrams for the electrical part of electrostatic harvesters with one mechanical degree of freedom. (**a**) Generic device with one electrical port. (**b**) Variable-capacitance device with two electrical ports and a constant, common priming voltage

and the voltage across the electrodes is

$$V = \frac{\partial E_e}{\partial Q} = V_0(x) + \frac{Q}{C(x)}. \tag{3.5}$$

E_e^0 is the electrostatic energy when the charge Q is zero and must be a quadratic form in the density of trapped charges. V_0 is the voltage between the electrodes for $Q = 0$ (open-circuit voltage) and is linearly related to the density of fixed charges. It changes when the electrode moves with respect to the fixed charges, for example, when a stripe patterned electrode moves above a stripe patterned electret as in Fig. 3.4. Finally C is the capacitance between the electrodes and is independent of the fixed charges. It is dependent on the displacement when capacitor-plate overlaps or gaps in the transducer change with displacement as in the examples in Fig. 3.3. As can be inferred from (3.5) or the corresponding circuit diagram in Fig. 3.5a, variations in electrode potential/current can arise from $V_0(x)$ when a load is connected. Hence, it is not necessary to have a variable capacitance in order to have transduction. For example, the rightmost design in Fig. 3.4 may have a negligible capacitance variation [18,46].

If we have instead two electrical ports, with related charges Q_1 and Q_2, the total electrostatic energy takes a quadratic form in these two charges. To be specific, consider the schematic representation of a harvester in Fig. 3.6 which corresponds to a device of the type shown in Fig. 3.3e. For this model, the electrostatic energy is

$$E_e(x, Q) = E_e^0(x) + V_e Q_1 + V_e Q_2 + \frac{1}{2C_1(x)} Q_1^2 + \frac{1}{2C_2(x)} Q_2^2 \tag{3.6}$$

In modeling the variable capacitances, the parallel plate capacitor formula for the inter-electrode capacitances $C_{c,1}$ and $C_{c,2}$ can be used together with stray capacitances $C_{p,1}$ and $C_{p,2}$ as indicated in the figure. Usually the two anti-phase capacitors are made equal, but due to asymmetries in the nonlinear springs, we consider different nominal overlaps l_1 and l_2 of the electrode fingers in the two transducers. The capacitances are then given by

$$C_{1/2}(x) = C_{p,1/2} + C_{1/2}^0 \left(1 \mp \frac{x}{l_{1/2}}\right), \tag{3.7}$$

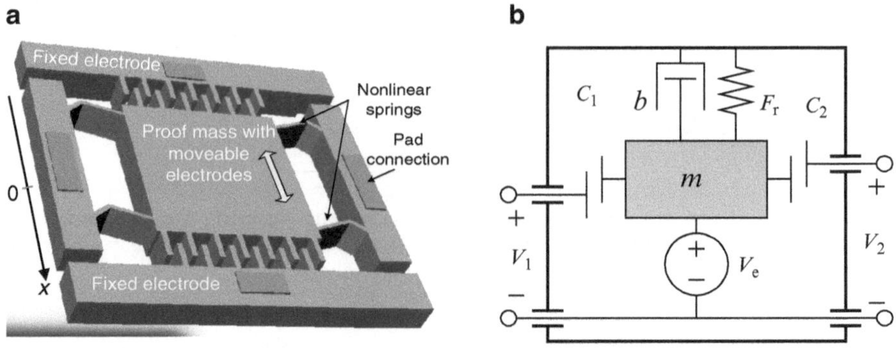

Fig. 3.6 Electrostatic energy harvester: (**a**) geometry (from [15]), (**b**) model

where $C^0_{1/2} = 2N\epsilon_0 t l_{1/2}/g$, N is the number of fingers in the comb structure, ϵ_0 is the permittivity of vacuum, t is the thickness of the fingers and g is the gap between the fingers. The stray capacitances in the model can account to some extent for fringing capacitances neglected in (3.7). If this model is too inaccurate at large displacements, the capacitances as a function of overlap length can always be calculated numerically. The port voltages in this model are

$$V_{1/2} = \frac{\partial E_e}{\partial Q_{1/2}} = \frac{Q_{1/2}}{C_{1/2}(x)} + V_e \tag{3.8}$$

represented by the circuit in Fig. 3.5b, and the electrical force is

$$F_e = \frac{\partial E_e}{\partial x} = \frac{1}{2}Q_1^2\frac{d}{dx}\frac{1}{C_1(x)} + \frac{1}{2}Q_2^2\frac{d}{dx}\frac{1}{C_2(x)}. \tag{3.9}$$

The mechanical equation of motion for a harvester, here with one mechanical degree of freedom, is simply Newton's second law for the proof mass, that is,

$$m\ddot{x} = -F_r - F_e - b\dot{x} - F_s + ma, \tag{3.10}$$

where m is the mass, a is the negative of the package acceleration, and b is the constant for linear mechanical damping representing parasitic mechanical loss. F_r, F_s, and F_e are the spring force, the end-stop force, and the electrical force, respectively.

The end-stop force can be modeled as a spring–damper system that is engaged at sufficiently large displacements. It is possible to use this as part of the design to exploit impacts as a means to improve the device bandwidth [47]. As we consider another route to that goal in this contribution, we will not pursue this further. It suffices to note that (a) end-stops will necessarily have to be included in a design, (b) they will tend to be much stiffer than the mechanical spring suspensions in order to restrict motion, and (c) for such rigid end-stops, the simple spring–dashpot model is adequate [5, 48].

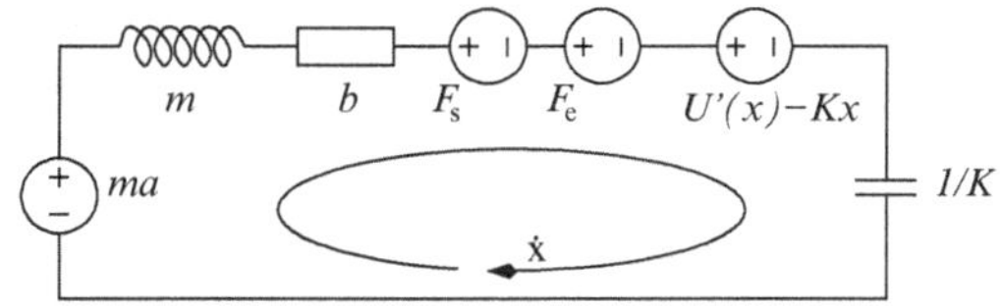

Fig. 3.7 Equivalent circuit for the mechanical part of an electrostatic energy harvester with nonlinear springs and nonlinear transducers

The spring force F_r is our primary concern and will be assumed to arise from an elastic potential energy U as $F_r = U'(x)$. The potential or the force can be modeled as a polynomial in x. Either on entirely phenomenologically grounds to get qualitative insight or as a polynomial fit to numerical calculations or to experimental data [10, 15].

Figure 3.7 shows an equivalent circuit for the mechanical part of the electrostatic energy harvester. Together with either of the two circuits in Fig. 3.5, it constitutes a completed harvester lumped model. To have a complete system model, one needs to add also a load that power can be delivered to. We will consider simple resistive loads, but other choices are possible and may even be desirable. That said, there are conversion circuits that appear as a resistive load as seen from the harvester electrical port [49].

With a resistive load, the system of equations is closed by Ohms law $V = -R\dot{Q}$ and (3.5) or $V_{1/2} = -R_{1/2}\dot{Q}_{1/2}$ and (3.8). The models can be solved by integrating the equations of motion in an ordinary differential equation solver [50] or in a standard circuit simulator [10]. The behavioral sources in, for example, SPICE can then be used to implement the variable capacitors and state-dependent sources in Figs. 3.5 and 3.7. For harmonic vibrations, the harmonic balance method [51] and multiscale analysis [52] are powerful alternatives to simulation.

For white-noise vibrations, the probability density p of the state of the energy harvester at any given time is given by the Fokker–Planck equation [53]. For the generic one-transducer device above, the equation for the stationary probability distribution was given in [23] and analyzed for some special cases. It is instructive to consider a model with linear transduction and no end-stop effects so that all nonlinearities arise from the springs. Formally this can be deduced by linearizing (3.4), (3.5) around the equilibrium operating point x_0, Q_0, that is, if we choose coordinates such that $x_0 = 0$ and set $Q = Q_0 + q$, then

$$m\ddot{x} = -[U'(x) - U'(0)] - K_e x - b\dot{x} - \Gamma q/C_0 + ma, \tag{3.11}$$

$$-R\dot{q} = \Gamma x/C_0 + q/C_0, \tag{3.12}$$

where $K_e = E_e^{0\prime\prime}(0) + V_0^{\prime\prime}(0)Q_0 + (1/2C(x))^{\prime\prime}|_{x=0}Q_0^2$, $C_0 = C(0)$, $\Gamma = C_0(V_0^{\prime}(x) + (1/C(x))^{\prime}Q_0|_{x=0})$, and R is the load resistance.

The leading asymptotics for the output power in the limits $\tau = RC_0 \to 0$ and $\tau \to \infty$ are, respectively, [23]

$$P \sim \Gamma^2\langle v^2\rangle\tau/C_0 \sim \tau\Gamma^2 m S_a/2C_0 b \tag{3.13}$$

and

$$P \sim \Gamma^2 \langle (x - \langle x \rangle)^2 \rangle / \tau C_0. \quad (3.14)$$

The statistical expectation in (3.14) is asymptotically given by the stationary probability distribution for the displacement and velocity with an open-circuit transducer

$$W_{\text{st}}(x, v) = \exp\left(-\frac{b}{m S_a} v^2 - \frac{2b}{m^2 S_a} \tilde{U}(x) \right) / Z_{\text{st}}, \quad (3.15)$$

where Z_{st} is a normalization constant and $\tilde{U} = U - U'(0)x + K_e x^2/2$.

Since the output power approaches zero in both limits $\tau \to 0$ and $\tau \to \infty$, the optimum load is at some finite, intermediate value. It is therefore misleading to consider either of the above two limits in judging whether or not a specific nonlinear stiffness would give more or less output power than another. From (3.13), one would then erroneously conclude that a nonlinear stiffness has no effect on the output power, while (3.14) would suggest that it is sufficient to maximize the mean square displacement. Clearly neither is true as there is a load dependence. Furthermore, in either limit, a much simpler optimization is to change τ, but that will eventually bring the system out of the asymptotic regimes where (3.13) are (3.14) valid. For small vibrations, when linear theory applies, the optimal choice is $\tau = 1/\omega_0$, where ω_0 is the open-circuit angular resonant frequency [23]. To reduce the workload in experiments, it is convenient to use the optimum τ for small vibrations over a range of vibration strengths. This makes sure that the load is optimal at least in one end of the range and that the asymptotic limits above are avoided at least for a range of vibration strengths. To the extent that it misrepresents the performance in the nonlinear regime, it underestimates the achievable output power.

3.4 Nonlinear Stiffness

In this section we discuss the background and motivation for using nonlinear springs. Then we argue for a special phenomenological form of the spring elastic potential and show that this potential is capable of giving increased bandwidth of the vibration spectrum. Finally we discuss bounds on the output power and what can be expected in terms of performance in general when utilizing mechanical nonlinearities.

3.4.1 Background and Motivation

Nonlinear stiffness can take a great variety of forms, but the archetypical examples come from the Duffing spring [54,55]. A Duffing spring has the force–displacement relationship

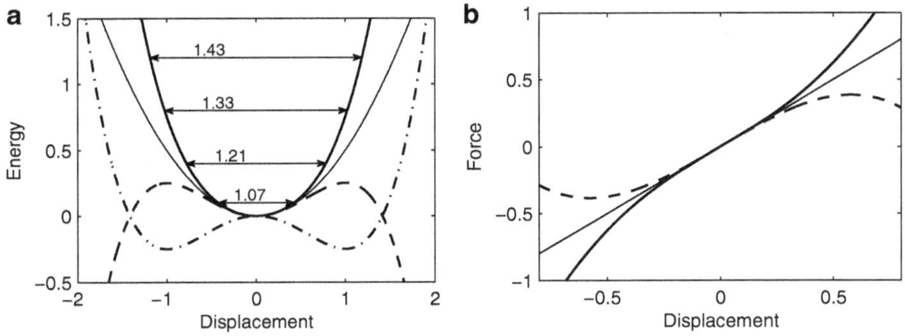

Fig. 3.8 Duffing spring examples. (**a**) Potentials. (**b**) Forces. *Solid line*: hardening spring. *Dashed line*: softening spring. *Dash-dotted line*: bistable system. *Thin solid line*: linear spring

$$F = Ax + Bx^3, \tag{3.16}$$

which follows from a symmetric quartic potential

$$U = \frac{1}{2}Ax^2 + \frac{1}{4}Bx^4. \tag{3.17}$$

Some typical potentials are shown in Fig. 3.8a: (a) $A > 0$ and $B = 0$, the linear spring; (b) $A < 0$ and $B > 0$ resulting in two equivalent stable minima, that is, a symmetric bistable potential; (c) $A > 0$ and $B > 0$ resulting in a spring tangential stiffness $U''(x)$ that is increasing with displacement and is therefore called a *hardening* (or *stiffening*) spring; and (d) the *softening* spring which is characterized by $A > 0$ and $B < 0$ and has a stiffness decreasing with displacement. The softening and hardening spring forces are shown in Fig. 3.8b.

It is intuitively clear that a spring with amplitude-dependent stiffness should also display amplitude-dependent characteristic frequencies. To make it precise, consider the possible natural frequencies of a proof mass m suspended in a general nonlinear spring with elastic energy U. For some selected energies, the corresponding displacement ranges and frequencies are calculated [56] and depicted for the hardening spring potential in Fig. 3.8a. It shows that a range of different frequencies are possible but that they correspond to different total energies. This is in contrast to a linear system which has a quadratic potential $U = Kx^2/2$ and oscillates at $\omega = \sqrt{K/m}$ no matter what the total energy is.

Typical frequency responses of a forced mass–spring–damper system with softening and hardening Duffing springs are shown in Fig. 3.9. If excited by a constant amplitude chirp vibration, a stable orbit will be followed in until a jump to the other stable orbit is made necessary by the present orbit ceasing to exist at a certain vibration frequency. The traces suggest that a bandwidth enhancement can be obtained for down-sweeps with a softening spring and for up-sweeps with a hardening spring.

Fig. 3.9 Proof mass displacement frequency responses for softening and hardening springs

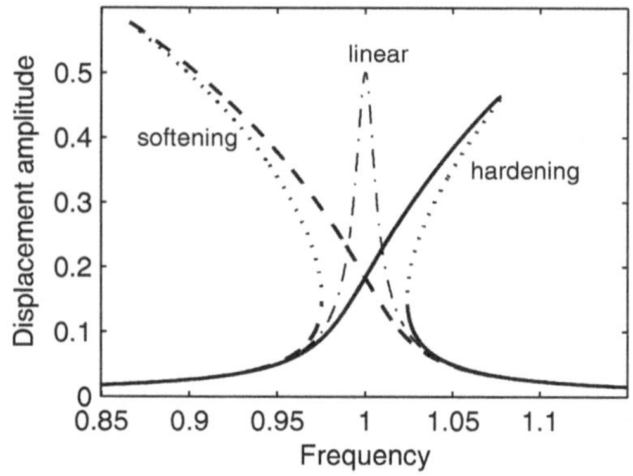

MEMS electrostatic energy harvesters are typically made from high aspect ratio etch of single crystal silicon. The material will therefore have little or no prestress except for possible packaging or mounting-induced stresses. Furthermore it is of interest to exploit the great freedom in layout geometry to obtain nonlinear behavior rather than adding extra features like magnets. For a spring suspension, we can then constrain the possible elastic energies $U = U(x)$. If we take $x = 0$ as the global equilibrium state in which the device is made, we have $U'(0) = 0$ and $U''(0) > 0$ for it to be locally stable. The absence of built-in stresses means $U(0) = 0$, and global stability requires $U(x) \geq 0\ \forall\ x$.

The softening Duffing spring violates the stability criterion $U(x) \geq 0$ for sufficiently large x but can represent an elastic suspension over a limited range of displacements. The symmetric bistable potential is not exactly realizable but can be approximated by an asymmetric potential as will be argued below. The hardening Duffing spring is typically used as a model for a clamped–clamped beam that experiences stretching at sufficient displacement amplitudes and is therefore quite simply realized [57].

3.4.2 Phenomenological Elastic Potential of Springs

With a white noise excitation, it is not possible to use the softening Duffing spring as a model, because a stationary state does not exist. It is therefore of interest to investigate a simple phenomenological potential that fulfills the stability criteria of a stable elastic suspension that is not prestressed [68]. The lowest-order polynomial potential that can fulfill criteria, and at the same time display nonlinearity, is a quartic polynomial on the form $U(x) = \frac{1}{2}Kx^2 + \frac{1}{3}K_3x^3 + \frac{1}{4}K_4x^4$, where $K > 0$, $K_4 > 0$, and $|K_3| < \frac{3}{2}\sqrt{2KK_4}$. To have a somewhat more transparent parametrization, define the length $l = \sqrt{K/K_4}$ which is such that the linear force equals the cubic force when $x = l$. Then introduce a nondimensional parameter

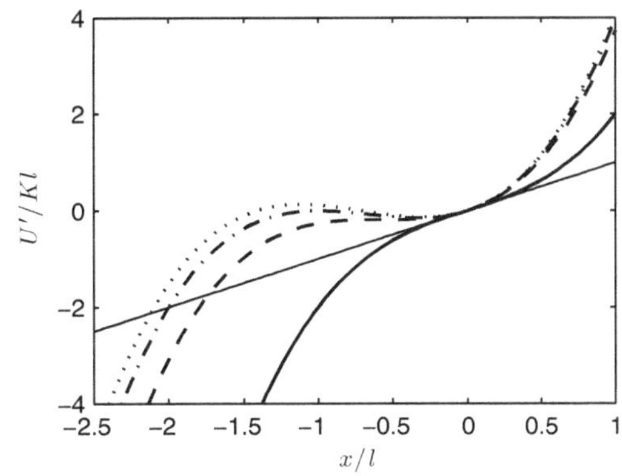

Fig. 3.10 Forces derived from phenomenological potentials for nonlinear springs for (*solid line*) $\eta = 0$, (*dashed line*) $\eta = \sqrt{2/3} \approx 0.8165$, (*dash-dotted line*) $\eta = 2\sqrt{2}/3 \approx 0.9428$, (*dotted line*) $\eta = 1$, and (*thin solid line*) linear spring

η such that $K_3 = 3\eta\sqrt{2KK_4}/2 = 3\sqrt{2}K\eta/2l$. The phenomenological quartic potential then reads

$$U(x) = \frac{1}{2}Kx^2 + \frac{K\eta}{\sqrt{2}l}x^3 + \frac{K}{4l^2}x^4. \tag{3.18}$$

We can think of the parameter l as a measure of the confinement provided by the quartic term functioning as a nonrigid end-stop. The parameter η makes sure that the stability criterion is fulfilled for $|\eta| < 1$ and gives us a one-parameter family of nonlinear potentials to be considered. If we take l as the unit of length and Kl^2 as the unit of energy, η is the only free parameter.

Some example force–displacement relations are given in Fig. 3.10. From the figures and some simple analysis of (3.18), we can identify several possible traits of the elastic potential. For $\eta = 0$, we have a hardening Duffing spring. For any $\eta \neq 0$, we will always have a hardening spring behavior in one direction. For $\eta > 0$, this is the case for $x > 0$ as shown in Fig. 3.10. In the other direction, the behavior is more complicated. As $|\eta|$ increases from zero, a displacement range of softening spring behavior exists up a certain magnitude of displacement. This range extends as $|\eta|$ increases. At $|\eta| = \sqrt{2/3}$, a range of negative tangential stiffness ($U''(x) < 0$) appears, and at $\eta = 2\sqrt{2}/3$, a bistability appears. The bistable potential is asymmetric but approaches a symmetric (about $x = \pm l/\sqrt{2}$) potential as $\eta \to \mp 1$. The case $|\eta| = \sqrt{2/3}$ also arises for the purely cubic Duffing oscillator with a constant force term [55].

Figure 3.11 shows the spectral density of velocity for spring–mass–damper systems with the elastic potentials discussed above. It was calculated using the matrix continued fraction method [53, 58]. Even for the hardening spring, we see that a certain bandwidth increase is obtained with increasing spectral density of acceleration. For nonzero η, we find considerably larger increases of bandwidth. The total spectral weight is the same for all varieties of the elastic potential and the same spectral density of acceleration because in the stationary state, the joint probability density has the simple form (3.15) with $\tilde{U} = U$.

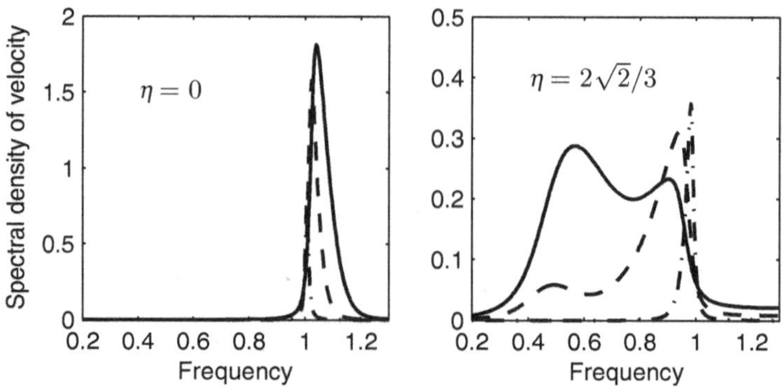

Fig. 3.11 Spectral density of velocity as a function of frequency

3.4.3 Power Bounds

Before proceeding to the actual design of nonlinear beams and the resulting harvester performance, it is instructive to consider what can be expected from a fundamental point of view without regard to the transduction method. Absolute upper bounds on the mean output power P are most easily achieved by studying the mean power P_{in} into the energy harvester as P obviously cannot exceed that. We will continue to restrict attention to devices with one mechanical degree of freedom.

For periodic vibrations, a simple and elegant derivation is possible as follows [59]. If the displacement limit is Z_l, the work done on the proof mass during a cycle cannot exceed $4Z_l \cdot \max_t m|a(t)| = 4mAZ_l$, where A is the peak acceleration amplitude. This is valid under the very weak assumption of the proof mass motion path during a period of vibration being no longer than that for motion back and forth between the extremes. Dividing by the period $T = 2\pi/\omega$, we get the bound

$$P \leq P_{\text{in}} \leq \frac{2}{\pi} m\omega A Z_l. \tag{3.19}$$

This bound makes no assumptions on the nature of the proof mass suspension and is valid regardless of whether it is linear, nonlinear, or at all present. For a sufficiently high $k^2 Q_m$, where k^2 is the electromechanical coupling factor and Q_m is the open-circuit mechanical Q, a resistively loaded linear device can be optimally operated at a low-load resistance [60], which means that it behaves as a velocity-damped generator. Such a device already has an input power that is $\pi/4 \approx 0.79$ of (3.19) [61].

For Gaussian white-noise vibrations, the input power is always given by $P_{\text{in}} = mS_a/2$. This was pointed out by Scruggs [62]. It is valid for any kind of suspensions and any kind of power conditioning circuitry. Hence, we have the bound

$$P \leq P_{\text{in}} = \frac{1}{2} m S_a. \tag{3.20}$$

In [23], it was shown that this limit is approached by a linear device when $k^2 Q_m \to \infty$. Hence, nonlinearities cannot improve an already efficient linear device.

In summary, the power bounds suggest that one cannot substantially improve the output power of a linear device with a high $k^2 Q_m$. It is rather for low $k^2 Q_m$ that we may expect an improvement in output power, but the greatest potential for improvement seems to lie in the increase of the harvester's frequency operating range. This can be important even for large $k^2 Q_m$, as it is k^2 that limits the maximum bandwidth of a linear device.

3.5 The Design of Nonlinear Springs

In designing nonlinear springs for microfabricated energy harvesters, there is great freedom in how to shape the springs. Almost any shape that can be drawn and which respects the minimum critical dimensions of the process at hand is feasible. In principle it should therefore be possible to determine by computational means what is the optimal force-deflection characteristic, and then use generic optimization methods [63] to generate a close approximation to this shape. Here, we will instead restrict attention to simple, parametrized spaces of beam shapes.

The analysis in Sect. 3.4.2 indicated that an asymmetric elastic potential with softening spring characteristics is promising with respect to increasing the operating bandwidth. To achieve this characteristic, we therefore consider two different types of spring designs that we will refer to as *angled* and *curved beams*, respectively.

3.5.1 Angled Beams

Figure 3.12 shows the geometry of an angled beam. When the spring is displaced in the negative x-direction, this geometry will initially experience, apart from bending, compressive axial forces along BC opposing the motion. With increasing displacement, these will relax and eventually become tensile. The resulting nonlinear characteristics is therefore first softening and then hardening. For displacements in the positive x-direction, tension builds up without an intermediate state of compressive forces. The spring therefore has an asymmetric force–displacement curve.

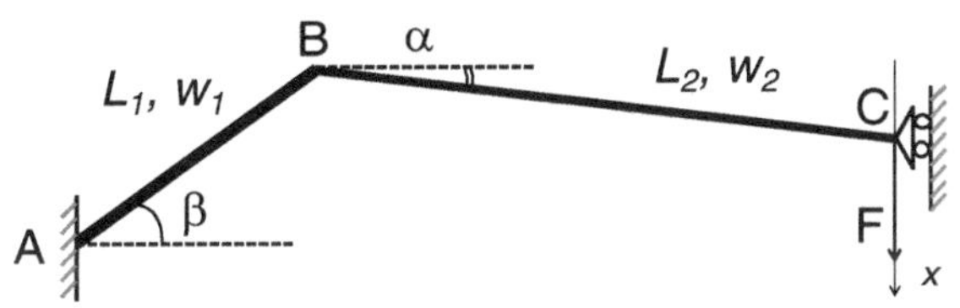

Fig. 3.12 The geometry of the angled beam. L_1, w_1 and L_2, w_2 are the lengths and widths of members AB and BC, respectively

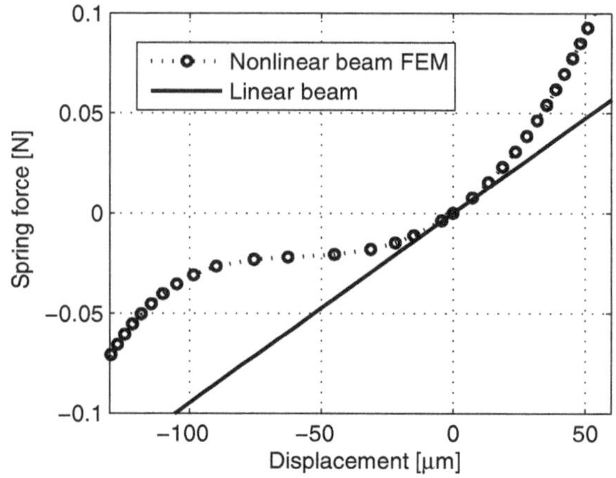

Fig. 3.13 The spring force vs. displacement for an angled beam obtained by FEM with $L_1 = 200\,\mu\text{m}$, $w_1 = 20\,\mu\text{m}$, $L_2 = 1,200\,\mu\text{m}$, $w_2 = 15\,\mu\text{m}$, $\alpha = 3.814^0$, and $\beta = 45^0$. The beam thickness $t = 300\,\mu\text{m}$, Young's modulus $E = 169\,\text{MPa}$. From [15]

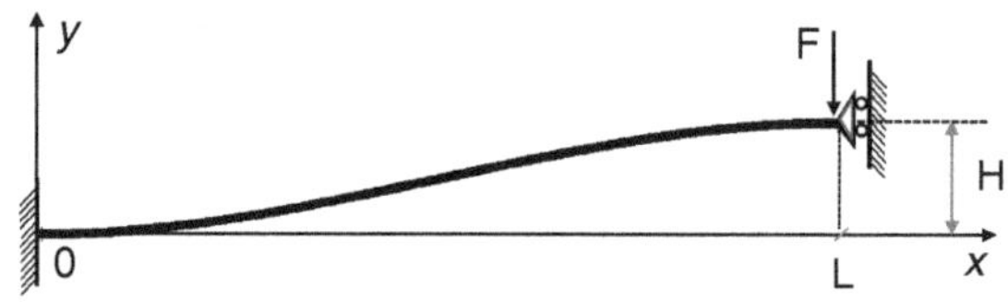

Fig. 3.14 Shape of the curved beam in its undeformed state

Figure 3.13 shows the spring force vs. displacement for an example spring design calculated using FEM. It is clearly asymmetric and qualitatively of the form previously arrived at in Fig. 3.10. It has a softening spring regime for a displacement down to about $-90\,\mu\text{m}$. In the opposite direction, it displays hardening spring behavior for the entire range of displacements.

We can optimize a design by varying the set of parameters L_1, L_2, β, and α and determine the performance from the resulting potential or force–displacement relation. The linear stiffness will decide the resonant frequency in the small vibration regime and, as seen in Fig. 3.11, approximately the high frequency cutoff in large vibration regime. To control the operating frequency range, it therefore makes sense to keep the linear stiffness constant in varying the parameters of the angled beam.

3.5.2 Curved Beams

An alternative to the angled beam design is a curved beam made to have the same undeformed shape as the deformed shape of a straight beam as shown in Fig. 3.14. The deviation from a straight beam is then given by

$$w(x) = H\frac{x^2}{L^2}\left(3 - 2\frac{x}{L}\right), \tag{3.21}$$

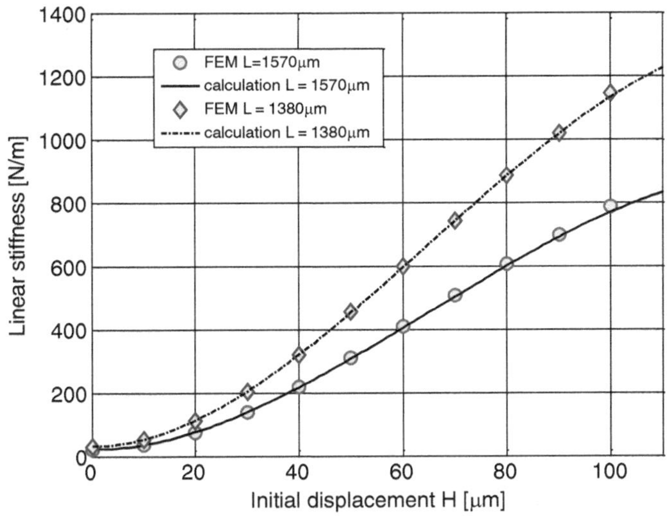

Fig. 3.15 The linear stiffness of the curved beam for the beam length of $L = 1,380\,\mu m$ and $L = 1,570\,\mu m$, the other parameters: $t = 150\,\mu m$, $w = 15\,\mu m$, $E = 169\,\text{GPa}$

where H is the initial distance between the base and tip along the y-axis, and L is the beam length in x-axis. Equation (3.21) is the deflection function of a straight clamped-guided beam at small deflection. This principle of nonlinear beam design was developed in [64] for actuator applications and was investigated for use in nonlinear energy harvesters in [65].

The linear stiffness k of the curved beam strongly depends on H, so to keep the linear stiffness while changing H, we adjust the length L at the same time. To accomplish that using a numerical solution, we approximate the linear stiffness k for a given cross-section as

$$k \approx \frac{Etw^3}{L^3} f(\widehat{H}), \tag{3.22}$$

where E is the Young's modulus, t is the beam thickness, w is the beam width, and $\widehat{H} = H/w$. Etw^3/L^3 is the stiffness of a straight clamped-guided beam at small deflection. We find $f(\widehat{H})$ by fitting a simple polynomial to the numerical calculation. An example fit is

$$f(\widehat{H}) = 1 + 1.56\widehat{H}^2 - 0.12\widehat{H}^3 \tag{3.23}$$

as shown in Fig. 3.15 which compares the fit and the finite element result for a selection of dimensions.

Figure 3.16 shows the spring force vs. displacement for different H. For moderate values of $H \leq 40\,\mu\text{m}$, the force–displacement curve looks qualitatively similar to what we already have discussed for the angled beam and in the phenomenological

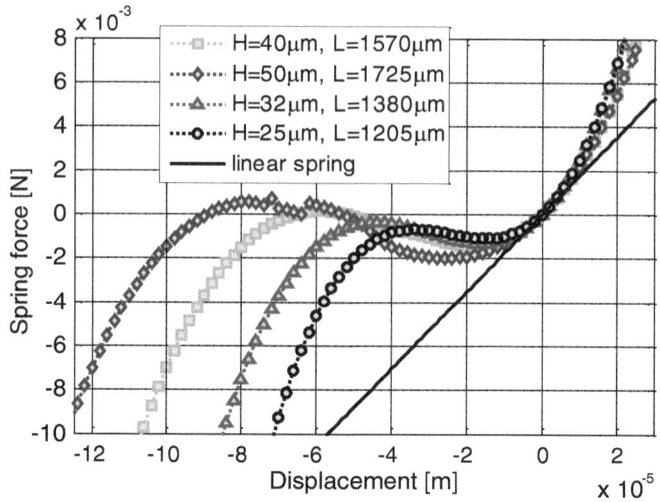

Fig. 3.16 Spring force vs. displacement for different initial displacements H, calculated by FEM [16]

treatment. For large enough initial displacement H, for example $H = 40\,\mu\text{m}$ and $H = 50\,\mu\text{m}$, the force is zero at three points representing two local minima and one local maximum of the corresponding potential energy, that is, we have bistability. The irregular features of the curve for $H = 50\,\mu\text{m}$ are due to a further elastic instability for the shape of the beam [66]. By varying H, the nonlinear spring design can thus be varied continuously from linear to asymmetrically bistable.

3.6 Measured Device Characteristics

In the light of the previous analysis, we now discuss measurements on three electrostatic energy harvesters with, respectively, straight beams, angled beams, and curved beams.

3.6.1 Harvester with Straight Beams

As mentioned in the previous section, the hardening Duffing spring is quite simply realized by a straight clamped-guided beam. An electrostatic device with straight beams was presented in [10]. The design is similar to that sketched in Fig. 3.6a except that the inertial mass is suspended in four straight beams, of which two has anchors also serving as electrical contact pads. The beams have no stress relief so

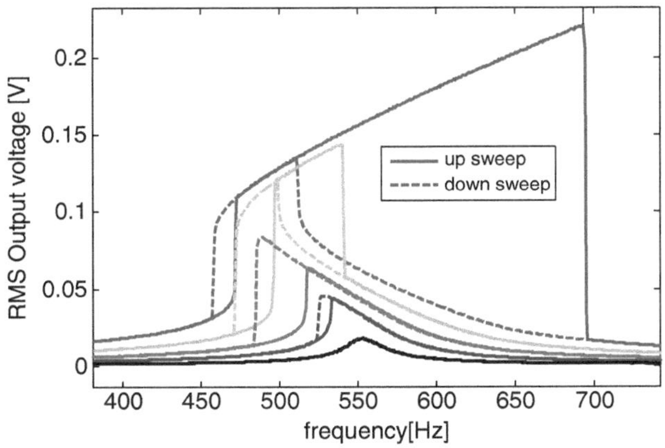

Fig. 3.17 Measurement of up and down frequency sweep excitation at 10 V bias with 0.015, 0.03, 0.06, 0.1, and 0.16 *g* RMS excitation (from lower to upper curve). From [10]

that they behave as hardening springs at sufficiently large displacement amplitudes. The anchors also function as mechanical stoppers at very large amplitude vibrations to avoid the capacitor fingers of the electrodes to collapse into each other. During test, the device had an external bias to emulate an electret.

The response of the energy harvester under linearly increasing frequency sweeps (up-sweeps) and decreasing frequency sweeps (down-sweeps) is shown in Fig. 3.17. At a small acceleration amplitude of 0.015 g RMS, the device behaves as a linear resonant harvester with a resonant frequency of 550 Hz and a 3-dB bandwidth of 21 Hz. For larger accelerations (0.03 g and 0.06 g RMS), the harvester shows a softening spring effect. For further increase in amplitude, the peak frequency drifts upward and develops yet another region of hysteresis or multivaluedness in the upper frequency range. Hence the device displays typical properties of both softening and stiffening springs.

The spring softening observed in this device was not by design. In [10], the analysis showed that it was a mechanical effect and was attributed to packaging or mounting stresses. A model of the type discussed in Sect. 3.3.2 with a symmetric spring potential exhibiting both softening and hardening regimes was able to account for all main features of the nonlinear response.

Figure 3.18 shows the output spectrum under white-noise vibrations. At $S_a = 1.6 \cdot 10^{-6}\,\mathrm{g}^2/\mathrm{Hz}$, the harvester response has a resonance at 550 Hz and bandwidth of about 19 Hz comparable to that measured with sinusoidal vibrations. The aforementioned softening behavior is seen upon increasing the vibrations, but at $S_a = 392 \cdot 10^{-6}\,\mathrm{g}^2/\mathrm{Hz}$, a 3-dB bandwidth of about 322 Hz is achieved due to the hardening spring effect. On output power, this device performed somewhat below that of the theoretical result for a corresponding linear device, but it demonstrated that a considerable bandwidth increase can be obtained from a hardening spring.

Fig. 3.18 Output PSD with several excitation levels 1.6×10^{-6}, 19×10^{-6}, 128×10^{-6}, and $392 \times 10^{-6} g^2/Hz$ at 10-V bias. Adapted from [10]

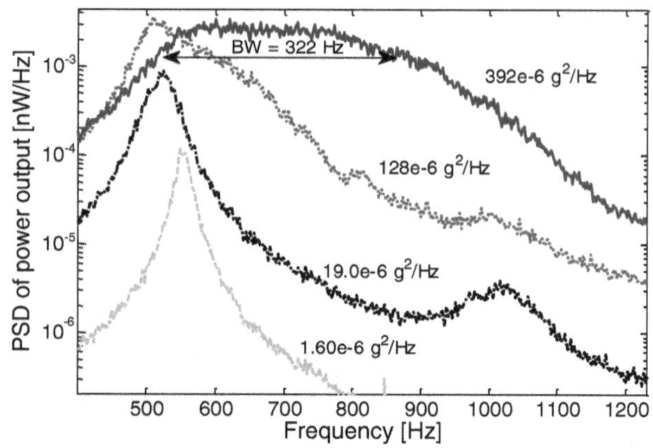

3.6.2 Harvesters with Angled Beams

A MEMS electrostatic energy harvester utilizing the angled beams as shown in Fig. 3.6 was presented in [15]. The force–displacement relation for the angled beam is designed as shown in Fig. 3.13.

When driven by vibrations, the asymmetric force–displacement relation causes the mean position of the proof mass to shift into the range of low compliance, giving an overall softening spring behavior. Therefore, the two variable-capacitor transducers are designed with different initial and maximum overlap. In the following, we refer to them as transducer 1 and transducer 2 for, respectively, the upper and lower spring drawn in Fig. 3.6a. The proof mass can move 125 μm in the negative direction of the x-axis corresponding to a softening spring regime and 45 μm in the other direction corresponding to a hardening spring regime. The maximum displacement of the proof mass is limited by mechanical end-stops. Again, the device had an external bias under test to emulate an electret.

A fabrication process specifically targeted towards prototyping MEMS devices with narrow springs and high aspect ratio spring gaps formed by through-wafer-thickness dry etch of a silicon wafer was developed for this device [15]. The process requires only three photolithography masks. A corner of a device sample is shown in Fig. 3.19 including one of the four angle beams, a part of the proof mass with its electrode fingers and a part of one transducer.

3.6.2.1 Harmonic Vibrations

The device was tested with load resistance 18 MΩ for transducer 1 and 17 MΩ for transducer 2. These values are optimal for harmonic vibrations at 560 Hz with a peak amplitude of 0.14 g and a bias voltage of 28.4 V and near the optimum for white-noise vibrations at $S_a = 4.7 \cdot 10^{-5} g^2$ Hz.

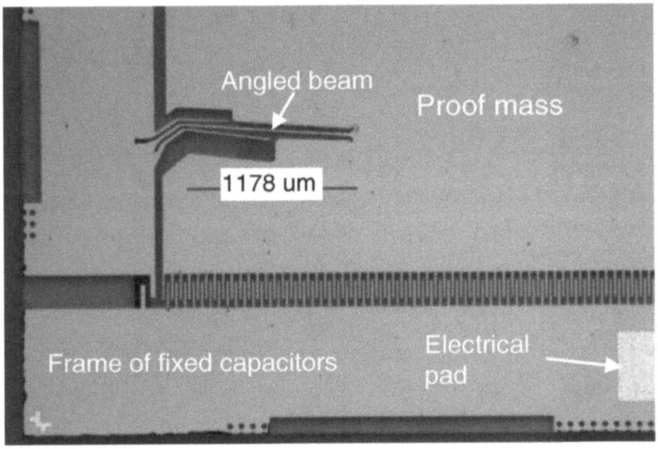

Fig. 3.19 A part of the fabricated energy harvester

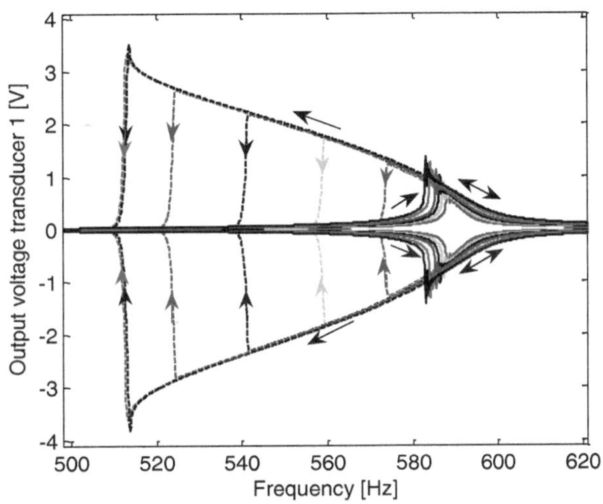

Fig. 3.20 Peak output voltages on transducer 1 as a function of frequency for frequency up-sweeps (*solid curves*) and down-sweeps (*dash-dot curves*). The sweep rate is 2.667 Hz/s, a bias voltage of 28.4 V. Peak acceleration amplitudes are 0.062 g, 0.088 g, 0.110 g, 0.135 g, 0.158 g, and 0.183 g (from lower to upper curve). Adapted from [15]

Figure 3.20 shows peak output voltages on transducer 1 as a function of frequency for frequency sweeps at different levels of acceleration amplitude. The jump phenomenon common to nonlinear systems is clearly seen in the traces, albeit somewhat smeared due to the finite sweep rate. The bandwidth of the harvester is considerably broadened for the frequency down-sweep and increases with acceleration amplitude. This phenomenon shows that the overall effect of the angled beams is to give a softening spring behavior. The bandwidth ceases to increase when the acceleration amplitude reaches 0.158 g.

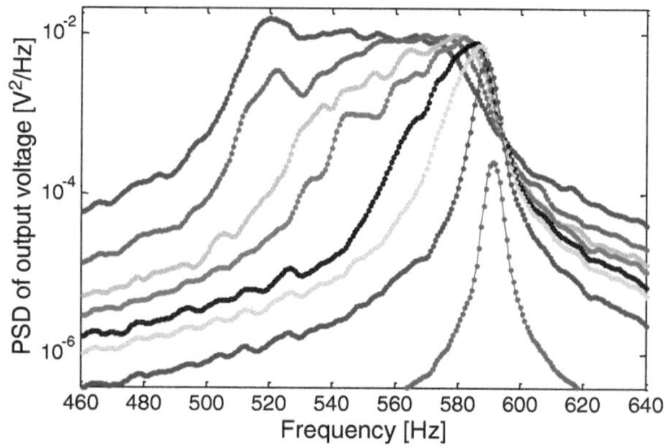

Fig. 3.21 The output spectrum of an angled beam harvester. The average PSDs of the acceleration are 1.5×10^{-6}, 4.6×10^{-5}, 1.310×10^{-4}, 1.8×10^{-4}, 2.7×10^{-4}, 3.4×10^{-4}, 4.8×10^{-4}, and $7.0 \times 10^{-4} g^2/\mathrm{Hz}$ (from lower to upper curve), bias voltage of 28.4 V. From [15]

3.6.2.2 White-Noise Vibrations

Figure 3.21 shows the output spectrum as a function of frequency for different spectral densities of random vibrations. When excited at a low level, $S_a = 1.5 \times 10^{-6} g^2/\mathrm{Hz}$, the energy harvester shows linear response with a resonant frequency of $f_0 = 591$ Hz and a 3-dB bandwidth of about 4.5 Hz. When the vibration increases, the softening behavior (broadening of bandwidth toward the lower frequency) can be observed to develop gradually. At the highest level of the broadband random vibration in Fig. 3.21, that is, $7.0 \cdot 10^{-4} g^2/\mathrm{Hz}$, the output spectrum is quite flat in a wide frequency range, corresponding to a bandwidth of about 60.2 Hz, which is more than 13 times wider than the bandwidth at the lowest level. Hence, the nonlinear spring design provides a wide bandwidth for sufficiently strong vibrations.

The total average output power of both transducers as a function of S_a is shown in Fig. 3.22. We observe that the average output power is linearly proportional to S_a at small values where the harvester behaves linearly (the straight line in Fig. 3.22) [32]. The average output power increases significantly when increasing S_a above $2.7 \cdot 10^{-4} g^2/\mathrm{Hz}$ consistent with the bandwidth increases shown in Fig. 3.21. It achieves 68% more harvested power than a linear energy harvester (152 nW and 90.4 nW) for $S_a = 7.0 \cdot 10^{-4} g^2/\mathrm{Hz}$. The harvester with softening springs not only widens bandwidth (Fig. 3.21) as could be expected but also harvests more output power than a linear energy harvester for sufficiently strong broadband random vibrations.

3.6.2.3 Band-Limited Random Vibrations

The superposition principle is not valid for nonlinear devices. Therefore, one cannot infer directly from frequency sweeps or white-noise excitation how a nonlinear device will respond to other types of excitations. It has to be checked. An energy harvester was tested with band-limited random vibrations generated with a spectral density of the form (3.2) in [67]. It had the same beam design and dimensions as the

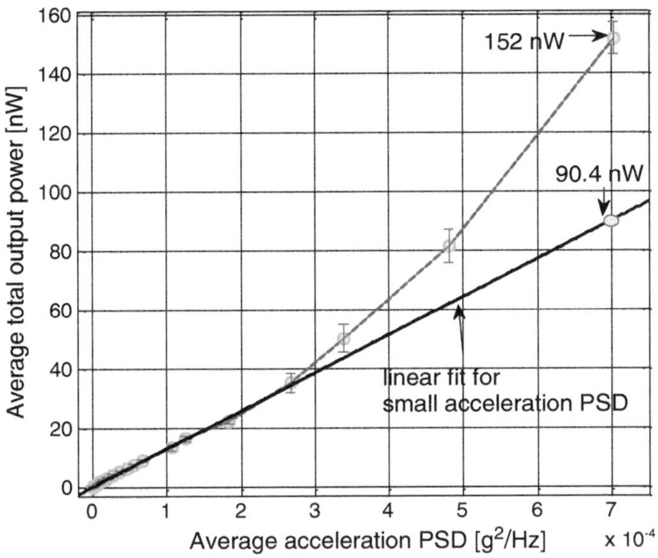

Fig. 3.22 The total average output power of the device as a function of the average PSD of the acceleration, bias voltage of 28.4 V. The straight line is linear fit for the average output power curve when the average PSD of the acceleration is small. From [15]

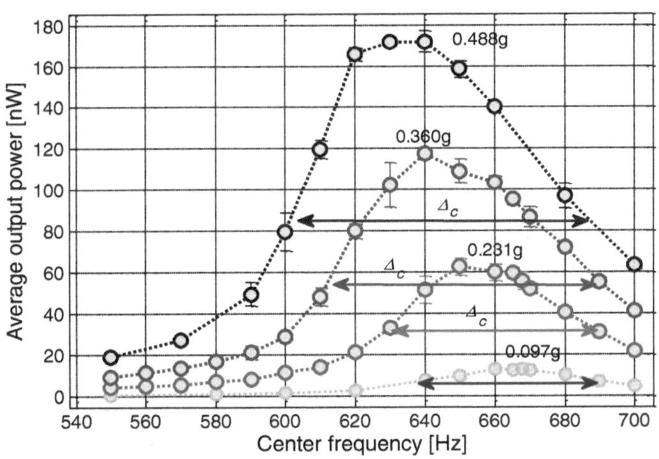

Fig. 3.23 Average output power vs. vibration center frequency for $\Delta = 50\,\text{Hz}$ and RMS accelerations of 0.097, 0.231, 0.360, and 0.488 g. Each average output power point is obtained by averaging the RMS values of three experiments of 40 s duration. From [67]

device shown in Fig. 3.19 except for slightly smaller proof mass and therefore higher resonance frequency of $f_0 = 668\,\text{Hz}$. The 3-dB bandwidth was $B = 1.37\,\text{Hz}$.

Figure 3.23 shows average output power vs. the vibration's center frequency f_c for a vibration bandwidth of $\Delta = 50\,\text{Hz}$ and at different RMS accelerations A.

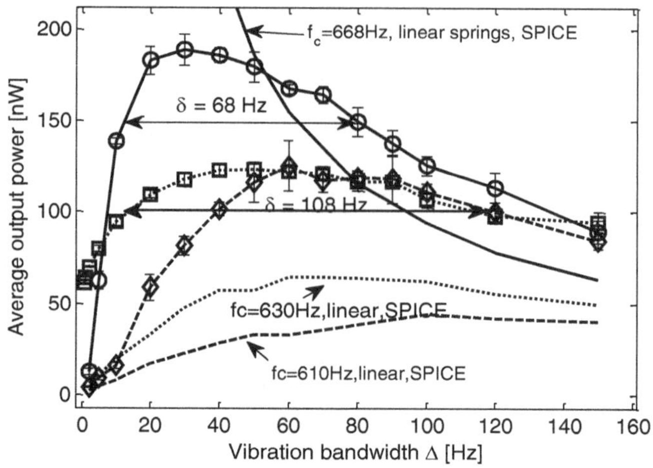

Fig. 3.24 Measured average output power of nonlinear spring device vs. vibration bandwidth Δ for different center frequencies, $f_c = f_0 = 668$ Hz, $f_c = 630$ Hz, and $f_c = 610$ Hz and an RMS acceleration of 0.488 g. SPICE simulations for a linear-spring harvester. From [67]

The 3-dB bandwidth Δ_c of the responses in Fig. 3.23 increases with increasing A beyond a certain value. For $A > 0.097\,g$, Δ_c exceeds Δ; specifically $\Delta_c = 60$ Hz, 75 Hz and 85 Hz at $A = 0.231\,g, 0.360\,g$, and $0.488\,g$, respectively. If the response bandwidth is taken instead at 1 dB below the peak power, its values are 30, 31, 40, and 50 Hz, respectively, for the four values of A. The 1-dB bandwidth of the vibration is 25.4 Hz. Hence, the nonlinear behavior gives increase in the tolerance towards frequency variations.

In the linear regime, the output power significantly reduces not only when f_c is far from f_0 but also when $\Delta > B$ because the spectral content of the vibration is filtered out and cannot be utilized. When the vibrations are strong enough to excite the harvester into its nonlinear regime, the situation improves as shown in Fig. 3.24. The nonlinear device can harvest the power when f_c is far, 38 Hz ($f_c = 630$ Hz) or 58 Hz ($f_c = 610$ Hz), from the resonance. In conclusion, the nonlinear spring design gives a device that has considerably higher tolerance also towards vibration bandwidth variations.

3.6.3 Harvester with Curved Beams

The energy harvester with angled beams showed a considerable bandwidth, for example, 60 Hz around 550 Hz in Fig. 3.21 or about 11% relatively. With real world vibration spectra in mind, such as, for example, the car tire or HVAC system in Fig. 3.1, it is interesting to see if this can be further improved. While this question is still a matter of ongoing investigations, early results on the curved beams designed as described in Sect. 3.5.2 are promising.

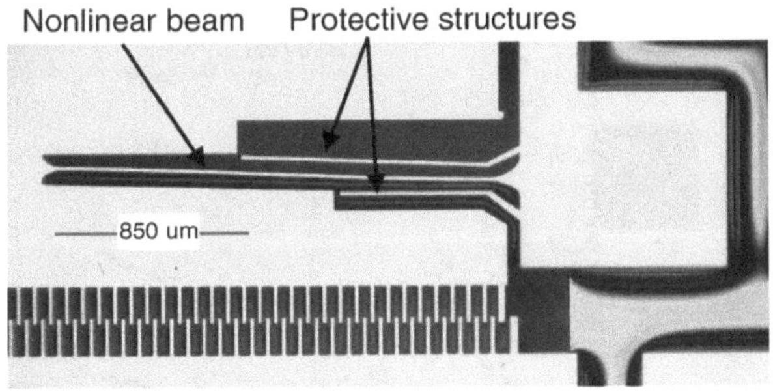

Fig. 3.25 Part of harvester with a curved beam, its dummy protective structures, and part of a transducer [16]

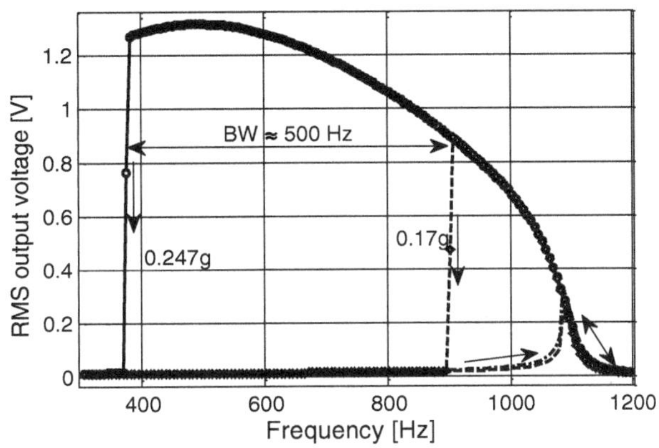

Fig. 3.26 Measured frequency sweeps of a curved-beam harvester at constant acceleration amplitude

The design and fabrication of a curved-beam energy harvester were presented in [16]. Again, the overall design followed the baseline shown in Fig. 3.6a. The important differences are that the device has curved beams as shown in Fig. 3.25 and that considerably more chip real estate was allocated to transducer structures.

The fabrication was made by high aspect ratio Silicon etch of a thick-device-layer Silicon-on-insulator wafer. SiO_2/Si_3N_4 layers were deposited on the vertical sidewalls of the capacitor fingers with the intent of subsequent charging to provide a self-bias, but an external DC voltage source was used for biasing in the experiments reported here.

The response of the harvester to frequency sweeps at very low acceleration amplitude displays the usual linear resonant behavior with a resonant frequency $f_0 = 1,104$ Hz and a narrow bandwidth of 4.1 Hz. Under frequency sweeps at high acceleration amplitude, as shown in Fig. 3.26, the harvester has a large down-sweep bandwidth. With an acceleration amplitude of 0.247 g, a bandwidth of 500 Hz is obtained.

Figure 3.27 shows the output spectral density for white-noise excitations. At the low-level acceleration of $0.32 \cdot 10^{-4} g^2/\text{Hz}$, the device has a very narrow bandwidth which subsequently increases as it progresses into the nonlinear regime

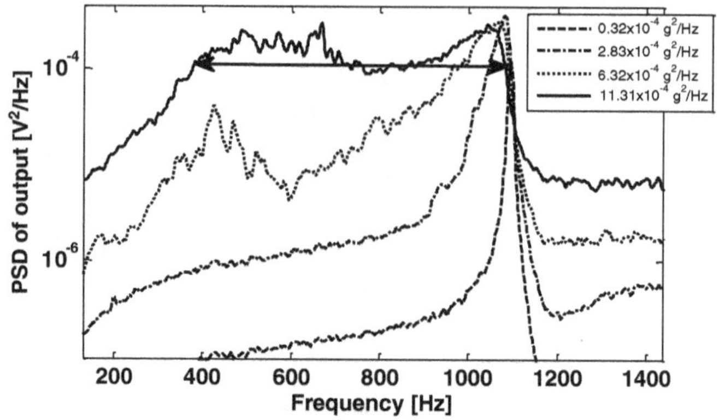

Fig. 3.27 Measured output voltage spectral density of a curved-beam harvester at several values of white-noise acceleration spectral densities

with increasing acceleration spectral density. At $S_a = 11.31 \cdot 10^{-4} \mathrm{g}^2/\mathrm{Hz}$, a wide bandwidth (at 3 dB) of about 481 Hz is achieved. Compared to the measured bandwidth in the linear regime, this bandwidth is about 117 times larger. The spectrum has two broad peaks. If it is possible to reduce the dip between them, one might hope for a bandwidth exceeding 600 Hz as indicated by the arrow in the figure.

3.7 Concluding Remarks

Fundamental properties of MEMS energy harvesters with nonlinear springs have been discussed based on theoretical considerations and recent experiments on microfabricated devices. The current status is that nonlinear springs allow for design of devices that have a much larger bandwidth during frequency sweeps or with white-noise vibrations than can be obtained from comparable linear devices. Compared to linear theory, a white-noise driven softening-spring harvester can even give more output power. The most important aspect of the increased bandwidth seen with these excitations is that it gives an indication of the robustness towards variation in the vibration spectrum. Recent experiments with varying bandwidth and center frequency of the vibrations have shown that this robustness is actually achieved. The use of nonlinear springs therefore provides a viable method to design energy harvesters that display tolerance towards variations in the vibration spectrum.

Acknowledgements This work was supported by the Research Council of Norway under grant no. 191282.

References

1. Burrow S, Clare L (2007) 2007 IEEE International electric machines drives conference, IEMDC '07, vol. 1. Antalya, Turkey, pp. 715–720
2. Burrow SG, Clare LR, Carrella A, Barton D (2008) Proc SPIE 6928:692807
3. Zhu D, Tudor MJ, Beeby SP (2010) Meas Sci Technol 21:022001
4. Soliman MSM, Abdel-Rahman EM, El-Saadany EF, Mansour RR (2008) J Micromech Microeng 18(115021):11p
5. Blystad LC, Halvorsen E (2011) Microsyst Technol 17:505
6. Matsumoto K, Saruwatari K, Suzuki Y (2011) In: PowerMEMS 2011, Seoul, The Republic of Korea, pp. 134–137
7. Liu H, Tay CJ, Quan C, Kobayashi T, Lee C (2011) J Microelectromech Syst 20:1131
8. Marinkovic B, Koser H (2009) Appl Phys Lett 94:103505
9. Marzencki M, Defosseux M, Basrour S (2009) J Microelectromech Syst 18:1444
10. Tvedt LGW, Nguyen DS, Halvorsen E (2010) J Microelectromech Syst 19:305. DOI 10.1109/JMEMS.2009.2039017
11. Stanton SC, McGehee CC, Mann BP (2009) Appl Phys Lett 95:174103
12. Cottone F, Vocca H, Gammaitoni L (2009) Phys Rev Lett 102:080601
13. Erturk A, Hoffmann J, Inman DJ (2009) Appl Phys Lett 94:254102
14. Barton DAW, Burrow SG, Clare LR (2010) J Vib Acoust 132:021009
15. Nguyen DS, Halvorsen E, Jensen GU, Vogl A (2010) J Micromech Microeng 20:125009. DOI 10.11088/0960-1317/20/12/125009
16. Nguyen SD, Tran NHT, Halvorsen E, Paprotny I (2011) In: Power MEMS 2011, Seoul, Republic of Korea, pp. 126–129
17. Löhndorf M, Kvisterøy T, Westby E, Halvorsen E (2007) In: PowerMEMS 2007 Technical Digest, Freiburg, pp. 331–334
18. Westby ER, Halvorsen E (2011) IEEE/ASME Trans Mechatronics 17:995. DOI 10.1109/TMECH.2011.2151203
19. Miller LM, Halvorsen E, Dong T, Wright PK (2011) J Micromech Microeng 21:045029.DOI 10.1088/0960-1317/21/4/045029
20. von Buren T, Mitcheson PD, Green TC, Yeatman EM, Holmes AS, Tröster G (2006) IEEE Sensor J 6:28
21. Blystad LCJ, Halvorsen E, Husa S (2010) IEEE Trans Ultrason Ferroelectr Freq Contr 57:908
22. Halvorsen E, Blystad LCJ, Husa S, Westby E (2007) In: International Conference on Perspective Technologies and Methods in MEMS Design, 2007. MEMSTECH 2007, pp. 117–122
23. Halvorsen E (2008) J Microelectromech Syst 17:1061
24. van Kampen NG (2007) Stochastic processes in physics and chemistry. Elsevier, Amsterdam
25. Gardiner CW (2004) Handbook of stochastic methods, 3rd edn. Springer, Berlin-Heidelberg
26. Kloeden PE, Platen E (2010) Numerical solution of stochastic differential equations. Springer, Berlin-Heidelberg
27. Blystad LCJ, Halvorsen E (2011) Smart Mater Struct 20:025011
28. Sebald G, Kuwano H, Guyomar D, Ducharne B (2011) Smart Mater Struct. 20:075022
29. Bao MH (2000) Micro Mechanical transducers: pressure sensors, accelerometers, and gyroscopes. Elsevier B.V., Amsterdam
30. Roundy S, Wright PK, Rabaey J (2003) Comput Commun 26:1131
31. Hoffmann D, Folkmer B, Manoli Y (2011) J Micromech Microeng 21:104002
32. Bartsch U, Gaspar J, Paul O (2010) J Micromech Microeng 20:035016
33. Meninger S, Mur-Miranda JO, Amirtharajah R, Chandrakasan A, Lang J (1999) In: ISLPED '99: Proceedings of the 1999 international symposium on low power electronics and design. ACM, New York, pp. 48–53
34. Mitcheson P, Sterken T, He C, Kiziroglou M, Yeatman E, Puers R (2008) Meas Contr UK 41:114
35. Yen BC, Lang JH (2006) IEEE Trans Circ Syst I 53:288

36. Kuehne I, Frey A, Marinkovic D, Eckstein G, Seidel H (2008) Sensor Actuator A 142:263
37. Ma W, Zhu R, Rufer L, Zohar Y, Wong M (2007) J Microelectromech Syst 16:29
38. Li Y, Cheng Z, San H, Duo Y, Chen X (2010) In: 5th IEEE International Conference on Nano/Micro Engineered and Molecular Systems. NEMS 2010, pp. 78–83
39. Suzuki Y (2011) IEEJ Trans Electr Electron Eng 6:101
40. Sessler GM (ed) (1999) Electrets, 3rd edn. Laplacian Press, California
41. Tsutsumino T, Suzuki Y, Kasagi N, Kashiwagi K, Morizawa Y (2006) In: PowerMEMS 2006 technical digest, Berkeley, USA, pp. 279–282
42. Sakane Y, Suzuki Y, Kasagi N (2008) J Micromech Microeng 18:104011
43. Sterken T, Baert K, Puers R, Borghs G, Mertens R (2003) In: Proceedings pan pacific microelectronics symposium, Hawaii, pp. 27–34
44. Gracewski SM, Funkenbusch PD, Jia Z, Ross DS, Potter MD (2006) J Micromech Microeng 16:235
45. Yamashita K, Honzumi M, Hagiwara K, Iguchi Y, Suzuki Y (2010) In: Proceedings of PowerMEMS 2010 oral presentations, Leuven, Belgium, pp. 165–168
46. Halvorsen E, Westby ER, Husa S, Vogl A, Østbø NP, Leonov V, Sterken T, Kvisterøy T (2009) In: The 15th international conference on solid-state sendors, actuators & microsystems, transducers 2009, Denver, Colorado, pp. 1381–1384
47. Soliman M, Abdel-Rahman E, El-Saadany E, Mansour R (2009) J Microelectromech Syst 18:1288
48. Le CP, Halvorsen E (2011) In: Proceedings of the 22nd micromechanics and microsystem technology europe workshop, Tønsberg, Norway
49. D'hulst R, Sterken T, Puers R, Deconinck G, Driesen J (2010) IEEE Trans Ind Electron 57:4170
50. Peano F, Tambosso T (2005) J Microelectromech Syst 14:429
51. Galayko D, Basset P (2011) IEEE Trans Circ Syst I 58:299
52. Mahmoud MAE, Abdel-Rahman EM, El-Saadany EF, Mansour RR (2010) Smart Mater Struct 19:025007
53. Risken H (1996) The Fokker–Planck equation: methods of solutions and applications. *Springer series in synergetics*, vol. 18, 2nd edn. Springer, New York
54. Rao SS (2004) Mechanical vibrations, Pearson Education, New Jersey
55. Kovacic I, Brennan MJ (eds) (2011) The duffing equation. Wiley, Chichester, West Sussex
56. Thomson WT, Dahleh MD (1998) Theory of vibration with applications, 5th edn. Prentice Hall, Upper Saddle River, New Jersey
57. Senturia SD (2001) Microsystem design, Kluwer Academic Publishers, Norwell, Massachusetts
58. Voigtländer K, Risken H (1985) J Stat Phys 40:397
59. Mitcheson PD, Yeatman EM, Rao GK, Holmes AS, Green TC (2008) Proc IEEE 96:1457
60. Renno JM, Daqaq MF, Inman DJ (2009) J Sound Vib 320:386
61. Ramlan R, Brennan M, Mace B, Kovacic I (2010) Nonlinear Dynam 59:545
62. Scruggs J (2009) J Sound Vib 320:707
63. Jutte CV, Kota S (2008) J Mech Des 130:081403
64. Krylov S, Bernstein Y (2006) Sensor Actuator Phys 130–131:497
65. Tran NHT, Nguyen SD, Halvorsen E (2011) In: Proceedings of the 22nd micromechanics and microsystem technology Europe workshop, Tønsberg, Norway
66. Tran NHT (2011) Design of nonlinear springs for MEMS vibration energy harvesting applications. Master's thesis, Vestfold University College
67. Nguyen SD, Halvorsen E (2011) J Microelectromech Syst 20:1225. DOI 10.1109/JMEMS.2011.2170824.
68. Halvorsen E (2012) Fundamental issues in nonlinear wide-band vibration energy harvesting (unpublished), arXiv:1209.3184

Chapter 4
Broadband Energy Harvesting from a Bistable Potential Well

B.P. Mann

Abstract This chapter describes recent research on the use of a bistable oscillator, in lieu of the more common choice of using a linear oscillator, for energy harvesting. The chapter progresses from the elementary theory for the potential well escape problem to analyzing several examples of using an oscillator with a bistable potential in electromagnetic and piezoelectric harvester applications. Select results are presented for harmonic and stochastic excitation.

4.1 Introduction

The success of portable electronics and remote sensing devices is dependent upon the availability of remote power. While batteries can sometimes fulfill this role over short-time intervals, they are often undesirable due to their finite life span, need for replacement, and environmental impact. Instead, researchers are now investigating methods of scavenging energy from the environment to eliminate the need for batteries or to prolong their life [1]. While solar, chemical, and thermal sources of energy transfer are sometimes viable, many have recognized the abundance of environmental disturbances that cause either rigid body motion or structural vibrations. This has led to a dramatic increase in the number of studies for vibration-based energy harvesting [2–7].

Most prior works have focused on the power harvested when the response behavior is adequately characterized as a linear oscillator being driven by harmonic excitation. For this type of design, the optimal performance is realized when the natural frequency of the oscillator is nearly identical to a dominant frequency in the ambient environment. Thus, the prototypical approach is to frequency match or

B.P Mann (✉)
Duke University, Durham, NC 27705, USA
e-mail: mannb@mac.com

N. Elvin and A. Erturk (eds.), *Advances in Energy Harvesting Methods*,
DOI 10.1007/978-1-4614-5705-3_4,

to design and fabricate energy harvesting devices to have a natural frequency that coincides with a dominant frequency in ambient environment [5, 8–10]. While this approach is relatively easy to implement, not to mention convenient for analysis, it suffers from the fact that the linear harvester is narrowband; thus, it can only take advantage of a narrow range of frequencies from the ambient environment.

While the research over the past decade has primarily focused on inertial generators that operate in a linear regime [8, 11–18], this chapter considers an alternative paradigm. More specifically, we consider the intentional use of nonlinearity and are interested in whether system nonlinearities can enhance performance. For example, the design-for-resonance approach places several performance limitations on the energy harvester. Specifically, a linear device is known to underperform when the system's resonance and excitation frequency do not coincide. Alternatively, the energy from multifrequency and/or broadband excitation may not be adequately captured. Problems also arise in application areas where the excitation frequency varies since the linear harvester will underperform, unless the complexity of the harvester is increased by adding more degrees of freedom, e.g., multiple oscillators [3, 4].

Several recent works have suggested the intentional use of nonlinearity might be beneficial to energy harvesting systems [13, 19–21]. More specifically, several studies have explored that the use of nonlinearities broaden the frequency spectrum, to extend the bandwidth to engage nonlinear resonances, and/or to facilitate tuning [13, 20–28]. These efforts take aim at overcoming the limitations of linear devices, which only perform well within a narrow band of frequencies [19]. The focus of this chapter is on a particularly promising type of nonlinear behavior commonly referred to as a bistable potential well.

The content of this chapter is organized as follows. The next section summarizes the limitations of a linear harvester by simply examining the response and uncertainty in the response of a linear oscillator. This is followed by a conceptual discussion of a bistable system and the examination of a potential well escape problem. Section 4.4 describes several examples where researchers have explored bistability in both piezoelectric and electromagnetic harvesters. The final section provides a summary of the findings and a discussion of potential future research avenues.

4.2 Limitations of a Linear Energy Harvester

Oscillators are often designed to operate in their linear regime when they are designed for vibratory energy harvesters. While restricting the oscillator to operate in a linear regime can greatly simplify the analysis of the system, it also limits the harvester performance in several ways. To illustrate these points, we have contrived a simple example. Consider the dimensionless linear oscillator

$$y'' + \mu y' + y = \Gamma \sin \eta\tau \,, \tag{4.1}$$

where y is the dimensionless displacement, a $()'$ denotes a derivative with respect to dimensionless time, μ is a damping coefficient, η is the ratio of the excitation frequency to the natural frequency, and Γ is the excitation level. For the typical case where $\mu > 0$, the steady-state response of Eq. (4.1) is given by

$$y = r\cos(\eta\tau - \phi)\,, \tag{4.2}$$

where the amplitude of the response is given by

$$r = \frac{\Gamma}{\sqrt{(1-\eta^2)^2 + (\mu\eta)^2}}\,. \tag{4.3}$$

Here, we note the power harvested is proportional to the amplitude of r. To both quantify and unveil the robustness of the linear oscillator's response to parameter variations, an expression for total uncertainty in the oscillator's response U_r, at a desired confidence level, is introduced

$$U_r^2 = \left(\frac{\partial r}{\partial \mu}\right)^2 U_\mu^2 + \left(\frac{\partial r}{\partial \eta}\right)^2 U_\eta^2 + \left(\frac{\partial r}{\partial \Gamma}\right)^2 U_\Gamma^2 \tag{4.4}$$

where U_{x_i} represents the uncertainty in the variable x_i at the same confidence level. It is common to express the uncertainty at the 95% confidence level (or 20:1 odds) and, consequently, 95% of the physical realizations can be expected to lie within the confidence intervals [29].

Figure 4.1 shows the response of the linear oscillator along with the associated confidence intervals. Here, the peak response occurs close to the natural frequency or, more precisely, at $\eta = \sqrt{1-\mu^2/2}$; this typically results in the design strategy of attempting to align the device natural frequency with a dominant frequency in the environment. Considering this strategy further, we have plotted the confidence intervals for the oscillator's response in Fig. 4.1. The confidence intervals were obtained by first determining the uncertainty in the response U_r; next, the upper and lower confidence intervals were determined from $r_u = r + U_r$ and $r_l = r - U_r$ where r_u is the upper confidence interval and r_l is the lower. In essence, the confidence intervals provide a measure of the robustness in the response of the system when parameter uncertainty is considered.

The results of Fig. 4.1 clearly show that an amplification effect is achieved when the oscillator frequency aligns with a frequency in the environment. However, the influence of the frequencies outside the narrowband peak is negligible. Furthermore, the confidence intervals of Fig. 4.1 highlight the lack of robustness in achieving a peak response since even small parameter variations or uncertainty can cause large differences in the expected response. For example, the upper and lower confidence intervals, dashed lines in Fig. 4.1, show the uncertainty in the oscillator response can sometimes be even larger than the nominal value (solid line). Armed with this understanding, we now focus our attention to an alternative strategy that could potentially address the limitations imposed by the linear oscillator.

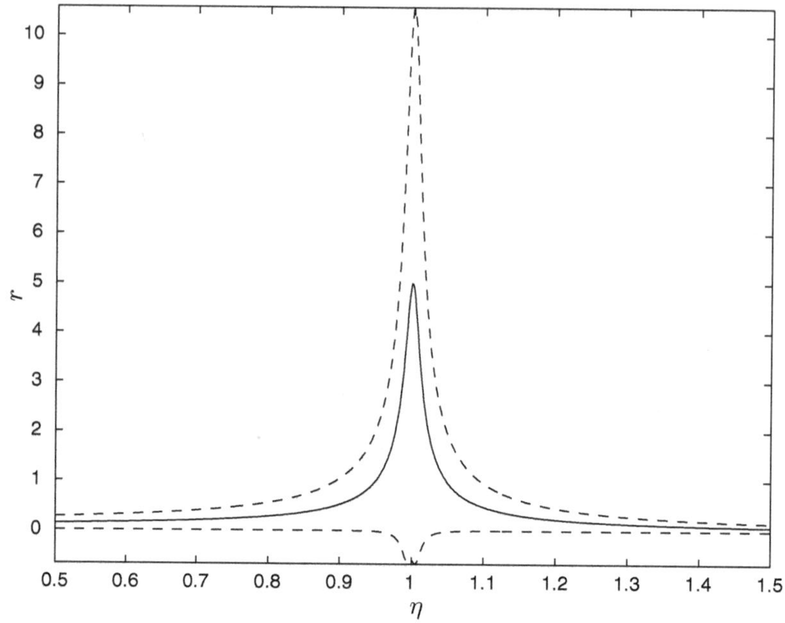

Fig. 4.1 Nominal response (*solid line*) and confidence intervals (*dashed lines*) showing the lack of robustness of a linear oscillator for $\mu = 0.02$ and $\Gamma = 0.1$ and parameter uncertainties $U_\mu = 0.5\mu$, $U_\eta = 0.01$, and $U_\Gamma = 0.1$. A negative value for a *dashed line* indicates the uncertainty in the response is even larger than the nominal r value

4.3 Bistable Potential Well Paradigm

The concept of interest can be brought into focus by considering the motion of a small ball rolling on the surface under the influence of gravity (see Fig. 4.2) where the ball height is proportional to the potential energy. Consider first the potential energy of a linear oscillator, shown in Fig. 4.2a. This system has a linear relationship between the restoring force and deflection which results in a quadratic potential energy well with a single equilibrium. Regardless of where the ball is placed, it will eventually come to rest at the bottom of the potential energy well. Shaking the parabola laterally yields the linear harmonic oscillator with the largest response occurring when it is shaken at its resonance frequency. As in the case of Fig. 4.1, the system's response is relatively small when nonresonant, stochastic, multifrequency, or broadband excitation is applied.

Consider next the same ball under the influence of a nonlinear restoring force where the potential energy description may be more complex—see Fig. 4.2b. Consider again the same ball under the influence of small lateral excitations. This results in a system that behaves linearly for small-amplitude motions with oscillations that remain confined to a single well. For increasingly large excitations, motion amplitudes grow until the threshold for a potential well escape occurs

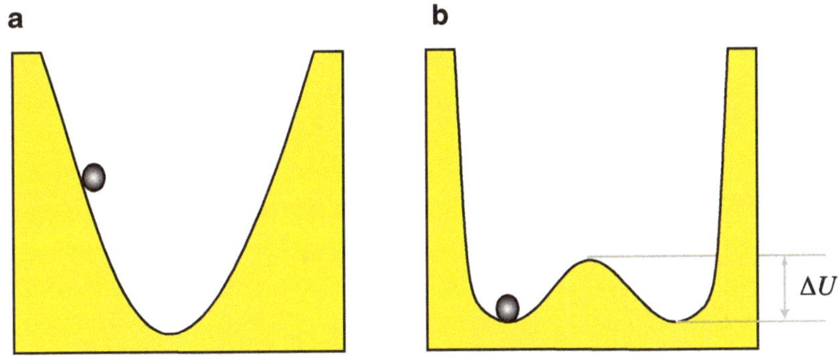

Fig. 4.2 Potential energy curves for (**a**) the quadratic potential well of a linear oscillator and (**b**) a nonlinear oscillator with two stable equilibria separated by an unstable equilibrium position. The energy difference between the potential energy barrier and the stable equilibria, labeled ΔU, is an important factor for determining the threshold for an escape

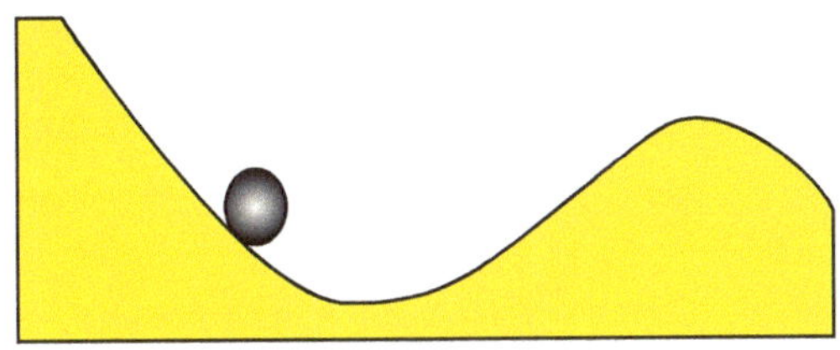

Fig. 4.3 Illustration of an oscillator with an asymmetric potential well. Escapes from this potential well result in an infinite response

(i.e., where an escape is imminent for energy levels above the threshold criteria ΔU in Fig. 4.2b). Once exceeding the threshold criteria, the small ball would then escape from the potential well and traverse both potential wells, sometimes called well-mixing behavior, with large-amplitude displacements and velocities. Acknowledging the dramatic increase in the energetic response of the oscillator in the post-escape regime [30], we depart from the qualitative discussion and focus our attention on more quantitative studies to ascertain whether this phenomena can yield a more broadband frequency response that is also robust to parameter variations.

4.3.1 Simple Escape Example

This section considers a simple escape problem motivated by the illustration of Fig. 4.3 where increasingly large motions of the base will cause the oscillating mass to escape from the potential well. While the present example provides a representative escape problem, we note that there exist distinct differences in the frequency response of this system and one with a bistable potential. Despite these differences, we can still gain insight into the escape behavior by examining this simpler system prior to studying the experimental realizations shown in Sect. 4.4.

The simplest governing equation for the forced oscillations of a mass oscillating in an asymmetric potential with damping is given by

$$m\ddot{x} + c\dot{x} + kx - k_b x^2 = f \sin \Omega t \,. \tag{4.5}$$

This equation can be nondimensionalized by introducing a dimensionless displacement $y = x/\ell$ and time $\tau = \omega t$, where $\omega = \sqrt{k/m}$ and $\ell = k/k_b$. The dimensionless equation takes the form

$$y'' + \mu y' + y - y^2 = \Gamma \sin \eta\tau \tag{4.6}$$

where a $()'$ denotes a derivative with respect to τ, $\mu = \dfrac{c}{m\omega}$ is a dimensionless dissipation coefficient, $\Gamma = k_b f/k^2$ is a dimensionless excitation level, and $\eta = \Omega/\omega$ is a dimensionless frequency ratio. If we assume a solution in the form of a truncated Fourier series with time-varying coefficients $y = p(\tau) + a(\tau)\cos\eta\tau + b(\tau)\sin\eta\tau$, the first and second derivates become

$$y' = p' + \left(a' + b\eta\right)\cos\eta\tau + \left(b' - a\eta\right)\sin\eta\tau \,, \tag{4.7a}$$

$$y'' = \left(2b' - a\eta\right)\eta\cos\eta\tau - \left(b\eta + 2a'\right)\eta\sin\eta\tau \,, \tag{4.7b}$$

where we have presumed that $p'' = a'' = b'' = 0$ which is in agreement with the presumption that the Fourier coefficients slowly vary with time. The expressions for y, y', and y'' are then returned to Eq. (4.6), and the constant coefficients, along with the coefficients of $\cos\eta\tau$ and $\sin\eta\tau$, are balanced on each side. Balancing the constant coefficients gives the top row of the following matrix:

$$\begin{bmatrix} \mu & 0 & 0 \\ 0 & \mu & 2\eta \\ 0 & -2\eta & \mu \end{bmatrix} \begin{bmatrix} p' \\ a' \\ b' \end{bmatrix} = \begin{bmatrix} p^2 - p + \frac{1}{2}\left(a^2 + b^2\right) \\ \left(\eta^2 - 1 + 2p\right)a - \mu\eta b \\ \left(\eta^2 - 1 + 2p\right)b + \mu\eta a + \Gamma \end{bmatrix} , \tag{4.8}$$

whereas balancing the coefficients of $\cos\eta\tau$ and $\sin\eta\tau$ gives the second and third rows, respectively.

The steady-state response of the systems requires $a' = b' = p' = 0$. After setting these terms to zero, we square and add the last two rows of Eq. (4.8) and solve for $a^2 + b^2$ to obtain

$$\tilde{r}^2 = \tilde{a}^2 + \tilde{b}^2 = \frac{\Gamma^2}{\left(\eta^2 - 1 + 2\tilde{p}\right)^2 + \left(\mu\eta\right)^2} \tag{4.9}$$

where $\tilde{r}$ is the amplitude of the periodic response. Here, a $(\tilde{\ })$ has been used to represent the equilibria or fixed-point solutions of our assumed solution form.

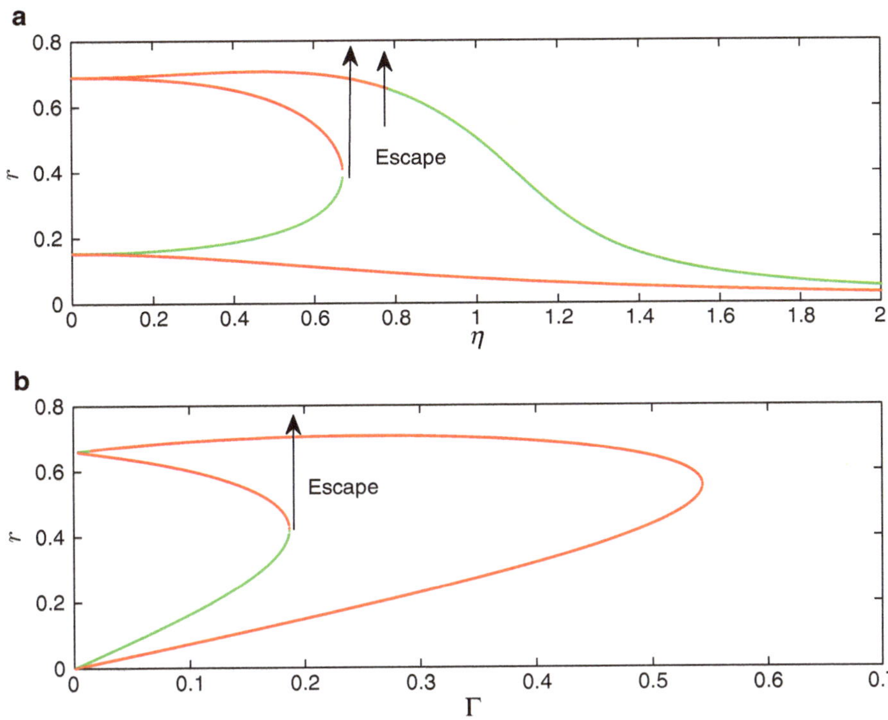

Fig. 4.4 Response amplitude as a function of the frequency ratio η (**a**) and the excitation amplitude Γ (**b**). Graph (**a**) shows that an escape to infinity will occur when $0.68 < \eta < 0.78$, $\mu = 0.01$, and $\Gamma = 0.15$. Similarly, graph (**b**) shows that steady-state escapes will occur when $\Gamma > 0.19$, $\mu = 0.01$, and $\eta = 0.6$

Equation (4.9) is then inserted into the top row of Eq. (4.8), and a polynomial in $\tilde{p}$ is obtained:

$$\tilde{p}^2 - \tilde{p} + \frac{1}{2}\frac{\Gamma^2}{(\eta^2 - 1 + 2\tilde{p})^2 + (\mu\eta)^2} = 0\,. \tag{4.10}$$

The roots of this polynomial in $\tilde{p}$ are then solved and inserted into Eq. (4.9) to obtain the amplitude of the periodic solution.

For relatively small levels of excitation, the motions of the mass remain confined within the potential well. However, escapes are more likely to occur as the level of excitation increases. For example, a representative case of the forced response is shown in Fig. 4.4. For this intermediate level of excitation, the frequency response of Fig. 4.4a indicates that the stable periodic oscillations gradually change for changes in η—at least for most of the response; however, the middle of this diagram shows that the stable solution branches vanish, near the frequency range $0.68 < \eta < 0.78$, thus indicating the absence of a stable periodic motion. Within this frequency range, a potential well escape occurs and the response of the system jumps to infinity.

Similarly, holding $\eta = 0.6$ constant and slowly increasing the excitation level shows that the mass will escape from the potential well escape when $\Gamma > 0.19$, as none of the neighboring solution branches are stable.

4.4 Physical Realizations of Bistable Potential Wells

This section describes piezoelectric and electromagnetic harvesters that have used a bistable oscillator. We provide both numerical and analytical treatment of the governing equations to help elucidate the behavior of the harvester. Results are presented for both harmonic and stochastic excitation.

4.4.1 Example Bistable Electromagnetic System

A schematic of the experimental system, which was studied in reference [13], is shown in Fig. 4.5. The device consisted of a series of magnets that were positioned to make the system bistable. For instance, the top and bottom magnets were

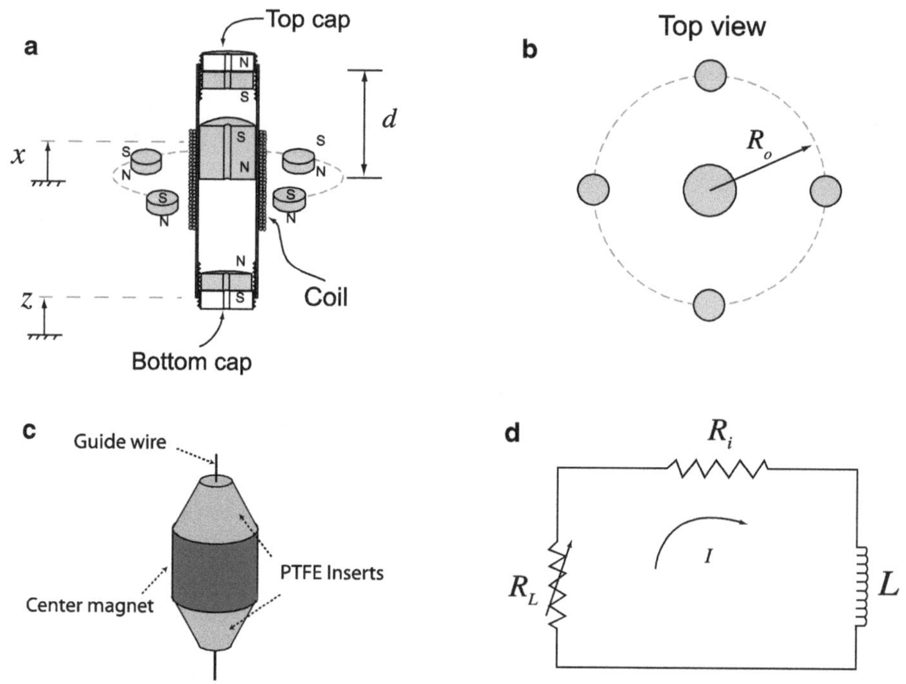

Fig. 4.5 Illustrations of the nonlinear generator (schematics (**a**) and (**b**)), the inserts used to reduce mechanical damping (**c**), and a schematic of the coupled electrical circuit (**d**)

mechanically attached to top and bottom caps that were inserted into a tube that allowed the half-spacing d (shown in Fig. 4.5a) to be adjusted. A center magnet was placed between these magnets and oriented to repel both outer magnets—thereby suspending the oscillating center magnet within the tube [19]. Four magnets were placed symmetrically around the tube's midpoint; the purpose of these magnets was to repel the center magnet away from the midpoint and to make the inertial generator bistable (see Fig. 4.5a,b).

To account for the interactions between the oscillating and fixed magnets, the potential energy was derived from a dipole model (see [31–34]); this section describes the salient features of the derivation by first introducing the expressions for two interacting magnets and then applying these expressions to multiple magnets. The magnetic flux density or **B**-field at the location $\mathbf{r}_p$ due to a magnet located at $\mathbf{r}_s$ is defined by

$$\mathbf{B} = -\frac{\mu_o}{4\pi}\nabla\frac{\mathbf{m}_s\cdot\mathbf{r}_{p/s}}{|\mathbf{r}_{p/s}|^3}\,, \tag{4.11}$$

where μ_o is the permeability of free space, ∇ is the vector gradient, $\mathbf{r}_{p/s} = \mathbf{r}_p - \mathbf{r}_s$ is a vector to the point of interest (the point $P(x, y, z)$) with respect to the location of the source magnet, and $|\mathbf{r}_{p/s}|$ is the distance between the two magnets. The magnetic moment of the source magnet, which is located at $\mathbf{r}_s$, is given by $\mathbf{m}_s = \mathbf{M}_s v_s$, where $\mathbf{M}_s$ and v_s are the magnetization and volume of the source magnet, respectively. The potential energy of the magnet at $\mathbf{r}_p$ in the field generated by the magnet at $\mathbf{r}_s$ is

$$U = -\mathbf{m}_p\cdot\mathbf{B}\,, \tag{4.12}$$

where $\mathbf{m}_p$ is the magnetic moment of the magnet located at $\mathbf{r}_p$. An expression for the interaction force between the magnets can be obtained by taking the gradient of Eq. (4.12).

The potential energy of the center magnet due to the interactions with the other six magnets can be obtained using Eqs. (4.11) and (4.12). For this purpose, we introduce the parameters v_c, v_o, and v_t to define the volumes of the center magnet, each of the outer-four magnets, and the identical top and bottom magnets, respectively; the corresponding magnetization amplitudes of these magnets are defined by M_c, M_o, and M_t, where the subscripts match those applied to the magnet volumes. Introducing d as the half-spacing between the top and bottom magnets (see Fig. 4.5a), the expression for the potential energy becomes

$$\begin{aligned} U = \frac{\mu_o M_c v_c}{2\pi}\Bigg[& v_t M_t\left(\frac{1}{(d+y)^3}+\frac{1}{(d-y)^3}\right) \\ & -\frac{M_o v_o N}{2}\left(\frac{y^2}{(y^2+R_o^2)^{5/2}}-\frac{1}{(y^2+R_o^2)^{3/2}}\right)\Bigg], \end{aligned} \tag{4.13}$$

where $y = x - z$ is the relative position of the center magnet with respect to the device's outer housing and the number of outer-ring magnets was $N = 4$. These

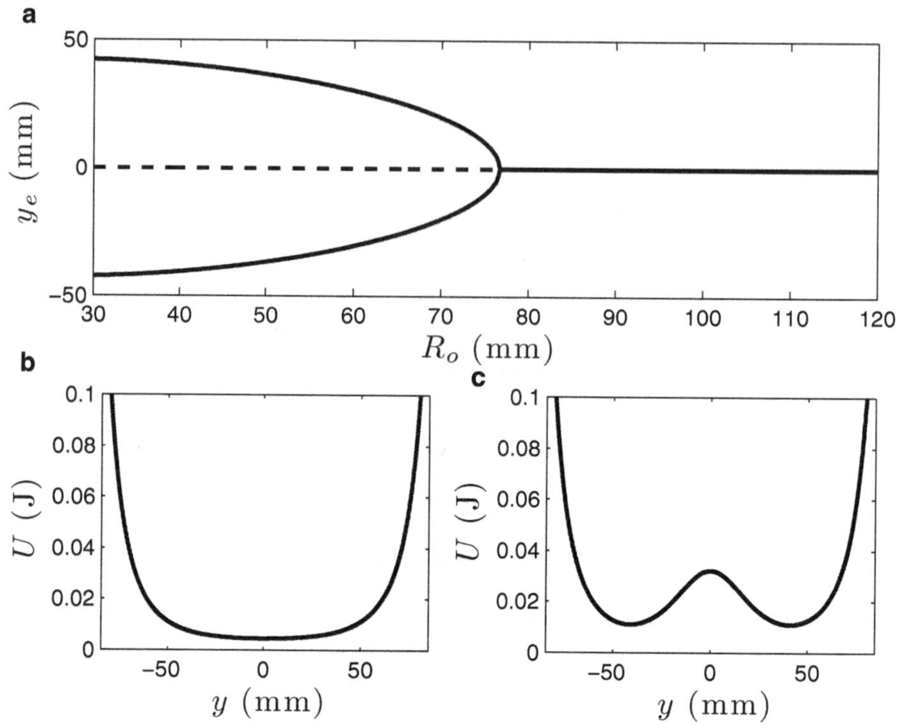

Fig. 4.6 Bifurcation diagram (**a**) and potential energy curves (**b**)–(**c**) for various radial spacings of the outer ring of magnets. Stable (*solid line*) and unstable (*dashed*) equilibria are marked in graph (**a**). Graphs (**b**) and (**c**) demonstrate that either a monostable or bistable potential can be realized for different values of R_o [13]

magnets were located at radial distance R_o from the central axis and on the plane of symmetry that divides the top from the bottom half of the device.

A bifurcation diagram and potential energy curves for this system are shown in Fig. 4.6. This bifurcation diagram shows the changes in the static equilibria y_e for changes in the radial distance R_o of the outer ring of magnets. Furthermore, the bifurcation diagram shows that changes in R_o give rise to a supercritical pitchfork bifurcation; this reveals the values of R_o where the generator is monostable and bistable. For instance, Fig. 4.6b,c shows a monostable and bistable system, respectively; these cases were obtained by simply using two different R_o values.

4.4.1.1 Numerical Investigations

This section investigates the response behavior of the electromagnetic harvester and the power delivered to an electrical load. Since the generator directly powered a

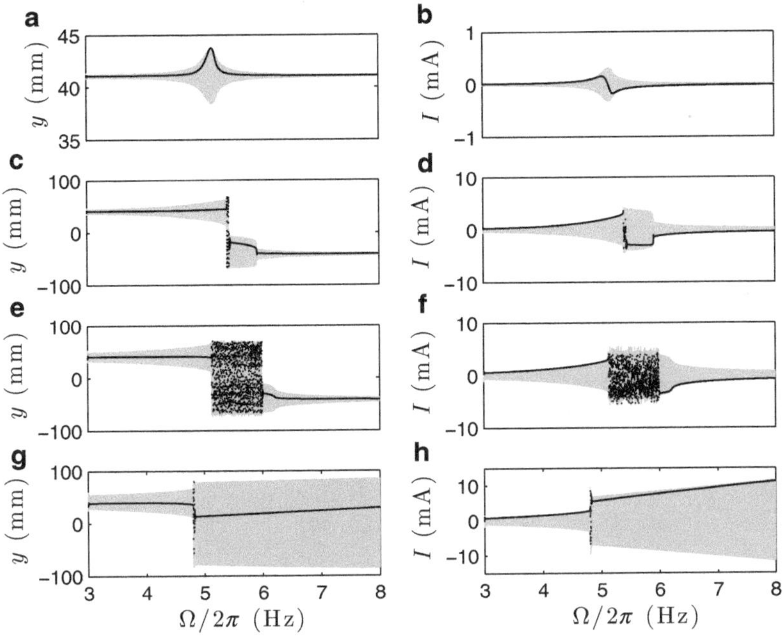

Fig. 4.7 Forward frequency sweep responses of the center magnet displacement (*left column*) and the current (*right column*) for different base excitations. Both the continuous time series (*shaded*) and stroboscopic samples (*black dots*) are shown. The following base accelerations were used: (**a,b**) $A = 0.1$ m/s^2, (**c,d**) $A = 3$ m/s^2, (**e,f**) $A = 6$ m/s^2, and (**g,h**) $A = 10$ m/s^2

resistive load, the instantaneous power delivered to the electrical load was given by $P = I^2 R_L$. The investigations that follow consider single-frequency excitation $\ddot{z} = A\cos\theta(t)$ where A is the acceleration amplitude and $\dot{\theta}(t) = \Omega(t)$ is the excitation frequency in rad/s. Frequency sweep results, where the excitation frequency is slowly increased or decreased while holding the value of A constant, are shown in the results.

Broadening the range of frequencies from which a relatively large amount of energy can be extracted is of considerable interest in energy harvesting. To provide insight into the harvester's capability to achieve this goal, we describe selective results from [13]. The graphs of Figs. 4.7–4.9 show an array of frequency responses that were predicted for different levels of base excitation. More specifically, the forward (or linearly increasing) frequency sweep results of Fig. 4.7 show the time series for the magnet displacement and the circuit current along with their respective stroboscopic points, i.e., each time series sampled at $\left\lfloor \dfrac{\theta(t)}{2\pi} \right\rfloor$ where $\theta(t) = \left(\Omega_o + \frac{1}{2}\Omega_r t\right) t$ is a function of the sweep rate Ω_r, start frequency Ω_o, and time. While the time series shows the amplitude of each state at a given frequency, the stroboscopic samples of these states reveal the periodicity of these responses

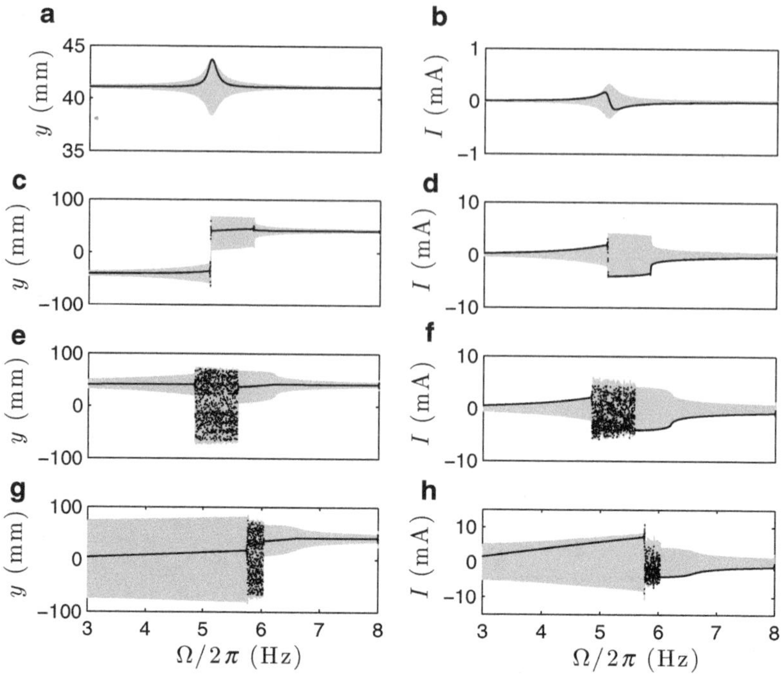

Fig. 4.8 Reverse frequency sweep responses of the center magnet displacement (*left column*) and the current (*right column*) for different base excitations. Both the continuous time series (*shaded*) and stroboscopic samples (*black dots*) are shown. The following base accelerations were used: **(a,b)** $A = 0.1$ m/s^2, **(c,d)** $A = 3$ m/s^2, **(e,f)** $A = 6$ m/s^2, and **(g,h)** $A = 10$ m/s^2

with respect to the excitation frequency, e.g., see periodic responses in Fig. 4.7a,b or chaotic responses in the middle of the frequency range for Fig. 4.7e,f.

The forward sweep cases of Fig. 4.7 can also be compared with the reverse sweep (or linearly decreasing) cases of Fig. 4.8. For instance, for small excitation levels, the harvester responds identically for forward and reverse frequency sweeps with a resonant peak near the linear natural frequency. In contrast, the responses at larger levels of excitation reveal a broadening of the peak response, hysteresis in the frequency responses, the presences of multiple attractors (coexisting solutions), and chaotic attractors. Furthermore, the largest level of excitation also unveils a significant broadening in the frequency response.

The frequency responses of the system can be used to ascertain the frequency dependence of the power delivered to an electrical load. In particular, the graphs of Fig. 4.9 provide the power over the range of frequencies that were studied in the previous two figures. For the lowest level of base excitation, the power graph shows a rather narrow peak near the natural frequency of the system. This type of response, which is reminiscent of classical linear behavior, no longer dominates the responses

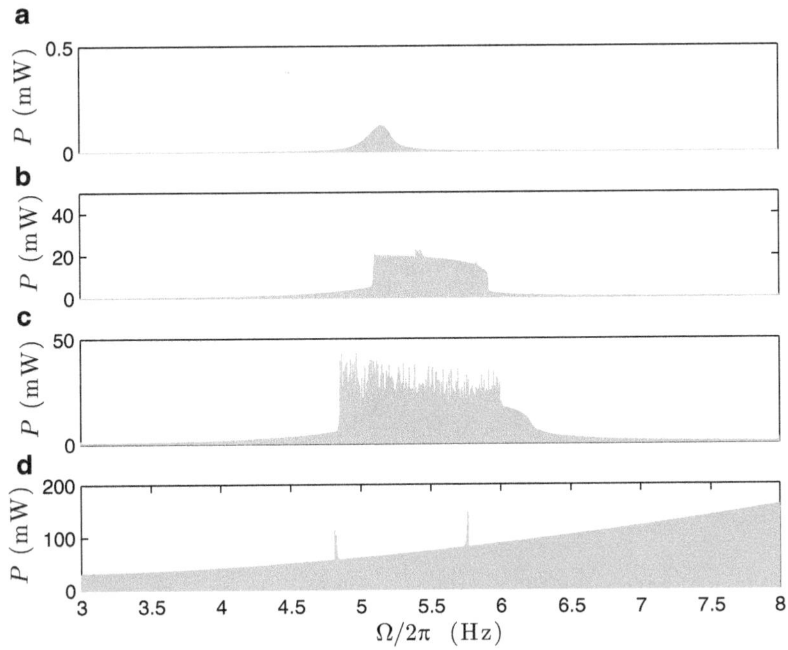

Fig. 4.9 Instantaneous power delivered to an electrical load for the forward and reverse frequency sweeps of Figs. 4.7 and 4.8. The following base accelerations were used: (**a**) $A = 0.1$ m/s^2, (**b**) $A = 3$ m/s^2, (**c**) $A = 6$ m/s^2, and (**d**) $A = 10$ m/s^2

for increased levels of base excitation; instead, the frequency response begins to broaden for higher levels of base excitation.

4.4.1.2 Theoretical Investigations

A dimensionless version of the governing equations for the system shown in Fig. 4.5 were derived in [35]. These equations take the form

$$I' + \rho I + \theta y' = 0\,, \tag{4.14a}$$

$$y'' + \mu y' - y + \beta y^3 - \epsilon\theta I = \Gamma \sin \eta\tau\,, \tag{4.14b}$$

where ρ is a dimensionless electric parameter for the electrical circuit and β is a dimensionless constant for the bistable system. This bistable case has multiple stable equilibria, which are located at $y_e = \pm\sqrt{1/\beta}$ when $\beta > 0$. We therefore expect that small levels of excitation will cause relatively small oscillations around one of the stable equilibria. However, past research on bistable mechanical oscillators has shown a dramatic increase in the energy-level of the response once exceeding the threshold for a potential well escape [30, 36].

The approximate analytical solution can be derived by assuming a solution in the form of a truncated Fourier series with time-varying coefficients $y = p(\tau) + a(\tau)\cos\eta\tau + b(\tau)\sin\eta\tau$ and $I = c(\tau)\cos\eta\tau + d(\tau)\sin\eta\tau$. The expressions for y, y', y'', I, and I', where we have taken $a'' = b'' = p'' = 0$ in agreement with the slowly varying presumption, were inserted into Eqs. (4.14a) and (4.14b) where the constant coefficients, along with the coefficients of $\cos\eta\tau$, $\sin\eta\tau$, are balanced on each side. Balancing the constant coefficients gives

$$\mu p' = p\left[1 - \beta\left(p^2 + \frac{3}{2}r^2\right)\right], \tag{4.15}$$

whereas balancing the coefficients of $\cos\eta\tau$ and $\sin\eta\tau$ gives

$$\mu a' + 2\eta b' = a\left(\eta^2 + 1 - 3\beta p^2 - \frac{3}{4}\beta r^2\right) - \mu\eta b + \epsilon\theta c\,, \tag{4.16a}$$

$$-2\eta a' + \mu b' = b\left(\eta^2 + 1 - 3\beta p^2 - \frac{3}{4}\beta r^2\right) + \mu\eta a + \epsilon\theta d + \Gamma\,. \tag{4.16b}$$

The steady-state response of the systems requires $a' = b' = p' = c' = d' = 0$. After setting these terms to zero, the expressions for $\tilde{c}$ and $\tilde{d}$ were then returned Eqs. (4.16a) and (4.16b) for c and d, respectively, to obtain

$$\left(\left(1 - \frac{\epsilon\theta^2}{\rho^2 + \eta^2}\right)\eta^2 + 1 - 3\beta\tilde{p}^2 - \frac{3}{4}\beta\tilde{r}^2\right)\tilde{a} - \left(\mu + \frac{\epsilon\theta^2\rho}{\rho^2 + \eta^2}\right)\eta\tilde{b} = 0\,, \tag{4.17a}$$

$$\left(\left(1 - \frac{\epsilon\theta^2}{\rho^2 + \eta^2}\right)\eta^2 + 1 - 3\beta\tilde{p}^2 - \frac{3}{4}\beta\tilde{r}^2\right)\tilde{b} + \left(\mu + \frac{\epsilon\theta^2\rho}{\rho^2 + \eta^2}\right)\eta\tilde{a} = -\Gamma\,. \tag{4.17b}$$

where a ($\tilde{}$) has been used to represent the fixed point solutions. Squaring and adding these two equations gives an expression for the frequency response of the mechanical system

$$\left[\left(1 - \frac{\epsilon\theta^2}{\rho^2 + \eta^2}\right)\eta^2 + 1 - 3\beta\tilde{p}^2 - \frac{3}{4}\beta\tilde{r}^2\right]^2\tilde{r}^2 + \left[\mu + \frac{\epsilon\theta^2\rho}{\rho^2 + \eta^2}\right]^2\eta^2\tilde{r}^2 = \Gamma^2 \tag{4.18}$$

where $\tilde{r}^2 = \tilde{a}^2 + \tilde{b}^2$ and $\tilde{p}$ may take on multiple values. More specifically, after setting $p' = 0$ in Eq. (4.15) and solving for the steady-state solution, the fixed points $\tilde{p} = 0$ and $\tilde{p}^2 = \dfrac{1}{\beta} - \dfrac{3}{2}\tilde{r}^2$ are obtained, where the latter solution restricts the values of $\tilde{r}$ that provide a physical solution for $\tilde{p}$ to $\tilde{r}^2 \leq \dfrac{2}{3\beta}$. For the plots that follow,

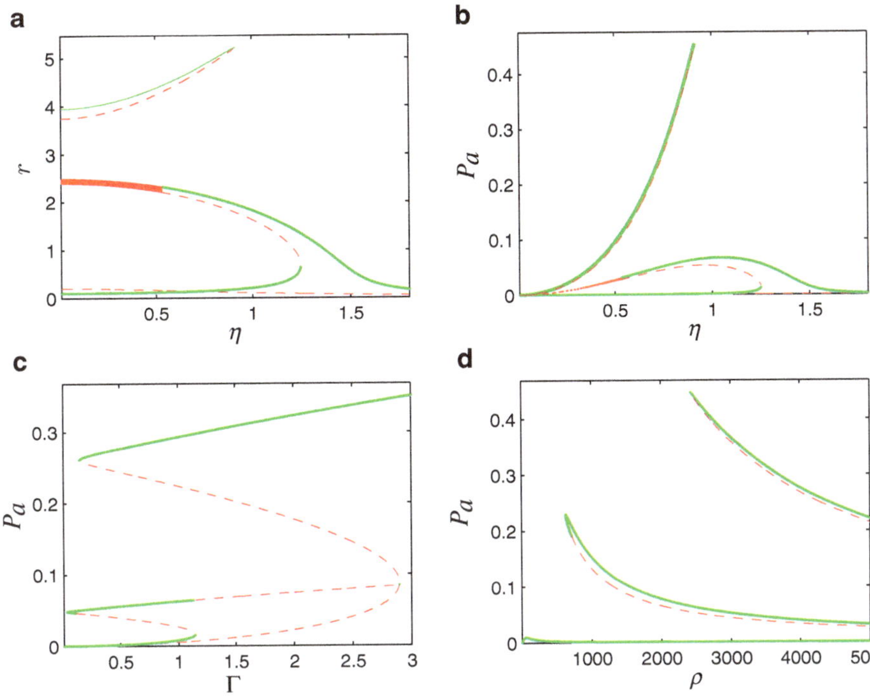

Fig. 4.10 Plots showing the stable (*solid line*) and unstable (*broken line*) response trends for the harvester with a bistable potential well. Graphs show frequency responses for (**a**) the oscillation amplitude of the mechanical system and (**b**) the dimensionless average power; graphs (**c**) and (**d**) plot the dimensionless average power for changes in Γ and ρ, respectively. In addition to $\mu = 0.01$, $\beta = 0.09$, $\epsilon = 0.8$, and $\theta = 10$, the following parameters were applied: both (**a**) and (**b**) used $\Gamma = 0.2$ and $\rho = 2,500$, (**c**) $\eta = 0.75$ and $\rho = 2,500$, (**d**) $\eta = 0.9$ and $\Gamma = 0.2$

an expression for the average power was formed by integrating the dimensionless form of the instantaneous power $P = \rho I^2$ over the period of the excitation

$$P_a = \frac{1}{T} \int_0^T \rho I^2 \, \mathrm{d}t, \tag{4.19}$$

where $T = 2\pi/\eta$ is the period of the excitation source.

Figure 4.10 shows a series of response trends for the bistable system. An interesting feature in the response of the bistable system is the emergence of additional solution branches. More specifically, these solutions are associated with the oscillations within a single potential well and those that cross the center potential well barrier. This system can exhibit P_a values similar to those of the linear system but displays more complex scaling in its response behavior as Γ is increased. In addition, the plots of ρ vs. P_a show the system can have even more local maxima.

4.4.2 Example Bistable Piezoelectric System

The bistable piezoelectric inertial generator has recently emerged as a potential solution to one of the foremost challenges in vibratory energy harvesting, i.e., consistent performance in complex spectral environments. More specifically, a dramatic increase in the design, validation, and analysis of such devices has appeared in the literature over the past few years. Shahruz [37], Ramlan et al. [38], and McInnes et al. [39] were among some of the first to propose exploiting a snap action instability for enhanced vibratory harvesting. While all these works provided novel ideas, such as energetic oscillations in the presence of noise [37], robustness to mistuning and low-frequency advantages [38], and stochastic resonance [39], each investigation chose to neglect the coupled electrical network modeling to emphasize the potential mechanical advantages.

Cottone et al. [40] and Gammaitoni et al. [41] provided experimentally validated models with electromechanical coupling considerations for bistable piezoelectric inertial generators. Cottone et al. [40] showed the clear advantage of oscillations about a double well potential in the presence of exponentially correlated noise, while Gammaitoni et al. extended the analysis and showed the optimal nonlinear parameters could yield marked improvement over the standard linear approach for both bistable and monostable nonlinear systems.

At the same time, Erturk et al. [20] and Stanton et al. [24] theoretically and experimentally studied harmonically forced bistable piezoelectric generators with an emphasis on broadband response. The device explored by Erturk et al. [20], which was inspired by the experiment of Moon and Holmes [42], used a ferromagnetic structure buckled by adjacent attracting magnets; Stanton et al. studied the system shown in Fig. 4.11, where a permanent magnet proof mass was oriented opposite to a fixed magnet to create a bistable potential. Both studies indicate that a harmonic cross-well attractor exists over a broad frequency range and can achieve similar marked improvement over the linear oscillator as discussed in references [40, 41]. More recently, Sneller et al. [43] demonstrated the use of a post-buckled beam harvester with piezoelectric patches. Erturk and Inman [44] have also provided a phenomenological discussion on the high-energy orbits of the bistable piezoelectric harvesters to include ideas for alternative transduction mechanisms. Ferrari et al. subjected the proof mass magnet bistable design to white noise stochastic excitation both numerically and experimentally and found an 88% improvement in the root-mean-square voltage response [45]. In addition, Ando et al. fabricated and validated all of the aforementioned advantages in a true MEMS-scale device, thus extending the advantages predicted and observed at the macroscopic scale towards the microscale regime [46].

This section considers the behavior of physical system shown in Fig. 4.12. Rather than derive the governing equations for this system, which was undertaken and refined over a series of articles, our focus is on the response of the dimensionless equations of motion. However, we note that the dimensionless equations have been derived from fundamental principles and point the interested reader towards

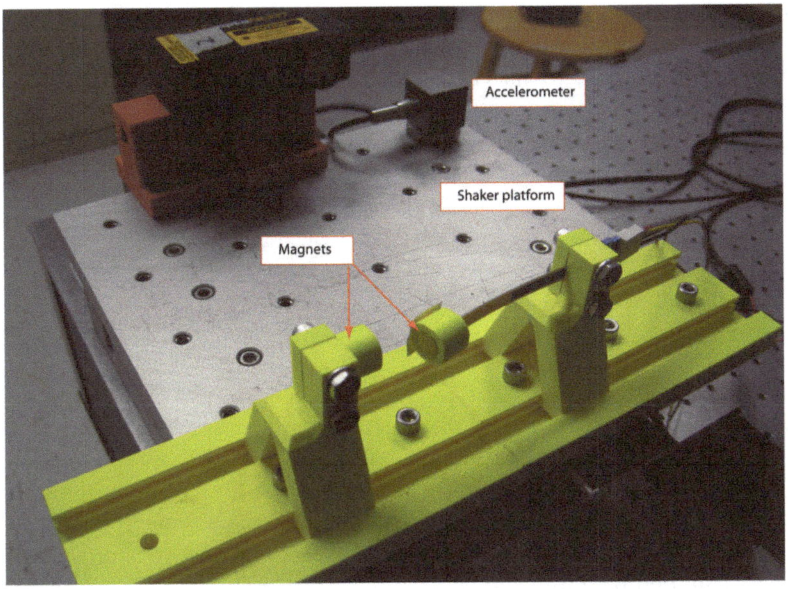

Fig. 4.11 Picture of the piezoelectric inertial generator studied in [24]

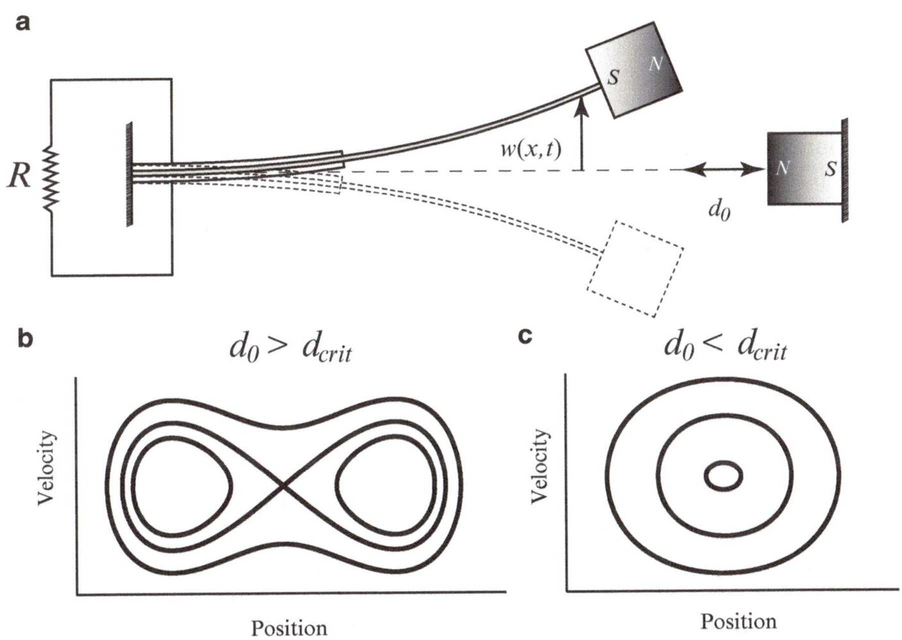

Fig. 4.12 Illustration of the bistable harvester concept. (**a**) The nonlinear boundary condition induces a pitchfork bifurcation in the continuum electroelastic harvester. Phase portraits upon surpassing the critical bifurcation distance d_{crit} are shown in (**b**) where the system is bistable, while in (**c**) the system is monostable with an adjustable hardening potential well due to the axial magnetic forces

[24,47,48] for more detailed information, which includes experimental comparisons. The remainder of this section will investigate the response behavior of the bistable piezoelectric energy harvester to harmonic and random excitations.

4.4.2.1 Harmonic Excitation Case

The dimensionless governing equations for a piezoelectric bistable harvester, as taken from [49, 50], are

$$\ddot{x} + \left(\mu_a + \mu_b x^2\right)\dot{x} - x + x^3 - \theta v = f\cos\omega t\,, \tag{4.20a}$$

$$\dot{v} + \mu_c v + \theta\dot{x} = 0\,, \tag{4.20b}$$

where x is a dimensionless modal displacement, v is a dimensionless voltage, μ_a is the linear dissipation coefficient, μ_b is the nonlinear dissipation coefficient, θ is the electromechanical coupling, f is the excitation level, ω is a dimensionless excitation frequency, and μ_c is the electrical dissipation coefficient.

In the method of harmonic balance, the harvester's response is presumed to be accurately captured by a truncated Fourier series; thus, we assume solutions in the following form:

$$x = c(t) + a_1(t)\sin\omega t + b_1(t)\cos\omega t\,, \tag{4.21a}$$

$$v = a_2(t)\sin\omega t + b_2(t)\cos\omega t\,. \tag{4.21b}$$

Similar to the previous studies on electromagnetic energy harvesting, the solution procedure requires the assumed solutions to be substituted into Eq. (4.20b) and the harmonics to be balanced. As shown in [50], it is then feasible to obtain an analytical expression for the frequency response of the bistable piezoelectric harvester. While the derivation of the frequency response equation is omitted, the remainder of this section describes some of the primary trends.

Oscillations within one of the two symmetric potential wells are known to be of the softening type [13, 36]. For the two most common piezoceramics utilized in harvester design, PZT-5A and PZT-5H, material elasticity is strongly softening as well, so much so that normal first mode hardening oscillations for large-amplitude piezoelectric beam motions rarely offset the piezoelectric softening effect [47, 48]. Therefore, it stands to reason that the intrawell motion of a bistable piezoelectric harvester will exhibit marked softening nonlinear resonance curves and exhibit linear response for nearly infinitesimal excitation ranges. Of equal importance to the higher-order piezoelectric elasticity influence is nonlinear material damping. The influence of these combined effects is shown in Fig. 4.13. As the nonlinear damping parameter μ_b is increased, a predicted saddle-node bifurcation is suppressed and is in agreement with experimental studies in references [47, 48]. In the inset of Fig. 4.13, the trend of a weakening jump phenomena can also be seen for a forcing frequency equal to the intrawell linear resonance, i.e., $\omega = \sqrt{2}$. For a dimensionless electrical impedance of $\mu_c = 0.0078$ (near open circuit conditions) and no nonlinear

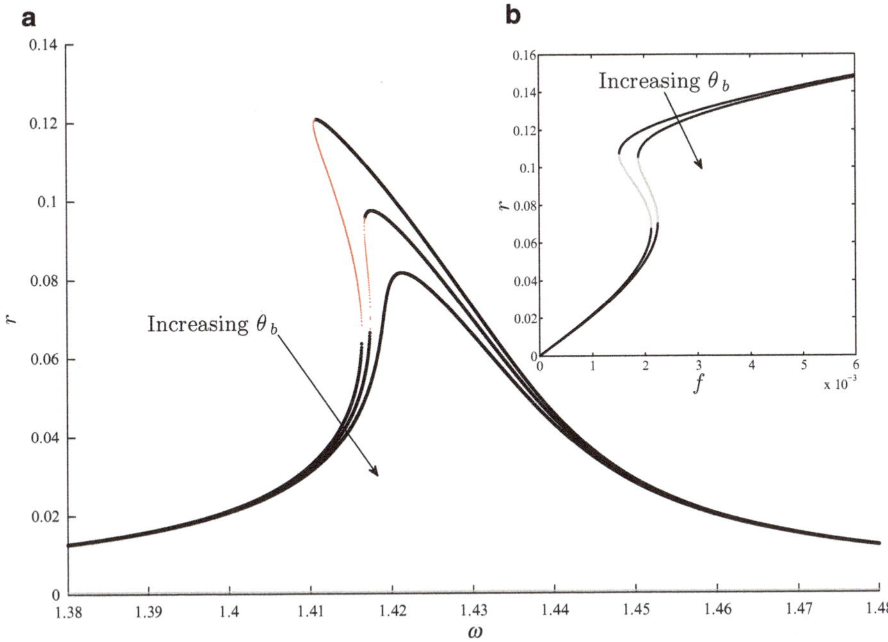

Fig. 4.13 Influence of nonlinear damping on the suppressing (**a**) the intrawell frequency response and (**b**) the intrawell force response curve

damping, the intrawell solution goes unstable and escapes over the potential barrier when $f = 0.4276$. However, for $\mu_b = 0.0025$, which corresponds to the nonlinear resonance curve in Fig. 4.13 with no multi-valued solutions, the escape amplitude increases to $f = 0.4376$. Therefore, the presence of nonlinear damping must be taken into consideration in designing a harvester to exhibit cross-well response to a harmonic excitation.

Erturk and Inman [44] experimentally demonstrated that there exists an optimal impedance load for the steady state large orbit dynamics of the bistable generator (see Fig. 13 of [44]). This operating condition can also be theoretically predicted by the method of harmonic balance [50]. Figure 4.14 focuses on the high-energy orbit solutions for forcing amplitudes that drive the system above the potential well barrier. A sample frequency response is shown in Fig. 4.14 for $f = 0.1222$, which corresponds to a realistic excitation level [50]. The distinct advantage in the bistable design lies in the existence of large-orbit attractor over a very wide frequency range, which in this case is shown over $\omega = 0.2$ to $\omega = 2$. For a range of frequencies, however, the intrawell harmonic solution does not coexist with the large orbit solution which would ensure the desired vigorous response in this parameter range. When there are coexisting solutions, the major design challenge is to realize large-orbit solutions amidst sensitivity to initial conditions, basins of attraction which may be extremely restricted in some instances and easily accessible in others, and

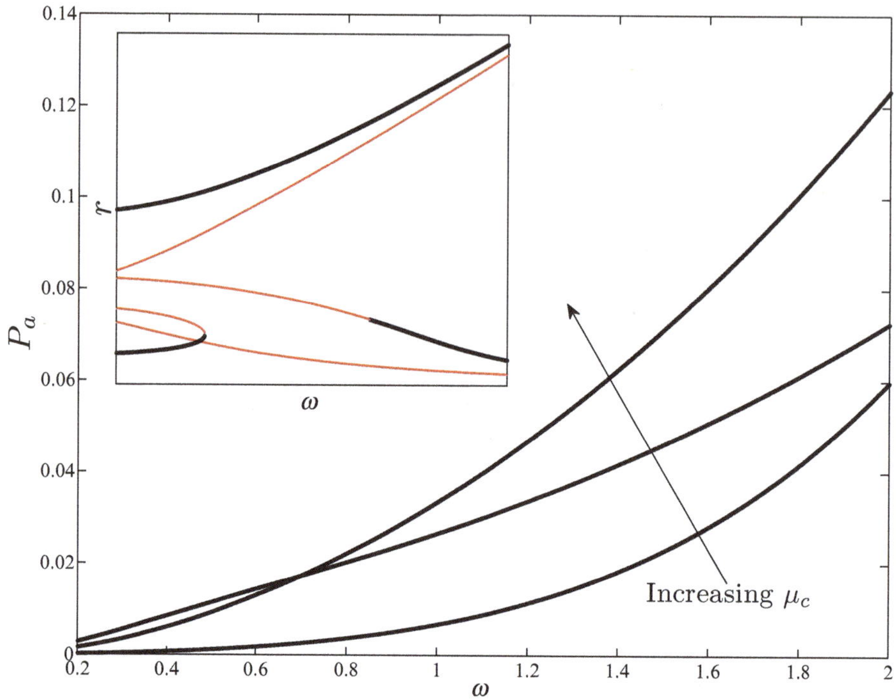

Fig. 4.14 Inset: a typical frequency response for large forcing such that a large-orbit solution manifests. The rest of the graph shows the effect of increasing electrical impedance on average power generated by the large-orbit motion

even interwell chaotic responses. The inset of Fig. 4.14 shows the average power for the same large-orbit response in main Fig. 4.14 but for three different values of μ_c. Again, the existence of an optimal impedance load is inferred by a region $\omega < 0.75$ where a lesser value of μ_c is shown to exceed the average power of a higher value of μ_c that otherwise clearly dominates the other parameter choices for μ_c. Fortunately, the harmonic balance method provides a method for finding such optimal solutions, see further details in [50].

4.4.2.2 Random Excitation Case

While the previous section examined the harvester's response to single frequency excitation, this section summarizes the investigations of [49] which considered the systems response under broadband excitation. More specifically we considered excitation of the form

$$f(t) = \sum_{n=1}^{N} f_n \sin(\omega_n t + \phi_n), \qquad (4.22)$$

where f_n is the amplitude of the nth harmonic with the phase ϕ_n. Following the detailed analysis from [49], the Melnikov necessary condition for chaos is

$$0 < -\Gamma + \gamma\xi(t), \qquad (4.23)$$

where γ the noise intensity and $\xi(t)$ is a realization of a stochastic process approximated by the sum $\xi(t) = \sum_{n=1}^{N} f_n \sin(\omega_n t + \phi_n)$ where $\omega_n = n\Delta\omega$, $\Delta\omega = \omega_{\rm cut}/N$, and $\omega_{\rm cut}$ is a cutoff frequency beyond the spectral distribution of $\xi(t)$ vanishes [51]. The nth phase ϕ_n is uniformly distributed over the interval $[0, 2\pi]$. The amplitudes f_n are determined from a spectral distribution similar to a Ornstein–Uhlenbeck colored noise process

$$f_n = \gamma\sqrt{\frac{4}{\gamma N(1 + \tau^2\omega^2)}}, \qquad (4.24)$$

where τ is the correlation time. As τ approaches zero, the spectral distribution has uniform frequency content and approximates a white noise spectrum over a finite frequency range.

The bistable generator was subjected to a broadband signal comprised of over 500 harmonics and for varying values of the correlation time τ. Both the nonlinear harvester and equivalent linear piezoelectric harvester (with no opposing magnet) were excited by a near-constant spectrum broadband excitation with $\tau = 0$ and $\omega_{\rm cut} = 5$. Figure 4.15 displays results for the RMS power across the electrical impedance load for increasing values of γ which correspond to physical driving amplitudes from zero up to 0.5 g. For all excitation levels, it was verified that the stochastic Melnikov inequality was satisfied, which indicates the likelihood of a large-amplitude response. In the course of collecting each data point, a total of 20 time-series simulations were performed for discrete values of γ, and the RMS power for each run was averaged over all 20 simulations. Near open-circuit conditions can be seen in Fig. 4.15 to surpass the RMS power generated by the equivalent linear oscillator for very weak excitations, while introducing the maximum Melnikov electrical damping ($\mu_c = 0.66$) suppressed this effect. For both impedance loads, however, a larger γ shows the linear oscillator is actually a more efficient harvester when the broadband constant spectrum is nearly constant. To verify the near open-circuit results at low driving amplitude, both the bistable system and its linear equivalent were simulated over a long time series. As indicated by the power spectrum shown in [49], the linear oscillator gathers a significant amount of energy from the excitation signal for frequencies near its fundamental resonance, while the bistable oscillator will draw larger amounts of power over the interval $[0, \omega_{\rm cut}]$ as well as at higher harmonics beyond this region.

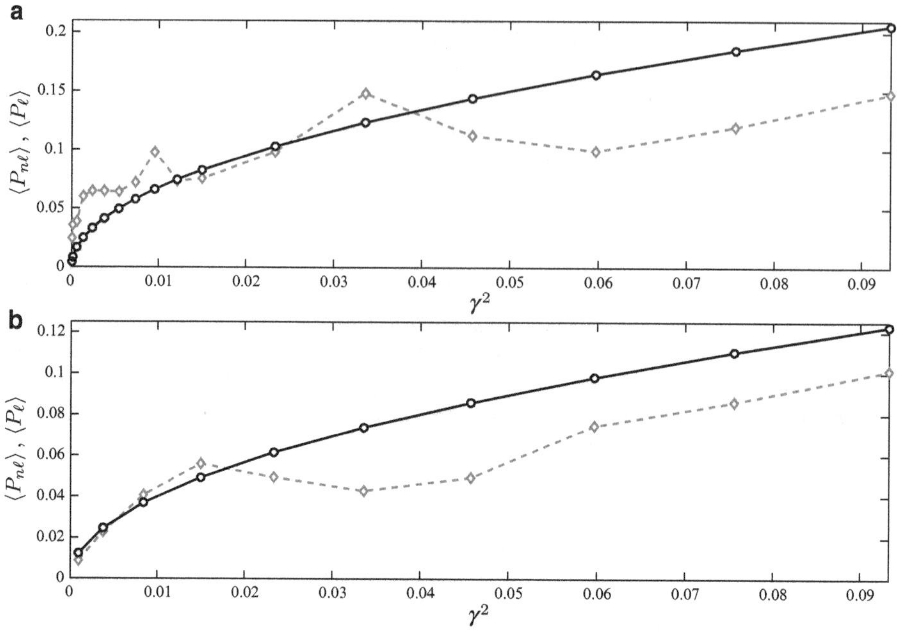

Fig. 4.15 (**a**) Constant spectrum broadband excitation for near open-circuit conditions $\mu_c = 0.035$ in the nonlinear harvester (*gray line*) can exceed the equivalent linear harvester (*black line*) for low-acceleration drives, but for (**b**) $\mu_c = 0.66$, the linear harvester tends to do as well or better than the nonlinear harvester

The same simulation procedure was carried out for $\tau \neq 0$ to investigate the performance of the harvester in a colored noise environment. Figure 4.16 illustrates a clear advantage of the nonlinear compliance over that of a linear oscillator for a wide range of γ. As the spectral content near the fundamental frequency of the equivalent linear system is tapered due to increasing values of τ, the linear system's response will degrade. In the bistable system, however, the stochastic Melnikov inequality continues to be satisfied, and the oscillator responds to a broad range of excitation frequencies with highly energetic motions. Figure 4.16 illustrates averaged RMS power values, plotted as a ratio of the nonlinear RMS power to the linear RMS power $(\langle P_{n\ell}\rangle/\langle P_{n\ell}\rangle)$ for several correlation times. For a more complex, and possibly more realistic noise spectrum, the advantage of a strong nonlinearity emerges.

4.5 Summary

This chapter described recent research on the use of a bistable potential well for energy harvesting applications using piezoelectric and electromagnetic devices. For

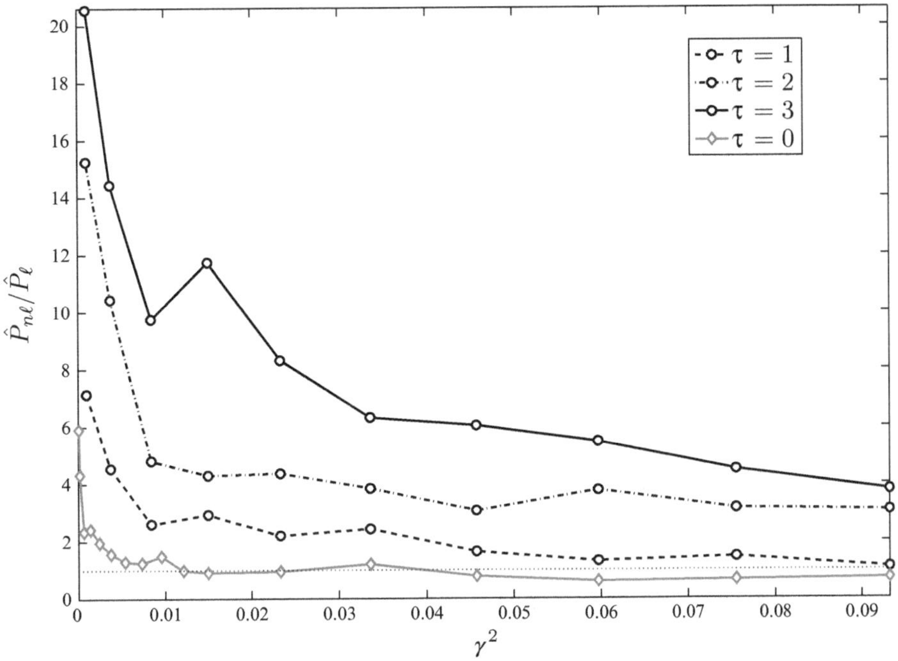

Fig. 4.16 Comparisons of the averaged power from the bistable $\hat{P}_{n\ell}$ and linear $\hat{P}_{\ell}$ for different types of noise. The variable γ represents the noise intensity and $\tau = 0$ represents broadband white noise. Apart from the $\tau = 0$ case, the bistable system harvests more energy than the linear system for colored noise excitation [49]

single harmonic excitation, both the piezoelectric and electromagnetic harvesters possess a large-amplitude response for sufficient levels of excitation. As shown in [35], through the use of uncertainty propagation, this large-amplitude response is robust to parameter variations and can offer potential advantages in several environments.

Beyond the case of single harmonic excitation, we have shown results for different forms of random excitation. In particular, it is interesting to note that the bistable harvester will, on average, harvest more energy than a linear harvester from colored noise. However, for the case of broadband white noise, the average power from both the linear and bistable devices is nearly equivalent.

Acknowledgements The author of this chapter would like to thank the Army Research Office for financial support and program manager Dr. Ronald Joslin from the Office of Naval Research for financial support.

References

1. Roundy S, Wright PK, Rabaey JM (2003) Energy scavenging for wireless sensor networks. Springer, New York
2. Saha CR (2006) Optimization of and electromagnetic energy harvesting device. IEEE Trans Magn 42(10):3509–3511
3. Shahruz SM (2006) Limits of performance of mechanical band-pass filters used in energy scavenging. J Sound Vib 293(1–2):449–461
4. Shahruz SM (2006) Design of mechanical band-pass filters for energy scavenging. J Sound Vib 292(3–5):987–998
5. Stephen NG (2006) On energy harvesting from ambient vibration. J Sound Vib 293:409–425
6. Yang B, Lee C, Xiang W, Xie J, He JH, Kotlanka RK, Low SP, Feng H (2009) Electromagnetic energy harvesting from vibrations of multiple frequencies. J Micromech Microengineering 19(035001):1–8
7. Yen BC, Lang JH (2005) A variable-capacitance vibration-to-electric energy harvester. IEEE Trans Circ Syst 1 Fund Theor Appl 53(2):288–295
8. Lesieutre GA, Ottman GK, Hofmann HF (2004) Damping as a result of piezoelectric energy harvesting. J Sound Vib 269(3–5):991–1001
9. Sodano HA, Inman DJ, Park G (2005) Generation and storage of electricity from power harvesting devices. J Intell Mater Syst Struct 16:67–75
10. Sodano HA, Inman DJ, Park G (2005) Comparison of piezoelectric energy harvesting devices for recharging batteries. J Intell Mater Syst Struct 16:799–807
11. Beeby SP, Torah RN, Tudor MJ, Glynne-Jones P, O'Donnell T, Saha CR, Roy S (2007) A micro electromagnetic generator for vibration energy harvesting. J Micromechanics Microengineering 17:1257–1265
12. Horowitz SB, Sheplak M, Cattafesta LN, Nishida T (2006) A mems acoustic energy harvester. J Micromechanics Microengineering 16:174–181
13. Mann B, Owens B (2010) Investigations of a nonlinear energy harvester with a bistable potential well. J Sound Vib 329:1215–1226
14. Beeby SP, Tudor MJ, White NM (2006) Energy harvesting vibration sources for microsystems applications. Meas Sci Tech 17:175–195
15. Leland ES, Wright PK (2006) Resonance tuning of piezoelectric vibration energy scavenging generators using compressive axial load. Smart Mater Struct 15:1413–1420
16. Poulin G, Sarraute E, Costa F (2004) Generation of electrical energy for portable devices comparative study of an electromagnetic and piezoelectric system. Sensor Actuator A 116:461–471
17. Renno JM, Daqaq MF, Inman DJ (2009) On the optimal energy harvesting from a vibration source. J Sound Vib 320:386–405
18. Roundy S (2005) On the effectiveness of vibration based energy harvesting. J Intell Syst Struct 16:809–823
19. Mann B, Sims N (2009) Energy harvesting from the nonlinear oscillations of magnetic levitation. J Sound Vib 319:515–530
20. Erturk A, Hoffmann J, Inman DJ (2009) A piezomagnetoelastic structure for broadband vibration energy harvesting. Appl Phys Lett 94(254102):1–4
21. Stanton SC, McGehee CC, Mann BP (2009) Reversible hysteresis for broadband magnetopiezoelastic energy harvesting. Appl Phys Lett 95:174103–3
22. Triplett A, Quinn DD (2009) The effect of nonlinear piezoelectric coupling on vibration-based energy harvesting. J Intell Mater Syst Struct 20(16):1959–1967
23. Soliman MS, Abdel-Rahman EM, El-Saadany EF (2008) A wideband vibration-based energy harvester. J Micromechanics Microengineering 18:1–11
24. Stanton SC, McGehee CC, Mann BP (2010) Nonlinear dynamics for broadband energy harvesting: Investigation of a bistable piezoelectric inertial generator. Phys D Nonlinear Phenom 239:640–653

25. Challa VR, Prasad MG, Shi Y, Fisher FT (2008) A vibration energy harvesting device with bidirectional resonance frequency tunability. Smart Mater Struct 17(1):015035
26. Barton DAW, Burrow SG, Clare LR (2010) Energy harvesting from vibrations with a nonlinear oscillator. J Vib Acoust 132(2):021009
27. Cammarano A, Burrow SG, Barton DAW (2011) Modelling and experimental characterization of an energy harvester with bi-stable compliance characteristic. J Syst Contr Eng 225:475–484
28. Owens BAM, Mann BP (2012) Linear and nonlinear electromagnetic coupling models in vibration-based energy harvesting. J Sound Vib 331:922–937
29. Coleman HW, Steele WG (1999) Experimentation and uncertainty analysis for engineers, 2 edn. Wiley
30. Mann BP (2009) Energy criterion for potential well escapes in a bistable magnetic pendulum. J Sound Vib 323:864–867
31. Yung KW, Landecker PB, Villani DD (1998) An analytical solution for the force between two magnetic dipoles. Magn Electr Separ 9:39–52
32. Landecker PB, Villani DD, Yung KW (1999) An analytical solution for the torque between two magnetic dipoles. Magn Electr Separ 10:29–33
33. Furlani EP (2001) Permanent magnet and electromechanical devices, 1 edn. Academic Press, New York
34. Vanderlinde J (2004) Classical electromagnetic theory, 2 edn. Springer, Dordrecht
35. Mann BP, Barton DAW, Owens BAM (2012) Uncertainty in performance for linear and nonlinear energy harvesting strategies. J Intell Mater Syst Struct 23(13):1448–1457
36. Virgin LN (2000) Introduction to experimental nonlinear dynamics. Cambridge University Press, Cambridge
37. Shahruz S (2004) Increasing the efficiency of energy scavengers by magnets. J Comput Nonlinear Dynam 3:1–12
38. Ramlan R, Brennan M, Mace B, Kovacic I (2010) Potential benefits of a non-linear stiffness in an energy harvesting device. Nonlinear Dynam 59:545–558
39. McInnes C, Gorman D, Cartmell M (2008) Enhanced vibration energy harvesting using nonlinear stochastic resonance. J Sound Vib 318:655–662
40. Cottone F, Vocca H, Gammaitoni L (2009) Nonlinear energy harvesting. Phys Rev Lett 102:1–4
41. Gammaitoni L, Neri I, Vocca H (2009) Nonlinear oscillators for vibration energy harvesting. Appl Phys Lett 94(164102)
42. Moon F, Holmes P (1979) A magnetoelastic strange attractors. J Sound Vib 65:275–296
43. Sneller AJ, Cette P, Mann BP (2011) Experimental investigation of a post-buckled piezoelectric beam with an attached central mass used to harvest energy. J Syst Contr Eng 225(4):497–509
44. Erturk A, Inman D (2011) Broadband piezoelectric power generation on high-energy orbits of the bistable Duffing oscillator with electromechanical coupling. J Sound Vib 330:2339–2353
45. Ferrari M, Ferrari V, Guizzettia M, Ando B, Bagliob S, Trigonab C (2010) Improved energy harvesting from wideband vibrations by nonlinear piezoelectric converters. Sensor Actuator A 162:425–431
46. Ando B, Baglio S, Trigona C, Dumas N, Latorre L, Nouet P (2010) Improved energy harvesting from wideband vibrations by nonlinear piezoelectric converters. J Micromechanics Microengineering 20(125020)
47. Stanton SC, Erturk A, Mann BP, Inman DJ (2010) Nonlinear piezoelectricity in electroelastic energy harvesters: Modeling and experimental identification. J Appl Phys 108:1–9
48. Stanton SC, Erturk A, Mann BP, Inman DJ (2010) Resonant manifestation of intrinsic nonlinearity within electroelastic micropower generators. Appl Phys Lett 97:1–3
49. Stanton SC, Mann BP, Owens BAM (2012) Melnikov theoretic methods for characterizing the dynamics of the bistable piezoelectric inertial generator in complex spectral environments. Phys D 241:711–720
50. Stanton SC, Mann BP, Owens BAM (2012) Harmonic balance analysis of the bistable piezoelectric inertial generator. J Sound Vib 331:3617–3627
51. Simiu E (2002) Chaotic transitions in deterministic and stochastic dynamical systems. Applied Mathematics. Princeton University Press, Princeton, NJ

Part III
Non-harmonic and Spectral Excitation

Chapter 5
Plucked Piezoelectric Bimorphs for Energy Harvesting

Michele Pozzi and Meiling Zhu

Abstract In this chapter, the plucking technique of frequency up-conversion is introduced, modelled and applied to energy harvesting. The aim of the technique is to bridge the gap between the high-frequency response of piezoelectric energy harvesters and the low-frequency input that is most often available from the ambient environment. After covering the general principle of plucking excitation, the plucking action is modelled analytically as well as with finite element analysis. Finally, the application of plucked piezoelectric bimorphs to a wearable knee-joint harvester (the pizzicato energy harvester) is discussed in some depth to show the potential of the plucking technique of frequency up-conversion.

5.1 Introduction

The modern drive towards mobility and wireless devices is motivating the growth of research efforts in energy harvesting (EH) technologies. The replacement of batteries with the renewable source of energy afforded by energy harvesters has the twofold advantage of reducing maintenance (battery replacement or recharging) and of increasing the reliability of the devices by removing the risk of being left without the power necessary for operation. Both benefits are of particular importance in medical applications, where any downtime might be life-threatening and where battery replacement may require surgery. This motivates the intense research in wearable energy harvesters.

Among the EH technologies that have proven effective for motion energy harvesting, piezoelectric transducers probably have the longest history and have attracted the greatest interest. There is a good selection of materials exhibiting

M. Pozzi • M. Zhu (✉)
Department of Manufacturing and Materials, Cranfield University, Cranfield, UK
e-mail: m.zhu@cranfield.ac.uk

N. Elvin and A. Erturk (eds.), *Advances in Energy Harvesting Methods*,
DOI 10.1007/978-1-4614-5705-3_5,

significant piezoelectric properties. Currently, the piezoelectric material of choice for macroscopic energy harvesters is undoubtedly lead zirconate titanate (PZT) in one of its many varieties.[1] The polymeric piezoelectric material PVDF offers low density and high compliance but is also characterised by modest piezoelectric activity. Although PZT (7.6 g/cm^3) is over four times more dense than PVDF (1.77 g/cm^3), it still affords the development of compact and light weight energy harvesters, thanks to its high-energy conversion efficiency. The advantages of compactness and light weight are arguably superior to many other energy harvesting technologies.

When used as an EH material, the high stiffness of PZT is often a hindrance, as it takes very large compressive forces to fully strain a PZT transducer by a few hundred micro-strain. In other words, the stiffness of piezoelectric ceramics is a source of problems in matching the mechanical impedance of the transducer with that of the rest of the system, including the environment that supplies the latent harvestable energy. For this reason, the piezoelectric transducer normally favoured in EH is the PZT bimorph, which works in bending as a cantilevered beam, therefore resulting in a much more compliant structure.

The PZT bimorph actuators resemble bimetallic strips: differential strains at either side of the midplane cause bending. With PZT this can be achieved in two ways: in a series bimorph the two PZT layers at either side of an internal shim are poled in the opposite directions, so that the electric field that builds up when the external electrodes are energised causes expansion of one layer and contraction of the other, yielding an overall bending of the device. In parallel[2] bimorphs, instead, the layers are poled in the same direction, which means that opposite electric fields must be applied to them to have a differential strain. This requires access to the middle electrode (the metal shim, if present[3]), so that the two external electrodes can be charged with the same polarity, opposite to that of the internal electrode. In bimorph actuators the parallel connection is normally preferable because it permits the permanent application of a bias voltage between external electrodes, whilst the middle electrode has a voltage ranging between these two; this prevents depoling of the PZT material. As energy generators, parallel bimorphs should have the two external electrodes connected together to form one terminal; as harvesters they have the advantage of higher capacitance, hence lower impedance, at the expense of increased wiring complexity. For this latter reason, series bimorphs may be preferable in compact EHs.

Piezoelectric devices which harvest energy from environmental vibrations encounter a frequency mismatch problem: even the more compliant bimorph structures

[1]The properties of PZT can be tailored by adjusting the relative concentration of zirconium and titanium as well as by the addition of doping elements.

[2]The names *series* and *parallel* derive from the configuration of the electrical connections, for example, in a parallel bimorph, the two layers are electrically connected in parallel.

[3]The internal shim is optional but normally added to increase the mechanical robustness of the device. Typical materials are stainless steel or brass.

have resonant frequencies above typical environmental vibrations. A traditional solution is the use of slender bimorphs and/or the attachment of a seismic mass at their end to reduce the system's fundamental frequency. On the other hand, the high fundamental frequency intrinsic to piezoelectric transducers remains desirable as, given that a small amount of mechanical energy can be converted into charge in each strain cycle, higher frequencies mean the potential of greater power generation. The frequency mismatch issue is even more severe within the scope of human-based energy harvesting, given that human motions are intrinsically slow, never exceeding a few hertz (with the notable exception of processes related to vocalisation). In this context, the application of the traditional approach, that is, longer bimorphs and additional mass to lower the response frequency, would unacceptably lead to increased encumbrance.

Therefore, the need for a technique capable of bridging between the high-frequency response of piezoelectric energy harvesters and the low-frequency input that is most often available in the environment and on the human body is clearly needed. The plucking excitation technique addresses this issue.

5.1.1 Mechanical Plucking: Principle of Operation

Mechanical plucking is made up of three main phases, as represented schematically in Fig. 5.1. In the *approach* phase, the distance between bimorph and plectrum is reduced until they come into contact. Immediately follows the *loading* phase, during which both elastic elements are deflected, according to their mechanical compliance: mechanical energy is input in the system plectrum-bimorph in the form of strain energy. As the deflection progresses, the overlap between the two elements is reduced so that their contact area becomes gradually smaller, until contact is lost (*release point*). From this instant on, the bimorph *vibrates* at its resonance frequency around its rest position as a cantilever beam. As the bimorph vibrates, the stored strain energy is converted by the direct piezoelectric effect into electrical energy and transferred to the external circuit; part of the energy is dissipated through various forms of damping, like air damping, dielectric losses and material internal damping. The outcome of plucking is frequency up-conversion, as by one single slow movement of the plectrum, a large number of vibrations are produced at high frequency.

It is worthwhile to highlight the differences between plucking and another technique of energy harvesting: impact excitation [13, 15]. From the physical point of view, impact involves the transfer of momentum; mathematically, the system's initial conditions have non-zero velocity; the principle is similar to hammers striking the wires in a piano. In plucking excitation, the piezoelectric devices are slowly deformed and then released; mathematically, the system's initial conditions feature a non-zero displacement; the principle is equivalent to the plucking of chords in a guitar (or a violin with the *pizzicato* technique).

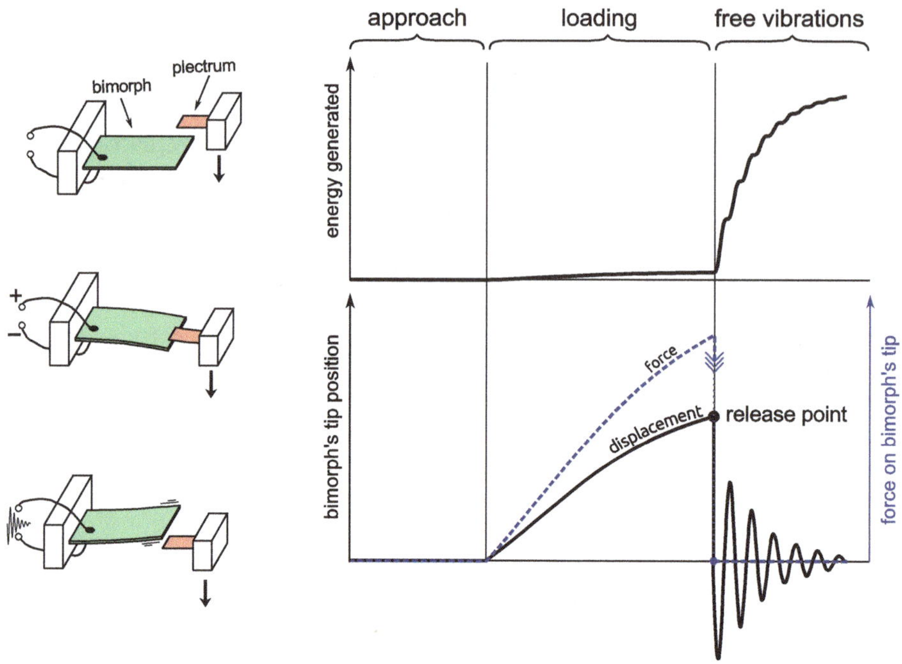

Fig. 5.1 The principle behind plucked bimorphs energy harvesting. On the *left*, the three main phases of approach, loading and free vibration after release. On the *right*, the time history plots during plucking for energy generation, tip displacement and external force on the bimorph

5.1.2 Possible Configurations and Applications

Frequency up-conversion techniques can be used in situations where the vibrational energy available is at much lower frequency than the transducers' natural frequency. For example, Kulah and Najafi [5, 6] have developed a microscale electromagnetic harvester where the low-frequency vibration of a suspended magnet excites the high-frequency vibration of a number of coil-carrying cantilevers placed around it. The relatively heavy magnet is suspended on a polymeric film and resonates with environmental vibrations at tens of hertz. At one end of its travel, it attracts the tip of the cantilevers until contact is made; as it moves towards the other extreme, the elastic restoring force of the cantilevers eventually overcomes the magnetic attraction and they spring back resonating at over 1 kHz. Frequency up-conversion techniques can also be applied to scavenge energy from slowly rocking platforms or slow rotary motions [12] or also from oceanic waves [8]. The plucking action for frequency up-conversion can also be applied to wearable energy harvesters, as human movements are always very low frequency [10]. The latter application will be discussed in detail in the second part of this chapter.

5.2 Modelling the Plucking Action

Although the principle behind the plucking technique of frequency up-conversion is rather simple, it is important to carry out some modelling of the process to understand its potential and its dominating features as an energy harvesting technique. It is important to know the maximum force needed to fully deflect the bimorph, otherwise, depending on the mechanical configurations, the plucking action may not be completed (if the mechanism providing the input energy stalls at a limiting force) or the process not exploited to the full (if the plectra are too compliant and contact is lost before the bimorph is deflected sufficiently). Another piece of information that can be gathered from modelling and which is important in the energy harvester design stage is the optimal plucking frequency (this is a rather involved topic that will be addressed in detail in a following section). On the electronics side, information on the output impedance of the harvester is needed to optimise energy extraction and can be readily found with device modelling. Modelling also allows the designer to try out different bimorph geometries and varieties of PZT material to optimise the performance of the energy harvester to the operating parameters of the intended application.

The following sections are therefore dedicated to the presentation of an analytical model, useful for an initial estimate of performance as well as ideally suited for some design optimisation techniques and a finite elements (FE) model, particularly useful for complex geometries.

5.2.1 *Analytical Modelling*

In this section the classical treatment of a vibrating clamped-free beam is adapted to a plucked piezoelectric bimorph; for more details on the beam-related derivations, the reader is referred to a book on vibrations such as [3].

The governing differential equation for a piezoelectric bimorph[4] (Fig. 5.2) is

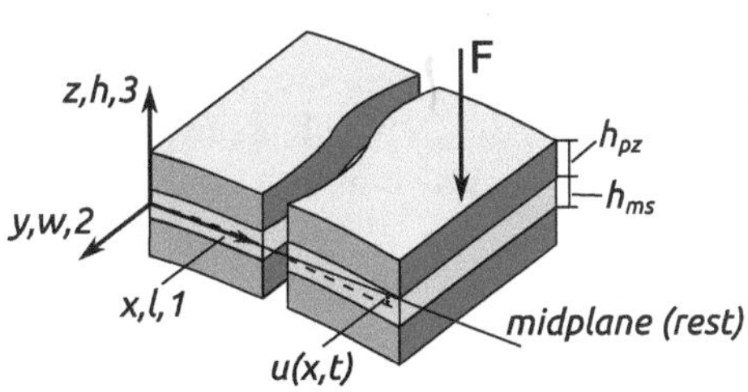

Fig. 5.2 Sketch of the piezoelectric bimorph with relevant dimensions and notations used in the analytical model

[4]It is assumed that the geometrical parameters of the bimorph permit the use the Euler–Bernoulli beam approximation.

$$\frac{\partial^2 M}{\partial x^2} + c_a \frac{\partial u}{\partial t} + c_s I \frac{\partial^5 u}{\partial x^4 \partial t} + m \frac{\partial^2 u}{\partial t^2} = f(x,t) \tag{5.1}$$

where c_s is the coefficient of internal damping, c_a is the coefficient of viscous damping (air), m is the linear mass density and M is made of two terms, M_{mech} and M_{el}, which are the bending moments of mechanical and electrical origin, respectively:

$$M = M_{\mathrm{mech}} + M_{\mathrm{el}} = B \frac{\partial^2 u}{\partial x^2} + \theta\, V \tag{5.2}$$

where θ will be derived later and the bending stiffness is

$$\begin{aligned} B &= 2Y_{\mathrm{pz}} I_{\mathrm{pz}} + Y_{\mathrm{ms}} I_{\mathrm{ms}} \\ &= 2Y_{\mathrm{pz}} \left[\frac{w h_{\mathrm{pz}}^3}{12} + w h_{\mathrm{pz}} \left(\frac{h_{\mathrm{pz}} + h_{\mathrm{ms}}}{2} \right)^2 \right] + Y_{\mathrm{ms}} \frac{w h_{\mathrm{ms}}^3}{12} \end{aligned} \tag{5.3}$$

where the pz and ms subscripts stand for piezoelectric layer and metal shim, respectively, Y is the Young's modulus in the x direction[5], I is the second moment of an area (the parallel axis theorem was applied to calculate the I for the two piezoelectric layers) and w is the width of the bimorph.

The parameter θ couples the mechanical and the electrical fields and can be derived from the following constitutive equations of piezoelectricity, which gives the stress[6] T as a function of the applied electric field:

$$T_1 = c_E S_1 - d_{31} Y_{\mathrm{pz}} E \tag{5.4}$$

We ignore its first term as it is already taken care of by Eq. (5.3); the second term leads to

$$\begin{aligned} M_{\mathrm{el}} &= 2 \int z \,\mathrm{d}a\, T = 2 \int_{h_{\mathrm{ms}}/2}^{h_{\mathrm{ms}}/2 + h_{\mathrm{pz}}} z\, (\mathrm{d}z\, w)\, d_{31}\, Y_{\mathrm{pz}} \frac{V}{2h_{\mathrm{pz}}} \\ &= \frac{w\, d_{31}\, Y_{\mathrm{pz}} (h_{\mathrm{pz}} + h_{\mathrm{ms}})}{2} V = \theta\, V \end{aligned} \tag{5.5}$$

where we have used the relationship between electric field E and voltage V ($E = -V/2h_{\mathrm{pz}}$), which assumes a series bimorph.

Once a plucked bimorph is released, it is not subjected to any external force, so we can set $f(x,t) = 0$. Substitution of Eq. (5.2) into (5.1) followed by the normalisation of the longitudinal coordinate (we define $\xi = x/L$) gives

---

[5]PZT ceramics are transversely isotropic as the poling direction is different from the other two directions.

[6]We are here adhering to the convention, common in piezoelectricity, of naming the stress T and the strain S.

$$\frac{B}{mL^4}\frac{\partial^4 u}{\partial \xi^4} + \frac{c_a}{m}\frac{\partial u}{\partial t} + \frac{c_s I}{mL^4}\frac{\partial^5 u}{\partial x^4 \partial t} + \frac{\theta}{mL^2}\frac{\partial^2 V}{\partial \xi^2} + \frac{\partial^2 u}{\partial t^2} = 0 \tag{5.6}$$

Without introducing an explicit dependence of V on x, this term would disappear in the following manipulations; so we use the Heaviside step function [2] to represent that V is constant over the electrodes, which we assume extend over the whole bimorph:

$$V(\xi, t) = V(t)\,[H(\xi) - H(\xi - 1)] \tag{5.7}$$

We can now begin solving Eq. (5.6) by separation of variables, assuming that the solution is given by

$$u(\xi, t) = \sum_{n=1}^{\infty} \phi_n(\xi) q_n(t) \tag{5.8}$$

where the spatial components ϕ_n are an orthonormal base in the functional space $L^2([0, 1])$ and can be chosen, as for a standard Euler–Bernoulli beam, as

$$\phi_n(\xi) = \cosh(k_n\xi) - \cos(k_n\xi) - \sigma_n\,[\sinh(k_n\xi) - \sin(k_n\xi)] \tag{5.9}$$

where k_n are the solutions to $1 + \cos k \cosh k = 0$ and

$$\sigma_n = \frac{\sinh(k_n) - \sin(k_n)}{\cosh(k_n) + \cos(k_n)} \tag{5.10}$$

Now Eqs. (5.7), (5.8) and (5.9) are substituted into Eq. (5.6), then we calculate the internal product with a single ϕ_m for every m (in other words, we project onto every vector in the functional vector space); in the process we use the orthonormal relationship:

$$\int_0^1 \phi_n(\xi)\phi_m(\xi) = \delta_{mn} \tag{5.11}$$

and the relationship:

$$\int_0^1 \phi_n(\xi)\frac{\partial^4 \phi_m(\xi)}{\partial \xi^4} = k_m^4 \delta_{mn} \tag{5.12}$$

After the projection, we obtain (with the compact dot-notation for time derivatives):

$$\ddot{q}_n + 2\zeta_n\omega_n\dot{q}_n + \omega_n^2 q_n - \chi_n V = 0 \tag{5.13}$$

with

$$\omega_n = \frac{k_n^2}{L^2}\sqrt{\frac{B}{m}} \tag{5.14}$$

$$\zeta_n = \frac{c_a}{2m\omega_n} + c_s\frac{I}{2mL^4\omega_n}k_n^4 = \frac{c_a}{2m\omega_n} + c_s\frac{I}{2B}\omega_n$$

$$\chi_n = \frac{d_{31}Y_{\mathrm{pz}}w(h_{\mathrm{pz}} + h_{\mathrm{ms}})}{2mL^2}\frac{\partial \phi_i(\xi = 1)}{\partial \xi}$$

Equation (5.13) is essentially the same that would be found for a standard Euler–Bernoulli beam, with the addition of the last term, which represents the converse piezoelectric coupling, as it states that the voltage on the electrodes acts as a force to determine the time evolution of the mode shape.

The direct piezoelectric effect is still to be included. To take into account the effect of the electrical circuit connected to the bimorph, we have to relate the voltage to the current, which is the time derivative of the charge Q, which in turn is given by the surface integral of the electric displacement D over the electrode area:

$$\begin{aligned} I(t) = \dot{Q}(t) &= \frac{\mathrm{d}}{\mathrm{d}t}\int_0^1 D_3(\xi,t)wL\,\mathrm{d}\xi \\ &= -\frac{d_{31}Y_{\mathrm{pz}}(h_{\mathrm{ms}}+h_{\mathrm{pz}})w}{2L}\int_0^1 \frac{\partial^3 u}{\partial t\,\partial^2\xi}\mathrm{d}\xi - \frac{wL\epsilon_{33}^S}{2h_{\mathrm{pz}}}\dot{V}(t) \end{aligned} \tag{5.15}$$

Here D_3 is given by another constitutive equation of piezoelectricity:

$$D_3(\xi,t) = d_{31}Y_{\mathrm{pz}}S_1(\xi,t) - \epsilon_3^S\frac{V(t)}{2h_{\mathrm{pz}}} \tag{5.16}$$

where we have focussed on D_3, which is parallel to the polarisation and so gives the charge density on the electrodes. We have also made explicit the spatial and temporal dependences: whereas D_3 and S_1 depend on both coordinates, the last term only depends on time as the electrodes ensure that the voltage is uniform on the surfaces of the piezoelectric material.

After the substitution, we decompose $u(\xi,t)$ as in expression (5.8), so that we find (using the dot-notation for time derivatives):

$$I(t) = -\frac{d_{31}Y_{\mathrm{pz}}w(h_{\mathrm{ms}}+h_{\mathrm{pz}})}{2L}\sum_n \dot{q}_n\frac{\mathrm{d}\phi(\xi=1)}{\mathrm{d}\xi} - \frac{\epsilon_3^S wL}{2h_{\mathrm{pz}}}\dot{V}(t) \tag{5.17}$$

Combining this with Ohm's law for a resistor R placed across the electrodes of the bimorph, we have

$$\dot{V} + \frac{V}{\tau} + \sum_{n=1}\Phi_n\dot{q}_n = 0 \tag{5.18}$$

with

$$\Phi_n = \frac{d_{31}Y_{\mathrm{pz}}h_{\mathrm{pz}}(h_{\mathrm{pz}}+h_{\mathrm{ms}})}{\epsilon_3^S L^2}\frac{\partial\phi_n(\xi=1)}{\partial\xi} \tag{5.19}$$

$$\tau = \frac{\epsilon_3^S RwL}{2h_{\mathrm{pz}}} \tag{5.20}$$

Substituting $s = \dot{q}$ to reduce the order, the system of differential equations to be solved can be written, for an arbitrary number of modes n:

$$\begin{aligned}
\dot{q}_1 &= s_1 \\
\dot{q}_2 &= s_2 \\
&\dots \\
\dot{q}_n &= s_n \\
\dot{s}_1 + 2\zeta_1 s_1 + \omega_1^2 q_1 - \chi_1 V &= 0 \\
\dot{s}_2 + 2\zeta_2 s_2 + \omega_2^2 q_2 - \chi_2 V &= 0 \\
&\dots \\
\dot{s}_n + 2\zeta_n s_n + \omega_n^2 q_n - \chi_n V &= 0 \\
\dot{V} + \frac{V}{\tau} + \sum_{i=1}^{n} \Phi_i \dot{q}_i &= 0
\end{aligned} \tag{5.21}$$

This is to be solved with initial conditions that take into account the fact that at the time of release the bimorph is deflected. For simplicity, we assume that the bimorph starts from a static condition; therefore, its shape is that of a statically deflected cantilever, all time derivatives are zero and the voltage across its electrodes is zero, assuming it has fully discharged across the resistor since it was deflected. Mathematically,

$$\begin{aligned}
q_i(0) &= \int_0^1 \phi_i(\xi) \frac{FL^3}{2B} \xi^2 \left(1 - \frac{\xi}{3}\right) \mathrm{d}\xi \qquad \forall\, i = 1 \dots n \\
s_i(0) &= 0 \qquad \forall\, i = 1 \dots n \\
V(0) &= 0
\end{aligned} \tag{5.22}$$

As evident from the formulation of the initial conditions, the force on the tip of the bimorph is inserted, using formulae for the static deflection of cantilevers, to give the initial deflection.

Equations (5.21) and (5.22), together with the symbols definitions given earlier can be directly used in mathematical software applications such as Maple® or Mathematica®, to solve the system for the desired number of mode shapes.

5.2.2 *Finite Element Modelling*

Finite element (FE) modelling permits the analysis of complex geometries, which may require the complete re-derivation of the differential equations used in the analytical models. On the downside, even when expertly employed, FE techniques have some drawbacks compared to analytical modelling, for example, the lack of insight offered and the large computational overhead required by parametric studies—although the latter is mitigated by the huge computational power of

Table 5.1 Geometrical parameters of the model

Parameter	Value
Length	25 mm
Width	10 mm
Total thickness	400 μm
PZT thickness	150 μm

modern computers, even of the desktop variety. Even so, FE software packages allow the designer to quickly assess the performance of a bimorph subject to plucking excitation.

This section describes a FE model developed in ANSYS® 11. ANSYS® is chosen for its ability to combine structural and piezoelectric analysis, in addition to the possibility of connecting very simple electric circuits to the electrodes of the piezoelectric components. In the FE model, the piezoelectric volumes are meshed with SOLID5 elements, which have the necessary four degrees of freedom (DoFs): electrical (VOLT) and structural displacements (UX, UY and UZ). The internal metallic shim can be meshed with SOLID45 elements, which are compatible with a piezoelectric analysis and only have structural DoFs: the three displacements UX, UY and UZ. Due to the simplicity of the harvester, the entire 3-D geometry is modelled; for more complex situations or for parametric analyses requiring many solution runs, it would be advisable to exploit the symmetry of the problem with respect to the sagittal plane. A region at the root of the cantilever bimorph is assumed clamped; hence, the structural displacements of the nodes on the top and bottom surface in that area are constrained to zero. The electrical DoFs (VOLT) of the nodes on each electrode are coupled together, ensuring equipotential condition across the electrode. For this example, the bimorph is assumed to be a series bimorph made of PZT-5H, a rather soft[7] variety of PZT. As a result of being a series bimorph, the two PZT layers must be meshed with oppositely poled materials; this can be achieved either by flipping the local coordinate system between the two meshing operations or by using two different material models. The material properties and dimensions used in the model are reported in Tables 5.1 and 5.2. In an ANSYS® piezoelectric analysis, it is possible to obtain electrical results beyond the structural results; of particular interest for a harvester is the voltage at the electrodes, from which it is also possible to calculate the output power. A long-standing common practice in the EH community is to measure the electrical energy dissipated across a resistor (R) as a yardstick for the performance of the device. This can easily be achieved in ANSYS® 11 by configuring a CIRCU94 element as a resistor and connecting its terminals to the electrodes of the bimorph.

In a harmonic analysis, the most meaningful measure of power output is usually the average power, calculated from the voltage amplitude V_0:

[7] Soft PZT is characterised by larger piezoelectric coupling coefficients, which are useful for energy harvesting; hard PZT has lower piezoelectric activity but has also higher-quality factor, which reduces the energy dissipated and is essential in high-frequency resonators.

Table 5.2 Materials' parameters of the model

Substrate material (brass)	
Density	8,530 kg/m^3
Young's modulus	110 GPa
Poisson's ratio	0.307
PZT-5H[a]	
Density	7,500 kg/m^3
s_{11}^E	16.5 10^{-12} 1/Pa
s_{33}^E	20.7 10^{-12} 1/Pa
s_{12}^E	−4.78 10^{-12} 1/Pa
s_{13}^E	−8.45 10^{-12} 1/Pa
s_{44}^E	43.5 10^{-12} 1/Pa
s_{66}^E	42.6 10^{-12} 1/Pa
ϵ_{11}^T	3,130
ϵ_{33}^T	3,400
Q_m	65
d_{31}	−274 10^{-12} C/N
d_{33}	593 10^{-12} C/N
d_{15}	741 10^{-12} C/N

[a] from [1]

$$P_{\text{ave}} = \frac{V_0^2}{2R} \tag{5.23}$$

In a transient analysis, the instantaneous power $P(t_i)$ can be calculated at each substep i from the voltage across the resistor at the same substep:

$$P(t_i) = \frac{V^2(t_i)}{R} \tag{5.24}$$

The instantaneous power can then be used to approximate the cumulative energy produced up to time t_n (substep n):

$$E(t_n) = \sum_{i=1}^{n} P(t_i)\Delta t_i \tag{5.25}$$

where Δt_i is the duration of substep i and $E(t_n)$ will be a monotonic function of time.

A significant source of energy loss is represented by damping; hence, it is advisable to include Rayleigh damping, which has the form expressed by the second equation in (5.14). In ANSYS® this form of damping is represented with the damping matrix C given by

$$C = \alpha M + \beta K \tag{5.26}$$

where α is the mass matrix multiplier, β is the stiffness matrix multiplier, M is the mass matrix and K is the structure stiffness matrix.

The α and β parameters are calculated solving the following system:

$$\xi_1 = \frac{\alpha}{2\omega_1} + \frac{\beta\omega_1}{2}$$
$$\xi_2 = \frac{\alpha}{2\omega_2} + \frac{\beta\omega_2}{2} \quad (5.27)$$

The damping of the first mode is related to the mechanical quality factor Q_m normally supplied by the manufacturer of the bimorph:

$$\xi_1 = \frac{1}{2Q_m} \quad (5.28)$$

If Q_m is not readily available, it can be measured by plucking the bimorph in open circuit conditions and recording the exponential decay of the vibration. The damping of the second mode is usually slightly larger than the first mode. The resonant frequency will depend on the free length of the bimorph and can be calculated analytically or via a FE modal analysis.

5.2.2.1 The Loading Phase

As we shall see, the speed at which the bimorph is initially deflected influences the energy it generates in the loading phase. In its turn, the deflection speed is determined by the operating conditions and by the design parameters. For example, in the case of the knee-wearable harvester discussed in the second part of this chapter, the angular velocity of the joint changes considerably during the gait cycle, directly affecting the speed of deflection of the bimorph. Besides this, the range of possible speed is determined at the design level, for example, by the diameter of the device, as this transforms angular velocity into relative speed between plectra and bimorph (see Sect. 5.3 below). Therefore, modelling the loading phase can feed useful information into the design process. Also, the reaction forces calculated by the model during deflection are used in the design of the plectra of the correct stiffness.

For maximal energy harvesting, the input impedance of the electrical load should optimally match the electromechanical characteristics of the harvester and the characteristics of the mechanical loading. Whereas this optimal impedance may in principle be different in the two phases of loading and free vibration, it is appropriate to set it to an optimal constant value as it is not feasible to change it at the release point. Modelling results (this section and the next) show that the energy generated in the loading phase is typically much lower than in the release phase; hence, it is advisable to optimise for the latter rather than the former.

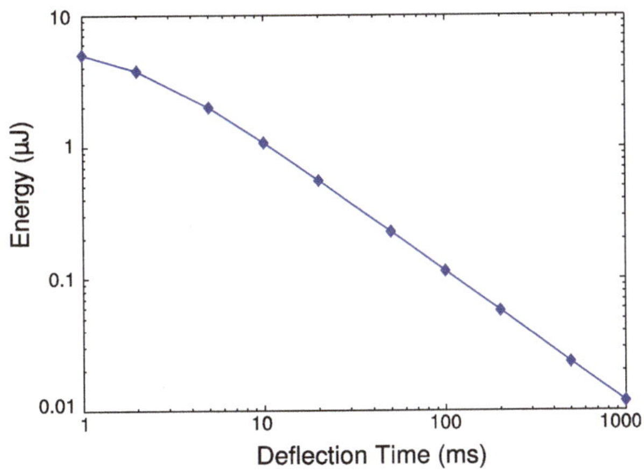

Fig. 5.3 FE predictions of the effect of the deflection time on the energy generated across a 25 kΩ resistor by a series bimorph as in Tables 5.1 and 5.2

Example results showing the dependence of energy on the time interval between initial contact and release point (*deflection time*, Δt) are plotted in Fig. 5.3. These results were obtained with an electrical load of 25 kΩ, which yields the best performance during the release phase (Sect. 5.2.2.2).

What emerges from a first look at Fig. 5.3 is that the energy produced decreases as the deflection time increases. This is not surprising. If we assume that the total charge Q produced by deflecting the piezoelectric material depends only on the final strain (i.e. Q does not depend on Δt), we have

$$I = \frac{Q}{\Delta t} \qquad \text{and so :} \qquad P = RI^2 = \frac{RQ^2}{\Delta t^2} \qquad \text{or also :} \qquad E = \frac{RQ^2}{\Delta t} \tag{5.29}$$

This relation between the energy E and the deflection time Δt, is asymptotically observed by the modelling results for slow deflections; however, when the deflection time is short, less energy is produced than would be expected from Eq. (5.29). The explanation for this effect is revealed by the time-resolved displacement and voltage output from the model [10], which shows that when the deflection is very fast (Δt equal to a few milliseconds) higher mode of vibrations are induced in the bimorph. These are detrimental because at higher modes different sections of the bimorph are under opposite strains, which induces opposite charges on the same electrode; this electrical energy is simply dissipated within the bimorph.

In conclusion, the designer must aim for fast deflections to maximise energy generation, but with the knowledge that extremely fast deflections are not as beneficial as they could be, because of the negative effect of the higher modes of vibration.

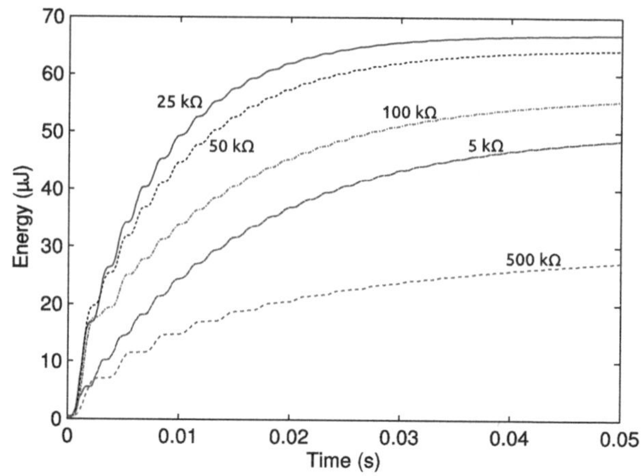

Fig. 5.4 Energy generation curves in the release phase for a plucked bimorph as in Tables 5.1 and 5.2

5.2.2.2 The Release Phase

The majority of mechanical energy is converted into electricity during the release phase. At the beginning of this phase, the bimorph is flexed, which means that it contains strain energy. As it oscillates around its rest position at the resonant frequency, all energy that is not mechanically dissipated is eventually converted into electrical energy. In the following, although the exact values presented are specific for the bimorph modelled, the qualitative behaviours discussed have general validity.

From simple considerations of electronic impedance matching, it is known that the optimal electrical load is approximately $R = 1/\omega C$, where ω is the operating frequency and C is the device capacitance (this approximation is less valid the stronger the piezoelectric coupling). By running the model with a selection of electrical loads, its value can be identified more accurately. Typical energy results are shown in Fig. 5.4 for the bimorph previously discussed, where the optimal load of 25 kΩ harvests 66 μJ of energy in under 40 ms.

A distinctive feature of these curves is the fast initial rise followed by asymptotic approach to a constant value, a sign that most of the energy is converted in the first few oscillations. This is more true for values of resistors close to the optimal, as optimal energy extraction means quicker electrical damping of oscillations. The fact that energy is produced by the oscillation of the bimorph is indicated by the slight oscillatory character of the curve. As the amplitude of the oscillations decreases rapidly so do the voltage and the power outputs. Interestingly, whereas the highest conversion efficiency is achieved with 25 kΩ, there is a time interval up to about 3 ms in which the energy produced is greater for 50 kΩ. Therefore, the chosen output impedance should take into account not only the final energy generated but also the frequency of plucking, depending on other operational parameters of the EH. The FE model presented here was validated by comparing its predictions with experimental results [10].

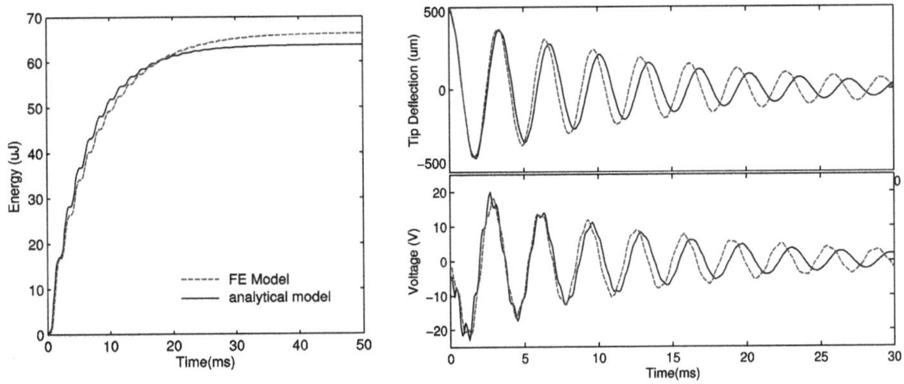

Fig. 5.5 Comparison of predictions obtained with the FE and the analytical models. A bimorph as in Tables 5.1 and 5.2 is connected to a 25 kΩ series resistor

5.2.3 Comparison of Analytical and FE Models

In this section the analytical and the FE models are compared. As can be seen in Fig. 5.5, the predictions of the two models are very close together, with a major difference being the fundamental resonance frequency, which is 305 Hz for the FE model and 297 Hz for the analytical model. Energy prediction is very similar, with a difference of 4% in total energy generated.

On the other hand, the analytical model is vastly superior to the FE model in terms of computational cost: the analytical model was solved with four mode shapes (results in Fig. 5.5) several thousand times faster than the FE model, on the same computer.

5.2.4 General Considerations

The models include only viscous damping within the material as a form of energy dissipation. As a result, conversion efficiency is in the order of 80%: the analytical model predicts that 64 μJ are produced at infinite time, when a total strain energy of 79 μJ is present in the bimorph just before release. A real EH faces other forms of energy loss, such as internal material dissipation within the plectra and friction in the bearings and in other mechanisms which are needed to transfer external energy into the harvester. These losses are device specific and should be dealt with as part of the detailed design of each energy harvester.

Whilst the total energy generated increases with the time interval in which the bimorph is free to vibrate, the average power clearly decays[8] over long periods, giving rise to an optimal plucking frequency, as shown above. Normally, the designer has some scope to determine the plucking frequency during operation, which should be optimised in the light of such considerations.

Complex circuitry for electrical power management, such as SSHI in one of its variants [4], is very difficult to model analytically and cannot be modelled within ANSYS®. Their non-linear behaviour would alter the mechanical response of the bimorph to some extent.

As will be highlighted in the following sections, whilst parameter optimisation is important, great thought should be given to the quality of manufacturing. Normally, the design of a plucking-based EH can be very tolerant of manufacturing uncertainties (such as the exact overlapping of bimorph and plectra when they come into contact). However, the sharpness and cleanliness of release have much bearing on performance: experimental results show that a considerable part of energy can be lost if the plectrum accompanies the bimorph over even a short distance, rather than releasing it swiftly [11].

5.3 Case Study: The Knee-Joint Pizzicato Energy Harvester

Among the areas on the human body where mechanical energy is available for harvesting, the most interesting ones are the foot and the knee-joint [14]. The biomechanical advantage of a knee-mounted brace is that during normal walking the angular displacement of the knee-joint is large and it delivers significant angular velocities at typical walking speeds. In addition, the attachment of such braces is simple, stable and repeatable. An electromagnetic device for harvesting energy from the knee-joint was presented in [7]; the power generation was very high at 4.8 ± 0.8 W during walking; however, due to the nature of the electromagnetic generator, the prototype included gears and other mechanisms, which negatively impacted on its size, mass and complexity. The plucking technique can be used to design a knee-joint EH which is lightweight and mechanically much simpler. In this second part of the chapter, the design, manufacture and testing of a knee-joint EH based on the plucking technique is discussed. The harvester is named "knee-joint pizzicato energy harvester" from the musical term of Italian origin *pizzicato*, meaning plucked.

[8] The average power is given by $P_{\text{av}} = E(t)/t$ (see Eq. (5.25)), where $E(t)$ tends to an asymptotic value for $t \rightarrow \infty$.

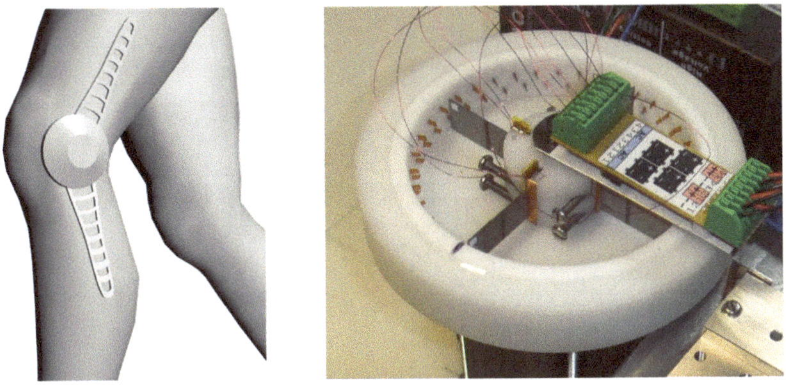

Fig. 5.6 The knee-joint pizzicato energy harvester

5.3.1 *Design Considerations*

The knee-joint pizzicato energy harvester is based on a design that ensures the repetition of the plucking actions to achieve sustained power production (sketched in Fig. 5.6). During normal gait, for example, while walking, the knee alternatively flexes and extends once per second, covering a rotation angle of approximately 70° in each direction. The pizzicato EH is held on the outside of the knee by braces. As the wearer walks, the inner hub and the outer ring rotate relatively to each other, so that the bimorphs (mounted on the hub) are forced to pass in front of the plectra (embedded in the outer ring). This simple mechanism is therefore capable of converting the slow motion of walking into high-frequency vibrations of many piezoelectric bimorphs.

5.3.2 *Testing of a Prototype*

A prototype of the pizzicato EH (Fig. 5.6) was realised based on the principle outlined above. To simplify manufacture and assembly, which was performed manually, the harvester features only four bimorphs fixed to the hub with a removable clamping system. The piezoelectric bimorphs are off-the-shelf components, so, in particular they have rectangular shape of nominal dimensions $31.8 \times .7 \times 0.38\,\text{mm}^3$ even if it is well known that triangular/trapezoidal bimorphs would yield a higher power/mass ratio. These bimorphs have a $130\,\mu\text{m}$ thick brass shim sandwiched between two layers of $125\,\mu\text{m}$ thick PZT and are poled for series operation, which simplifies the electrical connections. The plectra, of dimensions $3 \times 2\,\text{mm}^2$ ($l \times w$), are cut out of a $125\,\mu\text{m}$ thick Kapton® polyimide film. Along the outer ring, plectra are typically spaced 3.5 mm apart.

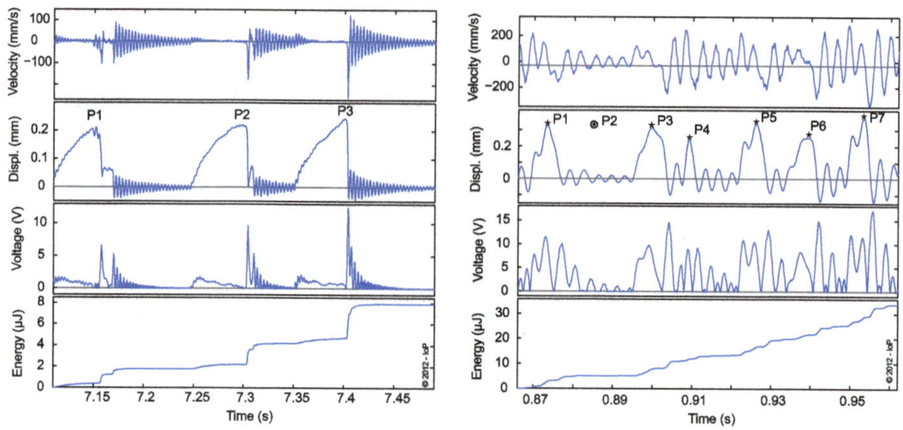

Fig. 5.7 Mechanical and electrical time-domain data for one of the bimorphs in the EH during slow rotation ($\pi/5$ rad/s, left) and fast rotation (2π rad/s, right) [11]

For testing purposes, the prototype is mounted on a knee-joint simulator (partly visible in the photograph), which uses a stepper motor to reproduce the kinematics of the knee-joint of a human subject. A typical measurement involves driving the simulator with the desired velocity profile and sampling the voltage drop across a resistor; for some measurements, full rectifying bridges are placed between bimorphs and loads.

5.3.2.1 Constant Speed Tests

The models predict that faster plucking frequencies yield higher average power. Therefore, once the optimal electrical load for power extraction is identified, the harvester is operated at a selection of rotational speeds, covering the range of speeds observed during normal gait, to investigate the vibrational characteristics of the bimorphs at several plucking frequencies and the impact these have on energy generation.

The results for the two extreme velocities investigated show remarkable differences (Fig. 5.7). At low speed, the plucking frequency is sufficiently low that the individual plucking actions can be identified in all measured and derived quantities: in particular, the displacement of the bimorph's tip clearly shows the loading phase and the vibrations following release, whilst the energy generated is obviously a stepwise build up with distinct contributions from each plucking action (P1 to P3). At high speed, the points of maximum deflection (P1 to P7[9]) can only be located with difficulty as the bimorph is vibrating almost continuously at

[9]Point P2 identifies the expected location of a plectrum which, due to manufacturing imperfections, was actually too short to interact with the bimorph.

high amplitude and the energy curve progressively rises with many small steps. The complete data, collected during full revolutions with a rectifying bridge interconnected, show that fast speeds yield almost twice the energy as low speeds; this ratio becomes almost three without the rectifying bridge.

Attentive observation of the curves in Fig. 5.7 gives insight in the release process. Focusing on the displacement data collected at low speed, we notice that whereas the release point at P3 is sharp, with the displacement steadily ramping up towards a maximum before a steep drop, at P1 and P2 the release is not "clean", suggesting that plectrum and bimorph rub against each other before contact is definitely lost. An unclean release, which becomes associated to only moderate spring-back velocities, is highly detrimental to energy production, as can be gathered by the energy curve, and generates considerably lower voltage peaks. A statistical analysis shows that unclean release is a phenomenon strongly correlated to the plectra, with the same set of plectra yielding the highest energy at different speeds. The effect of an unclean release cannot be observed visually at high speed (Fig. 5.7), but a statistical analysis showed that at high speed the quality of plectra is even more important, with an estimated 46% energy loss, as opposed to an estimated 36% loss at low speed.

5.3.2.2 Gait Cycle Tests

The constant speed tests summarised in the previous subsection are greatly useful for the characterisation of the EH and, as was discussed, have highlighted the importance of good quality manufacturing. Sapient use of the constant speed data can help predict the performance of the harvester in a more complex application such as the knee-joint, where revolution speed is not constant but varies significantly during the gait cycle. This is a valid approach in itself, until accurate ambient data become available and full simulation of the environment can be performed.

Cranfield University developed the knee-joint pizzicato energy harvester within the scope of the battery-free soldier initiative sponsored by the DSTL, part of the Ministry of Defence of the UK, and the EPSRC. Within this collaboration framework, the University of Liverpool (with support from the University of Salford) collected and analysed gait cycle data of human subjects carrying a selection of backpack loads up to 24 kg. In the following, the energy generation performance of the pizzicato energy harvester when driven with such biomechanical data will be briefly discussed [9].

The first peak in the knee angle vs. time curve (Fig. 5.8a) reaches approximately 20° and represents the flexion of the knee-joint immediately following heel-strike, when the leg is loaded with the body weight; the second peak, well over 50°, is associated with lift-off, when the leg has lost contact with the ground and is carried forward in preparation of the following heel-strike. As the knee-joint goes through a wide range of angular velocity during normal gait, both extremes of low and high speed discussed before are observed during gait simulation testing: next to direction reversals the speed is very low and individual plucking actions can be identified, during lift-off the bimorph is plucked multiple times in rapid succession generating an almost continuous vibration (Fig. 5.8a).

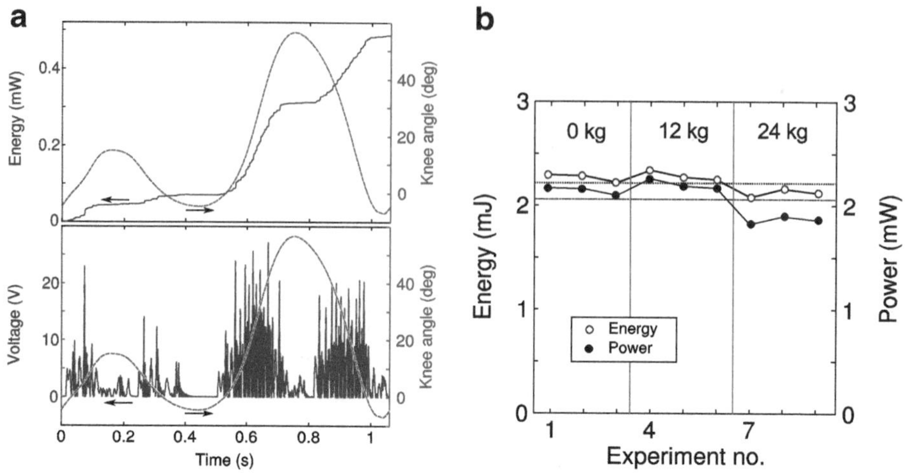

Fig. 5.8 (**a**) Typical voltage and energy output by one bimorph (left axis) when the pizzicato EH is driven with real gait cycle data (right axis). (**b**) Total energy and power produced by the pizzicato EH during a series of experiments with real biomechanical data from a human subject carrying a selection of backpack loads (indicated in the figure)

The biomechanical data collected show that the main effects of the backpack load on gait are an increase in its duration by 0.1 s (the wearer walks more slowly) and a change in the general pattern with, among others, an increase in the maximum knee angle when the foot is in contact with the ground (the knee bends more under the extra weight). In the light of constant speed tests, it is not surprising that the overall energy produced is therefore lower with the highest load (Fig. 5.8b), even if more plucking actions occur due to some peak angles being larger. Naturally, the average power is also lower, because of the lower energy produced and also the fact that the gait cycle lasts longer.

On average, this prototype of the knee-joint pizzicato energy harvester has proven the capability of producing a sustained power of 2 mW, while the wearer walks at a normal pace; very large backpack loads (24 kg) can bring about a power penalty of approximately 0.3 mW compared to the load-free or smaller load conditions [9]. With more bimorphs and overall design optimisation, it is not unreasonable to expect that the pizzicato EH can exceed 30 mW of continuous power generation.

5.4 Conclusions and Outlook

This chapter focused on providing the motivation, modelling methodology and application potential of plucking-based energy harvesters. There is currently great focus within the EH community on broadband harvesters, which are not constrained to operate in a very narrow frequency range. Frequency up-conversion by plucking is offered as a solution to the frequency mismatch problem: albeit not suitable for all applications and environments, it has been shown to offer promising results in different contexts.

The chapter has presented a framework for analytical modelling of plucked bimorphs, which can be used as is or adapted for other specific configurations. The discussion of the finite element model has strived to give sufficient information and insights in the technique to allow the reader to build its own model, taking full advantage of the flexibility and quick adaptability of FE techniques. Modelling results of general applicability include:

- Faster deflection leads to higher energy generation, although higher vibrational modes induced in the bimorph limit this advantage at very high speeds.
- The energy produced during the bimorph deflection is normally dwarfed by the subsequent release, so impedance matching should focus on the free vibration of the bimorph.
- Although the maximum energy is produced after a relatively long time, average power performance is optimised by a faster plucking repetition rate. This has direct consequences on the geometry of the design.

The knee-joint pizzicato energy harvester (concept and a prototype) are discussed as a case study in plucking-based EH. It permits useful power generation from sub-hertz movements of the human body. The experimental study of the pizzicato EH has highlighted some issues which could not be caught by the modelling; in particular, by measuring the energy loss during the release process, it shows the importance of a clean release and, consequently, of good quality plectra. Beyond sizing of harvester and bimorphs, choice of materials and spacing of plectra, the designer must pay great attention to the interaction between plectra and bimorphs by choosing the right materials and shape of plectra to attain the cleanest release together with a minimisation of the wear of the parts in contact. Following this line of thoughts, other interaction mechanisms may deserve investigation, for example, where the plectra are replaced with permanent magnets.

In conclusion, energy harvesting by plucking is still in its infancy and significant improvements in the technique and in the performance of the harvesters can be expected in the near future.

Acknowledgements The authors acknowledge the financial support dstl(MoD) and EPSRC (EP/H020764/1) which sponsored this work.

M. Pozzi would like to thank his wife Wimonrat for her support during the research work that led him to the pizzicato energy harvester and the modelling of the plucking excitation—*M. Pozzi*

References

1. Berlincourt D, Krueger HA (2000) Properties of morgan ElectroCeramic ceramics. Tech Rep TP-226, Morgan ElectroCeramics
2. Erturk A, Inman DJ (2008) A distributed parameter electromechanical model for cantilevered piezoelectric energy harvesters. J Vib Acoust 130(4):041,002–15. DOI 10.1115/1.2890402. URL http://link.aip.org/link/?VAJ/130/041002/1

3. Ferrari V, Gatti PL (2007) Applied structural and mechanical vibrations: theory, methods and measuring instrumentation, Taylor & Francis; 2nd Revised edition
4. Guyomar D, Lallart M (2011) Recent progress in piezoelectric conversion and energy harvesting using nonlinear electronic interfaces and issues in small scale implementation. Micromachines 2:274–294. DOI 10.3390/mi2020274. URL http://www.mdpi.com/2072-666X/2/2/274/
5. Kulah H, Najafi K (2004) An electromagnetic micro power generator for low-frequency environmental vibrations. In: Micro electro mechanical systems, 2004. 17th IEEE International Conference on. (MEMS), pp. 237–240. IEEE. DOI 10.1109/MEMS.2004.1290566
6. Kulah H, Najafi K (2008) Energy scavenging from Low-Frequency vibrations by using frequency Up-Conversion for wireless sensor applications. IEEE Sensor J **8**(3), 261–268. DOI 10.1109/JSEN.2008.917125
7. Li Q, Naing V, Donelan J (2009) Development of a biomechanical energy harvester. J NeuroEngineering Rehabil 6(1):22. DOI 10.1186/1743-0003-6-22
8. Murray R, Rastegar J (2009) Novel two-stage piezoelectric-based ocean wave energy harvesters for moored or unmoored buoys. In: Proceedings of SPIE, pp. 72, 880E–72, 880E–12. San Diego, CA, USA. DOI 10.1117/12.815852
9. Pozzi M, Aung M, Zhu M, Jones R, Goulermas J (2012) The pizzicato knee-joint energy harvester: characterization with biomechanical data and effect of backpack load. Smart Mater Struct 21(6):075023. DOI 10.1088/0964-1726/21/7/075023
10. Pozzi M, Zhu M (2011) Plucked piezoelectric bimorphs for knee-joint energy harvesting: modelling and experimental validation. Smart Mater Struct 20(5):055,007. DOI 10.1088/0964-1726/20/5/055007. URL http://iopscience.iop.org/0964-1726/20/5/055007/
11. Pozzi M, Zhu M (2012) Characterization of a rotary piezoelectric energy harvester based on plucking excitation for knee-joint wearable applications. Smart Mater Struct 21(5):055,004. DOI 10.1088/0964-1726/21/5/055004. URL http://iopscience.iop.org/0964-1726/21/5/055004
12. Rastegar J, Murray R (2010) Novel two-stage piezoelectric-based electrical energy generators for low and variable speed rotary machinery. pp. 76, 430C–76, 430C–8. San Diego, CA, USA. DOI 10.1117/12.847755
13. Renaud M, Fiorini P, van Schaijk R, van Hoof C (2009) Harvesting energy from the motion of human limbs: the design and analysis of an impact-based piezoelectric generator. Smart Mater Struct 18(3):035,001. DOI 10.1088/0964-1726/18/3/035001. URL http://iopscience.iop.org/0964-1726/18/3/035001
14. Riemer R, Shapiro A (2011) Biomechanical energy harvesting from human motion: Theory, state of the art, design guidelines, and future directions. J NeuroEngineering Rehabil 8(1):22. DOI 10.1186/1743-0003-8-22
15. Umeda M, Nakamura K, Ueha S (1996) Analysis of the transformation of mechanical impact energy to electric energy using piezoelectric vibrator. Jpn J Appl Phys 35(Part 1, No. 5B):3267–3273. DOI 10.1143/JJAP.35.3267. URL http://jjap.jsap.jp/link?JJAP/35/3267/

Chapter 6
Energy Harvesting with Vibrating Shoe-Mounted Piezoelectric Cantilevers

Denis Benasciutti and Luciano Moro

Abstract This chapter presents a study on energy harvesting from human walking via piezoelectric vibrating bimorphs. Heel accelerations are measured and compared with data from literature. All relevant features are summarized in a typical (standard) acceleration signal, used as a reference input in numerical simulations. The transient electromechanical response (beam deflection, output voltage, and average output power) of a shoe-mounted rectangular scavenger excited by the standard acceleration is calculated by numerical simulations. Step-by-step numerical integration is used, as the input is a non-sinusoidal signal and explicit analytical solution is not available. Results from simulations are also validated with measurements on a real shoe-mounted device. A sensitivity analysis is finally performed to find alternative scavenger configurations that could provide increased power levels. Acceptability criteria based on imposed geometrical constraints and material strength limits are also checked. This analysis allows a rapid screening of harvesting performance among a wide set of different scavenger configurations, which allows finding the one providing the largest output power.

6.1 Introduction

In the area of pervasive technologies, the development of wearable sensors for data measurement, transmission, and processing increasing has received attention especially in the recent years, both by industrial and academic research communities. Very interesting seems the possibility to realize distributed sensors networks in elder care technology, for monitoring of vital biometric parameters (e.g., acceleration, temperature, or blood pressure) of chronically ill patients [1–4].

D. Benasciutti (✉) • L. Moro
DIEGM, Dipartimento di Ingegneria Elettrica Gestionale Meccanica, Università di Udine, via delle Scienze 208, 33100 Udine, Italy
e-mail: denis.benasciutti@uniud.it; luciano.moro@uniud.it

N. Elvin and A. Erturk (eds.), *Advances in Energy Harvesting Methods*, DOI 10.1007/978-1-4614-5705-3_6,

A great challenge would be the design of wearable autonomous devices that could obtain the required energy from the surrounding environment, thus (optimistically) overcoming the use of traditional chemical batteries. A promising solution, especially for the power supply of wearable sensors, is represented by the conversion of human-generated power during everyday activity. This goal could be further facilitated by the continuous decrease in the power requirements of both portable electronic and biomedical devices. In elder care technology, this would make available completely autonomous sensors for real-time monitoring of patients [5, 6] or even the possibility to supply artificial organs without external energy sources [7].

A number of studies in the literature have documented how power can be generated from the human body, in the form of breathing, body heat, blood transport, arm motion, walking, etc. [5, 8–10]. Among them, walking has been viewed as particularly convenient, due to large deformation and stretching, as well as acceleration experienced by the foot during gait.

Several published papers (e.g., [11–13]) specifically investigated the design of wearable devices to harvest energy from human motion, with special focus on piezoelectric materials, due to their high efficiency in strain-to-energy conversion. Two types of piezoelectrics are commonly used: piezoceramics (e.g., lead zirconate titanate or PZT) and polymers (e.g., polyvinylidene fluoride or PVDF). Although characterized by lower energy conversion efficiency, PVDF is mechanically more flexible than PZT and it can thus withstand larger amounts of strains, which can explain its versatility in shoe-mounted inserts stretched by foot deformation. A typical configuration consists of bimorph in "31 mode," which gives large strain even with small applied forces and displacements.

While the literature has largely studied the behavior of flexible PVDF materials [12, 14], very little attention has been paid to the analysis of PZT performance as candidate to design vibrating shoe-mounted scavengers, excited by foot acceleration. In addition, most of the existing studies on PZT-based energy scavengers have been focused almost entirely on purely sinusoidal excitations, while non-sinusoidal excitations (as, for example, foot acceleration during walking) have not been considered at all. On the other hand, some existing literature [15, 16] shows that foot accelerations during normal walking could be sufficiently high to excite shoe-mounted vibrating scavengers.

In light of the above arguments, this chapter aims to evaluate the feasibility of energy harvesting from human walking by using shoe-mounted vibrating PZT scavengers. Heel accelerations during walking are first measured and the relevant features are next synthesized into a typical (standard) acceleration signal, which is then used as a reference excitation in numerical simulations. A distributed parameter model proposed in the literature [17] is used to evaluate the transient electromechanical response of a shoe-mounted bimorph cantilever excited by the standard acceleration signal. Step-by-step numerical integration of the equations of motion is required, as the acceleration input is a non-sinusoidal excitation and an explicit analytical solution is not available. The electrical response (beam deflection, output voltage, and average power dissipated across a resistive load) are calculated for a reference configuration; results are also compared with measurements on a real shoe-mounted prototype.

In order to seek alternative bimorph configurations that could give improved power levels, a sensitivity analysis is performed on several design parameters. Geometrical and mechanical constraints (e.g., maximum space available, material strength) are also introduced in the analysis, to check the feasibility and strength resistance of those configurations, which would give the largest harvested power.

6.2 Energy Harvesting from Human Walking

In recent years, a considerable effort has been spent by academic research to study the feasibility of wearable devices that are capable to harvest energy from different types of biomechanical motions (e.g., upper limbs, legs, ankle, knee, foot).

Since the early work of Starner [5], a number of publications have provided more or less detailed overviews on possible human activities that could be exploited as favorable energy source; see for example [18–21]. Such studies outlined that a promising power source could be represented by walking activity, due to the mechanical energy generated by shoe stretching and deformation under foot/ground contact, to the potential energy given by fall of the heel during each step, as well as to the ease of incorporating a scavenger into the shoe. In addition, shoe inserts are quite attractive for their high efficiency in converting everyday human activity into useful energy, as well as for their ease of implementation.

Theoretical studies have tried to estimate, under some simplifying assumptions, the maximum amount of power that could be harvested from walking. For example, the simple free fall motion of the heel during normal walking would generate about 67 W of power at a pace of two steps per second [5, 6]. Later studies, however, considered the above estimation as rather optimistic, considering that heel strike motion is not a real free fall. They also pointed out that only a fraction of total deformation energy (consequent to elastic compression of shoe heel) is actually available for power generation. They proposed to compute the net mechanical work produced during heel strike to get a more realistic estimate of available power. This gave a net energy of approximately 0.4–1.0 J/step and a maximum harvested power of about 2 W for a gait cycle of 1 Hz (two steps per second) [20, 21].

The estimated theoretical power levels must be properly reassessed and scaled when considering the specific transduction mechanism, material, and geometrical configuration of shoe-mounted devices. Suitable physical constraints (as, for example, material strength or maximum space available in the shoe) must be also considered in their design, as well as proper evaluation of possible changes in comfort during gait caused by their presence. For example, restriction to the maximum deflection of cantilevers in accordance to space available in a shoe cavity was investigated in [22]. Similarly, the thickness of a shoe insert is limited by the height of the sole, while the area must comply with footprint dimension; electrical criteria (breakdown electric field) could be considered as well [20].

Practical realization of shoe-mounted harvesting systems should also evaluate and compare the relative advantages offered by the two types of commercially available piezoelectric materials (flexible PVDF, brittle PZT). For example, piezoceramic PZT is characterized by an energy conversion efficiency larger than PVDF, but its extremely brittle nature imposes severe limitations in the maximum strain that it can safely sustain without failure. On the opposite, the piezoelectric polymer PVDF is very flexible and easy to shape, and it can also withstand larger strain before fracture.

Although in piezoelectric materials the most efficient energy conversion would occur from compression along the poling direction ("33" mode), the harvested power resultant from human-generated compression would be very small, since the required force would be impractically large [12].

The mechanical work produced by shoe deformation can be conveniently converted into energy through, for example, stretching of a piezoelectric shoe insert (PVDF would be a good candidate). Theoretical considerations have shown that the compression mode generation would be unfeasible in practice because of the geometrical constraints imposed (e.g., maximum thickness) and the large force necessary to induce enough mechanical strain. Harvested power levels of about 14–16 μW were estimated in this mode [20].

Another approach could convert energy using the piezoelectric material in the bending mode ("31" mode), which gives a higher strain-to-energy conversion efficiency, even if at the expense of more complicated device configurations. Theoretical estimates have been found in the order of 0.4 mW with PZT and 36 mW with PVDF from a cantilever-mounted bender [20]. A limitation is, however, that during heel strike the foot deformation is concentrated at the bending point of the foot, rather than distributed evenly along the whole foot.

A noteworthy practical example of shoe-mounted scavenger is given by Shenck and Paradiso [12], who designed a prestressed PZT unimorph for energy harvesting during heel strike. A so-called PVDF "stave," implanted in the front of a sole, delivered an average power of about 1.3 mW on a 250 kΩ load at a 0.9 Hz walking pace. In comparison, a so-called PZT dimorph, consisting of two curved unimorphs arranged in a clam shell configuration, produced 8.4 mW of power through a 500 kΩ resistive load.

The above survey has highlighted that past and current research in human-powered applications has been mainly interested on studying the performance of piezoelectric devices under stretching (static deformation), while no much attention has been dedicated to the feasibility analysis of human generated power by vibrating shoe-mounted devices.

On the other hand, common experience suggests that foot movement is characterized by large displacements and accelerations, which sounds very promising as possible vibrating source for a shoe-mounted device. This is supported by several literature studies [15, 16], which observed high velocities and accelerations in foot movement, especially during the foot/ground contact phase. As an example, Fig. 6.1 plots a typical acceleration time history measured in the calcaneous of one limb, in one gait during normal walking (the gravitation acceleration has been

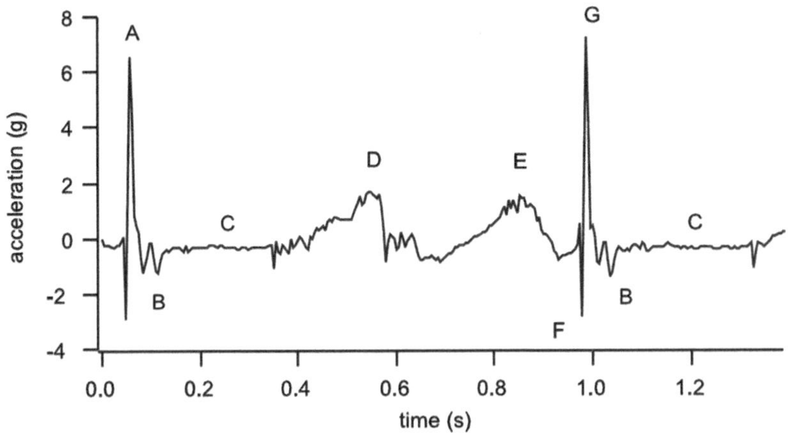

Fig. 6.1 Acceleration signal of the calcaneous for one complete gait (heel strike to heel strike of the same limb). Accelerations are normalized to acceleration of gravity $g = 9.81$ ms^{-2} (Reprinted from [16], with permission from Elsevier)

subtracted already). According to [16], some significant features are noteworthy: the acceleration peak at heel strike (A, G) followed by some oscillation after heel strike (B), the baseline during stance phase (C), acceleration during foot movement (D–E), and finally the downward acceleration at heel strike (F).

It is then the aim of next sections to investigate in more detail the feasibility of energy harvesting with shoe-mounted vibrating scavengers. More precisely, the goal is to investigate the acceleration levels during human gait as potential vibration sources and the resultant performance of vibrating bimorphs mounted inside the heel of the shoe.

6.3 Heel Acceleration in Human Gait

A gait cycle is characterized by a well-defined sequence of foot movements. Within a sequence of two consecutive heel strikes ("contact phase") of the same limb, the gait can be divided into a "stance phase" when the foot is at rest, followed by a "swing phase" when the foot is moving [23].

While the contact phase is practically instantaneous, the stance and swing phase occupy approximately 60 and 40% of a single gait cycle [24]. The heel starts moving at approximately 40% of the stride (while the foot is still in contact with the ground) and reaches its maximum displacement just after toe-off, when the velocity of the heel is approximately zero. During the upward and downward heel movement in the swing phase, the velocity has a sinusoidal trend, while the acceleration shows some oscillation consequent to velocity change [15].

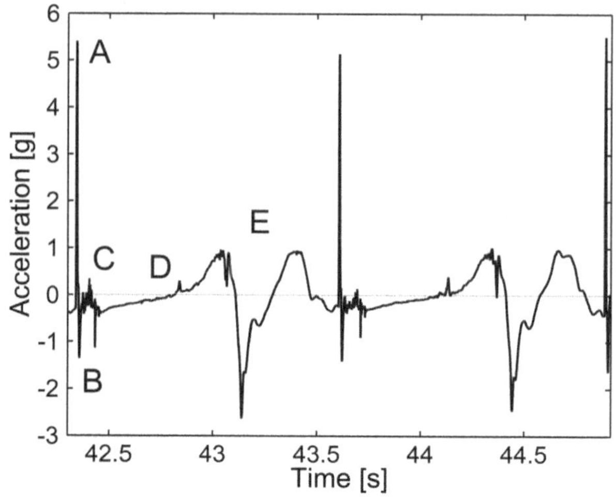

Fig. 6.2 Heel acceleration signal measured in normal walking (a two-gait sequence is shown). Accelerations are normalized to acceleration of gravity $g = 9.81\ \mathrm{ms}^{-2}$ (Reprinted from [26], with permission from IOP Science)

The vertical and horizontal acceleration in the foot are of interest, as a possible vibration source for a vibrating shoe insert. The acceleration signal shown in Fig. 6.1 has confirmed that large acceleration spikes occur at every heel strike, followed by some oscillation during the swing phase. Observed acceleration spikes are also confirmed by ground reaction forces measured under the foot [25].

It should be noted that heel acceleration may vary according to several factors, including ground surface, gait condition, shoe types, walker characteristics, etc.; in principle, different walking conditions are expected to give acceleration signals with different characteristics. The literature, however, seems to emphasize that a change in the above-mentioned walking conditions would produce only a change in measured values, while the overall trend would be generally preserved [16].

A first step in this analysis is the experimental measurement of gait acceleration time histories, in order to identify all the relevant features and to summarize them in a standardized acceleration signal, used as a reference excitation in the numerical evaluation of the electromechanical response of a shoe-mounted vibrating device.

A custom-made clamp system was used in the experiments to fix on the heel pad an accelerometer aligned along the tibial axis. Gait accelerations were measured during five repeated tests, consisting of a male subject (1.75 m height, 70 kg weight) walking on a level walkway at his natural pace (about 0.8 steps/s for the same limb) [26]. A segment of a typical acceleration time history for two consecutive gait cycles (the signal is virtually periodic) is shown in Fig. 6.2. It confirms all the relevant features already observed in the acceleration time history of Fig. 6.1: at contact phase, a large acceleration spike (A) is followed by negative valley (B), probably caused by mid-foot and toe impact to the ground. During the stance phase, acceleration is almost zero (D), while it shows a sinusoidal-shaped oscillation (E) during the swing phase.

Based on previous measurements and also on information gathered from the literature, it was possible to synthesize all the relevant features of gait acceleration into a

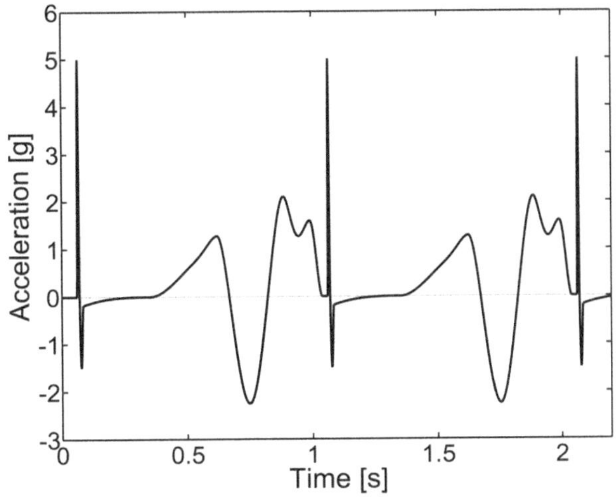

Fig. 6.3 Standard acceleration signal (footstep frequency is 1 Hz) (Reprinted from [26], with permission from IOP Science)

periodic acceleration signal, used as a reference input in numerical simulations of an electromechanical bimorph. As a first attempt, the standard acceleration signal can be established by elaborating and combining the values of vertical and horizontal displacement, velocity and acceleration, as well as the typical time variation of the foot-ground angle during walking, which are available from several studies on this topic. As an example, the average step distance is 65 cm [27] and the maximum vertical heel elevation is about 25 cm [15], where the foot-ground angle reaches its maximum of 46 degrees [25, 28]. Further aspects are also taken into account, according to existing literature [16, 29, 30]. Subsequently, experimental measurements were taken into account to validate the synthesized data, especially for the definition of acceleration during the contact phase.

The periodic standard acceleration signal is plotted in Fig. 6.3. Some typical features can be recognized: two sharp positive/negative peaks in the contact phase, an oscillation in the swing phase, followed by a small double peak, which results from processing literature data of vertical and horizontal accelerations. The small vibrations in the contact phase (see for example Fig. 6.2) are not included into the standard acceleration curve, since they are observed to vary among different measurements.

6.4 Numerical Simulations

This section analyzes the electromechanical response of a shoe-mounted vibrating scavenger excited by heel acceleration during human gait. As shown in Fig. 6.4, the device studied is a rectangular bimorph, made of two piezoelectric layers bounded to a metallic shim. One end is clamped, a tip mass is mounted on the other end, which reduces the natural frequency of the bender.

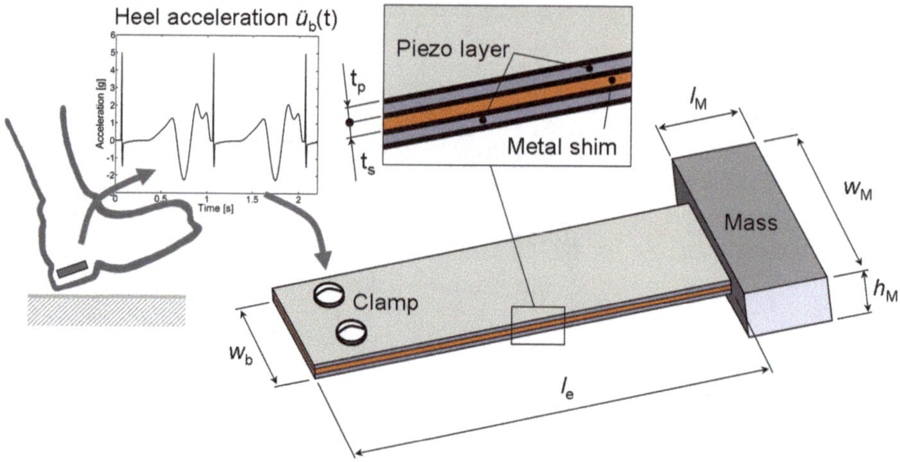

Fig. 6.4 Rectangular bimorph excited by heel acceleration at its base

Table 6.1 Geometrical and material parameters characterizing the reference rectangular bimorph of Fig. 6.1 ($\varepsilon_0 = 8.854$ pF m^{-1} is the permittivity of free space)

Parameter	Value	Parameter	Value
Electrode length, l_e (mm)	20	Thickness of piezo. layer, t_p (mm)	0.1 (each)
Beam width, w_b (mm)	14	Thickness of metal shim, t_s (mm)	0.2
Density ρ (kg m^{-3})[a]	7,800	Elastic modulus of piezoelectric, Y_{11} (GPa)[b]	66
Length of tip mass, l_M (mm)	10	Elastic modulus of metal shim, Y_s (GPa)	206
Width of tip mass, w_M (mm)	35	Electromechanical coupling coefficient, k_{31}	0.3
Height of tip mass, h_M (mm)	3.3	Permittivity at constant stress, ε^T (pF m^{-1})	1,800ε_0

[a]The same value of density is assumed for piezoelectric layers, metal shim, and tip mass
[b]Elastic modulus Y_{11} refers to longitudinal beam direction

The geometrical dimensions and material properties of the bimorph are listed in Table 6.1; they refer to a configuration that has been also used in previous works [26, 31]. Material properties in Table 6.1 are typical of PZT-5A piezoceramic and stainless steel, as data in reports of PiezoSystem Inc.

An electromechanical distributed parameter model from the literature [17] is used to evaluate the dynamic response and voltage output of the bimorph in Fig. 6.4 excited by heel acceleration at its base. This model includes the effects of both translational and rotational inertia of the tip mass, which is assumed to act at a single point. In a previous study [31] this model has been successfully applied to simulate the dynamic behavior of scavengers with similar geometrical parameters. Other models [32], including for example the rotational inertia effect caused by tip mass overhang, can also be used.

Only the fundamental equation of the model summarizing the bimorph electromechanical response is briefly discussed here; for its explicit derivation, the interested reader may refer to [17].

With reference to r-th mode with natural frequency ω_r, the bimorph electromechanical response (for piezoelectric layers connected in series) is synthesized by the two mechanical and electrical coupled equations:

$$\begin{cases} \ddot{\eta}_r(t) + 2\zeta_r\omega_r\dot{\eta}_r(t) + \omega_r^2\eta_r(t) + \chi_r v(t) = f_r(t) \\ \dfrac{C_\mathrm{p}}{2}\dot{v}(t) + \dfrac{v(t)}{R_\mathrm{L}} = i(t) \end{cases}, \quad (6.1)$$

where $\eta_r(t)$ is the modal mechanical coordinate (the dot stands for time derivative), ζ_r is the modal mechanical damping ratio, χ_r is a backward electromechanical modal coupling term correlated to mode shape $\phi_r(x)$, and C_p is the internal capacitance of the piezoelectric layer. The current $i(t)$ can be evaluated by modal expansion. Explicit expressions of some constant parameters (ω_r, χ_r, ζ_r, C_p) as well as the equations of modal shape $\phi_r(x)$ can be found in [17].

Symbol $f_r(t)$ in Eq. (6.1) represents the modal mechanical forcing function:

$$f_r(t) = -\ddot{u}_\mathrm{b}(t)\left[m\int_0^l \phi_r(x)\mathrm{d}x + M\phi_r(l_\mathrm{e})\right] \quad (6.2)$$

which is proportional to base vertical acceleration $\ddot{u}_\mathrm{b}(t)$; symbol M is the tip mass, m is the mass per unit length of the beam, while $l = l_\mathrm{e} + l_\mathrm{M}/2$.

In the electrical part of expression in Eq. (6.1), $v(t)$ and $i(t)$ are the instantaneous voltage and current across the resistive load R_L connected to piezoelectric layers. Considering a pure resistive load is actually a strong simplification in the model, although it is a common assumption in modeling vibrating scavengers as it allows an effective comparison of harvested power levels.

The possibility to obtain analytical solutions of Eq. (6.1) strictly depends on the particular expression of the input base acceleration $\ddot{u}_\mathrm{b}(t)$ that defines the modal forcing function $f_r(t)$. For example, for a simple harmonic input $u_\mathrm{b}(t) = U_0 e^{j\omega t}$ with frequency ω, Eq. (6.1) allows a closed-form analytical solution for both the transverse displacement response and electric voltage across the piezoelectric layers, as shown in [17]. However, for a nonharmonic excitation such as the heel acceleration in Fig. 6.3, Eq. (6.1) does not have a closed-form analytical solution and therefore a numerical integration must be used to obtain the beam response. As explained in [26], the numerical solution of Eq. (6.1) can be easily obtained by using the ode45 function in the "ordinary differential equation" (ODE) MatLab® toolbox (default options of function ode45 are generally sufficient for accurate analysis results, although a low absolute tolerance, as 10^{-10}, is suggested for more reliable results).

The numerical procedure detailed in [26] is then used to compute the dynamic deflection and voltage output of the rectangular bimorph of Fig. 6.4, using the parameter values of Table 6.1. Simulations assume a modal mechanical damping ratio $\zeta_1 = 0.02$ for the first mode (only the first resonant mode in bending has been considered in these simulations). As a preliminary validation, in [26] the response under a harmonic base acceleration of 0.2 ms^{-2}, with frequency close to $\omega_1 = 23$ Hz, was first calculated and compared with the analytical closed-form solution given by Eq. (48) in [17]. After a short transient, the numerically simulated response was shown to converge to the analytical solution, thus confirming the validity of the proposed numerical procedure.

As a second analysis step, the numerical procedure is applied to study the scavenger response under the standardized heel acceleration signal of Fig. 6.3. A resistive load $R_L = 14$ kΩ is assumed; note that this is only an attempt value, not the optimal resistance that assures the maximum harvested power for the given scavenger configuration. The calculated output voltage for a sequence of three footsteps is shown in Fig. 6.5a. The maximum beam deflection at the free end has a similar trend (not shown). In the figure, damped oscillating response of the bimorph to each heel strike is clearly observed in response to pulse excitation caused by walking.

6.5 Experimental Measurements

Experimental tests were carried out to validate the numerically calculated transient response, shown in Fig. 6.5a, of a shoe-mounted bimorph excited by the nonharmonic heel acceleration.

A bimorph prototype, manufactured by PiezoSystem Inc. according to dimensions and properties of Table 6.1, was fixed in the heel cavity by a custom clamp system, see Fig. 6.6. An accelerometer was placed on the base (clamped end) to monitor the heel acceleration signal during gait. The clamping system is well integrated into the shoe, and it does not penalize the comfort during walking.

Similarly to the numerical simulations, a resistive load $R_L = 14$ kΩ was connected to the bimorph. The same test conditions already described in Sect. 6.3 (male subject in normal walking) were then replicated.

A series of gait cycles was performed, and both the input acceleration and the output voltage were measured in real time during the test. A sample of output voltage is plotted in Fig. 6.5b; it is compared with the voltage waveform calculated by simulations under the standard acceleration input.

In both figures, the time axis has been normalized to footstep period T_{step}, because the footstep frequency of the measured signal slightly differs from that of standard signal (which was fixed at 1 Hz). At each heel strike, the scavenger vibrates with a damped response.

Similarities between the simulation and the experiment, shown in Fig. 6.5, confirm that the standard acceleration signal, even if somewhat simplified, retains all the relevant features of gait acceleration to faithfully reproduce the real measured signal. Hence, it can be conveniently used as a surrogate of the measured signals to simulate the dynamic response of a shoe-mounted scavenger. This also gives an advantage from a computational point of view. Direct use of measured signal

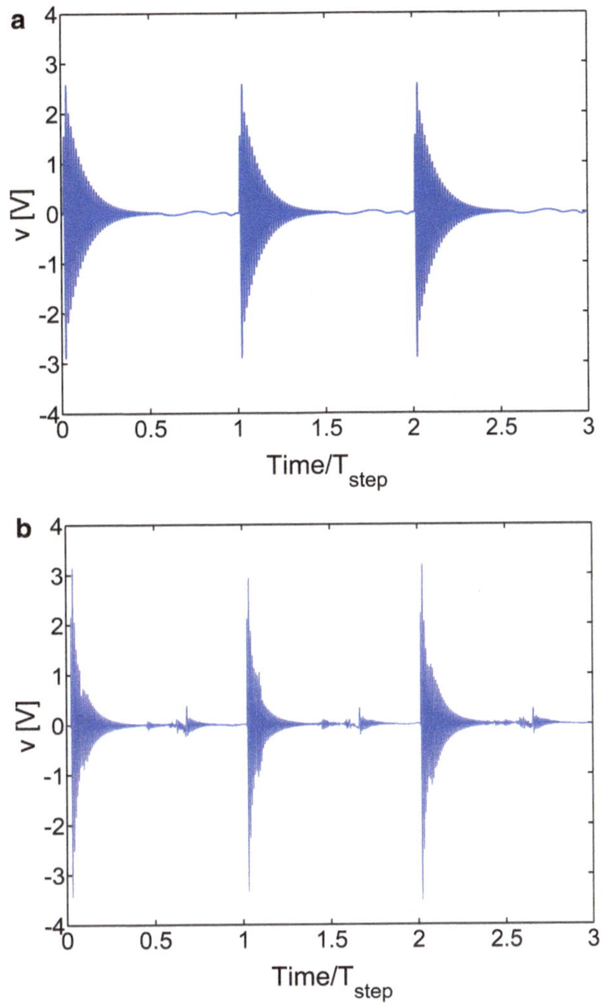

Fig. 6.5 Transient output voltage under heel acceleration for three heel strikes of the same foot: (**a**) numerical solution and (**b**) experimental measurement (Reprinted from [26], with permission from IOP Science)

in numerical simulation would be impractical; in fact, any measured acceleration signal (see for example Fig. 6.2) is usually characterized by many small oscillations (noise), which would greatly increase the computational time. As an example, Fig. 6.7 shows a comparison between experiments and simulations, in which measured acceleration is used as input instead of the standard signal. The numerical simulation shows a trend that is very close to the measured one (thus confirming the validity of the model) at the expense of a considerable increase of overall computational time. Nevertheless, the output power calculated with the standard input signal is very similar to that obtained experimentally, as will be discussed in the following section.

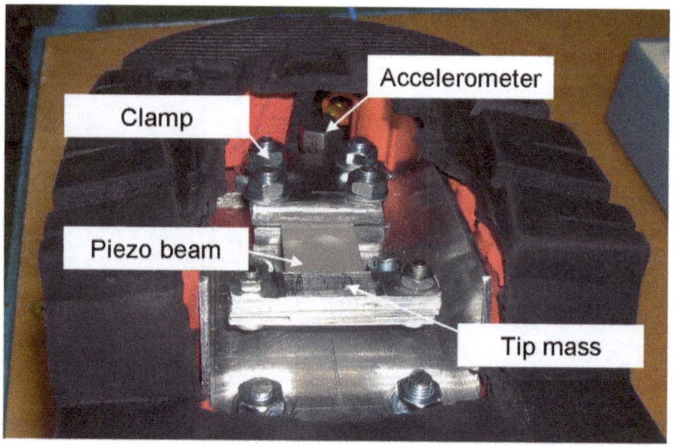

Fig. 6.6 Experimental setup: rectangular bimorph mounted inside the heel shoe (Reprinted from [26], with permission from IOP Science)

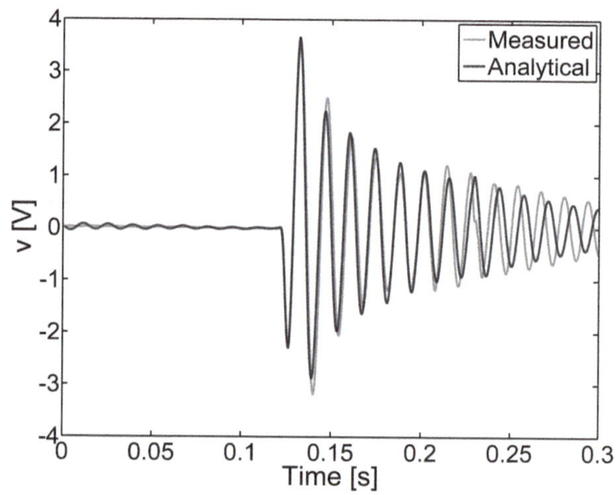

Fig. 6.7 Comparison between analytical and experimental trends with measured acceleration input (Reprinted from: [26], with permission from IOP Science)

As a matter of fact, the scavenger performance can be represented by the average power per footstep dissipated across the resistance R_L:

$$P_{\text{ave}} = \frac{1}{n_s} \int_0^{T_{\text{step}}} \frac{v^2(t)}{R_L} dt, \tag{6.3}$$

where n_s is the number of footsteps counted in time period T_{step}. This definition of power is actually independent of pace frequency or period. Power P_{ave} will be conveniently used to compare the performance of different scavenger configurations in the sensitivity analysis illustrated in the next section.

The voltage waveforms in Fig. 6.5 give comparable output power levels. The harvested power per step obtained from experimental measurements is approximately 13 μW, about 6% higher than the power calculated numerically with the standard acceleration input. This small difference may be attributed to unpredictable small oscillations observed in the measured signal and also to the fact that consecutive steps provide slightly different acceleration signals, even under stable walking conditions.

For practical applications, the obtained power levels are not very high. However, it should be emphasized that they are only indicative of the real bimorph performance, since the applied resistive load R_L (as mentioned before) is not optimal. Furthermore, a change in scavenger geometry could provide an additional increase in harvested power.

It is thus desirable to estimate the amount of power that could be effectively harvested, if other scavenger configurations were actually tested under the same input excitation. Each hypothetical configuration, however, should be also checked with some imposed external constraints (such as the available space in the shoe and the material strength limit), before manufacturing real prototypes. A rapid solution to screen among different possible scavenger configurations is proposed as a sensitivity analysis, described in the next section.

6.6 Sensitivity Analysis

The average harvested power P_{ave} calculated by Eq. (6.3) depends on the voltage $v(t)$ across the resistive load R_L. Each scavenger configuration has an optimum applied resistance R_L^{opt} that would provide the highest dissipated power. The voltage $v(t)$ and the optimum resistive load R_L^{opt} are mutually correlated through the electromechanical bimorph behavior, summarized by Eq. (6.1).

A change in scavenger geometry would then affect v(t) and R_L^{opt}, and hence the resultant harvested power. Different configurations are thus expected to supply different power levels, although the correlation with scavenger parameters may be nonintuitive because of the electromechanical coupling in the bimorph system.

The aim of the sensitivity analysis is to verify if other scavenger geometries could provide improved harvested power levels, compared to the reference rectangular configuration defined in Table 6.1.

The performance of hypothetical rectangular bimorphs obtained by changing some of the parameters in Table 6.1 is now tested. To ascertain that such configurations would be actually realizable in practice and would then behave properly when deployed, limit criteria are also checked. For example, scavenger maximum dimensions are compared to the space available within the shoe, and bimorph resistance is tested by verifying that the maximum stress is lower than allowable material strength limit.

A preliminary analysis shows that not all the parameters in Table 6.1 could be arbitrarily changed. For example, material properties (such as density, elastic

Table 6.2 Ranges of the variables used in sensitivity analysis

Parameter	Values Minimum	Maximum
Electrode length, l_e (mm)	15	40
Beam width, w_b (mm)	10	20
Tip mass thickness, h_M (mm)	2	6
Tip mass weight, M (grams)	5.5	16.4

modulus, permittivity) are actually fixed, once the type of metallic shim and piezoceramic layer are established. Moreover, thickness values t_p and t_s of piezoelectric and metallic layers, respectively, cannot be easily customized in practice and thus their values must be taken as constant as well (values of Table 6.1 are used in the study). Also the tip mass length $l_M = 10$ mm has to be considered as fixed in the present study, because for practical reasons the mass has been manufactured using commercial iron bars having a constant width of 10 mm.

A further reduction in the total number of variables can be achieved if the mass width w_M is increased to the maximum value that is compatible with the cavity depth in the shoe (accordingly, a constant value 35 mm is then chosen).

An additional parameter in the sensitivity analysis that requires special attention is the modal mechanical damping ratio ζ_r. On one hand, the damping is correlated to bimorph geometry, as for example in the Rayleigh damping model [33]; hence different scavenger configurations are expected to have different damping ratios. On the other hand, damping is generally estimated from experimental measurements [34–36]. However, the present sensitivity analysis assumes various scavenger configurations that are hypothetical and not yet physically realized; hence the damping would be hard to be estimated in advance (no experimental measurement is available). For practical reasons, the analysis then assumes for all configurations a constant modal damping ratio equal to $\zeta_r = 0.02$.

After the above screening, a total of three variables can be actually changed in sensitivity analysis: electrode length l_e, beam width w_b, tip mass thickness h_M. Their ranges are established according to physical constraints. For example, their maximum values cannot be chosen arbitrarily, because they must define scavenger geometries that are consistent with the available space in the heel of a shoe. The strategy adopted is to fix their maximum values by changing the geometry of the reference scavenger of Table 6.1, under the design constraint on the maximum space available. On the other hand, also the minimum values should define scavenger configurations that are physically reasonable. Such arguments are followed to establish the ranges of sensitivity parameters listed in Table 6.2. The range of mass weight M is also explicitly listed for more clarity, although it is not really an independent parameter, as it can be calculated from h_M values as $M = \rho \cdot l_M \cdot h_M \cdot w_M$, where $w_M = 35$ mm and $l_M = 10$ mm.

For each analyzed scavenger configuration, the optimum resistive load that dissipates the maximum power under the assumption of weak electromechanical coupling is found as [37]:

$$R_{\mathrm{L}}^{\mathrm{opt}} = \frac{1}{\omega_1 C_{\mathrm{p}}}, \tag{6.4}$$

where ω_1 is the first resonant frequency, C_{p} is the equivalent capacitance of piezoelectric layers, and this equation is valid for transient vibrations dominated by the first vibration mode. Given the parameter values within the ranges of Table 6.2, frequency ω_1 calculated by the analytical formula in [17] for all analyzed configurations is shown to vary approximately from 20 to 150 Hz.

As anticipated before, a change in scavenger geometry is expected to modify the harvested power level. Configurations maximizing the output power can be found by evaluating the dynamic response and output power for all the geometries defined by any possible combination of the parameters in Table 6.2. At the same time, a checking procedure is also introduced to find out only acceptable bimorph configurations, i.e., to verify if all those configurations that provide increased power levels are also physically realizable considering the prescribed limit criteria. For example, all hypothetical scavenger configurations that exceed strength limits characteristics of piezoelectric material should be clearly discarded.

Two constraints are checked in the analysis, the first on compatibility between overall bimorph deflection and available space, the second on the strength limit of the bimorph. The first criterion compares the maximum deflection of the tip mass, u_{max}, to the available cavity depth within insole space, u_{lim}, which is usually several millimeters. The second mechanical constraint is related to the maximum stress at the clamped end σ_{max}, which has not to exceed the strength limit $\sigma_{\mathrm{p,lim}}$ of piezoelectric material, to prevent either static or fatigue failure. A bending fatigue limit of $\sigma_{\mathrm{p,lim}} = 55$ MPa is assumed for PZT material, as estimated in [38, 39].

The maximum stress at the clamped end is estimated as a function of tip maximum displacement u_{max} by using the following formula for beam static deflection:

$$\sigma_{\mathrm{max}} = \frac{3Y_{11}}{l_{\mathrm{b}}^2}\left(\frac{t_{\mathrm{s}}}{2} + t_{\mathrm{p}}\right) u_{\mathrm{max}}, \tag{6.5}$$

where other symbols are defined in Table 6.1. Scavenger configurations where $\sigma_{\mathrm{max}} > \sigma_{\mathrm{p,lim}}$ were considered as critical (failure occurs) and then discarded, independent of the output power they harvested.

Compared to other approaches in the literature, the merit of the present analysis is that it takes also the material strength limit, and not simply the maximum beam deflection, as an acceptability criterion to check the behavior of analyzed bimorph configurations.

A simple approach is adopted to identify different scavenger geometries by changing the three parameters in Table 6.2: each parameter range is divided into a finite set of discrete values and a three-dimensional grid of points is constructed, each point defining a particular scavenger configuration. All points in the grid are next analyzed sequentially by numerical simulations. No particular sampling

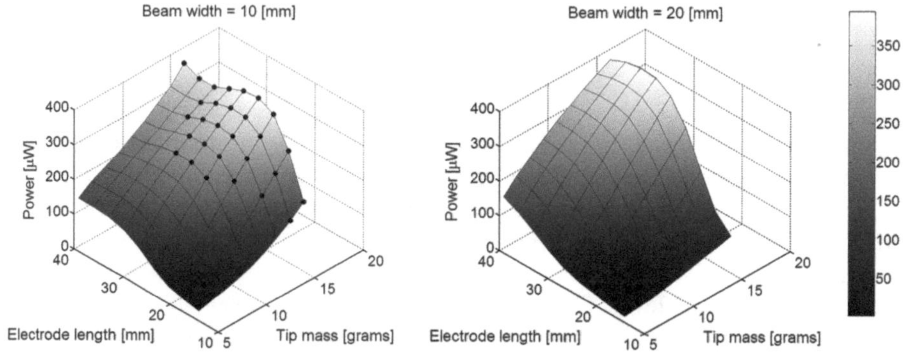

Fig. 6.8 Trend of harvested power, as a function of mass weight M, electrode length l_e, and beam width w_b. *Black markers* identify critical configurations (Reprinted from: [26], with permission from IOP Science)

technique for selecting inputs is necessary, as parameters in Table 6.2 have no statistical variability.

Each point in the grid defines a specific bimorph configuration, for which the dynamic response and average harvested power under the standard acceleration input can be evaluated by numerical simulations. A sequence of three footsteps is chosen as a reference in all simulations. The number of analyzed bimorph configurations (that is the amount of simulation runs) is equal to the grid size, which in turn depends on the number of levels dividing each parameter range. Selecting 9 equally spaced levels for 3 range in Table 6.2, it must be performed a total amount of $9^3 = 729$ simulation runs, which has been considered as the best balance between the desired number of results points and the required total computational time.

The proposed approach is really effective, as it assures that each range in Table 6.2 is totally spanned, while the total simulation time is kept to reasonable levels. Nevertheless, nothing prevents the analyst to use different grid points to get more closely spaced results, since it does not absolutely affect the accuracy of results in sensitivity analysis.

The average power defined by Eq. (6.3) is used as a figure of merit to compare the performance of the set of different scavenger configurations. A typical trend of harvested power, as a function of electrode length l_e for different values of mass weight M and two values of beam width w_b, is shown in Fig. 6.8. Those configurations that exceed material strength limit, $\sigma_{max} > \sigma_{p,lim}$, are identified by a black marker. Note that all bimorph configurations satisfy the constraint imposed by available insole space, as the calculated maximum deflection u_{max} is always lower than 2.5 mm, i.e., significantly below the available cavity depth.

As shown in Fig. 6.8, at small mass values, the power is roughly proportional to electrode length l_e. Instead, at larger mass values the power is not always proportional to l_e, as it has a minimum within an intermediate region of l_e values. For example, given a mass of 16.4 g, the maximum power (378 μW) is attained at

Table 6.3 Comparison of the output power for various bimorphs

Configuration	l_e (mm)	w_b (mm)	h_M (mm)	M (grams)	P_{ave} (μW)
A	24.4	10	6	16.4	378
B	33.8	10	6	16.4	336
C	24.4	10	5	13.7	312
D	33.8	10	5	13.7	293

an intermediate electrode length $l_e = 24.4$ mm, while a lower power (336 μW) is obtained at a greater length $l_e = 33.8$ mm (compare A and B in Table 6.3). A similar nonlinear trend is also observed for other tip mass values, as the examples C and D listed in Table 6.3.

This nonlinear trend has been explained in [26] with reference to the strong relationship between bimorph mechanical dynamic response and the applied standard acceleration excitation. In particular, the range of the instantaneous output power is shown to depend on the relative phase between beam oscillation and acceleration excitation. In fact, the acceleration signal of Fig. 6.3 is characterized by an irregular pattern with a close peak/valley sequence, which makes the scavenger dynamic response more complex than the simple response that would be given by a simple harmonic excitation. Specifically, it has been shown [26] that the bimorph response oscillation is amplified if its first negative oscillation is exactly in-phase with the first valley in the standard acceleration signal. On the contrary, a relatively damped response is observed if the bimorph oscillation is not exactly in-phase with excitation. However, bimorph oscillation also depends on its first resonant frequency, which is correlated to electrode length (or, equivalently, beam length) and tip mass weight.

The trend in Fig. 6.8 also shows that an increase of mass gives an increase in output power, because of larger tip deflection and thus mechanical strain, which, however, could induce fracture for those bimorphs where $\sigma_{max} > \sigma_{p,lim}$ at the clamped end (this critical condition is identified by a black marker in Fig. 6.8a).

Other configurations with larger beam width are characterized by a similar trend and comparable power levels. However, an increase in beam width gives a larger stiffness that reduces the stress/strain levels in the piezoelectric layers and thus reduces the possibility of fracture, without however significantly penalizing the harvested power. The configurations in Fig. 6.8b where maximum stress does not exceed the material strength should thus be preferred.

The most important result of this sensitivity analysis is the possibility to screen the performance of a wide set of scavenger configurations and to identify those characterized by the largest power levels. Five different configurations are listed in Table 6.4. The second column refers to the reference bimorph configuration defined by the parameters of Table 6.1. The third column identifies the configuration that supplies the largest output power, while the remaining columns on the right list other configurations characterized by power levels very close to the maximum.

Table 6.4 Comparison of the performance of different bimorph configurations, using the same acceleration time history as input

	Configuration Reference	Optimal	#1	#2	#3
Geometrical parameters					
l_e (mm)	20	33.8	27.5	36.9	30.6
w_b (mm)	14	20	16.3	18.8	13.8
h_M (mm)	3.3	6	6	6	6
M (grams)	9	16.4	16.4	16.4	16.4
Mechanical properties					
w_{max} (mm)	0.4	1.4	1.2	1.6	1.5
σ_{max} (MPa)	31.3	46.6	54.9	45.3	57.3[b]
Electrical properties					
$R_L{}^{opt}$ (kΩ)	220	199	254	226	337
C_p (nF)	10.2	24.5	16.2	25.1	15.3
P_{ave} (μW)[a]	61	$395_{(+547\%)}$	$381_{(+525\%)}$	$381_{(+525\%)}$	$381_{(+525\%)}$

[a]Average power per footstep
[b]Maximum stress exceeds material strength limit

For the reference bimorph, a maximum power of 61 μW is dissipated across a resistive load $R_L{}^{opt} = 220$ kΩ.[1] The optimal configuration gives an output power of 395 μW over a 199 kΩ resistance, i.e., an increase of 547% with respect to the reference geometry. Slightly lower power levels are provided by configurations #(1–3) having different geometries. A schematic picture of some bimorphs in Table 6.4 is given in Fig. 6.9, which confirms the nonobvious relationship between electrical performances and geometrical parameters under imposed constraints: Starting from the reference configuration in Fig. 6.9a, a rise in output power can be obtained with an increase in bimorph length and tip mass weight, as suggested by configuration #3 in Fig. 6.9b. As a consequence, the maximum stress could exceed the limit material strength (in fact geometry #3 should be discarded) and also a change in beam width should be evaluated in order to increase the flexural stiffness. The so-called optimal configuration, represented in Fig. 6.9b, provides a greater amount of power (its equivalent capacity is higher compared to those of the other bimorphs, see Table 6.4) with a tolerable value of bending stress.

This analysis shows that with a proper change in geometrical parameters, a significant increase in power output can be obtained with respect to the reference configuration of Table 6.1. The obtained improved power levels actually confirm the feasibility of energy harvesting with shoe-mounted vibrating scavengers excited by heel accelerations typical of human walking.

Nevertheless, different strategies can be devised to further improve the harvested power. As an example, the literature [22, 31, 40–42] has extensively investigated shape optimization of bimorph geometry, finding that trapezoidal shapes could

[1]This confirms that the resistive load $R_L = 14$ kΩ used in Sect. 6.4 is not the optimum one.

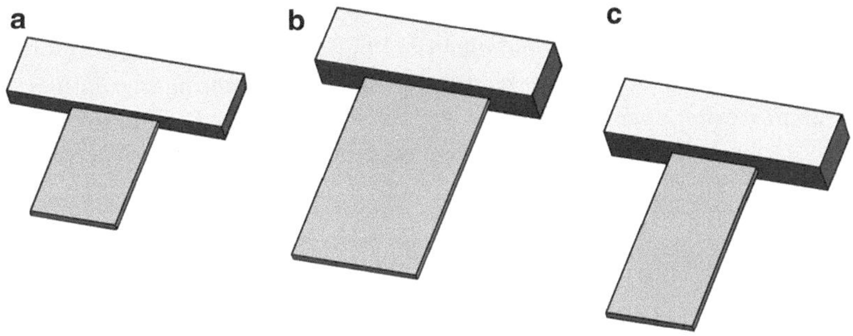

Fig. 6.9 Comparison of different bimorphs of Table 6.4: (**a**) reference; (**b**) optimal; (**c**) configuration #3 (fracture)

provide higher mechanical strains and in turn increased levels of specific power. Rather than changing bimorph geometry, optimization of beam deformation to get constant maximum strain in piezoelectric materials has been also recently proposed [43]. The idea is to constrain beam deflection onto supports with an optimal shape to obtain a constant curvature deformation.

Finally, proper optimization of the electrical circuitry, in place of the simple resistive load used in this study, could better represent electrical loads encountered in practical applications and could then further increase the harvested power [37, 44].

6.7 Conclusions

This chapter discussed the feasibility of energy harvesting with vibrating piezoelectric shoe-mounted devices, excited by gait acceleration. The typical features of heel acceleration, measured during normal walking, are first identified and compared with relevant data from the literature. A standard heel acceleration signal is next defined, which is used as reference excitation in the numerical simulations.

The electromechanical dynamic response of a shoe-mounted rectangular bimorph with tip mass, excited by the standard heel acceleration, is studied using a distributed parameter model from the literature. The beam deflection and the average output power are calculated. The results given by the numerical model are preliminary validated with experimental measurements on a real prototype.

A sensitivity analysis is finally performed to find hypothetical scavenger configurations that could provide enhanced power levels. Acceptability criteria on both geometrical constraints and strength limits are also checked to verify if optimal bimorphs could be actually realizable in practice.

The results of this study confirmed how the obtained power levels, although lower than values claimed by other studies [12, 18, 45], may however be sufficiently high for practical applications. Suitable electrical storage systems may also be designed, to accumulate energy for future use. Further improvements can be obtained by proper mechanical and electrical optimization strategies, as suggested in the literature. For example, mechanical optimization adopting trapezoidal shapes could maximize the average strain in piezoelectric material, which results in an increase of about 25–30% of specific power compared to classical rectangular geometries. An alternative that does not require any change in bimorph geometry is the optimization of bending deflection to give a constant maximum strain and thus increase power output [43]. Optimization of cavity depth inside the shoe, by using multiple harvesters, could be another strategy. Finally, optimized circuitries could further improve the harvested power.

Acknowledgements This work is a part of "Tech-Up" project for the development of ubiquitous and pervasive technologies, which is financially supported by the Friuli-Venezia Giulia Region, Italy. The authors wish also to acknowledge Mr. Elvio Castellarin for precious help in experimental measurements.

References

1. Paradiso JA, Starner T (2005) Energy scavenging for mobile and wireless electronics. IEEE Pervasive Comput 4(1):18–27
2. Beeby SP, Tudor MJ, White NM (2006) Energy harvesting vibration sources for microsystems applications. Meas Sci Technol 17:R175–R195
3. Park G, Farrar CR, Todd MD, Hodgkiss W, Rosing T (2007) Energy harvesting for structural health monitoring sensor networks. Los Alamos National Laboratory Report LA-14314-MS
4. Mitcheson PD, Yeatman EM, Kondala Rao G, Holmes AS, Green TC (2008) Energy harvesting from human and machine motion for wireless electronic devices. Proc IEEE 96(9):1457–86
5. Starner T (1996) Human powered wearable computing. IBM Syst J 35(3–4):618–29
6. Starner T, Paradiso JA (2004) Human generated power for mobile electronics. In: Piguet G (ed) Low power electronics design. CRC, Boca Raton, FL
7. Antaki JF, Bertocci GE, Green CG, Nadeem A, Rintoul T, Kormos RL, Griffith BP (1995) A gait-powered autologous battery charging system for artificial organs. ASAIO J 41(3):M588–M595
8. Granstrom J, Feenstra J, Sodano HA, Farinholt K (2007) Energy harvesting from a backpack instrumented with piezoelectric shoulder straps. Smart Mater Struct 16:1810. doi:10.1088/0964-1726/16/5/036
9. Donelan JM, Li Q, Naing V, Hoffer JA, Weber DJ, Kuo AD (2008) Biomechanical energy harvesting: generating electricity during walking with minimal user effort. Science 319:807–10. doi:10.1126/science.1149860
10. Wang Z, Leonov V, Fiorini P, Van Hoofa C (2009) Realization of a wearable miniaturized thermoelectric generator for human body applications. Sens Actuators A Phys 156(1):95–102
11. Kymissis J, Kendall C, Paradiso J, Gershenfeld N (1998) Parasitic power harvesting in shoes. 2nd IEEE international conference on wearable computing, pp 132–136
12. Shenck NS, Paradiso JA (2001) Energy scavenging with shoe-mounted piezoelectrics. IEEE Micro 21(3):30–42
13. Renaud M, Fiorini P, van Schaijk R, van Hoof C (2009) Harvesting energy from the motion of human limbs: the design and analysis of an impact-based piezoelectric generator. Smart Mater Struct 18:035001

14. Anton SR, Sodano HA (2007) A review of power harvesting using piezoelectric materials (2003–2006). Smart Mater Struct 16:R1–R21
15. Winter DA (1992) Foot trajectory in human gait: a precise and multifactorial motor control task. Phys Ther 72(1):45–53
16. Ledoux WR, Hillstrom HJ (2001) Acceleration of the calcaneous at heel strike in neutrally aligned and pes planus feet. Clin Biomech 16:608–13
17. Erturk A, Inman DJ (2009) An experimentally validated bimorph cantilever model for piezoelectric energy harvesting from base excitation. Smart Mater Struct 18(2):025009
18. Shenck NS (1999) A demonstration of useful electric energy generation from piezoceramics in a shoe. BS Thesis, Massachusetts Institute of Technology, MIT
19. González JL, Rubio A, Moll F (2002) Human powered piezoelectric batteries to supply power to wearable electronic devices. Int J Soc Mater Eng Resour 10:34–40
20. Niu P, Chapman P, Riemer R, Zhang X (2004) Evaluation of motions and actuation methods for biomechanical energy harvesting. 35th annual IEEE power electronics specialists conference, Aachen, Germany, pp 2100–2106
21. Riemer R, Shapiro A (2011) Biomechanical energy harvesting from human motion: theory, state of the art, design guidelines, and future directions. J Neuro Eng Rehabil 8, article n. 22
22. Mateu L, Moll F (2005) Optimum piezoelectric bending beam structures for energy harvesting using shoe inserts. J Intell Mater Syst Struct 16:835–45
23. Whittle WM (2007) Gait analysis, an introduction, 4th edn. Elsevier, Philadelphia, PA
24. Perry J (1992) Gait analysis: normal and pathological function. SLACK Incorporated, Thorofare, NJ
25. Giddings VL, Beauprè GS, Whalen RT, Carter DR (2000) Calcaneal loading during walking and running. Med Sci Sports Exerc 32(3):627–34
26. Moro L, Benasciutti D (2010) Harvested power and sensitivity analysis of vibrating shoe-mounted piezoelectric cantilevers. Smart Mater Struct 19:115011. doi:10.1088/0964-1726/19/11/115011
27. Tripathy BK (2004) A study on step distance and its relation with some morphometric features in adult male. Anthropologist 6(2):137–139
28. Ren L, Jones RK, Howard D (2007) Predictive modelling of human walking over a complete gait cycle. J Biomech 40:1567–74
29. Chi KJ, Schmitt D (2005) Mechanical energy and effective foot mass during impact loading of walking and running. J Biomech 38:1387–95
30. Danion F, Varraine E, Bonnard M, Pailhous J (2003) Stride variability in human gait: the effect of stride frequency and stride length. Gait Posture 18:69–77
31. Benasciutti D, Moro L, Zelenika S, Brusa E (2010) Vibration energy scavenging via piezoelectric bimorphs of optimized shapes. Microsyst Technol 16:657–668
32. Kim M, Hoegen M, Dugundji J, Wardle BL (2010) Modeling and experimental verification of proof mass effects on vibration energy harvester performance. Smart Mater Struct 19:045023. doi:10.1088/0964-1726/19/4/045023
33. Banks HT, Inman DJ (1991) On damping mechanisms in beams ASME. J Appl Mech 58:716–23
34. Clough RW, Penzien J (1993) Dynamics of structures, 2nd edn. McGraw-Hill, New York
35. Erturk A, Inman DJ (2008) A distributed parameter electromechanical model for cantilevered piezoelectric energy harvesters. ASME J Vib Acoust 130:041002
36. Goldschmidtboeing F, Woias P (2008) Characterization of different beam shapes for piezoelectric energy harvesting. J Micromech Microeng 18:104013
37. Badel A, Guyomar D, Lefeuvre E, Richard C (2005) Efficiency enhancement of a piezoelectric energy harvesting device in pulsed operation by synchronous charge inversion. J Intell Mater Syst Struct 16:889–901
38. Okayasu M, Aoki S, Mizuno M (2008) Effects of silver-based metal electroplate on fatigue properties of PZT ceramics. Int J Fatigue 30:1115–24
39. Okayasu M, Ozeki G, Mizuno M (2010) Fatigue failure characteristics of lead zirconate titanate piezoelectric ceramics. J Eur Ceram Soc 30:713–25

40. Benasciutti D, Brusa E, Moro L, Zelenika S (2008) Optimised piezoelectric energy scavengers for elder care. Proceedings of the European Society for Precision Engineering and Nanotechnology (EUSPEN) conference, Zurich, 18–22 May 2008, pp 41–45
41. Brusa E, Zelenika S, Benasciutti D, Moro L (2009) Analytical characterization and experimental validation of performance of piezoelectric vibration energy scavengers. In: Schmid U, Cané C, Shea H (eds) Proceedings of the SPIE conference, Dresden, Germany, 4–6 May 2009, vol 7362, pp 736204-1/736204-12
42. Baker J, Roundy S, Wright P (2005) Alternative geometries for increasing power density in vibration energy scavenging for wireless sensor networks. Proceedings of the 3rd international energy conversion engineering conference, San Francisco, CA, 15–18 Aug 2005, paper AIAA-2005-5617
43. Moro L, Benasciutti D (2011) On the optimal bending deflection for piezoelectric scavengers. EuroMech Colloquium 530, Structural Control and Energy Harvesting, Bristol, 25–27 July
44. Renno JM, Mohammed FD, Inman DJ (2009) On the optimal energy harvesting from a vibration source. J Sound Vib 320:386–405
45. Kendall CJ (1998) Parasitic power collection in shoe mounted devices. BS Thesis, Massachusetts Institute of Technology, MIT

Chapter 7
Role of Stiffness Nonlinearities in the Transduction of Energy Harvesters under White Gaussian Excitations

Mohammed F. Daqaq

Abstract This chapter dileneates the influence of stiffness-type nonlinearities on the transduction of vibratory energy harvesters (VEHs) under random excitations that can be approximated by a white Gaussian noise process. Both mono- and bi-stable Duffing-type harvesters are considered. The Fokker–Plank–Kolomogorov equation governing the evolution of the harvetser's transition probability density function is formulated and used to generate the moment differential equations governing the response statistics. The moment equations are then closed using a fourth-order cumulant-neglect closure scheme and solved for the relevant steady-state response statistics. The influence of the nonlinearity, time constant ratio (the ratio between the nominal period of the mechanical subsystem and the time constant of the harvesting circuit), and noise intensity on the mean square value of the electric output (voltage or current) and the average power is detailed. Results are then compared to those obtained by analytically solving the FPK equation for the linear resonant harvester. It is demonstrated that a Duffing-type monostable harvester with hardening nonlinearity can never outperform its linear counterpart. A bi-stable harvester, on the other hand, can outperform a linear harvester only when the time constant ratio is small and its potential energy function is optimized based on a known excitation intensity.

7.1 Introduction

Energy harvesting is the process by which ambient energy is captured and transformed into a useful form. Historically, mankind has relied on this process to fill its basic energy needs using windmills, sailing ships, and waterwheels. However, our

M.F. Daqaq (✉)
Clemson University, Clemson, SC 29634, USA
e-mail: mdaqaq@clemson.edu

N. Elvin and A. Erturk (eds.), *Advances in Energy Harvesting Methods*,
DOI 10.1007/978-1-4614-5705-3_7,

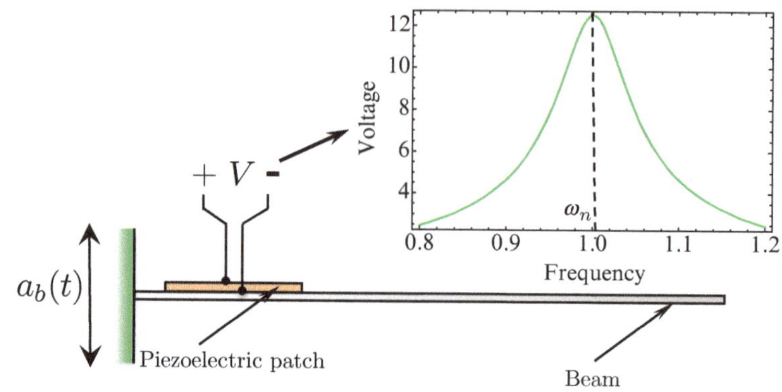

Fig. 7.1 A schematic of a linear piezoelectric harvester and its voltage–frequency response

ever-increasing energy demands and changing research trends are pushing these old concepts into newer directions. For instance, today, we continue to produce smaller and lower-power consumption devices that span different fields of technology. Wireless sensors, data transmitters, controllers, and implantable medical devices that require only sub-milliwatts of average power to function are being developed [1–4]. Unfortunately, further evolution of such technologies is currently being moderated by the lack of continuous scalable energy sources that can be used to power and maintain them. Batteries, which remain the most adequate power choice, have not kept pace with the devices' demands, especially in terms of energy density [5]. In addition, their finite life span which necessitates regular replacement can be a very costly and cumbersome process. Consider, for instance, the difficulty of replacing batteries for a spatially dense remotely located wireless sensor network or the risks involved in changing batteries for patients with implantable pacemakers.

In light of such challenges and the low-power consumption of many new critical technologies, this last decade has witnessed a new evolution in energy-harvesting technologies whereby the concept of micro-power generators (MPGs) was introduced [6–8]. MPGs are compact energy-harvesting devices that can transform the smallest amounts of available wasted ambient energy into electricity. When embedded with electronic devices, these generators can provide a continuous power supply permitting an autonomous operation process. Within the vast field of micro-power generation, vibratory energy harvesters (VEHs) have flourished as a major thrust area. Various devices have been developed to transform mechanical motions directly into electricity by exploiting the ability of active materials and some electromechanical mechanisms to generate an electric potential in response to mechanical stimuli and external vibrations [6–8].

The most prolific energy-harvesting design consists of a cantilever beam subjected to base excitations as shown in Fig. 7.1. Attached to the beam near the clamped end are piezoelectric patches. External environmental excitations set the beam in motion producing large strains near the clamped end, which, in turn, produce a voltage difference across the piezoelectric patches. By designing the proper circuitry, this electric potential can be used to create a current that transfers energy from the environment to an electric device.

Key for efficient energy transduction is the ability to set the beam into large-amplitude oscillations. For an environmental excitation exhibiting harmonic fixed-frequency characteristics, large-amplitude beam deflections can be excited by tuning one modal frequency of the beam, usually the first, to be equal to the excitation frequency (resonance condition). This tuning approach, however, can, in many instances, be very difficult to achieve in realistic vibratory environments for the following reasons:

1. Linear VEHs similar to the cantilever beam shown in Fig. 7.1 have a very narrow frequency bandwidth. Small drifts in the excitation frequency around the harvester's fundamental frequency can easily occur due to small variations in the nature of the excitation source and/or changes in the design parameters of the harvester around their nominal values. This drops the already small energy output of VEHs even further, making the energy-harvesting process inefficient.
2. Most environmental excitations are not harmonic but have broadband or nonstationary (time-dependent) characteristics in which the energy is distributed over a wide spectrum of frequencies or the dominant frequency vary with time. For instance, environmental excitations to which a bridge is subjected are generally random resulting from wind loadings whose frequency and intensity vary depending on the atmospheric conditions and moving vehicles whose number, speed, weight, etc., vary at different times during a given day. Common sources for oscillations in microsystems have white noise characteristics normally due to nonequilibrium thermal fluctuations as well as shot and low-frequency noise [9, 10].

7.1.1 Current Solutions

To remedy this problem, some initial solutions called for the design of vibratory energy harvesters with tunable characteristics. Tuning mechanisms use passive/active means to alter the fundamental frequency of the harvester to match the dominant frequency of the excitation [11–14, 14–17]. Following a number of research investigations, it became evident that tunable VEHs can only be utilized to account for slow drifts in the excitation's frequency and are not efficient under random or fast-varying frequency inputs [11]. In addition, tuning mechanisms usually require external power or complex designs. Others proposed solutions that utilize stacks of harvesters with different fundamental frequencies such that at least one will have a matching fundamental frequency and will, thereby, resonate and harvest energy from the corresponding excitation's component [14, 16, 17]. This, however, reduces the power density and adversely affects the scalability of the harvester.

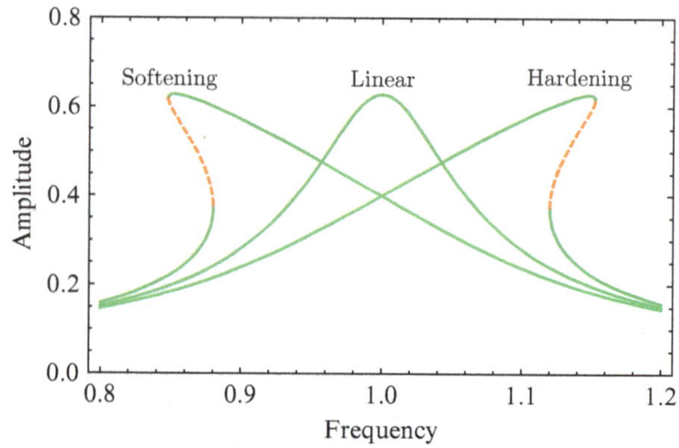

Fig. 7.2 A normalized steady-state (displacement or voltage) frequency-response curve of a monostable energy harvester. *Dashed lines* represent unstable solutions

A significant body of the current research on vibratory energy harvesting is focused on the concept of purposeful inclusion of stiffness nonlinearities for broadband transduction. When compared to their linear resonant counterparts, nonlinear energy harvesters have a wider steady-state frequency bandwidth, leading to the idea that they can be utilized to improve performance especially in random and nonstationary vibratory environments. The basic idea lies in altering the potential energy function of the harvesting system by introducing a nonlinear restoring force. In general, the nonlinearity can be introduced using external design means such as magnetic or mechanical forces [18–29]. There are two different classes of these harvesters. The first is designed to exhibit a nonlinear resonant behavior similar to that of a *monostable Duffing oscillator* with a hardening/softening nonlinearity [19–21]. As shown in Fig. 7.2, the nonlinearity bends the frequency-response curves to the left or to the right depending on its nature, thereby extending the bandwidth of the harvester to a wider range of frequencies. However, the enhanced bandwidth is accompanied with the presence of regions of multiple stable solutions with different basins of attraction. As such, a monostable harvester is not always guaranteed to operate on the large-amplitude branch of solutions.

Examples of monostable VEHs include, but are not limited to, the magnetically levitated inductive harvester proposed by Mann and Sims [20] and shown in Fig. 7.3a. This harvester comprises of two outer magnets to levitate an oscillating central magnet. The nonlinearity is introduced in the form of the magnetic restoring force, which also enables the system to be tuned to a specific resonant frequency. Per Faraday's law, energy is generated as a result of the relative motion between the coil and the center magnet. Barton et al. [19] proposed a monostable inductive VEH but in the form of a tip magnet attached to a cantilever beam, Fig. 7.3b. When the beam oscillates, the magnet moves relative to a coil wound around an iron core, generating a current in the coil.

Masana and Daqaq [21] also proposed a monostable axially loaded piezoelectric clamped–clamped beam harvester as shown in Fig. 7.3c. The axial preload, which is kept below the critical buckling load, serves to tune the natural frequency of

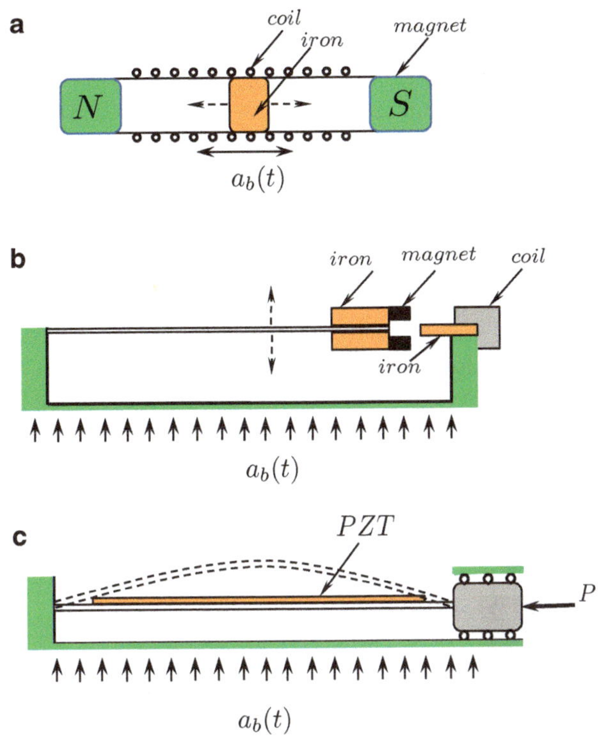

Fig. 7.3 Schematics of different nonlinear energy harvesters. (**a,b**) Inductive VEHs proposed in [20, 30] and [19], respectively; the linear stiffness and nonlinearity can be tuned by varying the distance between the magnets. (**c**) Piezoelectric VEH proposed in [21]; linear stiffness and nonlinearity can be tuned by varying an axial load, P. All devices can operate in the mono- and bistable configurations, and $a_b(t)$ refers to the environmental base excitation

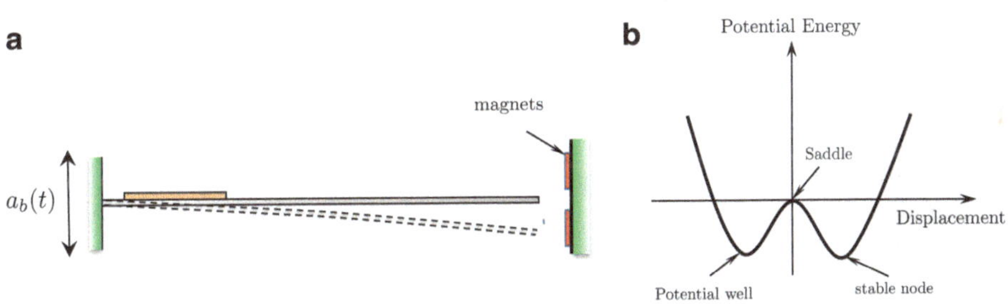

Fig. 7.4 (**a**) Schematic of a bistable piezoelectric harvester proposed by [23]. (**b**) Associated potential energy function

the beam and to introduce a cubic nonlinearity which depends on the magnitude of the axial load. The device harvests energy as a result of the excitation-induced deformation of a piezoelectric patch attached to the surface of the beam.

The second class of nonlinear harvesters is designed to have a double-well potential energy function exhibiting the response of a *bistable oscillator* as shown in Fig. 7.4b. The operation concept of this class, which has been initially proposed by Cottone et al. [24] and later studied by several researchers [18, 19, 21, 23, 26], is based on the dynamics of a bistable Duffing oscillator. While several variances of the device were proposed, the main concept of operation is very similar. As shown in Fig. 7.4a, a version of this harvester consists of a piezoelectric beam (harvester) with a ferroelectric tip oscillating between two magnets. For a certain separation range between those magnets, the system becomes bistable, having a *double-well* potential energy function with two stable equilibria and one unstable saddle. When such a device is subjected to small input excitations, the dynamics remain confined to one potential well exhibiting a nonlinear resonant behavior similar to that of a regular monostable Duffing oscillator. However, when enough energy is supplied to allow the dynamic trajectories to overcome the potential barrier and escape to the other potential well (inter-well motion), the harvester can exhibit complex dynamic responses which can, under some conditions, be favorable for energy harvesting [23, 26]. Masana and Daqaq [31] investigated the relative response of a mono- and bistable VEHs based on the axially loaded clamped–clamped piezoelectric beam design under harmonic excitations illustrating that the bistable harvester can *only* outperform the monostable design for some shapes of the potential energy function and for large-base excitation levels.

7.1.2 Motivations

While the purposeful introduction of nonlinearities has been aimed to resolve the issue of excitations' nonstationarities and randomness, the associated analyses and predicted power enhancements were, for the most part, based on the steady-state response which assumes a harmonic fixed-frequency excitation. As of today, we still do not have a clear understanding of how the nature of the excitation influences the output power or what role stiffness nonlinearities play in the transduction of energy harvesters under random excitations. Still, it is not even well understood whether the steady-state fixed-frequency analysis currently adopted in the literature is a valid performance indicator.

A few recent research studies have tried to address some of these unanswered questions by providing a clearer picture of how randomness and nonstationarities in the excitation influence the average power of nonlinear VEHs [32–38]. Along similar lines, this chapter aims to delineate the influence of stiffness nonlinearities on the response statistics when the harvester is excited by white Gaussian excitations. Both mono- and bistable Duffing-type VEHs are considered and compared to their linear resonant counterparts to identify the design conditions and excitation intensities under which a nonlinear harvester can actually outperform a linear one.

The rest of this chapter is organized as follows: Sect. 7.2 presents a generic electromechanical lumped-parameter model which can be used to capture the qualitative behavior of VEHs (capacitive and inductive). Section 7.3 formulates the problem in the Itô stochastic sense and presents the Fokker–Planck–Kolmogorov (FPK) equation governing the evolution of the transition probability density function (PDF). Section 7.4 solves the FPK equation analytically for the special case of a large time constant ratio. Section 7.5 studies the response of the linear harvester by obtaining analytical expressions for the relevant response statistics. Section 7.6 uses a non-Gaussian fourth-order cumulant-neglect closure scheme to approximately solve the FPK equation in the case of the mono- and bistable harvesters, respectively. Results are compared with those obtained in the linear scenario, and discussions are presented to delineate the relative performance and the influence of the nonlinearity on the average power. Finally, Sect. 7.7 presents the important conclusions.

7.2 Electromechanical Model

Without loss of generality, the basic physics of a nonlinear energy-harvesting device can be captured by considering an oscillating structure coupled to an electric circuit through an electromechanical coupling mechanism which is mainly either (a) capacitive (piezoelectric, electrostatic), Fig. 7.5a, or (b) inductive (electromagnetic, magnetostrictive), Fig. 7.5b. The equation of motion can then be written in the following form:

$$m\ddot{\bar{x}} + c\dot{\bar{x}} + \frac{\mathrm{d}\bar{U}(\bar{x})}{\mathrm{d}\bar{x}} + \theta\bar{y} = -m\ddot{x}_b,$$

$$C_\mathrm{p}\dot{\bar{y}} + \frac{\bar{y}}{R} = \theta\dot{\bar{x}}, \text{(capacitive)}, \qquad L\dot{\bar{y}} + R\bar{y} = \theta\dot{\bar{x}}, \text{(inductive)}, \tag{7.1}$$

where the dot represents a derivative with respect to time, τ. The variable $\bar{x}$ represents the relative displacement of the mass m, c is a linear viscous damping coefficient, θ is a linear electromechanical coupling coefficient, $\ddot{x}_b$ is the base acceleration, C_p is the capacitance of the piezoelectric element, L is the inductance of the harvesting coil, and $\bar{y}$ is the electric quantity representing the induced voltage in capacitive harvesters and the induced current in inductive ones. These are

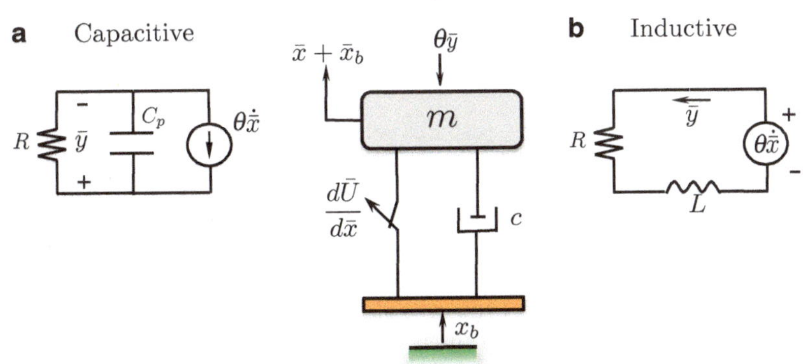

Fig. 7.5 A simplified representation of a vibratory energy harvester

measured across an equivalent resistive load, R . The function $\bar{U}(\bar{x})$ represents the potential energy of the mechanical subsystem and is given in the following *Duffing* form:

$$\bar{U}(\bar{x}) = \frac{1}{2}k_1(1-r)\bar{x}^2 + \frac{1}{4}k_2\bar{x}^4, \tag{7.2}$$

where k_1 and k_2 are, respectively, linear and nonlinear stiffness coefficients, and r is introduced to permit tuning of the linear stiffness around its nominal value using external design means such as the axial load in the axially loaded VEH proposed by Daqaq et al. [21], Fig. 7.3c, or the magnetic field in the magnetically tuned VEHs proposed by [20, 30, 34], Fig. 7.3a,b. Introduction of this constant is necessary to reflect the fact that, for physical VEHs, the nonlinear stiffness cannot be easily changed without altering the linear stiffness due to physical and design constraints.

The equations of motion can be further nondimensionalized by introducing the following dimensionless quantities:

$$x = \frac{\bar{x}}{l_c}, \qquad t = \tau\omega_n, \qquad y = \frac{C_\mathrm{p}}{\theta l_c}\bar{y}\ \text{(capacitive)}, \ y = \frac{L}{\theta l_c}\bar{y}\ \text{(inductive)}, \tag{7.3}$$

where l_c is a length scale, and $\omega_n = \sqrt{k_1/m}$ is the short-circuit natural frequency. With these transformations, the nondimensional equations of motion can be expressed as

$$\begin{aligned} &\ddot{x} + 2\zeta\dot{x} + \frac{\mathrm{d}U}{\mathrm{d}x} + \kappa^2 y = -\ddot{x}_b, \\ &\dot{y} + \alpha y = \dot{x}, \end{aligned} \tag{7.4}$$

where

$$\frac{\mathrm{d}U}{\mathrm{d}x} = (1-r)x + \delta x^3 \tag{7.5}$$

and

$$\begin{aligned} &\zeta = \frac{c}{2\sqrt{k_1 m}}, \qquad \kappa^2 = \frac{\theta^2}{k_1 C_\mathrm{p}}\ \text{(capacitive)}, \ \kappa^2 = \frac{\theta^2}{k_1 L}\ \text{(inductive)}, \\ &\delta = \frac{k_2 l_c^2}{k_1}, \qquad \alpha = \frac{1}{R C_\mathrm{p}\omega_n}\text{(capacitive)}, \ \alpha = \frac{R}{L\omega_n}\ \text{(inductive)}. \end{aligned} \tag{7.6}$$

Here, ζ is the mechanical damping ratio, κ is a linear dimensionless electromechanical coupling coefficient, and α is the ratio between the mechanical and electrical time constants of the harvester. Finally, δ is the coefficient of the cubic nonlinearity. The form of Eq. (7.4) permits classifying energy harvesters, regardless of their coupling mechanism, into three major categories based on the shape of their potential energy function, Fig. 7.6. The harvester is considered to be linear monostable when $\delta = 0$ and $r < 1$, nonlinear monostable when $\delta > 0$ and $r \leq 1$, and nonlinear bistable when $\delta > 0$ and $r > 1$. For the bistable harvester, when

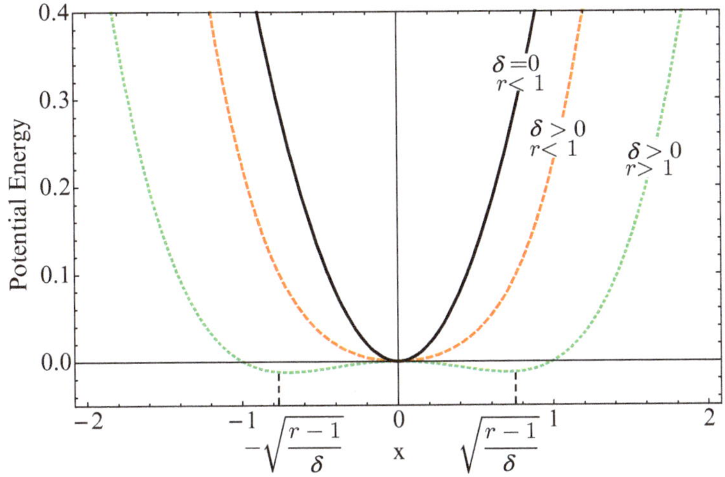

Fig. 7.6 Energy potentials of different vibratory energy harvesters

δ is increased, the separation distance between the wells which is defined by the location of the system's stable fixed points $x_0 = \pm\sqrt{\frac{r-1}{\delta}}$ decreases. The height of the potential barrier at the unstable saddle, $h_b = \frac{(r-1)^2}{4\delta}$, also decreases. This creates shallower potential wells, which, in turn, facilitates the transition of dynamic trajectories from one potential well to the other, a phenomenon commonly known as the inter-well dynamics.

7.3 Stochastic Dynamics

The environmental excitation, $\ddot{x}_b$, to which the harvester is subjected, is assumed to be a physical zero-mean Gaussian process with a very small correlation time which approaches zero. In such a case, $\ddot{x}_b$ can be approximated by a Gaussian white noise process such that[1]

$$\langle \ddot{x}_b(t) \rangle = 0, \qquad \langle \ddot{x}_b(t)\ddot{x}_b(s) \rangle = \sigma^2 \hat{\delta}(s-t), \tag{7.7}$$

where $\langle\rangle$ denotes the expected value, σ^2 is the variance of the process, and $\hat{\delta}$ is the Dirac-delta function. The response statistics associated with the stochastic dynamics of Eq. (7.4) can be generated by expressing the equations in the Itô stochastic form as [39, 40]

[1]The assumption of white noise is not as restrictive as it may appear. If the bandwidth of the excitation is sufficiently larger than that of the harvester's, then a random excitation can be safely considered to be white.

$$\mathrm{d}\mathbf{x}(t) = \mathbf{f}(\mathbf{x},t)\mathrm{d}t + \mathbf{G}(\mathbf{x},t)\mathrm{d}\mathbf{B}, \tag{7.8}$$

where $\mathbf{x} = (x_1, x_2, x_3)^{\mathrm{T}} \equiv (x, \dot{x}, y)^{\mathrm{T}}$, B is a Brownian motion process such that $\ddot{x}_b(t) = \mathrm{d}\mathbf{B}/\mathrm{d}t$, and

$$\mathbf{f}(\mathbf{x},t) = \begin{Bmatrix} x_2 \\ -2\zeta x_2 - \frac{\mathrm{d}U}{\mathrm{d}x_1} - \kappa^2 x_3 \\ -\alpha x_3 - x_2 \end{Bmatrix}, \qquad \mathbf{G}(\mathbf{x},t) = \begin{Bmatrix} 0 \\ -1 \\ 0 \end{Bmatrix}. \tag{7.9}$$

The solution of Eq. (7.8) is determined by the evolution of the transition PDF, $P(\mathbf{x},t)$, which, in turn, is governed by the following FPK equation:

$$\frac{\partial P(\mathbf{x},t)}{\partial t} = -\sum_{i=1}^{3} \frac{\partial}{\partial x_i}[P(\mathbf{x},t)\mathbf{f}_i(\mathbf{x},t)] + \frac{1}{2}\sum_{i=1}^{3}\sum_{j=1}^{3} \frac{\partial^2}{\partial x_i x_j}[P(\mathbf{x},t)(\mathbf{Q}\mathbf{G}\mathbf{G}^{\mathrm{T}})_{ij}],$$
$$P(\infty,t) = P(-\infty,t) = 0, \tag{7.10}$$

where

$$\mathbf{Q} = \begin{bmatrix} 0 & 0 & 0 \\ 0 & \sigma^2 & 0 \\ 0 & 0 & 0 \end{bmatrix}.$$

With the knowledge of $\mathbf{f}(\mathbf{x},t)$ and $\mathbf{G}(\mathbf{x},t)$, the FPK equation reduces to

$$\frac{\partial P(\mathbf{x},t)}{\partial t} = -x_2\frac{\partial P(\mathbf{x},t)}{\partial x_1} + 2\zeta\frac{\partial(x_2 P(\mathbf{x},t))}{\partial x_2} + \left(\frac{\mathrm{d}U}{\mathrm{d}x_1} + \kappa^2 x_3\right)\frac{\partial P(\mathbf{x},t)}{\partial x_2}$$
$$+\alpha\frac{\partial(x_3 P(\mathbf{x},t))}{\partial x_3} + x_2\frac{\partial P(\mathbf{x},t)}{\partial x_3} + \frac{\sigma^2}{2}\frac{\partial^2 P(\mathbf{x},t)}{\partial^2 x_2},$$
$$P(\infty,t) = P(-\infty,t) = 0. \tag{7.11}$$

Upon solving Eq. (7.11) for $P(\mathbf{x},t)$, the response statistics can then be obtained via

$$\left\langle \prod_{i=1}^{3} x_i^{k_i} \right\rangle = \int_{-\infty}^{\infty}\int_{-\infty}^{\infty}\int_{-\infty}^{\infty} \prod_{i=1}^{3} x_i^{k_i} P(\mathbf{x},t)\mathrm{d}x_1\mathrm{d}x_2\mathrm{d}x_3, \tag{7.12}$$

where $k_i = 0, 1, 2, \ldots$.

7.4 Response Statistics for Large Time Constant Ratio

We start by considering the response of the harvesters when the time constant ratio, α, is large. We aim to understand how the nonlinearity influences the response of the harvester in such a scenario. From a mathematical point of view, the circuit dynamics, Eq. (7.4), represents a first-order low-pass filter with the velocity being its input; the electric quantity, y, representing its output; and α characterizing the inverse of its time constant. When α is large, the bandwidth of the filter is large, and the circuit dynamics can be approximated via $\alpha y = \dot{x}$. This allows the dynamics of the coupled system, Eq. (7.4), to be reduced to the following form:

$$\ddot{x} + \zeta_{\text{eff}}\dot{x} + \frac{\mathrm{d}U}{\mathrm{d}x} = -\ddot{x}_b, \tag{7.13}$$

where $\zeta_{\text{eff}} = 2\zeta + \kappa^2/\alpha$. For the reduced system, the PDF of the response can be obtained by solving a reduced FPK equation in the form

$$\begin{aligned} \frac{\partial P(\mathbf{x},t)}{\partial t} &= -x_2\frac{\partial P(\mathbf{x},t)}{\partial x_1} + 2\zeta_{\text{eff}}\frac{\partial(x_2 P(\mathbf{x},t))}{\partial x_2} + \left(\frac{\mathrm{d}U}{\mathrm{d}x_1}\right)\frac{\partial P(\mathbf{x},t)}{\partial x_2} + \frac{\sigma^2}{2}\frac{\partial^2 P(\mathbf{x},t)}{\partial^2 x_2}, \\ P(\infty,t) &= P(-\infty,t) = 0, \end{aligned} \tag{7.14}$$

where $(x_1, x_2)^{\mathrm{T}} \equiv (x, \dot{x})^{\mathrm{T}}$. Since steady-state response statistics are of particular relevance for energy harvesting, we focus our attention on obtaining the stationary solutions of Eq. (7.14). In such a scenario, the transition probability function is time invariant, i.e., $\partial P(\mathbf{x},t)/\partial t = 0$ or $P(\mathbf{x},t) = P(\mathbf{x})$, and Eq. (7.14) admits the following stationary solution:

$$P(x_1,x_2) = A_1 \exp\left\{\frac{-2\zeta_{\text{eff}}}{\sigma^2}U(x_1)\right\} \times A_2 \exp\left\{\frac{-2\zeta_{\text{eff}}}{\sigma^2}\frac{x_2^2}{2}\right\}, \tag{7.15}$$

where $A_1^{-1} = \int_{-\infty}^{\infty} \exp\left\{\frac{-2\zeta_{\text{eff}}}{\sigma^2}U(x_1)\right\}\mathrm{d}x_1$ and $A_2^{-1} = \int_{-\infty}^{\infty} \exp\left\{\frac{-2\zeta_{\text{eff}}}{\sigma^2}\frac{x_2^2}{2}\right\}\mathrm{d}x_2$. Note that the resulting PDF can be factored into a function of the displacement, x_1, and velocity, x_2. This implies that the displacement and velocity can be treated as two independent random variables. In such a case, the expected mean square value of the velocity, $\langle x_2^2\rangle$, is independent of the displacement, nonlinearity, and the potential function altogether and is given by

$$\langle \dot{x}^2\rangle \equiv \langle x_2^2\rangle = A_2\int_{-\infty}^{\infty} x_2^2 \exp\left\{\frac{-\zeta_{\text{eff}}}{\sigma^2}x_2^2\right\}\mathrm{d}x_2 = \frac{\sigma^2}{2\zeta_{\text{eff}}}. \tag{7.16}$$

Next, using the relation $y = \dot{x}/\alpha$ in conjunction with Eq. (7.16), the expected mean square value of the electric quantity can be written as

$$\langle y^2\rangle \equiv \langle x_3^2\rangle = \frac{\sigma^2}{2\alpha^2\zeta_{\text{eff}}}. \tag{7.17}$$

Equation (7.17) reveals that the expected value of the electric quantity, voltage in the case of piezoelectric harvesters and current in the case of electromagnetic ones, is independent of the shape of the potential function leading to the conclusion that, under white Gaussian excitations and for large values of the time constant ratio

α, no matter how the potential function of the harvester is altered, it will have no influence on the average output power. This conclusion holds for harvesters with nonlinearities appearing in the restoring force.

7.5 Energy Harvesters with a Linear Restoring Force

Next, we investigate the response statistics for any arbitrary value of α. For the purpose of performance comparison, we start by studying the response of the linear harvester ($\delta = 0, r < 1$). Since for linear systems the response to a Gaussian input is also Gaussian, it is possible to obtain an exact stationary solution of Eq. (7.11) in the general Gaussian form

$$P(x_1, x_2, x_3) = A \exp\left(\sum_{i,j=1}^{3} a_{ij} x_i x_j \right), \tag{7.18}$$

where A is a constant obtained via the following normalization scheme:

$$\int_{-\infty}^{\infty} \int_{-\infty}^{\infty} \int_{-\infty}^{\infty} P(x_1, x_2, x_3) \mathrm{d}x_1 \mathrm{d}x_2 \mathrm{d}x_3 = 1, \tag{7.19}$$

and the a_{ij} are attained by substituting Eq. (7.18) into Eq. (7.11), then forcing the solvability conditions. This yields

$$a_{ij} = -\frac{1}{2} \frac{|R|_{ij}}{|R|}, \tag{7.20}$$

where

$$R = \frac{\sigma^2}{2\zeta(1 - r + \alpha^2 + 2\alpha\zeta) + \kappa^2(\alpha + 2\zeta)} \begin{bmatrix} \frac{1-r+\alpha^2+2\alpha\zeta}{1-r} & 0 & \frac{1}{2} \\ 0 & \frac{1-r+\alpha^2+2\alpha\zeta+\kappa^2}{2} & \frac{\alpha}{2} \\ \frac{1}{2} & \frac{\alpha}{2} & \frac{1}{2} \end{bmatrix}.$$

Here, $|R|$ and $|R|_{ij}$ are, respectively, the determinant and cofactors of R.

With the knowledge of the exact stationary probability function, the required response statistics can now be obtained using Eq. (7.12). Of special importance are the mean square values of the displacement, velocity, and electric quantity which can be expressed as

$$\left\langle x_1^2 \right\rangle = \frac{\sigma^2}{1 - r} \frac{1 - r + \alpha^2 + 2\alpha\zeta}{2\zeta(1 - r + \alpha^2 + 2\alpha\zeta) + \kappa^2(\alpha + 2\zeta)}, \tag{7.21a}$$

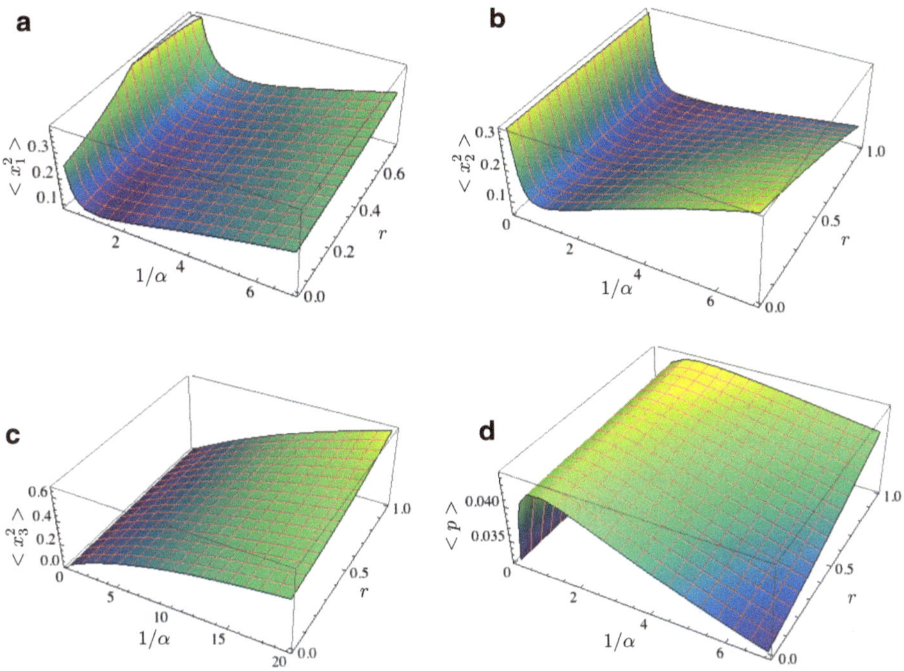

Fig. 7.7 Variation of the mean square values of (**a**) displacement, (**b**) velocity, (**c**) electric quantity, and (**d**) mean power with $1/\alpha$ and r for $\sigma = 0.2$, $\zeta = 0.01$, and $\kappa = 0.65$

$$\langle x_2^2 \rangle = \frac{\sigma^2}{2} \frac{1 - r + \alpha^2 + 2\alpha\zeta}{2\zeta(1 - r + \alpha^2 + 2\alpha\zeta) + \kappa^2(\alpha + 2\zeta)}, \tag{7.21b}$$

$$\langle x_3^2 \rangle = \frac{\sigma^2}{2} \frac{1}{2\zeta(1 - r + \alpha^2 + 2\alpha\zeta) + \kappa^2(\alpha + 2\zeta)}. \tag{7.21c}$$

Using Eq. (7.21c), a nondimensional expression for the average power can also be expressed in the simple form

$$\langle \mathrm{p} \rangle = \frac{\langle \bar{\mathrm{p}} \rangle}{\mathrm{p}_o} = \alpha \langle x_3^2 \rangle, \tag{7.22}$$

where $\mathrm{p}_o = k_1 \omega_n \kappa^2 l_c^2$.

Equations (7.21a)–(7.22) can be utilized to investigate the influence of the time constant ratio, α, and the stiffness tuning constant, r, on the average power as well as the mean square values of the velocity, displacement, and electric quantity. Figure 7.7d demonstrates that, similar to a deterministic excitation, the power exhibits a maximum at an optimal time constant ratio which can be obtained by finding the extrema of Eq. (7.22) with respect to α. This yields

$$\alpha_{\text{opt}} = \sqrt{1 - r + \kappa^2}, \qquad r \leq 1. \tag{7.23}$$

Since white noise has equal intensity in the spectral domain, the optimization results simplify significantly when compared to their deterministic harmonic counterparts as the optimal time constant ratio is only dependent on r and κ. Furthermore, while under deterministic excitations, there exist two optimal excitation frequencies known as the resonance and antiresonance frequency; such optimal frequencies vanish under white noise excitations due to the constant mean square value of the excitation in the frequency domain.[2]

The optimal value of the average power occurs when the mean square value of the velocity exhibits a minimum as depicted in Fig. 7.7b. This can be further confirmed by minimizing Eq. (7.21b) with respect to α which yields the same expression for α_{opt} as given in Eq. (7.23). One plausible explanation is that maximum electric damping occurs at the point where maximum energy is channeled from the mechanical subsystem. This yields a minimum in the mean square value of the velocity. It is worth mentioning that in the limiting case of α approaching infinity, the mean square value of the velocity reduces to $< x_2^2 > = \frac{\sigma^2}{4\zeta}$ which represents the same expression obtained in Eq. (7.16) for a general potential function with $\alpha \gg \kappa^2$. This, again, implies that both of the linear and nonlinear harvesters provide similar mean square velocities and, hence, electric output for large values of α.

It can also be noted that, in general, the output power increases as the stiffness decreases (r increases) for any value of α. When the stiffness decreases, the harvester becomes softer, experiencing larger motions for smaller input excitations. As a result, the mean square value of the displacement increases and approaches infinity as the stiffness approaches zero; see also Fig. 7.7a. The increased mean square value of the displacement aids in increasing the average output power. However, at the same time, the reduction in stiffness reduces the frequency of oscillation, thereby reducing the mean square value of the electric output and, hence, the average power. It turns out that the net increase in the average power due to the increase in the mean square value of the displacement overcomes the decrease due to the reduction in the frequency which causes the average power to increase with r.

7.6 Energy Harvesters with a Nonlinear Restoring Force

When the restoring force has a nonlinear dependence on the displacement, an exact solution of the FPK equation, Eq. (7.11), is not easily attainable even in the stationary sense. The reason being that, even when the input excitation is Gaussian,

[2]The reader can refer to Renno et al. [41] for more details on the optimization of energy harvesters under sinusoidal deterministic excitations.

the response PDF will deviate from the Gaussian distribution depending on the strength of the nonlinearity. To approximate the response statistics in such scenarios, it is common to seek an approximate solution of the FPK equation. One approach is based on discretizing the FPK equation into a set of ordinary differential equations governing the response statistics (moment differential equations). To that end, we multiply both sides of Eq. (7.11) by $\prod_{i=1}^{3} x_i^{k_i}$ and integrate over the whole domain to obtain

$$\begin{aligned}\int_{-\infty}^{\infty}\int_{-\infty}^{\infty}\int_{-\infty}^{\infty}\prod_{i=1}^{3} x_i^{k_i}\frac{\partial P(\mathbf{x},t)}{\partial t}\mathrm{d}\mathbf{x} = &-\int_{-\infty}^{\infty}\int_{-\infty}^{\infty}\int_{-\infty}^{\infty}\prod_{i=1}^{3} x_i^{k_i}\left(x_2\frac{\partial P(\mathbf{x},t)}{\partial x_1}\right)\mathrm{d}\mathbf{x}\\
&+2\zeta\int_{-\infty}^{\infty}\int_{-\infty}^{\infty}\int_{-\infty}^{\infty}\prod_{i=1}^{3} x_i^{k_i}\frac{\partial(x_2P(\mathbf{x},t))}{\partial x_2}\mathrm{d}\mathbf{x}\\
&+\int_{-\infty}^{\infty}\int_{-\infty}^{\infty}\int_{-\infty}^{\infty}\prod_{i=1}^{3} x_i^{k_i}\left(\frac{\mathrm{d}U}{\mathrm{d}x_1}+\kappa^2 x_3\right)\frac{\partial P(\mathbf{x},t)}{\partial x_2}\mathrm{d}\mathbf{x}\\
&+\alpha\int_{-\infty}^{\infty}\int_{-\infty}^{\infty}\int_{-\infty}^{\infty}\prod_{i=1}^{3} x_i^{k_i}\frac{\partial(x_3P(\mathbf{x},t))}{\partial x_3}\mathrm{d}\mathbf{x}\\
&+\int_{-\infty}^{\infty}\int_{-\infty}^{\infty}\int_{-\infty}^{\infty}\prod_{i=1}^{3} x_i^{k_i}\left(x_2\frac{\partial P(\mathbf{x},t)}{\partial x_3}\right)\mathrm{d}\mathbf{x}\\
&+\frac{\sigma^2}{2}\int_{-\infty}^{\infty}\int_{-\infty}^{\infty}\int_{-\infty}^{\infty}\prod_{i=1}^{3} x_i^{k_i}\frac{\partial^2 P(\mathbf{x},t)}{\partial^2 x_2}\mathrm{d}\mathbf{x},\end{aligned} \tag{7.24}$$

where $\mathrm{d}\mathbf{x} \equiv \mathrm{d}x_1\mathrm{d}x_2\mathrm{d}x_3$. Integrating each term in Eq. (7.24) by parts while taking into account the boundary conditions, $P(\infty,t) = P(-\infty,t) = 0$, yields

$$\begin{aligned}\frac{\mathrm{d}m_{k_1,k_2,k_3}}{\mathrm{d}t} = &\, k_1 m_{k_1-1,k_2+1,k_3} + k_3 m_{k_1,k_2+1,k_3-1} - \alpha k_3 m_{k_1,k_2,k_3} - \delta k_2 m_{k_1+3,k_2-1,k_3}\\
&- 2\zeta k_2 m_{k_1,k_2,k_3} - \kappa^2 k_2 m_{k_1,k_2-1,k_3+1} - k_2(1-r)m_{k_1+1,k_2-1,k_3}\\
&+ \frac{1}{2}(k_2-1)k_2\sigma^2 m_{k_1,k_2-2,k_3},\end{aligned} \tag{7.25}$$

where $m_{k_1,k_2,k_3} = \langle\prod_{i=1}^{3} x_i^{k_i}\rangle$ is the statistical moment of order $K = \sum_{i=1}^{3} k_i$. Equations (7.25) which are also known as the moment differential equations can be simultaneously integrated to study time evolution of the response statistics for any set of initial conditions. At steady state, the response statistics are time invariant which permits setting the left-hand side of Eq. (7.25) equal to zero and solving the resulting algebraic equations together for the steady-state response moments. The stability of the resulting solutions can be assessed by finding the eigenvalues of the associated Jacobian matrix.

For a linear system, moments of order lower than two, i.e., $K \leq 2$, are independent of higher-order moments. For nonlinear systems, however, the resulting moment equations are coupled to higher-order moments. In Eq. (7.25), the coupling occurs through the term associated with the nonlinearity δ. This yields an infinite hierarchy problem which should be closed in an appropriate manner so that the approximate response statistics can be calculated. A proper *closure scheme* truncates the problem into a finite set of equations while preserving the statistical moment properties and Schwarz's inequality ($\langle x\rangle^2 \leq \langle x^2\rangle < \infty$) as well as leading solutions that are close to those obtained by numerically solving the FPK equation [42].

Here, we implement a cumulant-neglect closure method [42, 43] to close the moment equations by assuming that the response cumulants above a certain threshold are too small when compared to lower-order cumulants and hence can be neglected.[3] This provides additional equations that relate higher-order moments (above the closure level) to lower-order moments (below the closure level). The response cumulants λ_K of order K can be obtained using the following relation:

$$\lambda_K[x_1^{k_1}x_2^{k_2}x_3^{k_3}] = \frac{1}{\mathrm{i}^K}\frac{\partial^K \ln(F_{\mathbf{x}}(\mathbf{u}))}{\partial u_1^{k_1}\partial u_2^{k_2}\partial u_3^{k_3}}|_{\mathbf{u}=0}, \tag{7.26}$$

where $\mathrm{i}^2 = -1$ and $F_{\mathbf{x}}(\mathbf{u})$ is the characteristic function given by

$$F_{\mathbf{x}}(\mathbf{u}) = \langle \exp(\mathrm{i}u_1x_1 + \mathrm{i}u_2x_2 + \mathrm{i}u_3x_3)\rangle = 1 + \sum_{j=1}^{\infty}\frac{1}{j!}\langle(\mathrm{i}u_1x_1 + \mathrm{i}u_2x_2 + \mathrm{i}u_3x_3)^j\rangle. \tag{7.27}$$

As shown in Fig. 7.8, based on comparisons with numerical simulations, a fourth-order cumulant-neglect closure scheme, also known as a non-Gaussian closure, yields acceptable results that reflect the general trends while balancing the computational cost. In this scheme, cumulants of order five and six (λ_5 and λ_6) are set equal to zero to relate fifth-and sixth-order moments ($m_{k_1,k_2,k_3}, K = 5, 6$) to lower-order moments ($m_{k_1,k_2,k_3}, K \leq 4$). This yields a total of 34 differential moment equations, Eq. (7.25), and a total of 49 algebraic equations from the cumulant-negligence scheme, Eq. (7.26). A higher-order closure scheme has been shown to yield even more accurate results but is, computationally, much more expensive.

To find the stationary response statistics, the differential moment equations are transformed into a set of algebraic equations by setting the time derivatives equal to zero. The resulting 83 coupled algebraic equations are solved together for the stationary response moments. These equations yield a nonunique set of solutions whose stability is assessed using the eigenvalues of the associated Jacobian matrix.

[3]Cumulants are used to provide a measure of correlation strength among different random variables [42].

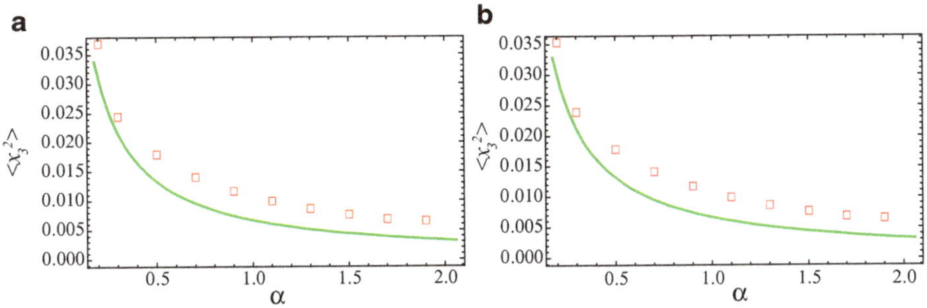

Fig. 7.8 (**a**) Variation of the mean square voltage with α for a monostable system with $r = 0.9$, $\delta = 4$, $\zeta = 0.01$, $\kappa = 0.75$, and $\sigma = 0.09$. (**b**) Variation of the mean square voltage with α for a bistable system with $r = 1.1$, $\delta = 4$, $\zeta = 0.01$, $\kappa = 0.75$, and $\sigma = 0.09$. Squares represent results obtained via numerical integration using Matlab's stochastic communication toolbox

Only one of these solutions yields stable, physically realizable, response moments. This represents the only actual and unique solution of the linear FPK equation. In what follows, we use this solution to investigate the influence of nonlinearity on the output power in nonlinear harvesters.

7.6.1 Nonlinear Harvesters with a Monostable Potential Function

Many of the nonlinear VEHs proposed in the literature exhibit monostable potential characteristics [18–29]. Their nonlinear behavior can exist even when the nonlinearity is not intentionally introduced. For instance, due to the nonlinear strain–displacement relation, the well-known piezoelectric cantilever beam harvester usually exhibits a nonlinear hardening response when undergoing large oscillations. Under similar conditions, the piezoelectric element can also produce softening nonlinearities as a result of its nonlinear constitutive (stress–strain) relation [44,45]. This section serves to delineate the influence of nonlinearity on the performance of nonlinear monostable VEHs under white Gaussian excitations.

Figure 7.9a depicts variation of the mean square value of the electric output with the nonlinearity coefficient δ for different values of r and a small value of the time constant ratio, $\alpha = 0.05$. It is evident that, as the nonlinearity increases, the mean square value of the electric quantity decreases regardless of the value of r. The decrease in the voltage follows a linear trend for the range of the nonlinearity studied. The same trend can also be seen for a larger value of the time constant ratio, $\alpha = 10$, as shown in Fig. 7.9a. Here, however, variation of the mean square value with the nonlinearity is very small to the extent that it can be considered negligible. This corroborates the findings discussed earlier in Sect. 7.4. The decrease in the

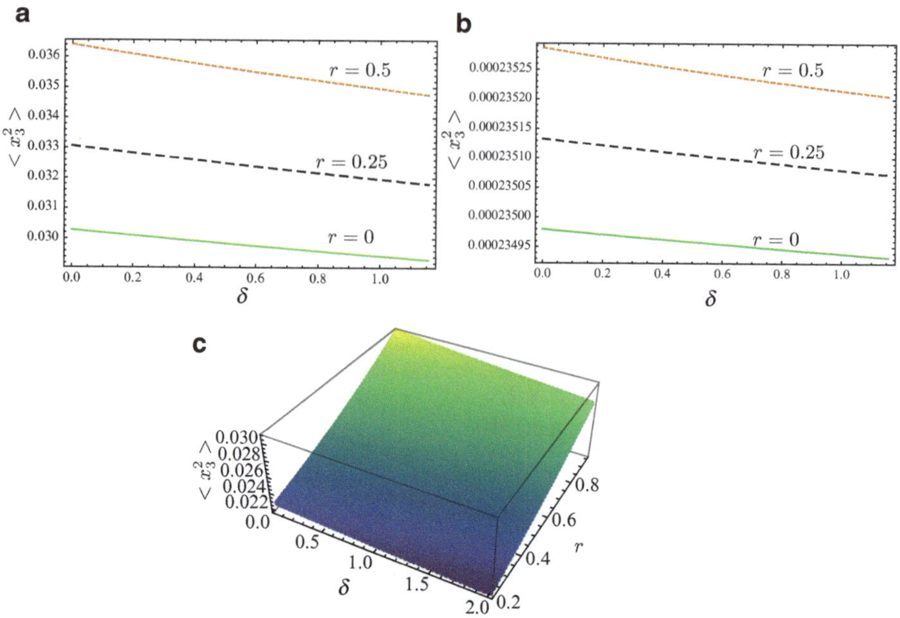

Fig. 7.9 Variation of the mean square value of the electric output with δ for different values of r and (**a**) $\alpha = 0.05$ and (**b**) $\alpha = 10$. (**c**) Surface showing variation of the mean square value of the electric quantity with r and δ for $\alpha = 0.05$. Results are obtained using $\zeta = 0.01$, $\kappa = 0.75$, and $\sigma = 0.06$

mean square of the electric quantity occurs for any value of $r < 1$ as shown in the surface depicted in Fig. 7.9c.

Figure 7.10 illustrates variation of the output power with the time constant ratio for the linear harvester, $\delta = 0$, as compared to the nonlinear monostable harvester with $\delta = 10, 20$. Again, it is evident that the nonlinearity has an adverse influence on the average output power. The reduction in the average power is mostly pronounced for smaller values of α. The two curves converge to one another as α increases, further emphasizing that the influence of the nonlinearity diminishes with α. This clearly demonstrates that intentional inclusion of hardening stiffness-type nonlinearities in monostable VEHs will adversely influence their performance under random excitations that can be approximated as a white Gaussian noise process. The larger the nonlinearity is, the smaller the average power will be. Reduction in the output power is mostly pronounced for smaller values of the time constant ratio.

Further inspection of Fig. 7.10a reveals that the optimal time constant ratio, which is inversely proportional to the load resistance in capacitive harvesters and directly proportional to it in inductive ones, increases with the nonlinearity. The optimal operating load resistance should be chosen such that it shifts the time constant ratio of the VEH away from the linear optimal value and towards larger values. This result can be of importance even for VEHs that do not intentionally

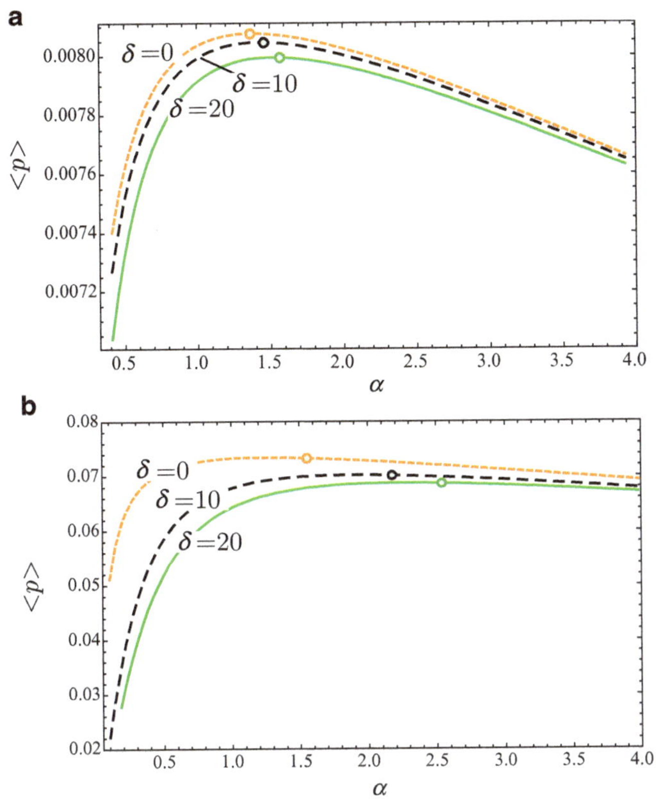

Fig. 7.10 Variation of the average power with α for different values of δ and $r = 0$, $\zeta = 0.01$, and $\kappa = 0.75$. Results are obtained for (**a**) $\sigma = 0.1$ and (**b**) $\sigma = 0.3$. *Circles* represent the optimal value of the time constant ratio, α_{opt}

incorporate nonlinearities as it will help increase the average power under random broadband excitations. Similar trends can be seen in Fig. 7.10b when the variance of the excitation is increased to $\sigma = 0.3$. The main difference is that the optimal time constant shifts even further towards larger values.

7.6.2 Harvesters with a Bistable Potential Function

In this section, we treat VEHs with bistable potentials. The goal is to analyze how the average power of the harvester depends on the potential's shape and excitation's intensity as compared to the linear harvester. Here, again, a fourth-order cumulant-neglect closure scheme is utilized to obtain the response statistics. We start by investigating how the potential shape influences the mean square value of the electric

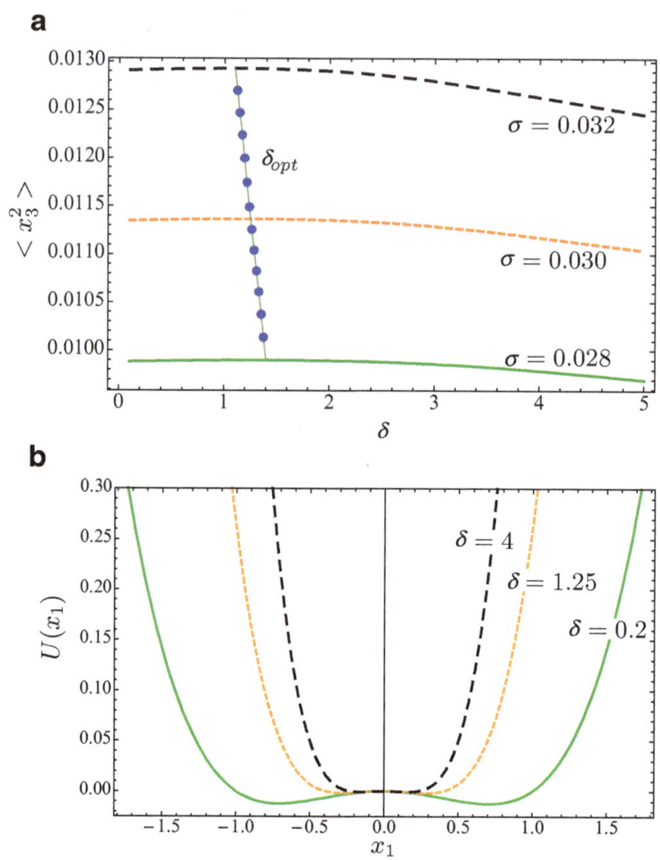

Fig. 7.11 (**a**) Variation of the mean square value of the electric quantity with δ for different values of σ and $\alpha = 0.05$, $\zeta = 0.01$, $\kappa = 0.75$, and $r = 1.1$. (**b**) Associated potential energy functions for different values of δ and $r = 1.1$

output. Figure 7.11a depicts variation of the mean square value of the electric output with δ for $r = 1.1$ and a time constant ratio $\alpha = 0.05$. Results indicate that the electric output increases initially with δ then decreases beyond an optimal value, δ_{opt}, which decreases as the noise intensity is increased. To put these results in a better perspective, we show the shape of the potential function for three different values of δ and $r = 1.1$ as depicted in Fig. 7.11b. As it can be clearly seen, when δ is increased, the separation distance between the wells which is defined by the location of the system's stable fixed points decreases. The height of the potential barrier at the unstable saddle also decreases which creates shallower potential wells that facilitate the transition of dynamic trajectories from one potential well to the other.

Let us consider, for instance, the case when $\delta = 4$. Here, the inter-well dynamics can be easily activated, and the frequency of trajectories' transitions is very high

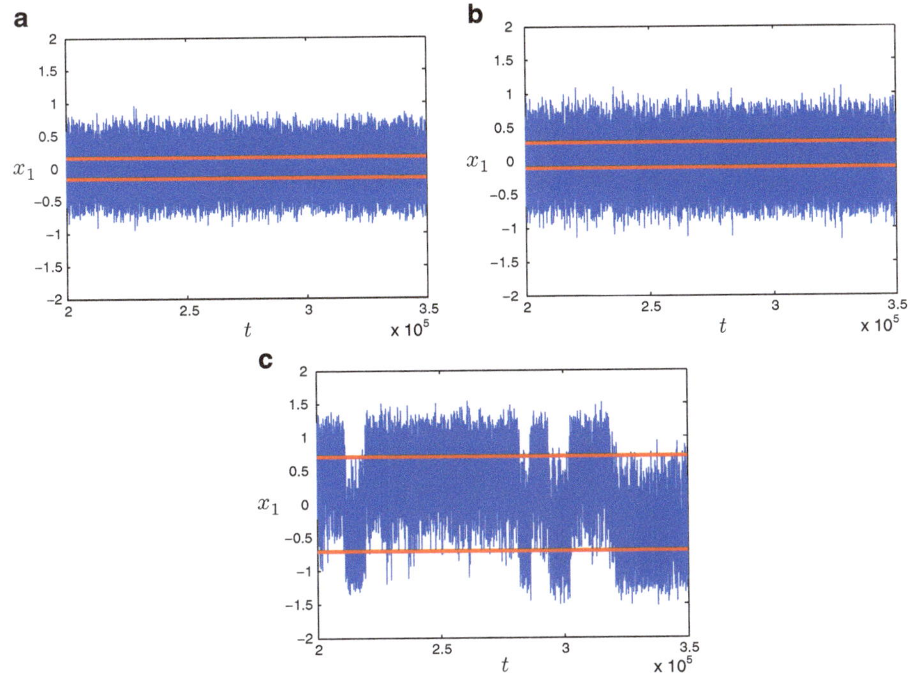

Fig. 7.12 Displacement time histories for (**a**) δ =4, (**b**) δ = 1.25, and (**c**) δ = 0.2. Results are obtained for σ = 0.09, α = 0.05, ζ = 0.01, κ = 0.75, and r = 1.1. *Solid lines* represent stable equilibria

as depicted in the time history shown in Fig. 7.12a. However, since the separation distance between the wells is too small, the mean square value of the electric quantity remains small at about $< x_3^2 >= 0.1046$. When δ is decreased to $\delta =$ 1.25, the inter-well dynamics can still be activated at a high transition frequency, Fig. 7.12b. In this case, however, because the separation distance is larger, each transition event produces a larger spike in the electric output which helps increase the mean square value. As such, the effective mean square value increases to $< x_3^2 >= 0.1134$. When the nonlinearity coefficient is decreased further to $\delta = 0.2$, transition of dynamic trajectories becomes very difficult, confining the dynamics mostly to one potential well (intra-well response) as shown in Fig. 7.12c. This has the adverse influence of reducing the mean square value to $< x_3^2 >= 0.0979$.

Based on the preceding discussion, it is evident that δ_{opt} represents the value of the nonlinearity which creates a potential function that balances the frequency of transition with the separation distance between the wells to maximize the mean square value of the electric output. This optimal value decreases with the noise intensity, because for a larger intensity, the balance will be maintained for a potential function with a higher barrier and larger separation distance between its wells. The

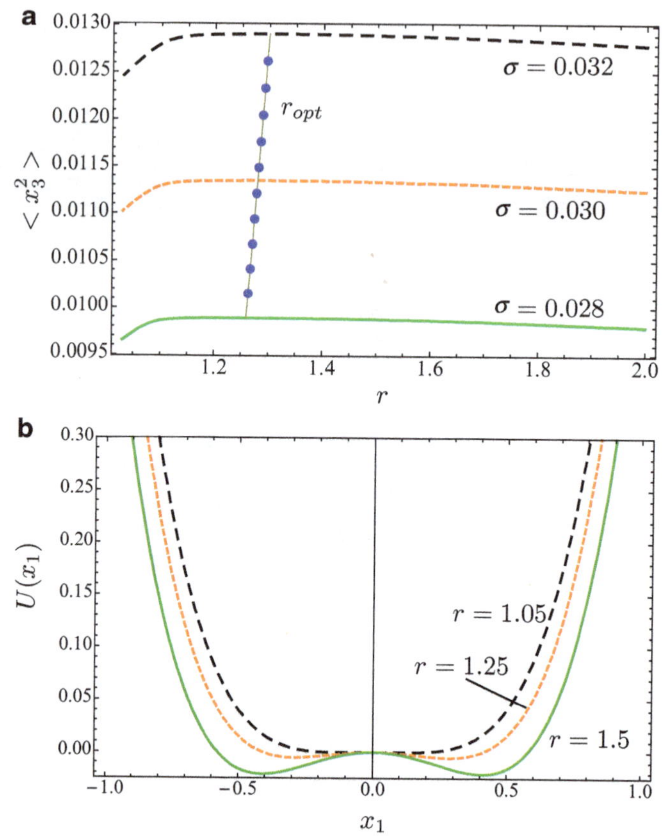

Fig. 7.13 (**a**) Variation of the mean square voltage with r for different values of σ and $\alpha = 0.05$, $\zeta = 0.01$, $\kappa = 0.75$, and $\delta = 3$. (**b**) Associated potential energy functions for different values of r and $\delta = 3$

same conclusions can also be deduced but in the opposite sense as r is increased for a given δ as can be seen in Fig. 7.13a,b.

A clearer picture illustrating the influence of the potential shape on the electric output can be drawn by inspecting the surfaces generated in Fig. 7.14 for increasing noise intensity. It can be noted that for an excitation with a small variance, Fig. 7.14a, the mean square value of the electric output is much more sensitive to variations in r than it is to variations in δ, at least for the range considered. For a given δ, the electric output decreases initially with r exhibiting an optimal value and then decreases again. This optimal r shifts towards smaller values as δ decreases, approaching unity when δ approaches zero. The decrease in the optimal r occurs to balance the increase in the height of the potential barrier and the separation between the potential wells due to the increase in δ. Hypothetically, the maximum mean square value of the electric output occurs when r approaches unity and δ approaches

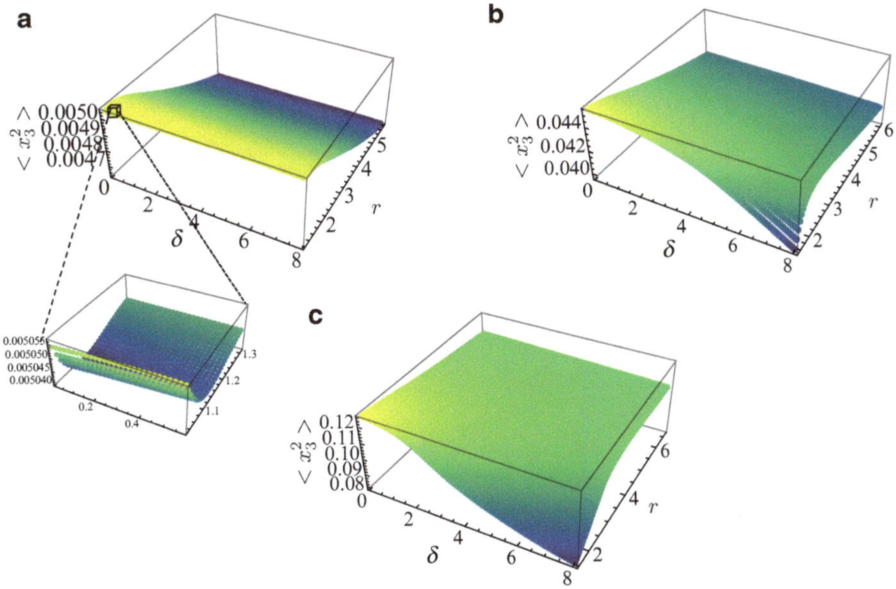

Fig. 7.14 Variation of the mean square voltage with r and δ for $\alpha = 0.05$, $\zeta = 0.01$, $\kappa = 0.75$, and (**a**) $\sigma = 0.02$, (**b**) $\sigma = 0.06$, (**c**) $\sigma = 0.1$

zero, i.e., a harvester without a restoring force. This, however, is not physically realizable because it implies that the mean square value of the displacement would approach infinity.

When the noise intensity is increased to $\sigma = 0.06$ as seen in Fig. 7.14b, the electric output becomes more sensitive to variations in δ. An optimal value of r can now be clearly identified for a given value of δ. The electric output decreases sharply below that optimal value especially for large values of δ. In this scenario, even though the inter-well dynamics can be easily activated, the separation distance between the wells is too small for this high noise intensity which reduces the velocity and the electric output. Beyond the optimal value, the power also decreases but with a smaller slope. Here, the inter-well dynamics of the harvester is activated less frequently, but the separation distance between the wells is large enough to balance this reduction in the frequency. Similar trends can be seen in Fig. 7.14c when the noise intensity is increased even further to $\sigma = 0.1$.

The preceding results emphasize an important conclusion concerning the design of a bistable VEH for a random excitation source. Specifically, the optimal design of the potential shape depends on the excitation intensity. A certain potential shape optimized for a given known noise intensity does not necessarily provide an acceptable performance for other noise intensities. This further complicates the process of designing bistable VEHs with desirable performance characteristics.

Figure 7.15 investigates whether designing a harvester with a bistable potential can enhance the transduction under white Gaussian excitations when compared to

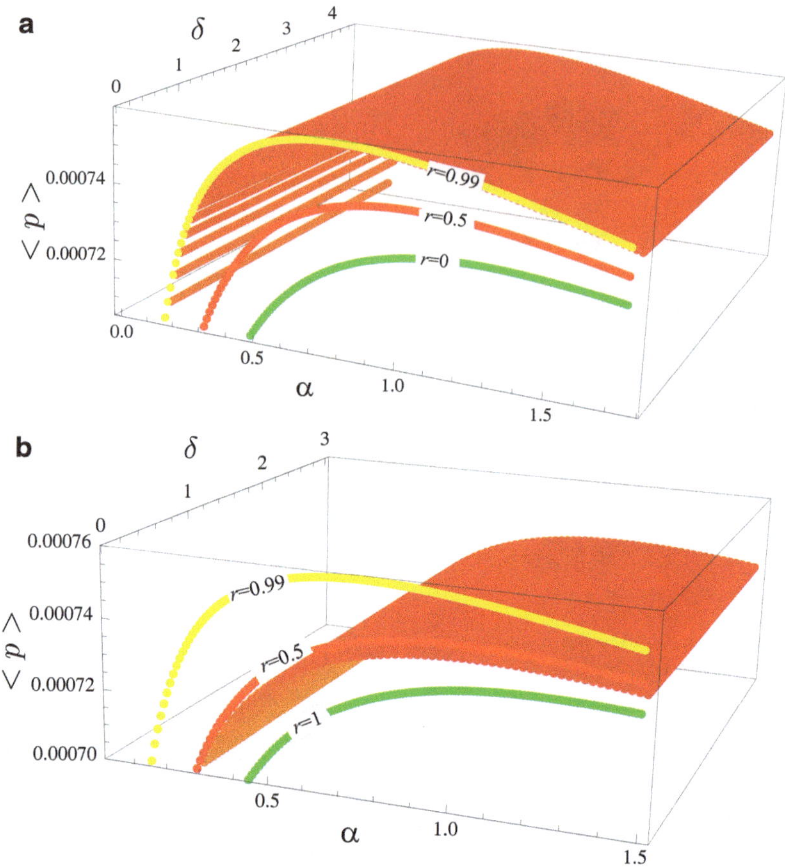

Fig. 7.15 Variation of the average power with α and δ for $\zeta = 0.01$, $\kappa = 0.75$, and $\sigma = 0.06$. (**a**) $r = 1.4$ and (**b**) $r = 8$

a linear harvester. The figure depicts variation of the average power with the time constant ratio, α, and the nonlinearity coefficient, δ. An r value of 1.4 was chosen to generate the surface based on the results shown in Fig. 7.14b which demonstrate that this value of r produces maximum power for most values of δ. As expected, the average power exhibits a maximum value for a given optimal α. Quite interestingly, the nonlinearity coefficient seems to have a very little influence on the average power for this value of r. The surface of average power is compared to average power curves obtained using a linear harvester with three different values of r, namely, $r = 0, 0.5$, and 0.99. It is evident that the bistable harvester with the optimal design parameters significantly outperforms the linear harvester for $r = 0$ and $r = 0.5$. However, when $r = 0.99$, the linear harvester provides slightly more power than the bistable one. It should be borne in mind, however, that designing a harvester very close to a linear instability, i.e., $r \to 1$ and $\delta \to 0$, is very hard to achieve in an experimental setting using tuning mechanisms similar to the ones discussed in the introduction. Furthermore, when r approaches one without the presence of amplitude-limiting nonlinearities, the amplitude of the response unrealistically approaches infinity.

While Fig. 7.15a demonstrates that an energy harvester with a bistable potential can outperform a linear harvester, this does not necessarily occur for any arbitrary potential shape. To further confirm this conclusion, Fig. 7.15b depicts variation of the average power with the time constant ratio, α, and the nonlinearity coefficient, δ, for $r = 8$, which is far from the optimal value. Here, we can clearly see that the average output power is reduced significantly and a linear harvester with values of r ranging between 0.5 and 1 can clearly outperform the bistable design. As such, extreme care should be taken when designing harvesters with bistable potential for excitation sources that exhibit the characteristic of white Gaussian noise.

7.7 Conclusions

This chapter studied the influence of stiffness-type nonlinearities on the transduction of VEHs under white Gaussian excitations. The analysis and simulation results culminated in the following conclusions:

- A nonlinear monostable Duffing-type harvester with hardening nonlinearity can never outperform its linear counterpart regardless of the linear stiffness or noise intensity. The larger the coefficient of the cubic nonlinearity is, the lower is the average output power.
- The time constant ratio, i.e., ratio between the period of the mechanical subsystem and the time constant of the harvesting circuit, plays a critical role in characterizing the influence of stiffness nonlinearities on the average power:
  - When the time constant ratio is large, the influence of the nonlinearity diminishes, and both a linear and a nonlinear harvester provide the same value of average power regardless of the noise intensity. In such a case, no matter how the potential function of the harvester is altered, it has no influence on the average output power of the device. This conclusion holds for any harvester with nonlinearities appearing in the restoring force.
  - When the time constant ratio is small, a monostable Duffing-type harvester always produces less power than the linear one. A bistable Duffing harvester with hardening nonlinearity can, for some optimal potential shapes, outperform the linear harvester. The potential function's optimal shape balances the transitional frequency of the dynamic trajectories between its wells and their separation distance to maximize the mean square value of the electric quantity. The shape of the optimal potential function is sensitive to the noise intensity further complicating the design of efficient bistable VEHs for random excitations with unknown or variable noise intensities.

Future research efforts should investigate other types of nonlinearities, e.g., nonlinearities in the harvesting circuit dynamics or the damping mechanisms to enhance the transduction of energy harvesters under random excitations that can be approximated by a white Gaussian noise process.

Acknowledgements This material is based upon work supported by the National Science Foundation under CAREER Grant No. 1055419. Any opinions, findings, and conclusions or recommendations expressed in this material are those of the author and do not necessarily reflect the views of the National Science Foundation.

References

1. Gregori S, Li Y, Li H, Liu J, Maloberti F (2004) 2.45 GHz power and data transmission for a low-power autonomous sensors platform. ISLPED 04, pp 269–273
2. Kim JW, Takao H, Sawada K, Ishida M (2007) Integrated inductors for RF transmitters in CMOS/MEMS smart microsensor systems. Sensors 7:1387–1398
3. Bracke W, Merken P, Puers R, Van Hoof C (2007) Generic architectures and design methods for autonomous sensors. Sensor Actuator A 135:881–888
4. Baerta K, Gyselinckxa B, Torfsa T, Leonova V, Yazicioglua F, Brebelsa S, Donnaya S, Vanfleterena J, Beyna E, Van Hoof C (2006) Technologies for highly miniaturized autonomous sensor networks. Microelectron J 37:1563–1568
5. Paradiso JA, Starner T (2005) Energy scavenging for mobile and wireless electronics. IEEE Pervasive Comput 4:18–27
6. Sodano H, Inman DJ, Park G (2004) A review of power harvesting from vibration using piezoelectric materials. Shock Vib Digest 36:197–205
7. Sodano H, Inman DJ, Park G (2005) Generation and storage of electricity from power harvesting devices. J Intell Mater Syst Struct 16:67–75
8. Roundy S (2005) On the effectiveness of vibration-based energy harvesting. J Intell Mater Struct 16:809–823
9. Yu P, Yuan Y, Xu J (2002) Study of double hopf bifurcation and chaos for an oscillator with time-delayed feedback. Comm Nonlinear Sci Numer Simulat 7:69
10. Chau HL, Wise KD (1987) Noise due to Brownian motion in ultrasensitive solid-state pressure sensors. IEEE Trans Electron Dev 34
11. Roundy S, Zhang Y (2005) Toward self-tuning adaptive vibration-based micro-generators. In: Smart materials, nano- and micro-smart systems, Sydney, Australia
12. Wu W, Chen Y, Lee B, He J, Peng Y (2006) Tunable resonant frequency power harvesting devices. In: Proceedings of smart structures and materials conference, SPIE, p 61690A, San Diego, CA
13. Challa V, Prasad M, Shi Y, Fisher F (2008) A vibration energy harvesting device with bidirectional resonance frequency tunability. Smart Mater Struct 75:1–10
14. Shahruz SM (2006) Design of mechanical band-pass filters for energy scavenging. J Sound Vib 292:987–998
15. Shahruz SM (2006) Limits of performance of mechanical band-pass filters used in energy harvesting. J Sound Vib 294:449–461
16. Baker J, Roundy S, Wright P (2005) Alternative geometries for increasing power density in vibration energy scavenging for wireless sensors. In: Proceedings of the third international energy conversion conference, p 959–970, San Francisco, CA
17. Rastegar J, Pereira C, Nguyen HL (2006) Piezoelectric-based power sources for harvesting energy from platforms with low frequency vibrations. In: Proceedings of smart structures and materials conference, SPIE, p 617101, San Diego, CA

18. McInnes CR, Gorman DG, Cartmell MP (2008) Enhanced vibrational energy harvesting using nonlinear stochastic resonance. J Sound Vib 318:655–662
19. Barton D, Burrow S, Clare L (2010) Energy harvesting from vibrations with a nonlinear oscillator. J Vib Acoust 132:0210091
20. Mann B, Sims N (2008) Energy harvesting from the nonlinear oscillations of magnetic levitation. J Sound Vib 319:515–530
21. Masana R, Daqaq MF (2011) Electromechanical modeling and nonlinear analysis of axially-loaded energy harvesters. J Vib Acoust 133:011007
22. Quinn D, Triplett L, Vakakis D, Bergman L (2011) Comparing linear and essentially nonlinear vibration-based energy harvesting. J Vib Acoust 133:011001
23. Erturk A, Hoffman J, Inman DJ (2009) A piezo-magneto-elastic structure for broadband vibration energy harvesting. Appl Phys Lett 94:254102
24. Cottone F, Vocca H, Gammaitoni L (2009) Nonlinear energy harvesting. Phys Rev Lett 102:080601–1–080601–4
25. Daqaq MF, Stabler C, Seuaciuc-Osorio T, Qaroush Y (2009) Investigation of power harvesting via parametric excitations. J Intell Mater Syst Struct 20:545–557
26. Stanton SC, McGehee CC, Mann BP (2010) Nonlinear dynamics for broadband energy harvesting: investigation of a bistable piezoelectric inertial generator. Phys D Nonlinear Phenom 239:640–653
27. Daqaq MF, Bode D (2010) Exploring the parametric amplification phenomenon for energy harvesting. J Syst Contr Eng 225:456–466
28. Abdelkefi A, Nayfeh AH, Hajj M (2012) Global nonlinear distributed-parameter model of parametrically excited piezoelectric energy harvesters. Nonlinear Dynam 67(2):1147–1160
29. Abdelkefi A, Nayfeh AH, Hajj M (2012) Effects of nonlinear piezoelectric coupling on energy harvesters under direct excitation. Nonlinear Dynam 67(2):1221–1232
30. Mann BP, Owens BA (2010) Investigations of a nonlinear energy harvester with a bistable potential well. J Sound Vib 329:1215–1226
31. Masana R, Daqaq MF (2011) Comparing the performance of a nonlinear energy harvester in mono- and bi-stable potentials. In: Proceedings of the ASME 2011 international design engineering technical conference and computers and information in engineering conference, Washington, DC
32. Adhikari S, Friswell MI, Inman DJ (2009) Piezoelectric energy harvesting from broadband random vibrations. Smart Mater Struct 18:115005
33. Seuaciuc-Osorio T, Daqaq MF (2010) Energy harvesting under excitations of time-varying frequency. J Sound Vib 329:2497–2515
34. Barton D, Burrow S, Clare L (2009) Energy harvesting from vibrations with a nonlinear oscillator. In: Proceedings of the ASME 2009 international design engineering technical conference and computers and information in engineering conference, San Diego, CA
35. Daqaq MF (2010) Response of uni-modal duffing type harvesters to random forced excitations. J Sound Vib 329:3621–3631
36. Gammaitoni L, Neri I, Vocca H (2009) Nonlinear oscillators for vibration energy harvesting. Appl PhysLett 94(16):164102
37. Daqaq MF (2011) Transduction of a bistable inductive generator driven by white and exponentially correlated gaussian noise. J Sound Vib 330:2554–2564
38. Daqaq MF (2012) On intentional introduction of stiffness nonlinearities for energy harvesting under White Gaussian excitations. Nonlinear Dynam. DOI: 10.1007/s11071-012-0327-0
39. Ito K (1944) Stochastic integral. Proc Imper Acad Tokyo 20:519–524
40. Jazwinski AH (1970) Stochastic processes and filtering theory. Academic, New York
41. Renno J, Daqaq MF, Inman DJ (2009) On the optimal energy harvesting from a vibration source. J Sound Vib 320:386–405
42. Ibrahim RA (1985) Parametric random vibrations. Research Studies Press, NY
43. Wojtkiewicz S, Spencer B, Bergman LA (1995) On the cumulant-neglect closure method in stochastic dynamics. Int J Non Lin Mech 95:657–684

44. Triplett A, Quinn D (2009) The effect of non-linear piezoelectric coupling on vibration-based energy harvesting. J Intell Mater Syst Struct 20(16):1959–1967
45. Mahmoodi N, Daqaq MF, Jalili N (2009) Modeling, nonlinear dynamics, and identification of a piezoelectrically actuated microcantilever sensor. IEEE/ASME Trans Mechatron 13(1):58–65

Chapter 8
Random Excitation of Bistable Harvesters

Sondipon Adhikari and Michael I. Friswell

Abstract This chapter considers nonlinear piezoelastic energy harvesters driven by stationary random noise. A range of devices that exhibit nonlinear dynamics have been proposed, and their response to sinusoidal excitation is often complex, with coexisting periodic solutions or even chaotic solutions. The response of nonlinear harvesters to random noise depends on the statistics of the excitation; the maximum response can occur at particular excitation variances, and this is called stochastic resonance. The stochastic linearisation method is proposed for the optimal design of bistable harvesters subjected to random excitation.

8.1 Introduction

The previous chapters have introduced energy harvesting using linear systems, for both harmonic and random base excitation. This chapter considers nonlinear piezoelastic energy harvesters driven by stationary random noise. Williams and Yates [46] described three basic vibration-to-electric energy conversion mechanisms: electromagnetic [3, 5, 8, 46], electrostatic [7, 24] and piezoelectric [1, 13, 36, 43]. In the last decade, these transduction mechanisms have been investigated by numerous researchers for vibration-based energy harvesting, and extensive discussions can be found in the existing review articles [4, 7, 36, 44]. Regardless of the transduction mechanism, a primary issue in vibration-based energy harvesting is that the best performance of a generator is usually limited to excitation at its fundamental resonance frequency. If the applied ambient vibration deviates slightly from the resonance condition, then the power output is drastically reduced. Hence a major issue in energy harvesting is the concept of broadband energy harvesters [19, 30].

S. Adhikari (✉) • M.I. Friswell
College of Engineering, Swansea University, Swansea SA2 8PP, UK
e-mail: s.adhikari@swansea.ac.uk; m.i.friswell@swansea.ac.uk

N. Elvin and A. Erturk (eds.), *Advances in Energy Harvesting Methods*,
DOI 10.1007/978-1-4614-5705-3_8,

A range of devices that exhibit nonlinear dynamics have been proposed. The key aspect of the nonlinear harvesters is the use of a double well potential function, so that the device will have two equilibrium positions [10, 28, 37, 38]. Masana and Daqaq [31] highlighted the advantages of a double well potential for energy harvesting, particularly when inter-well dynamics were excited. The simplest equation of motion with a double well potential is the well-known Duffing oscillator, which has been extensively studied, particularly for sinusoidal excitation. The dynamics is often complex, sometimes with coexisting periodic solutions and sometimes exhibiting a chaotic response. The Duffing oscillator model has been used for many energy harvesting simulations, with the addition of electromechanical coupling for the harvesting circuit. One popular implementation of such a potential well is a piezomagnetoelastic system based on the magnetoelastic structure that was first investigated by Moon and Holmes [33] as a mechanical structure that exhibits strange attractor motions. Erturk et al. [19] investigated the potential of this device for energy harvesting when the excitation is harmonic and demonstrated an order of magnitude larger power output over the linear system (without magnets) for nonresonant excitation. One problem with multiple solutions to harmonic excitation is that the system can respond in the low amplitude solution; Sebald et al. [42] proposed a method to excite the system to jump to the high amplitude solution at low energy cost. Stanton et al. [45] and Erturk and Inman [18] investigated the dynamic response, including the chaotic response, for such a system. Cottone et al. [10] used an inverted beam with magnets and also considered random excitation. Mann and Sims [29] and Barton et al. [6] used an electromagnetic harvester with a cubic force nonlinearity.

The literature on the random excitation of nonlinear harvesters is somewhat sparse. McInnes et al. [32] considered the mechanical energy available in a bistable harvester and introduced the notion of stochastic resonance [21] to the harvesting community. Gammaitoni et al. [22, 23] considered the random excitation of a double well and a single well potential, respectively. Litak et al. [27] and Ferrari et al. [20] considered the performance of nonlinear piezomagnetoelastic systems to random excitations by numerical integration and highlighted the stochastic resonance for these systems. Daqaq [11] solved the Fokker–Plank–Kolmogorov equation to obtain the statistics of the response of a harvester with a cubic stiffness nonlinearity. Ali et al. [2] investigated the stochastic linearisation approach to predict the performance of piezomagnetoelastic energy harvesters.

In this chapter we give an account of recent methods developed for nonlinear energy harvesters subjected to random excitations. Throughout this chapter the mechanical system is modelled using a single degree of freedom equation of motion and may be used for a range of harvesters based on different physical stiffness properties. In Sect. 8.2 we briefly review the response of linear harvesters subjected to stationary Gaussian excitation. The response of bistable harvesters to random excitation is discussed in Sect. 8.3 using direct simulation. In Sect. 8.4 a stochastic linearisation method is proposed for optimal design of bistable harvesters subjected to stationary random excitation.

8.2 The Response of Linear Harvesters to Random Excitation

8.2.1 Single Degree of Freedom Electromechanical Model

We consider stack-type piezoelectric harvesting as shown in Fig. 8.1. We have considered two types, namely, without and with an inductor as shown in Fig. 8.1a,b, respectively. Energy is harvested through base excitations, and the piezoceramic is operated in the {33} direction. Here we use a simple single degree of freedom model for the mechanical motion of the harvester, and the same equations may also be used to describe the dynamics of piezoelectric harvesters using beam bending. A more detailed model, along with correction factors for a single degree of freedom model that accounts for distributed mass effects, was given by Erturk and Inman [14, 15, 16, 17]. This enables the analysis described here to be used in a wide range of practical applications, providing that the broadband base acceleration does not excite high vibration modes of the harvester. The single degree of freedom model could be extended to multi degree of freedom mechanical systems by using a modal decomposition of the response. This chapter only considers a linear model of the piezoelectric material, which allows the application of linear random vibration theory. The relaxation of the linearity assumption would require the use of nonlinear random vibration theory which is considered later.

8.2.1.1 Circuit Without an Inductor

duToit and Wardle [12] expressed the coupled electromechanical behavior by the linear ordinary differential equations:

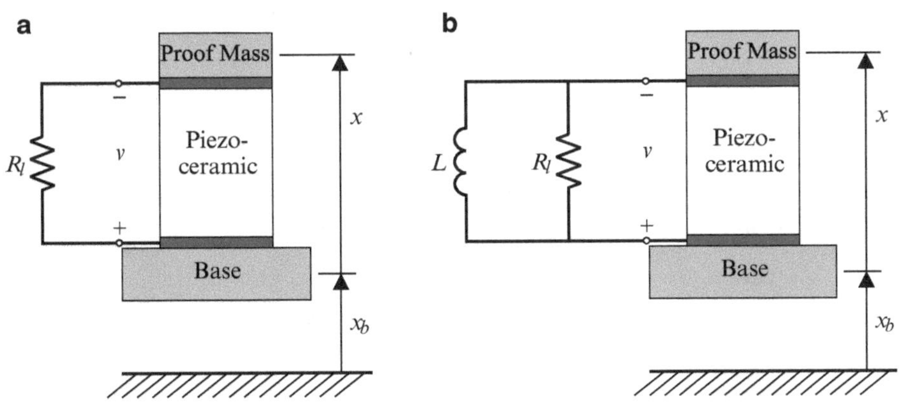

Fig. 8.1 Schematic diagrams of piezoelectric energy harvesters with two different harvesting circuits. (**a**) Harvesting circuit without an inductor, (**b**) Harvesting circuit with an inductor

$$m\ddot{x}(t) + c\dot{x}(t) + kx(t) - \theta v(t) = -m\ddot{x}_b(t) \tag{8.1}$$

$$\theta\dot{x}(t) + C_p\dot{v}(t) + \frac{1}{R_l}v(t) = 0. \tag{8.2}$$

Equation (8.1) is simply Newton's equation of motion for a single degree of freedom system, where t is the time; $x(t)$ is the displacement of the mass; m, c and k are respectively the mass, damping and stiffness of the harvester; and $x_b(t)$ is the random base excitation. In this chapter we consider the base excitation to be a random process. θ is the electromechanical coupling, and the mechanical force is modelled as proportional to the voltage across the piezoceramic, $v(t)$. Equation (8.2) is obtained from the electrical circuit, where the voltage across the load resistance arises from the mechanical strain through the electromechanical coupling, θ, and the capacitance of the piezoceramic, C_p. Transforming both the equations into the frequency domain and dividing the first equation by m and the second equation by C_p, we obtain

$$\left(-\omega^2 + 2\mathrm{i}\omega\zeta\omega_n + \omega_n^2\right)X(\omega) - \frac{\theta}{m}V(\omega) = \omega^2 X_b(\omega) \tag{8.3}$$

$$\mathrm{i}\omega\frac{\theta}{C_p}X(\omega) + \left(\mathrm{i}\omega + \frac{1}{C_pR_l}\right)V(\omega) = 0. \tag{8.4}$$

Here $X(\omega)$, $V(\omega)$ and $X_b(\omega)$ are respectively the Fourier transforms of $x(t)$, $v(t)$ and $x_b(t)$. The natural frequency of the harvester, ω_n, and the damping factor, ζ, are defined as

$$\omega_n = \sqrt{\frac{k}{m}} \quad \text{and} \quad \zeta = \frac{c}{2m\omega_n}. \tag{8.5}$$

Dividing the preceding equations by ω_n and writing in matrix form, one has

$$\begin{bmatrix} (1-\Omega^2) + 2\mathrm{i}\Omega\zeta & -\frac{\theta}{k} \\ \mathrm{i}\Omega\frac{\alpha\theta}{C_p} & (\mathrm{i}\Omega\alpha + 1) \end{bmatrix} \begin{Bmatrix} X \\ V \end{Bmatrix} = \begin{Bmatrix} \Omega^2 X_b \\ 0 \end{Bmatrix}, \tag{8.6}$$

where the dimensionless frequency and dimensionless time constant are defined as

$$\Omega = \frac{\omega}{\omega_n} \quad \text{and} \quad \alpha = \omega_n C_p R_l. \tag{8.7}$$

α is the time constant of the first-order electrical system, nondimensionalised using the natural frequency of the mechanical system. Inverting the coefficient matrix, the displacement and voltage in the frequency domain can be obtained as

$$\begin{Bmatrix} X \\ V \end{Bmatrix} = \frac{1}{\Delta_1} \begin{bmatrix} (\mathrm{i}\Omega\alpha + 1) & \frac{\theta}{k} \\ -\mathrm{i}\Omega\frac{\alpha\theta}{C_\mathrm{p}} & \left(1 - \Omega^2\right) + 2\mathrm{i}\Omega\zeta \end{bmatrix} \begin{Bmatrix} \Omega^2 X_b \\ 0 \end{Bmatrix} = \begin{Bmatrix} (\mathrm{i}\Omega\alpha + 1)\,\Omega^2 X_b/\Delta_1 \\ -\mathrm{i}\Omega^3\frac{\alpha\theta}{C_\mathrm{p}} X_b/\Delta_1 \end{Bmatrix}, \tag{8.8}$$

where the determinant of the coefficient matrix is

$$\Delta_1(\mathrm{i}\Omega) = (\mathrm{i}\Omega)^3\alpha + (2\,\zeta\,\alpha + 1)\,(\mathrm{i}\Omega)^2 + \left(\alpha + \kappa^2\alpha + 2\,\zeta\right)(\mathrm{i}\Omega) + 1 \tag{8.9}$$

and the nondimensional electromechanical coupling coefficient is

$$\kappa^2 = \frac{\theta^2}{kC_\mathrm{p}}. \tag{8.10}$$

8.2.1.2 Circuit with an Inductor

For this case, following [39], the electrical equation becomes

$$\theta\ddot{x}(t) + C_\mathrm{p}\ddot{v}(t) + \frac{1}{R_l}\dot{v}(t) + \frac{1}{L}v(t) = 0 \tag{8.11}$$

where L is the inductance of the circuit. Transforming Equation (8.11) into the frequency domain and dividing by $C_\mathrm{p}\omega_n^2$, one has

$$-\,\Omega^2\frac{\theta}{C_\mathrm{p}}X + \left(-\Omega^2 + \mathrm{i}\Omega\frac{1}{\alpha} + \frac{1}{\beta}\right)V = 0 \tag{8.12}$$

where the second dimensionless constant is defined as

$$\beta = \omega_n^2 L C_\mathrm{p} \tag{8.13}$$

and is the ratio of the mechanical to electrical natural frequencies. Similar to Eq. (8.6), this equation can be written in matrix form with the equation of motion of the mechanical system (8.3) as

$$\begin{bmatrix} \left(1 - \Omega^2\right) + 2\mathrm{i}\Omega\zeta & -\frac{\theta}{k} \\ -\Omega^2\frac{\alpha\beta\theta}{C_\mathrm{p}} & \alpha\left(1 - \beta\Omega^2\right) + \mathrm{i}\Omega\beta \end{bmatrix} \begin{Bmatrix} X \\ V \end{Bmatrix} = \begin{Bmatrix} \Omega^2 X_b \\ 0 \end{Bmatrix}. \tag{8.14}$$

Inverting the coefficient matrix, the displacement and voltage in the frequency domain can be obtained as

$$\begin{Bmatrix} X \\ V \end{Bmatrix} = \frac{1}{\Delta_2} \begin{bmatrix} \alpha\left(1 - \beta\Omega^2\right) + \mathrm{i}\Omega\beta & \frac{\theta}{k} \\ \Omega^2 \frac{\alpha\beta\theta}{C_\mathrm{p}} & \left(1 - \Omega^2\right) + 2\mathrm{i}\Omega\zeta \end{bmatrix} \begin{Bmatrix} \Omega^2 X_b \\ 0 \end{Bmatrix}$$

$$= \begin{Bmatrix} \left(\alpha\left(1 - \beta\Omega^2\right) + \mathrm{i}\Omega\beta\right)\Omega^2 X_b/\Delta_2 \\ \Omega^4 \frac{\alpha\beta\theta}{C_\mathrm{p}} X_b/\Delta_2 \end{Bmatrix} \tag{8.15}$$

where the determinant of the coefficient matrix is

$$\Delta_2(\mathrm{i}\Omega) = (\mathrm{i}\Omega)^4 \beta\,\alpha + (2\,\zeta\,\beta\,\alpha + \beta)\,(\mathrm{i}\Omega)^3 + (\beta\,\alpha + \alpha + 2\,\zeta\,\beta + \kappa^2\beta\,\alpha)(\mathrm{i}\Omega)^2$$
$$+ (\beta + 2\,\zeta\,\alpha)\,(\mathrm{i}\Omega) + \alpha. \tag{8.16}$$

8.2.2 A Brief Overview of Stationary Random Vibration

We consider that the base excitation $x_b(t)$ is a random process. It is assumed that $x_b(t)$ is a weakly stationary, Gaussian, broadband random process. Mechanical systems driven by this type of excitation have been discussed by Bolotin [9], Lin [26], Nigam [35], Roberts and Spanos [40] and Newland [34] within the scope of random vibration theory. To obtain the samples of the random response quantities such as the displacement of the mass $x(t)$ and the voltage $v(t)$, one needs to solve the coupled stochastic differential equations (8.1) and (8.2) or (8.1) and (8.11). However, analytical results developed within the theory of random vibration allows us to bypass numerical solutions because we are interested in the average values of the output random processes. Here we extend the available results to the energy harvester. Since $x_b(t)$ is a weakly stationary random process, its autocorrelation function depends only on the difference in the time instants, and thus

$$\mathrm{E}\left[x_b(\tau_1)x_b(\tau_2)\right] = R_{x_b x_b}(\tau_1 - \tau_2). \tag{8.17}$$

This autocorrelation function can be expressed as the inverse Fourier transform of the spectral density $\Phi_{x_b x_b}(\omega)$ as

$$R_{x_b x_b}(\tau_1 - \tau_2) = \int_{-\infty}^{\infty} \Phi_{x_b x_b}(\omega) \exp[\mathrm{i}\omega(\tau_1 - \tau_2)]\mathrm{d}\omega. \tag{8.18}$$

We are interested in the average harvested power given by

$$\mathrm{E}\left[P(t)\right] = \mathrm{E}\left[\frac{v^2(t)}{R_l}\right] = \frac{\mathrm{E}\left[v^2(t)\right]}{R_l}. \tag{8.19}$$

For a damped linear system of the form $V(\omega) = H(\omega)X_b(\omega)$, it can be shown that [26, 35] the spectral density of V is related to the spectral density of X_b by

$$\Phi_{VV}(\omega) = |H(\omega)|^2 \Phi_{x_b x_b}(\omega). \tag{8.20}$$

Thus, for large t, we obtain

$$\mathrm{E}\left[v^2(t)\right] = R_{vv}(0) = \int_{-\infty}^{\infty} |H(\omega)|^2 \Phi_{x_b x_b}(\omega)\, \mathrm{d}\omega. \tag{8.21}$$

This expression will be used to obtain the average power for the two cases considered. We assume that the base acceleration $\ddot{x}_b(t)$ is Gaussian white noise so that its spectral density is constant with respect to frequency.

The calculation of the integral on the right-hand side of Eq. (8.21) in general requires the calculation of integrals of the following form:

$$I_n = \int_{-\infty}^{\infty} \frac{\Xi_n(\omega)\, \mathrm{d}\omega}{\Lambda_n(\omega)\Lambda_n^*(\omega)} \tag{8.22}$$

where the polynomials have the form

$$\Xi_n(\omega) = q_{n-1}\omega^{2n-2} + q_{n-2}\omega^{2n-4} + \cdots + q_0, \tag{8.23}$$

$$\Lambda_n(\omega) = p_n(\mathrm{i}\omega)^n + p_{n-1}(\mathrm{i}\omega)^{n-1} + \cdots + p_0. \tag{8.24}$$

Following [40], this integral can be evaluated as

$$I_n = \frac{\pi}{p_n} \frac{\det[\mathbf{D}_n]}{\det[\mathbf{N}_n]}. \tag{8.25}$$

Here the $m \times m$ matrices are defined as

$$\mathbf{D}_n = \begin{bmatrix} q_{n-1} & q_{n-2} & & \cdots & & & q_0 \\ -p_n & p_{n-2} & -p_{n-4} & p_{n-6} & \cdots & 0 & \cdots \\ 0 & -p_{n-1} & p_{n-3} & -p_{n-5} & \cdots & 0 & \cdots \\ 0 & p_n & -p_{n-2} & p_{n-4} & \cdots & 0 & \cdots \\ 0 & \cdots & & & \cdots & 0 & \cdots \\ 0 & 0 & & & \cdots & -p_2 & p_0 \end{bmatrix} \tag{8.26}$$

and

$$\mathbf{N}_n = \begin{bmatrix} p_{n-1} & -p_{n-3} & p_{n-5} & -p_{n-7} & & & \\ -p_n & p_{n-2} & -p_{n-4} & p_{n-6} & \cdots & 0 & \cdots \\ 0 & -p_{n-1} & p_{n-3} & -p_{n-5} & \cdots & 0 & \cdots \\ 0 & p_n & -p_{n-2} & p_{n-4} & \cdots & 0 & \cdots \\ 0 & \cdots & & & \cdots & 0 & \cdots \\ 0 & 0 & & & \cdots & -p_2 & p_0 \end{bmatrix}. \tag{8.27}$$

8.2.3 Mean Power for Systems Without an Inductor

From Eq. (8.8) we obtain the voltage in the frequency domain as

$$V = \frac{-\mathrm{i}\Omega^3 \frac{\alpha\theta}{C_p}}{\Delta_1(\mathrm{i}\Omega)} X_b. \tag{8.28}$$

Following [12], we are interested in the mean of the normalised harvested power when the base acceleration is Gaussian white noise, that is, $|V|^2/(R_l \omega^4 \Phi_{x_b x_b})$. Note that $\omega^4 \Phi_{x_b x_b}$ is the spectral density of the acceleration and is assumed to be constant. After some algebra, from Eq. (8.28), the normalised power is

$$\widetilde{P} = \frac{|V|^2}{(R_l \omega^4 \Phi_{x_b x_b})} = \frac{k\alpha\kappa^2}{\omega_n^3} \frac{\Omega^2}{\Delta_1(\mathrm{i}\Omega)\Delta_1^*(\mathrm{i}\Omega)}. \tag{8.29}$$

Using Eq. (8.21), the average normalised power can be obtained as

$$\mathrm{E}\left[\widetilde{P}\right] = \mathrm{E}\left[\frac{|V|^2}{(R_l \omega^4 \Phi_{x_b x_b})}\right] = \frac{k\alpha\kappa^2}{\omega_n^3} \int_{-\infty}^{\infty} \frac{\Omega^2}{\Delta_1(\mathrm{i}\Omega)\Delta_1^*(\mathrm{i}\Omega)} \mathrm{d}\omega. \tag{8.30}$$

From Eq. (8.9) observe that $\Delta_1(\mathrm{i}\Omega)$ is third-order polynomial in $(\mathrm{i}\Omega)$. Noting that $\mathrm{d}\omega = \omega_n \mathrm{d}\Omega$, from Eq. (8.9), the average harvested power can be obtained from Eq. (8.30) as

$$\mathrm{E}\left[\widetilde{P}\right] = \mathrm{E}\left[\frac{|V|^2}{(R_l \omega^4 \Phi_{x_b x_b})}\right] = m\alpha\kappa^2 I^{(1)} \tag{8.31}$$

where

$$I^{(1)} = \int_{-\infty}^{\infty} \frac{\Omega^2}{\Delta_1(\mathrm{i}\Omega)\Delta_1^*(\mathrm{i}\Omega)} \mathrm{d}\Omega. \tag{8.32}$$

Comparing $I^{(1)}$ with the general integral in Eq. (8.22), we have

$$n = 3, \quad q_2 = 0, \quad q_1 = 1, \quad q_0 = 0,$$

$$p_3 = \alpha, \quad p_2 = (2\zeta\alpha + 1), \quad p_1 = \left(\alpha + \kappa^2\alpha + 2\zeta\right), \quad p_0 = 1. \tag{8.33}$$

Now using Eq. (8.25), the integral can be evaluated as

$$I^{(1)} = \frac{\pi}{\alpha} \frac{\det\begin{bmatrix} 0 & 1 & 0 \\ -\alpha & \alpha + \kappa^2\alpha + 2\zeta & 0 \\ 0 & -2\zeta\alpha - 1 & 1 \end{bmatrix}}{\det\begin{bmatrix} 2\zeta\alpha + 1 & -1 & 0 \\ -\alpha & \alpha + \kappa^2\alpha + 2\zeta & 0 \\ 0 & -2\zeta\alpha - 1 & 1 \end{bmatrix}}. \tag{8.34}$$

Combining this with Eq. (8.31), we finally obtain the average harvested power due to white-noise base acceleration as

$$\mathrm{E}\left[\widetilde{P}\right] = \mathrm{E}\left[\frac{|V|^2}{(R_l \omega^4 \Phi_{x_b x_b})}\right] = \frac{\pi\, m\, \alpha\, \kappa^2}{(2\,\zeta\,\alpha^2 + \alpha)\,\kappa^2 + 4\,\zeta^2\alpha + (2\,\alpha^2 + 2)\,\zeta}. \tag{8.35}$$

Since α and κ^2 are positive, the average harvested power is monotonically decreasing with damping ratio ζ. Thus, the mechanical damping in the harvester should be minimised. For fixed α and ζ, the average harvested power is monotonically increasing with the coupling coefficient κ^2, and hence the electromechanical coupling should be as large as possible. Maximising the average power with respect to α gives the condition

$$\alpha^2\left(1 + \kappa^2\right) = 1 \tag{8.36}$$

or in terms of physical quantities

$$R_l^2 C_\mathrm{p}\left(k C_\mathrm{p} + \theta^2\right) = m. \tag{8.37}$$

8.2.4 Mean Power for Systems with an Inductor

From Eq. (8.15) we obtain the voltage in the frequency domain as

$$V = \frac{\Omega^4 \frac{\alpha\beta\theta}{C_\mathrm{p}}}{\Delta_2(\mathrm{i}\Omega)} X_b. \tag{8.38}$$

Following [39] the average normalised harvested power can be obtained as

$$\mathrm{E}\left[\widetilde{P}\right] = \mathrm{E}\left[\frac{|V|^2}{(R_l \omega^4 \Phi_{x_b x_b})}\right] = m\alpha\beta^2\kappa^2 I^{(2)} \tag{8.39}$$

where

$$I^{(2)} = \int_{-\infty}^{\infty} \frac{\Omega^4}{\Delta_2(\mathrm{i}\Omega)\Delta_2^*(\mathrm{i}\Omega)}\,\mathrm{d}\Omega. \tag{8.40}$$

Using the expression of $\Delta_2(\mathrm{i}\Omega)$ in Eq. (8.16) and comparing $I^{(2)}$ with the general integral in Eq. (8.22), we have

$$n = 4,\quad q_3 = 0,\quad q_2 = 1,\quad q_1 = 0,\quad q_0 = 0,\quad p_4 = \beta\alpha,\quad p_3 = (2\,\zeta\,\beta\,\alpha + \beta),$$
$$p_2 = \left(\beta\,\alpha + \alpha + 2\,\zeta\,\beta + \kappa^2\beta\,\alpha\right),\quad p_1 = (\beta + 2\,\zeta\,\alpha),\quad p_0 = \alpha. \tag{8.41}$$

Now using Eq. (8.25), the integral can be evaluated as

$$I^{(2)} = \frac{\pi}{\beta\alpha}\frac{\det\begin{bmatrix} 0 & 1 & 0 & 0 \\ -\beta\,\alpha & \beta\,\alpha+\alpha+2\,\zeta\,\beta+\kappa^2\beta\,\alpha & -\alpha & 0 \\ 0 & -2\,\zeta\,\beta\,\alpha-\beta & \beta+2\,\zeta\,\alpha & 0 \\ 0 & -\beta\,\alpha & \beta\,\alpha+\alpha+2\,\zeta\,\beta+\kappa^2\beta\,\alpha\,\alpha \end{bmatrix}}{\det\begin{bmatrix} 2\,\zeta\,\beta\,\alpha+\beta & -\beta-2\,\zeta\,\alpha & 0 & 0 \\ -\beta\,\alpha & \beta\,\alpha+\alpha+2\,\zeta\,\beta+\kappa^2\beta\,\alpha & -\alpha & 0 \\ 0 & -2\,\zeta\,\beta\,\alpha-\beta & \beta+2\,\zeta\,\alpha & 0 \\ 0 & -\beta\,\alpha & \beta\,\alpha+\alpha+2\,\zeta\,\beta+\kappa^2\beta\,\alpha\,\alpha \end{bmatrix}}. \tag{8.42}$$

Combining this with Eq. (8.31), we finally obtain the average normalised harvested power as

$$\begin{aligned} \mathrm{E}\left[\widetilde{P}\right] &= \mathrm{E}\left[\frac{|V|^2}{(R_l\omega^4\Phi_{x_bx_b})}\right] \\ &= m\alpha\beta\kappa^2\pi\,(\beta+2\alpha\zeta)/\left[\left(4\beta\alpha^3\zeta^2+2\beta\alpha^2\,(\beta+1)\,\zeta+\beta^2\alpha\right)\kappa^2\right. \\ &\qquad \left.+8\beta\alpha^2\zeta^3+4\beta\alpha\,(\beta+1)\,\zeta^2+2\left(\beta^2\alpha^2+\beta^2-2\beta\alpha^2+\alpha^2\right)\zeta\right] \\ &= \frac{m\alpha\beta\kappa^2\pi\,(\beta+2\alpha\zeta)}{\beta\,(\beta+2\alpha\zeta)\,(1+2\alpha\zeta)\,(\alpha\kappa^2+2\zeta)+2\alpha^2\zeta\,(\beta-1)^2}. \end{aligned} \tag{8.43}$$

This is the complete closed-form expression of the normalised harvested power under Gaussian white noise base acceleration.

Since α, β and κ^2 are positive, the average harvested power is monotonically decreasing with damping ratio ζ. Thus, the mechanical damping in the harvester should be minimised. For fixed α, β and ζ, the average harvested power is monotonically increasing with the coupling coefficient κ^2, and hence the electromechanical coupling should be as large as possible. These are the same conclusions as for the case without an inductor, although slightly more difficult to prove.

We can also determine optimum values for α and β. Dividing both the numerator and denominator of the last expression in Eq. (8.43) by $\beta\,(\beta+2\alpha\zeta)$ shows that the optimum value of β for all values of the other parameters is $\beta = 1$. This value of β implies that $\omega_n^2 LC_\mathrm{p} = 1$, and thus the mechanical and electrical natural frequencies are equal. With $\beta = 1$ the average normalised harvested power is

$$\mathrm{E}\left[\widetilde{P}\right] = \frac{m\alpha\kappa^2\pi}{(1+2\alpha\zeta)\,(\alpha\kappa^2+2\zeta)}. \tag{8.44}$$

If κ and ζ are fixed, then the maximum power with respect to α is obtained when $\alpha = 1/\kappa$.

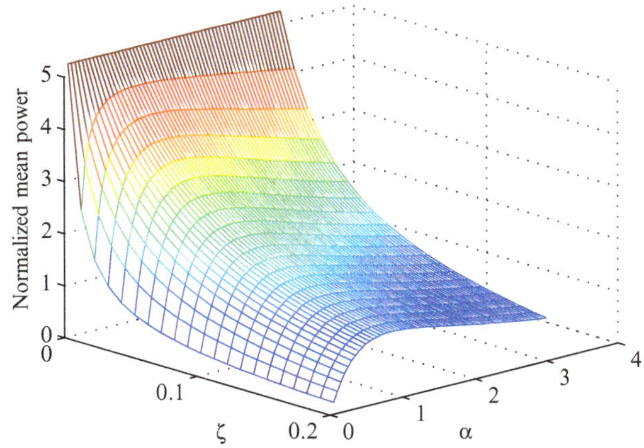

Fig. 8.2 The normalised mean power of a harvester without an inductor as a function of α and ζ, $\kappa = 0.6$

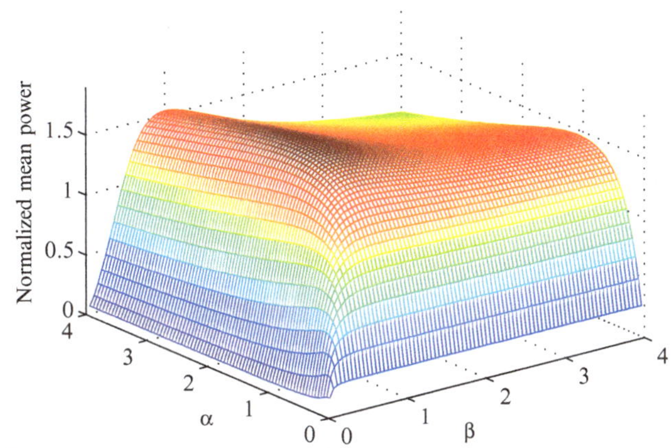

Fig. 8.3 The normalised mean power of a harvester with an inductor as a function of α and β, with $\zeta = 0.1$ and $\kappa = 0.6$

8.2.5 Numerical Illustrations

The expressions of Sects. 8.2.3 and 8.2.4 are now illustrated numerically for a system with unit mass. In Fig. 8.2 the normalised mean power of a harvester without an inductor, as given by Eq. (8.35), is shown as function of α and ζ. For illustration, the value of the coupling coefficient κ is kept fixed at 0.6. The increased harvested energy as the damping ratio ζ decreases is clearly seen. Also there is a maximum in the harvested energy for $\alpha = 0.86$, corresponding to the optimum predicted by Eq. (8.36).

In Fig. 8.3 the normalised mean power of a harvester with an inductor, as given by Eq. (8.43), is shown as function of α and β. For illustration, the value of the coupling coefficient κ is again kept fixed at 0.6, while the value of damping factor

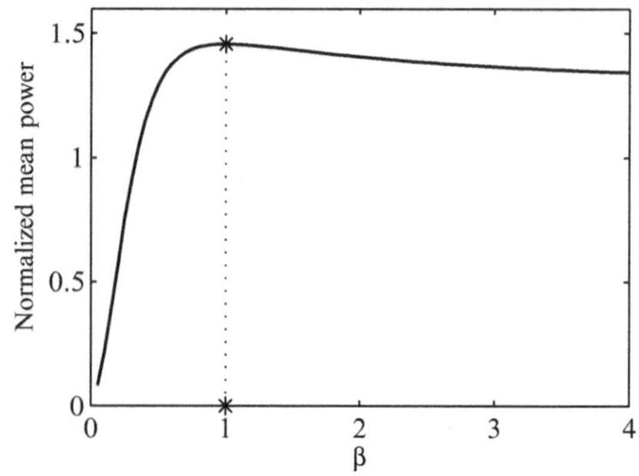

Fig. 8.4 The normalised mean power of a harvester with an inductor as a function of β for $\alpha = 0.6$, $\zeta = 0.1$ and $\kappa = 0.6$. The * corresponds to the optimal value of β for the maximum mean harvested power

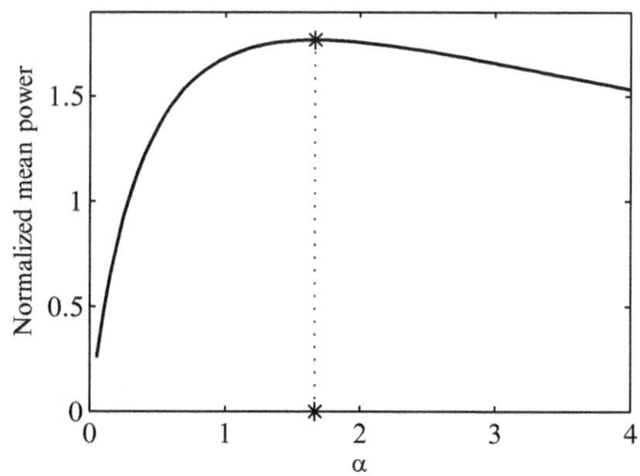

Fig. 8.5 The normalised mean power of a harvester with an inductor as a function of α for $\beta = 1$, $\zeta = 0.1$ and $\kappa = 0.6$. The * corresponds to the optimal value of $\alpha (= 1.667)$ for the maximum mean harvested power

ζ is kept fixed at 0.1. There is clearly a well-defined maximum harvested energy at $\beta = 1$ and $\alpha = 1/\kappa = 1.667$, as predicted in Sect. 8.2.4, although this is better illustrated by taking sections through this 3D surface.

The mean harvested power as a function of β is shown in Fig. 8.4, with the other parameters fixed at $\alpha = 0.6$, $\zeta = 0.1$ and $\kappa = 0.6$. The optimum value occurs at $\beta = 1$, which is shown by the star in Fig. 8.4. It was highlighted in Sect. 8.2.4 that this value of β is optimum for all values of α. The mean harvested power as a function of α is shown in Fig. 8.5 for $\beta = 1$ and clearly shows the maximum at $\alpha = 1.667$.

Vibration energy-based piezoelectric energy harvesters are expected to operate under a wide range of ambient environments. Analytical expressions of the normalised mean harvested power due to stationary Gaussian white noise base excitation have been derived. The resulting two-dimensional stochastic differential equations are solved using the theory of linear random vibrations. Two cases, namely, the harvesting circuit with and without an inductor, have been considered. For both cases exact closed-form expressions of the harvested power involving the nondimensional time constants, the nondimensional electromechanical coupling coefficient and the mechanical viscous damping factor have been derived. Optimal values of the parameter for which the harvested power is maximum have been discussed. It was shown that in order to maximise the mean of the harvested power, (a) the mechanical damping in the harvester should be minimised, and (b) the electromechanical coupling should be as large as possible. The electrical circuits may also be optimised to obtain the maximum mean power, and the expressions for these optima have been given. For the circuit with an inductor, the maximum mean power occurs when the natural frequency of the electrical circuit is equal to that of the mechanical system.

Equation (8.43) may be used to (a) design energy harvesters subject to random excitation and (b) to provide insight into the physical nature of harvesting when subject to random ambient energy. The approach described can be extended to filtered white noise and non-Gaussian excitation that may be described as a rational fraction polynomial in the frequency domain. Such excitation will simulate more realistic excitation spectra compared to the pure Gaussian white noise base acceleration, although the derived expressions will be complicated. The higher-order moments of the harvested power, such as the standard deviation, could also be derived, in addition to the mean power derived here.

8.3 The Response of Bistable Harvesters to Random Excitation

In this section we consider the numerical integration of the equations of motion of a bistable energy harvester. The device consists of a ferromagnetic cantilever beam that is excited at the support (see Fig. 8.6). Two permanent magnets are located symmetrically on the base near the free end. The distance between the beam and the magnets determines the stable equilibrium points. Here we are interested in the case when the system has three equilibrium positions, two of which are stable, and the mechanical system is characterised by the classical double well potential.

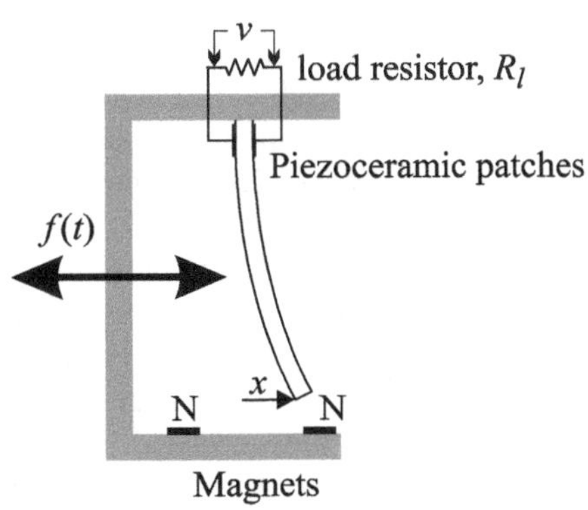

Fig. 8.6 Schematic diagram of the magnetopiezoelastic device [19]

The nondimensional equations of motion for this system are

$$\ddot{x} + 2\zeta\dot{x} - \frac{1}{2}x(1 - x^2) - \chi v = f(t), \tag{8.45}$$

$$\dot{v} + \lambda v + \eta\dot{x} = 0, \tag{8.46}$$

where x is the dimensionless transverse displacement of the beam tip, v is the dimensionless voltage across the load resistor, χ is the dimensionless piezoelectric coupling term in the mechanical equation, η is the dimensionless piezoelectric coupling term in the electrical equation, $\lambda \propto 1/R_l C_{\mathrm{p}}$ is the reciprocal of the dimensionless time constant of the electrical circuit, R_l is the load resistance and C_{p} is the capacitance of the piezoelectric material. The excitation $f(t)$ is proportional to the base acceleration on the device and is assumed to be band-limited white noise, with zero mean and specified variance σ_f^2.

The system parameters have been taken as [19]: $\zeta = 0.01$, $\chi = 0.05$ and $\eta = 0.5$, while λ was varied between 0.01 and 10. The white noise excitation is assumed to have a bandwidth of 2 Hz, and Eqs. (8.45) and (8.46) are integrated for 2621 s, with the first 200 s discarded. The standard deviations of the displacement x and the voltage v are calculated for a range of excitation noise amplitudes σ_f. For each value of σ_f the integration is repeated 10 times.

Figure 8.7a shows the standard deviation of the displacement relative to that of the excitation for $\lambda = 0.01$, and the pronounced peaks correspond to the stochastic resonance phenomenon [21]. This is a nonlinear effect; for the linear system the output is proportional to the random excitation, and the corresponding plot would be a horizontal line. The sudden increase of the displacement amplitude x is associated with the escape from the single well potential in the presence of stochastic excitation. This may be highlighted by considering the phase plane for three values of σ_f shown in Fig. 8.8, for $\lambda = 0.01$. For low excitation levels the response remains in one of the potential wells, with a low response variance. For high excitation levels the system easily hops between the potential wells and the system approximates a cubic stiffness nonlinearity. For intermediate excitation levels, the occurrence of hops between potential wells is stochastic, and this is clearly shown in Fig. 8.7a by the stochastic nature of the response variance. Figure 8.7b,c shows the displacement response for $\lambda = 1$ and $\lambda = 10$, respectively. The value of λ is related to the electrical time constant; the changes in the mechanical response are small because the mechanical and electrical systems are only weakly coupled. However, the hopping between potential wells tends to occur at slightly higher σ_f for $\lambda = 1$.

The effect of λ is greater in the electrical response, shown in Fig. 8.7d,e, and the stochastic resonance phenomenon is clearly seen for $\lambda = 0.01$. As the mechanical system is coupled to the electrical circuit through the piezoelectric transducers, the increase in mechanical displacement amplitude causes the voltage amplitude to rise. This also causes the harvested power to increase; Fig. 8.7f shows the variance of the voltage multiplied by λ, which is proportional to the mean harvested power. The electrical constant λ significantly affects the voltage produced and hence the

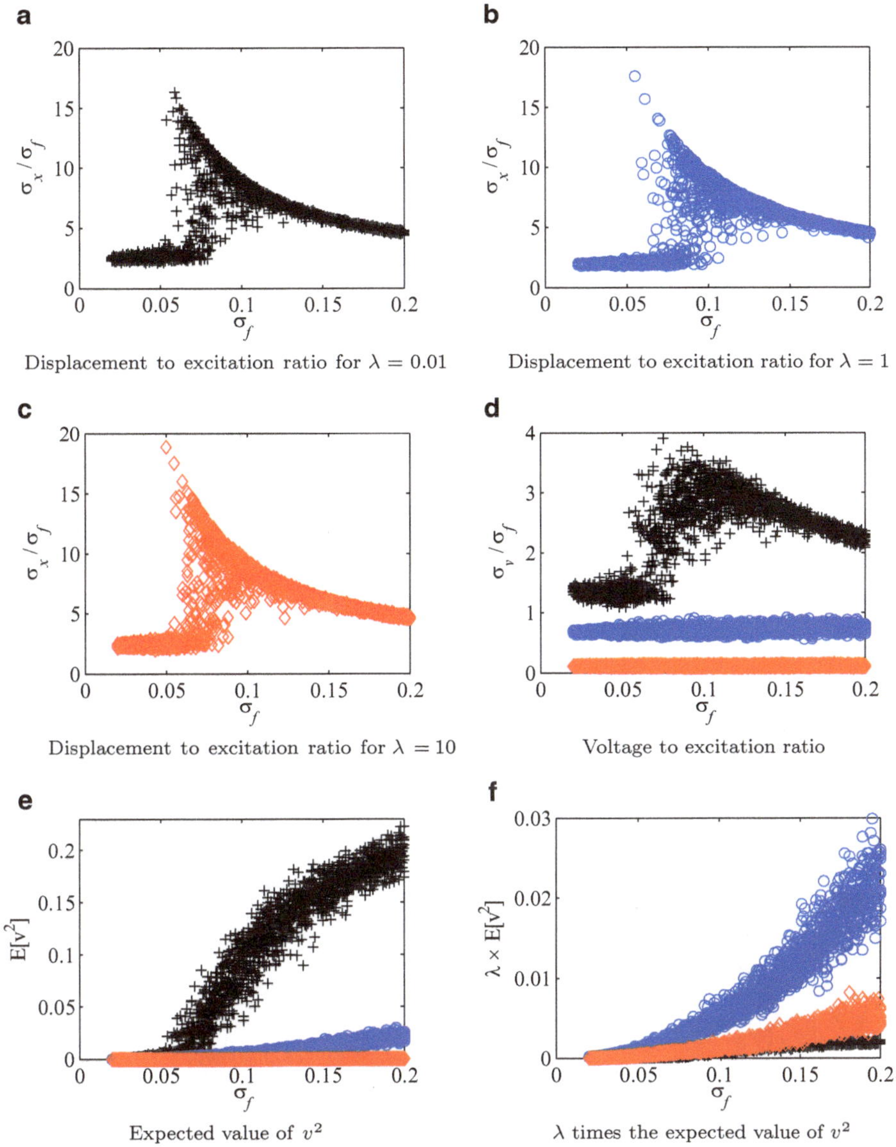

Fig. 8.7 Simulated responses of the piezomagnetoelastic oscillator in terms of the standard deviations of displacement and voltage (σ_x and σ_v) as the standard deviation of the random excitation σ_f varies, for $\lambda = 0.01$ (*black cross*), $\lambda = 1$ (*blue circle*), and $\lambda = 10$ (*red diamond*)

power generated. Note that this increase in power above the stochastic resonance is due to the changes in the electrical system, since the mechanical responses shown in Fig. 8.7a–c are almost identical. Decreasing λ increases the electrical time constant,

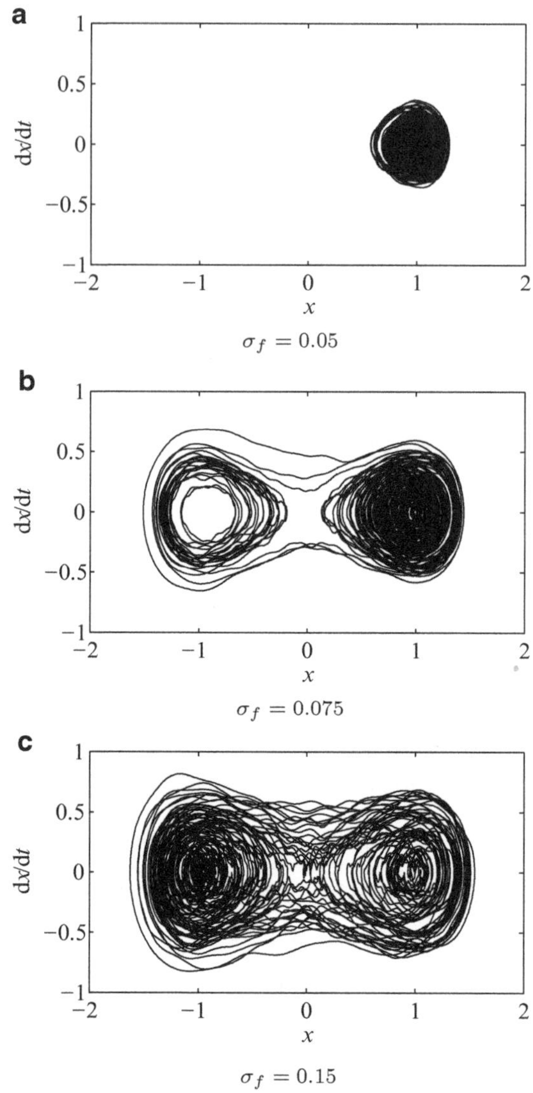

Fig. 8.8 Phase portraits for $\lambda = 0.01$. Note that the increasing noise level overcomes the potential barrier resulting in a significant increase in the displacement x

and this allows the load to capture energy from the slow scale dynamics, where the mechanical states hop between the two potential wells. However, a value of λ close to 1 matches the excitation frequency range very well and therefore appears to capture most energy. The practical implementation of an optimum value of λ requires the selection of the electrical resistance and capacitance; this selection should be included in the trade-offs required for the design of a real system.

Figure 8.9 shows the probability density functions for the displacement x and voltage v for the responses shown in Fig. 8.8 for $\lambda = 0.01$. The bimodal distributions

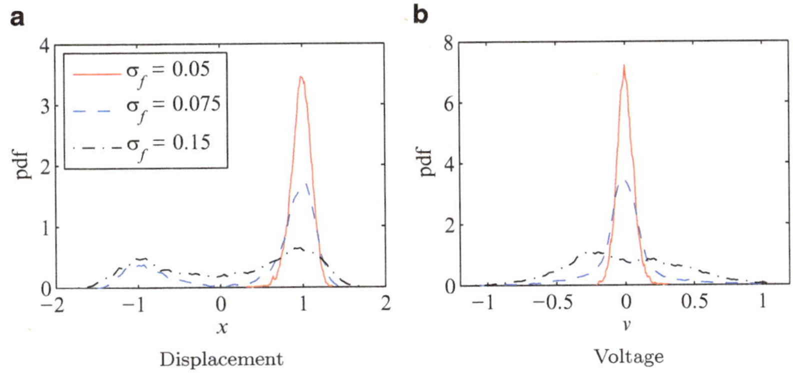

Fig. 8.9 Probability density functions of the mechanical displacement x and the voltage across the piezoelectric patch, v, for $\lambda = 0.01$: $\sigma_f = 0.05$ (*solid*), $\sigma_f = 0.075$ (*dashed*), and $\sigma_f = 0.15$ (*dash dot*)

in Fig. 8.9a clearly show that the mechanical system spends most of its time close to the two equilibrium positions. The bimodal distribution is not visible in the voltage, Fig. 8.9b, because the electrical circuit acts as a first-order filter.

This analysis indicates that the energy harvested from a piezomagnetoelastic device is most efficient for a certain range of the noise intensity. In practice the noise intensity, or noise variance, would be known and the device designed so that it operates at the stochastic resonance condition. Essentially this design process would tailor the height of the energy barrier in the double well potential. The understanding of the slow dynamics of this bistable system can be used to improve the coupling to the electrical system and hence the performance of the energy harvester.

8.4 Stochastic Linearisation of Bistable Harvesters

The theoretical analysis of piezomagnetoelastic systems as in Fig. 8.6 is absent in the literature. The exact analysis of nonlinear piezomagnetoelastic energy harvesting system under random excitation requires the solution of the multidimensional Fokker–Planck equation to obtain the governing the probability density function of the harvested power. A different approach is considered here; the system is stochastically linearised and an equivalent model of the overall system is developed. The linear model is then analysed to determine probability density functions of the system response and the power scavenged by the system.

Equation (8.45) is a nonlinear equation with nonlinearity in the stiffness term, whereas Eq. (8.46) is a linear equation. To facilitate the linearisation process, Eq. (8.45) can be rewritten as

$$\ddot{x} + 2\zeta\dot{x} + g(x) - \chi v = f(t). \tag{8.47}$$

The nonlinear stiffness is represented as $g(x) = -\frac{1}{2}(x - x^3)$. The linearised model for the Duffing equation (8.47) is developed using the stochastic linearisation approach [41].

Assuming a non-zero mean random excitation (i.e. $f(t) = f_0(t) + m_f$) and a non-zero mean system response (i.e. $x(t) = x_0(t) + m_x$), the following equivalent linear system is considered:

$$\ddot{x_0} + 2\zeta \dot{x_0} + a_0 x_0 + b_0 - \chi v = f_0(t) + m_f \tag{8.48}$$

where $f_0(t)$ and $x_0(t)$ are zero mean random processes. m_f and m_x are the mean of the original processes $f(t)$ and $x(t)$, respectively. a_0 and b_0 are the constants to be determined with $b_0 = m_f$, and a_0 represents the square of the natural frequency of the linearised system $\omega_{\rm eq}^2$. The stochastic moments of the nonlinear and the linearised model outputs will be matched by minimising the expectation of the error norm (i.e., $\mathbb{E}\left[\epsilon^2\right]$, with$\epsilon = g(x) - a_0 x_0 - b_0$). To determine the constants a_0 and b_0 in terms of the statistics of the response x, we take partial derivatives of the error norm *w.r.t.* a_0 and b_0 and equate them to zero individually:

$$\frac{\partial}{\partial a_0}\mathrm{E}\left[\epsilon^2\right] = \mathrm{E}\left[g(x)x_0\right] - a_0\mathrm{E}\left[x_0^2\right] - b_0\mathrm{E}\left[x_0\right] \tag{8.49}$$

$$\frac{\partial}{\partial b_0}\mathrm{E}\left[\epsilon^2\right] = \mathrm{E}\left[g(x)\right] - a_0\mathrm{E}\left[x_0\right] - b_0. \tag{8.50}$$

Equating Eqs. (8.49) and (8.50) to zero, we get

$$\begin{aligned} a_0 &= \frac{\mathrm{E}\left[g(x)x_0\right]}{\mathrm{E}\left[x_0^2\right]} \\ &= \frac{\mathrm{E}\left[g(x)x_0\right]}{\sigma_x^2} \end{aligned} \tag{8.51}$$

$$\begin{aligned} b_0 &= \mathrm{E}\left[g(x)\right] \\ &= m_f. \end{aligned} \tag{8.52}$$

As a special case if we assume that $x(t)$ is a Gaussian random process, the expressions in Eqs. (8.51) and (8.52) can be further simplified [25, 41] as

$$\begin{aligned} a_0 &= \mathrm{E}\left[\frac{\mathrm{d}}{\mathrm{d}x}g(x)\right] \\ &= -\frac{1}{2}\left(1 - 3\mathrm{E}\left[x^2\right]\right) \\ &= -\frac{1}{2}\left\{1 - 3\sigma_x^2 - 3m_x^2\right\} \end{aligned} \tag{8.53}$$

which gives

$$3m_x^2 + 3\sigma_x^2 - 2a_0 - 1 = 0, \tag{8.54}$$

and

$$\begin{aligned} m_f &= \mathrm{E}\left[g(x)\right] \\ &= -\frac{1}{2}\left(\mathrm{E}\left[x\right] - \mathrm{E}\left[x^3\right]\right). \end{aligned} \tag{8.55}$$

Note that for a non-zero mean Gaussian process x with mean m_x and its zero mean part x_0, we have the following relations:

$$\mathrm{E}\left[x_0^3\right] = \mathrm{E}\left[(x - m_x)^3\right] = \mathrm{E}\left[x^3\right] - 3m_x\sigma_x^2 - m_x^3. \tag{8.56}$$

Since for a zero mean Gaussian process $\mathrm{E}\left[x_0^3\right] = 0$, using the relation in Eq. (8.56), we get

$$\mathrm{E}\left[x^3\right] = 3m_x\sigma_x^2 + m_x^3. \tag{8.57}$$

Substituting $\mathrm{E}\left[x^3\right]$ from Eq. (8.57) into Eq. (8.55), we get

$$m_f = -\frac{1}{2}\left\{m_x - \left(3m_x\sigma_x^2 + m_x^3\right)\right\} \tag{8.58}$$

which gives

$$m_x\left(m_x^2 + 3\sigma_x^2 - 1\right) - 2m_f = 0 \tag{8.59}$$

where m_x and σ_x are the mean and standard deviation of the system response x.

The process of statistical linearisation reduces to finding three unknowns a_0, m_x and σ_x with only two equations (8.54) and (8.59). Another expression for σ_x can be obtained from the linearised system equations.

Equation (8.48) along with Eq. (8.46) can be rewritten as

$$\ddot{x}_0 + 2\zeta\dot{x}_0 + a_0x_0 - \chi v = f_0(t) \tag{8.60}$$

$$\dot{v} + \lambda v + \eta\dot{x}_0 = 0. \tag{8.61}$$

The analysis is identical to that of Sect. 8.2.1.1 but is repeated here briefly because the notation and the meaning of the parameters is different.

Taking the Fourier transform of Eqs. (8.60) and (8.61), we get

$$\begin{bmatrix} (a_0 - \Omega^2) + 2\mathrm{i}\Omega\zeta & -\chi \\ \mathrm{i}\Omega\eta & (\mathrm{i}\Omega + \lambda) \end{bmatrix} \begin{Bmatrix} X(\Omega) \\ V(\Omega) \end{Bmatrix} = \begin{Bmatrix} F_0(\Omega) \\ 0. \end{Bmatrix} \tag{8.62}$$

Inverting the coefficient matrix, the displacement and voltage in the frequency domain can be obtained as

$$\begin{Bmatrix} X \\ V \end{Bmatrix} = \frac{1}{\Delta}\begin{bmatrix} (\mathrm{i}\Omega + \lambda)\,\chi \\ -\mathrm{i}\Omega\eta\left(a_0 - \Omega^2\right) + 2\mathrm{i}\Omega\zeta \end{bmatrix}\begin{Bmatrix} F_0 \\ 0 \end{Bmatrix} \tag{8.63}$$

$$= H(\Omega)\begin{Bmatrix} F_0 \\ 0, \end{Bmatrix} \tag{8.64}$$

where $H(\Omega)$ is the 2×2 matrix of frequency response functions and the determinant of the coefficient matrix is

$$\Delta(\mathrm{i}\Omega) = (\mathrm{i}\Omega)^3 + (2\zeta + \lambda)\,(\mathrm{i}\Omega)^2 + (2\zeta\lambda + \eta\chi + a_0)\,(\mathrm{i}\Omega) + \lambda a_0. \tag{8.65}$$

We consider the excitation $f_0(t)$ to be a zero mean weakly stationary, Gaussian, broadband random process. To obtain the statistics of the random response quantities such as the displacement of the mass $x(t)$ and the voltage $v(t)$, one needs to solve the coupled stochastic differential equations (8.60) and (8.61). Analytical results developed within the theory of random vibration discussed in Sect. 8.2.2 allow us to bypass numerical solutions because we are interested in the average values of the output random processes. The standard deviation of the mechanical response may be obtained as

$$\sigma_x^2 = \Phi_{f_0 f_0}\int_{-\infty}^{\infty}\frac{\lambda^2 + \Omega^2}{\Delta(\Omega)\Delta^*(\Omega)}\mathrm{d}\Omega. \tag{8.66}$$

The calculation of the integral on the right-hand side of Eq. (8.66) can be performed in closed form as

$$I_n = \int_{-\infty}^{\infty}\frac{\Xi_n(\Omega)\,\mathrm{d}\Omega}{\Lambda_n(\Omega)\Lambda_n^*(\Omega)} = \frac{\pi(a_0 + 2\lambda\zeta + \lambda^2)}{a_0\left(4\lambda\zeta^2 + 2\zeta\eta\chi + 2\zeta a_0 + 2\lambda^2\zeta + \lambda\eta\chi\right)}. \tag{8.67}$$

Combining Eq. (8.67) with Eq. (8.66) and then simplifying the resulting expression, we obtain the final relation between a_0 and σ_x and which is given as

$$a_0\sigma_x^2\left(4\lambda\zeta^2 + 2\zeta\eta\chi + 2\zeta a_0 + 2\lambda^2\zeta + \lambda\eta\chi\right) - \pi\Phi_{f_0 f_0}\left(a_0 + 2\lambda\zeta + \lambda^2\right) = 0. \tag{8.68}$$

Equation (8.68) along with Eqs. (8.54) and (8.59) provides three equations to solve for the unknown variables a_0, σ_x and m_x. Analytical solutions of Eqs. (8.54), (8.59) and (8.68) are not possible and one should make use of numerical schemes. In the next section, solutions for zero mean white noise excitation are shown. To summarise, one has to solve for m_x, σ_x and a_0 from the following three nonlinear coupled equations:

$$\begin{gathered} 3m_x^2 + 3\sigma_x^2 - 2a_0 - 1 = 0, \\ m_x\left(m_x^2 + 3\sigma_x^2 - 1\right) - 2m_f = 0, \\ a_0\sigma_x^2\left(4\lambda\zeta^2 + 2\zeta\eta\chi + 2\zeta a_0 + 2\lambda^2\zeta + \lambda\eta\chi\right) - \pi\Phi_{f_0 f_0}\left(a_0 + 2\lambda\zeta + \lambda^2\right) = 0\,. \end{gathered} \tag{8.69}$$

8.4.1 Zero Mean White Noise Excitation

8.4.1.1 Determination of σ_x, a_0 and m_x

Without loss of generality the external excitation can be assumed to be a zero mean white noise process, that is, $m_f = 0$. This largely simplifies our analysis and provides a simple relation between the mean (m_x) and the standard deviation (σ_x) of the response. Putting $m_f = 0$ in Eq. (8.59), we get

$$m_x\left(m_x^2 + 3\sigma_x^2 - 1\right) = 0 \tag{8.70}$$

which gives either $m_x = 0$ or $m_x = \sqrt{1 - 3\sigma_x^2}$.

Substituting $m_x = 0$ in Eq. (8.54), we get

$$3\sigma_x^2 = 1 + 2a_0. \tag{8.71}$$

Equation (8.71) shows that for $m_x = 0$ and for any real $a_0 \geq 0$, we have $\sigma_x^2 \geq \frac{1}{3}$.

Substituting $m_x = \sqrt{1 - 3\sigma_x^2}$ in Eq. (8.54), we get

$$\begin{aligned} 3(1 - 3\sigma_x^2) + 3\sigma_x^2 - 2a_0 - 1 &= 0 \\ 1 - 3\sigma_x^2 - a_0 &= 0 \end{aligned} \tag{8.72}$$

which gives

$$3\sigma_x^2 = 1 - a_0. \tag{8.73}$$

Equation (8.73) shows that for $m_x^2 = 1 - 3\sigma_x^2$ and real, positive a_0, we have $\sigma_x^2 \in [0, \frac{1}{3}]$, and this bounds $a_0 \in [0, 1]$. A further analysis will show that $a_0 = m_x^2$.

Substituting Eq. (8.71) and Eq. (8.73) separately in Eq. (8.68), we obtain the following two cubic polynomial equations:

$$\begin{aligned} &4\zeta {a_0}^3 + \left(8\lambda\zeta^2 + 4\lambda^2\zeta + 4\eta\chi\zeta + 2\zeta + 2\lambda\eta\chi\right){a_0}^2 \\ &\quad + \left(4\lambda\zeta^2 + 2\lambda^2\zeta + 2\eta\chi\zeta + \lambda\eta\chi - 3\Phi_{f_0 f_0}\pi\right)a_0 - 3\Phi_{f_0 f_0}\pi\lambda\left(\lambda + 2\zeta\right) = 0 \end{aligned} \tag{8.74}$$

$$\begin{aligned} &2\zeta {a_0}^3 + \left(4\lambda\zeta^2 + 2\lambda^2\zeta + 2\eta\chi\zeta - 2\zeta + \lambda\eta\chi\right){a_0}^2 \\ &\quad - \left(4\lambda\zeta^2 + 2\lambda^2\zeta + 2\eta\chi\zeta + \lambda\eta\chi - 3\Phi_{f_0 f_0}\pi\right)a_0 + 3\Phi_{f_0 f_0}\pi\lambda\left(\lambda + 2\zeta\right) = 0. \end{aligned} \tag{8.75}$$

Numerical solutions show that Eq. (8.74) has a single real positive root and Eq. (8.75) has two real positive roots for any given value of $\Phi_{f_0 f_0}$.

8.4.1.2 Determination of E $[v^2]$

The spectral density of the voltage generated across the harvester can be related to the excitation as

$$\Phi_{vv}(\Omega) = |H_{21}(\Omega)|^2\, \Phi_{f_0 f_0}(\Omega), \tag{8.76}$$

where $H_{21}(\Omega)$ is the element in the second row and first column of the matrix $H(\Omega)$ (see Eq. (8.63)). Thus:

$$\mathrm{E}\left[v^2\right] = \int_{-\infty}^{\infty} |H_{21}(\Omega)|^2 \Phi_{f_0 f_0} \mathrm{d}\Omega, \tag{8.77}$$

where $\Phi_{f_0 f_0}$ is a constant for weakly stationary, white noise process.

Combining Eqs. (8.63), (8.64), (8.65) and (8.77), we get

$$\mathrm{E}\left[v^2\right] = \Phi_{f_0 f_0} \int_{-\infty}^{\infty} \frac{\eta^2 \Omega^2}{\Delta(\Omega)\Delta^*(\Omega)} \mathrm{d}\Omega. \tag{8.78}$$

Proceeding in a similar way to Sect. 8.2.2, we find an expression for E $[v^2]$ as

$$\mathrm{E}\left[v^2\right] = \frac{\eta^2 \lambda}{(4\lambda\zeta^2 + 2\zeta\eta\chi + 2\zeta a_0 + 2\lambda^2\zeta + \lambda\eta\chi)} \pi \Phi_{f_0 f_0} \tag{8.79}$$

8.4.2 Numerical Analysis

The stochastic linearisation approach is demonstrated using system parameters identical to those in Sect. 8.3 [19]: $\zeta = 0.01$, $\chi = 0.05$ and $\eta = 0.5$, while λ was varied between 0.01 and 10. The excitation $f(t)$ is considered stationary band-limited white noise with standard deviation σ_f. The excitation bandwidth is assumed to be 2 Hz, and hence the spectral density is given by $\Phi_{f_0 f_0} = \sigma_f^2/8\pi$. As discussed in Sect. 8.4.1.1, the analytical solution of Eqs. (8.74) and (8.75) is not feasible. Numerical methods are used to solve for σ_x, a_0 and m_x for various values of standard deviation of the excitation, σ_f.

The nonlinear piezomagnetoelastic system given by Eq. (8.45) represents a Duffing-type equation of motion. The nonlinear system has three equilibrium points: two stable equilibrium points at $x = \pm 1$ and one unstable equilibrium point at the origin ($x = 0$). Numerical simulations show that Eq. (8.74) has a single real positive solution for a_0, whereas Eq. (8.75) has two real positive solutions. The solutions for different standard deviations of excitation are shown in Fig. 8.10. *Solution 1* and *Solution 2* are the solutions for Eq. (8.75), with $m_x = \sqrt{1 - 3\sigma_x^2}$, and *Solution 3* is obtained by solving Eq. (8.74) and represents the zero mean response. a_0 is the square of the natural frequency of the equivalent linear system (see Eq. (8.48)) and

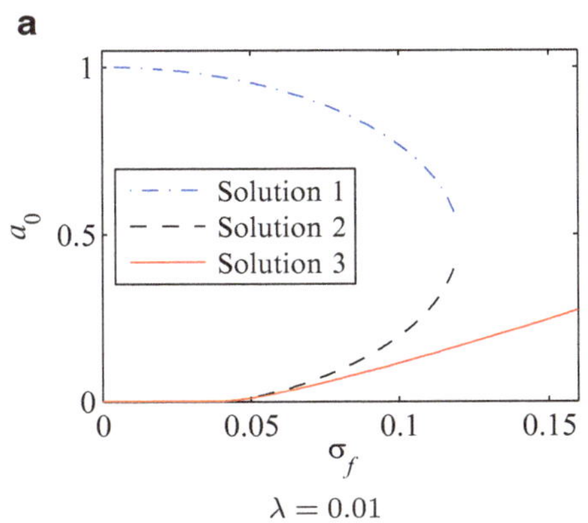

Fig. 8.10 Square of the natural frequency of the equivalent linear system, a_0, for different standard deviations of the excitation, σ_f

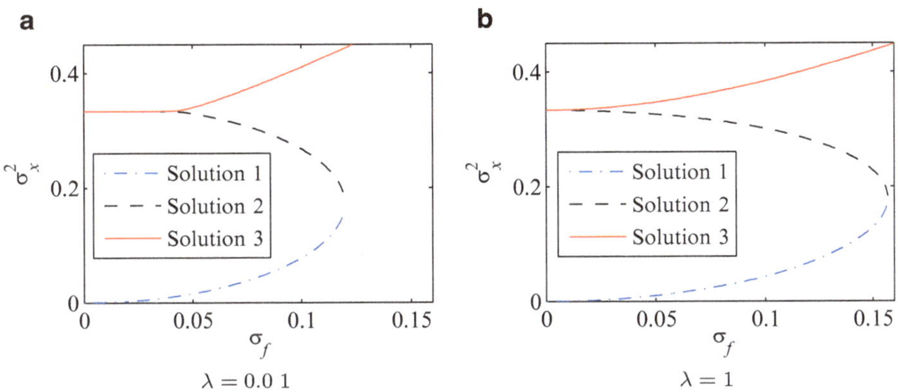

Fig. 8.11 Variance of the response, σ_x^2, for different standard deviations of the excitation, σ_f

therefore cannot be negative. Consistent line types for the three solutions are used for all figures in this section.

Figure 8.11 shows the square of the standard deviation of the response as a function of the standard deviation of the excitation. It can be observed (and as described in Sect. 8.4.1.1) that two real positive solutions of σ_x^2 exist for $m_x \neq 0$ which have values less than $\frac{1}{3}$. Another real positive solution is obtained solving Eq. (8.74) which results in $\sigma_x \geq \frac{1}{3}$.

Figure 8.12 shows the ratio of standard deviations of the response and the excitation. Note that Solutions 2 and 3 give high ratios for low σ_f but are unlikely to occur in practice. Fig. 8.13 shows the variation of the square of the mean of the system responses for different values of σ_f. Solution 3 has a zero mean mechanical response. The variance of the voltage is a measure of the mean power harvested and is shown in Fig. 8.14 for $\lambda = 0.01$ and $\lambda = 1$. The electrical constant λ

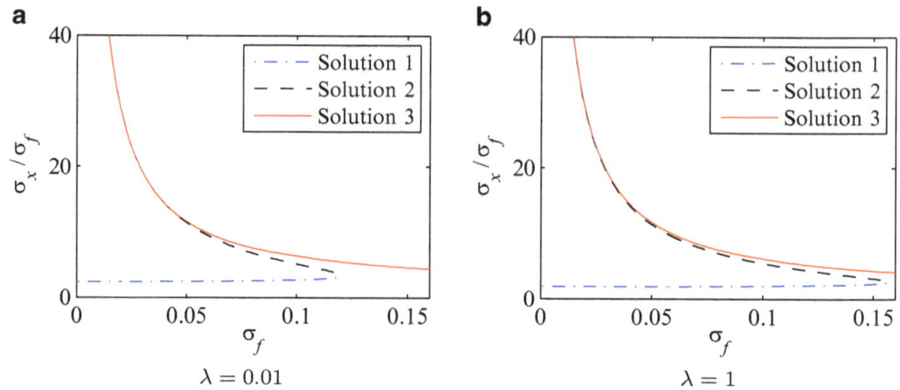

Fig. 8.12 Ratio of the standard deviations of the displacement to the excitation, σ_x/σ_f, for different standard deviations of the excitation, σ_f

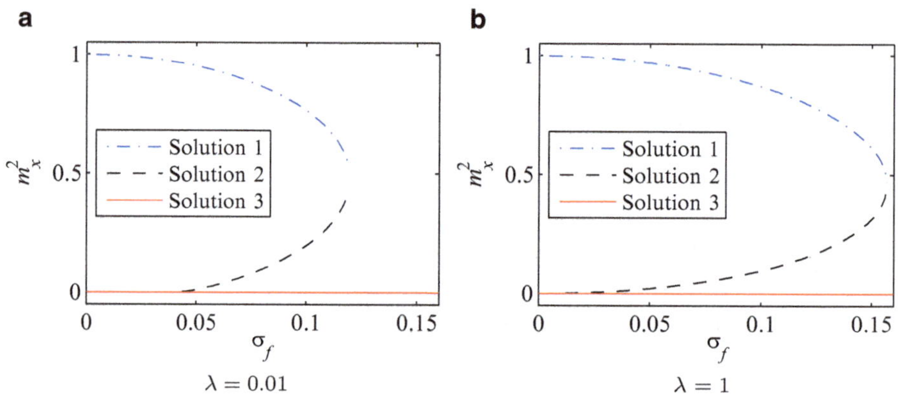

Fig. 8.13 Square of the mean displacement response for different standard deviations of the excitation, σ_f

significantly affects the voltage produced and hence the power generated, even though the mechanical responses shown in Figs. 8.10–8.13 are almost identical. The practical implementation of an increase in λ requires a reduction in either electrical resistance or capacitance. This matches reality, where power increases as the resistance across the circuit decreases. These requirements should be included in trade-off studies required for the design of a real system.

For all values of the excitation amplitude, σ_f, the stochastic linearisation produces three solutions. The form of the solutions is very similar as λ varies, with the solutions scaling along the σ_f axis. However, these solutions are not equally likely to occur; for small σ_f Solutions 1 and 2, about the equilibrium position, are most likely to occur, whereas for large σ_f Solution 3 is most likely. This was demonstrated by simulation in Sect. 8.3. For very large values of σ_f, only Solution

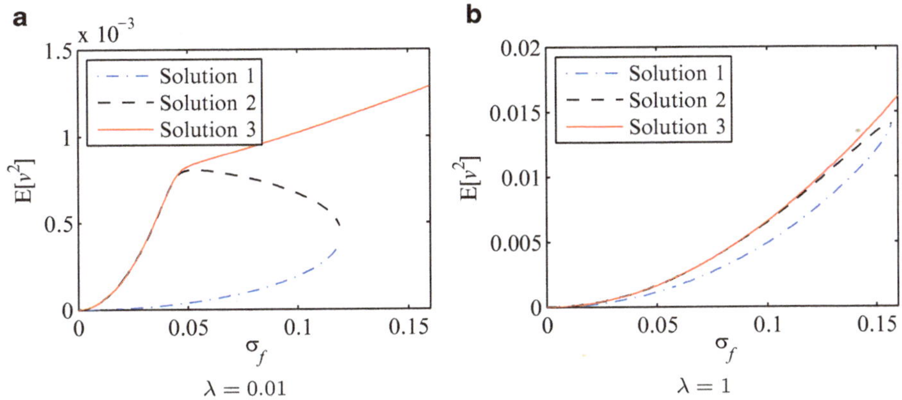

Fig. 8.14 Expected value of the voltage squared for different standard deviations of the excitation, σ_f

Fig. 8.15 Excitation standard deviation for various λ to give $\sigma_x^2 = 0.03$

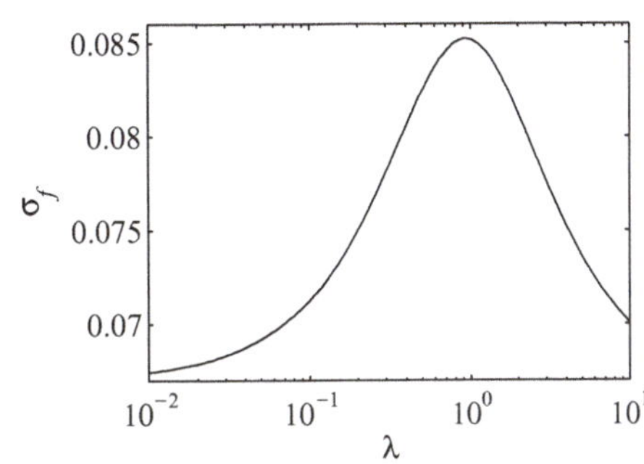

3 exists. The simulations suggest that there is a range of σ_f where the actual displacement response changes from a local response (within single potential well) to a global response (hopping between potential wells). The range of σ_f where this occurs may be estimated by considering the probability that Solution 1 escapes from the potential well. This produces a threshold on σ_x and hence for particular system a corresponding value of σ_f. Figure 8.15 shows how this value of σ_f varies with λ for $\sigma_x^2 = 0.03$ and shows that $\lambda = 1$ delays the onset of hopping between potential wells to higher values of σ_f, as shown in Sect. 8.3.

8.5 Conclusion

Piezomagnetoelectric harvesters are well suited to broadband energy harvesting. These devices are nonlinear, and their mechanical counterpart is represented by Duffing-type oscillator with a double well potential. The dynamic response of the system to broadband excitation depends on the amplitude of the excitation. For low excitation levels the response remains within one of the potential wells with a low response variance. For high excitation levels the system easily hops between the potential wells, and the system approximates a cubic stiffness nonlinearity. For intermediate excitation levels, the occurrence of hops between potential wells is stochastic, and this produces a stochastic response variance. The result is a stochastic resonance and the optimum mechanical response occurs at these intermediate excitation amplitudes.

For many energy harvesting systems, including those analysed in the chapter, the coupling between the mechanical and electrical systems is small, and hence the electrical time constant only has a small effect on the mechanical response. However, the electrical time constant has a significant impact on the voltage output and thus the power generated. The analysis of the linear harvester excited by broadband excitation highlighted that the electrical time constant should be matched to the response statistics and in particular the resonance frequency of the single degree of freedom system. For the nonlinear piezomagnetoelectric harvester, there are several significant frequencies in the response. The first is related to the natural frequency of the equivalent linear model of the nonlinear system, and this chapter has used stochastic linearisation to estimate this frequency. The second is the frequency of hopping between the potential wells, which is approximately related to the probability that a response about one of the equilibrium positions escapes from the potential well. Tuning the electrical time constant to the main system response appears to give the highest power output, which increases with excitation amplitude. However, if the electric time constant is tuned to the hopping frequency, then the stochastic resonance phenomenon is also observed in the voltage and power output. Thus, the analytical expressions for the equivalent linear system developed in this chapter may be used to aid the design of nonlinear energy harvesters.

References

1. Ali S, Adhikari S, Friswell MI (2010) Piezoelectric energy harvesting with parametric uncertainty. Smart Mater Struct 19(105010)
2. Ali SF, Adhikari S, Friswell MI, Narayanan S (2011) The analysis of magnetopiezoelastic energy harvesters under broadband random excitations. J Appl Phys 109(074904)
3. Amirtharajah R, Chandrakasan A (1998) Self-powered signal processing using vibration-based power generation. IEEE J Solid State Circ 33(5):687–695
4. Anton SR, Sodano HA (2007) A review of power harvesting using piezoelectric materials (2003–2006). Smart Mater Struct 16(3):R1–R21
5. Arnold DP (2007) Review of microscale magnetic power generation. IEEE Trans Magn 43(11):3940–3951
6. Barton DAW, Burrow SG, Clare LR (2010) Energy harvesting from vibrations with a nonlinear oscillator. J Vib Acoust 132(021009)
7. Beeby SP, Tudor MJ, White NM (2006) Energy harvesting vibration sources for microsystems applications. Meas Sci Tech 17(12):175–195

8. Beeby SP, Torah RN, Tudor MJ, Glynne-Jones P, O'Donnell T, Saha CR, Roy S (2007) A micro electromagnetic generator for vibration energy harvesting. J Micromechanics Microengineering 17(7):1257–1265
9. Bolotin VV (1984) Random vibration of elastic systems. Martinus and Nijhoff Publishers, The Hague
10. Cottone F, Vocca H, Gammaitoni L (2009) Nonlinear energy harvesting. Phys Rev Lett 102(080601)
11. Daqaq M (2010) Response of a uni-modal Duffing-type harvesters to random force excitations. J Sound Vib 329:3621–3631
12. duToit NE, Wardle BL (2007) Experimental verification of models for microfabricated piezoelectric vibration energy harvesters. AIAA J 45(5):1126–1137
13. Dutoit NE, Wardle BL, Kim SG (2005) Design consideration for mems scale piezoelectric mechanical vibration energy harvesters. Integrated Ferroelectrics Int J 71(1):121–160
14. Erturk A, Inman DJ (2008a) A distributed parameter electromechanical model for cantilevered piezoelectric energy harvesters. J Vib Acoust 130(041002)
15. Erturk A, Inman DJ (2008b) Issues in mathematical modeling of piezoelectric energy harvesters. Smart Mater Struct 17(065016)
16. Erturk A, Inman DJ (2008c) On mechanical modeling of cantilevered piezoelectric vibration energy harvesters. J Intell Mater Syst Struct 19(11):1311–1325
17. Erturk A, Inman DJ (2009) An experimentally validated bimorph cantilever model for piezoelectric energy harvesting from base excitations. Smart Mater Struct 18(025009)
18. Erturk A, Inman DJ (2011) Broadband piezoelectric power generation on high-energy orbits of the bistable duffing oscillator with electromechanical coupling. J Sound Vib 330:2339–2353
19. Erturk A, Hoffmann J, Inman DJ (2009) A piezomagnetoelastic structure for broadband vibration energy harvesting. Appl Phys Lett 94(254102)
20. Ferrari M, Ferrari V, Guizzetti M, Ando B, Baglio S, Trigona C (2010) Improved energy harvesting from wideband vibrations by nonlinear piezoelectric converters. Sensor Actuator Phys 162:425–431
21. Gammaitoni L, Hanggi P, Jung P, Marchesoni F (1998) Stochastic resonance. Rev Mod Phys 70(1):223–287
22. Gammaitoni L, Neri I, Vocca H (2009) Nonlinear oscillators for vibration energy harvesting. Appl Phys Lett 94:164102
23. Gammaitoni L, Neri I, Vocca H (2010) The benefits of noise and nonlinearity: extracting energy from random vibrations. Chem Phys 375:435–438
24. Halvorsen E (2008) Energy harvesters driven by broadband random vibrations. J Microelectromech Syst 17(5):1061–1071
25. Kazakov IE (1965) Generalization of methods of statistical linearizarion to multidimensional systems. Autom Rem Contr 26:1201–1206
26. Lin YK (1967) Probabilistic theory of structural dynamics. McGraw-Hill, New York
27. Litak G, Friswell MI, Adhikari S (2010) Magnetopiezoelastic energy harvesting driven by random excitations. Appl Phys Lett 96(21):214,103
28. Mann BP, Owens BA (2010) Investigations of a nonlinear energy harvester with a bistable potential well. J Sound Vib 329:1215–1226
29. Mann BP, Sims ND (2009) Energy harvesting from the nonlinear oscillations of magnetic levitation. J Sound Vib 319(1–2):515–530
30. Marinkovic B, Koser H (2009) Smart sand-a wide bandwidth vibration energy harvesting platform. Appl Phys Lett 94(103505)
31. Masana R, Daqaq MF (2011) Relative performance of a vibratory energy harvester in mono- and bi-stable potentials. J Sound Vib 330:6036–6052
32. McInnes C, Gorman D, Cartmell M (2010) Enhanced vibrational energy harvesting using nonlinear stochastic resonance. J Sound Vib 318(4–5):655–662
33. Moon FC, Holmes PJ (1979) A magnetoelastic strange attractor. J Sound Vib 65(2):275–296

34. Newland DE (1989) Mechanical vibration analysis and computation. Longman, Harlow and Wiley, New York
35. Nigam NC (1983) Introduction to random vibration. MIT, Cambridge, MA
36. Priya S (2007) Advances in energy harvesting using low profile piezoelectric transducers. J Electroceramics 19(1):167–184
37. Quinn DD, Triplett AL, Bergman LA, Vakakis AF (2011) Comparing linear and essentially nonlinear vibration-based energy harvesting. J Vib Acoust 133(011001)
38. Ramlan R, Brennan MJ, Mace BR, Kovacic I (2010) Potential benefits of a non-linear stiffness in an energy harvesting device. Nonlinear Dynam 59:545–558
39. Renno JM, Daqaq MF, Inman DJ (2009) On the optimal energy harvesting from a vibration source. J Sound Vib 320(1–2):386–405
40. Roberts JB, Spanos PD (1990) Random vibration and statistical linearization. Wiley, Chichester
41. Roberts JB, Spanos PD (2003) Random vibration and statistical linearization. Dover Publications, Mineola, New York
42. Sebald G, Kuwano H, Guyomar D, Ducharne B (2011) Experimental Duffing oscillator for broadband piezoelectric energy harvesting. Smart Mater Struct 20(102001)
43. Sodano HA, Inman DJ, Park G (2004a) A review of power harvesting from vibration using piezoelectric materials. Shock Vib Digest 36(3):197–205
44. Sodano HA, Park G, Inman DJ (2004b) Estimation of electric charge output for piezoelectric energy harvesting. Strain 40(2):49–58
45. Stanton SC, McGehee CC, Mann BP (2010) Nonlinear dynamics for broadband energy harvesting: investigation of a bistable piezoelectric inertial generator. Phys D 640–653
46. Williams C, Yates R (1996) Analysis of a micro-electric generator for microsystems. Sensor Actuator Phys 52:8–11

Part IV
Fluidic Energy Harvesting

Chapter 9
Energy Harvesting from Fluids Using Ionic Polymer Metal Composites

Maurizio Porfiri and Sean D. Peterson

Abstract In this chapter, we discuss energy harvesting from steady, oscillatory, and unsteady water flows using ionic polymer metal composites (IPMCs). After a brief description of this new class of active materials and their ability to transduce strain energy into electrical form, we present three case studies spanning this range of flow environments. First, we examine energy harvesting from a heavy flapping flag hosting IPMCs in a steady flow water channel; second, we consider base excitation of a cantilevered IPMC to simulate the effect of an oscillatory flow; and finally, we investigate the impact of a vortex ring with an IPMC. Analytical insight on the mechanics of the coupled fluid–structure problem is used to interpret experimental results and provide design guidelines for energy harvesters based on active compliant materials in fluids.

9.1 Introduction

Recent progress in lightweight smart materials has opened the door to scientific and technological advancements in the area of energy harvesting [30]. Energy extracted from the surroundings can be potentially used to power low consumption electronic devices [20, 35, 39, 40, 47, 66, 82]. The main advantages over traditional batteries of powering devices via energy harvesting are extended device lifetime, limited

M. Porfiri (✉)
Department of Mechanical and Aerospace Engineering, Polytechnic Institute of New York University, 6 Metrotech Center, Brooklyn, NY 11201 USA
e-mail: mporfiri@poly.edu

S.D. Peterson
Department of Mechanical and Mechatronics Engineering, University of Waterloo, 200 University Avenue West, Waterloo, ON N2L 3G1 Canada
e-mail: peterson@mme.uwaterloo.ca

N. Elvin and A. Erturk (eds.), *Advances in Energy Harvesting Methods*,
DOI 10.1007/978-1-4614-5705-3_9,

maintenance, and reduced onboard weight. Comprehensive overviews of energy harvesting methods and implementations can be found in [49, 51, 68, 87].

Mechanical vibration has received particular attention as a simple and viable source for energy harvesting using active materials. In this context, piezoelectrics have seen considerable research efforts in the conversion of mechanical energy into electrical energy via deformation (see the reviews in [7, 75]). Energy harvesting from the base excitation of cantilevered piezoelectric beams is analytically studied and experimentally validated in [28, 29]. The feasibility of enhancing the energy harvesting capabilities of piezoelectrics through shunting and nonlinear circuit elements is presented in [38] and strategies for optimization of design parameters are discussed in [43, 67]. Methods for expanding the frequency range and storing electric energy from piezoelectric harvesters are discussed in [46, 76, 77].

The momentum carried by research on piezoelectric energy harvesting of mechanical vibration has sparked into emerging interest for exploring alternative uses of piezoelectrics for energy scavenging. Specifically, a few implementations of piezoelectrics in fluidic energy harvesting, wherein the fluid is used as a source of energy instead of a passive damper, have been recently presented. Drawing inspiration from traditional windmills, the design of a small scale piezoelectric-based windmill is presented in [53]. Energy harvesting from a piezoelectric strip placed in the wake of a cylinder in the direction of the cross flow is studied in [6, 81]. Therein, the strip is subject to the time-varying pressure loading caused by the Kármán vortex street shed from the cylinder. A similar configuration placed in a turbulent boundary layer is considered in [2]. Structural instabilities induced by the fluid coupling are explored in [21, 31, 78]. Specifically, fluttering structures for energy harvesting are described in [21, 31] and flow-induced self-oscillations of structures hosting piezoelectrics are considered in [78].

While piezoelectrics offer several advantages over other active materials in terms of energy harvesting, including a good energy conversion rate and reliable performance, their implementation in harvesting fluid energy as miniature devices can be limited by their stiffness, brittleness, and challenges associated with working in wet environments. Ionic polymer metal composites (IPMCs), sometimes referred to as ionic polymer transducers, are a novel class of electro-active polymers that are currently receiving considerable attention for their propitious attributes for both sensing and actuation. An IPMC is a porous ionic polymer saturated with an electrolytic solution and plated by noble metal electrodes (see for example [70]). Mechanical deformation yields a voltage difference across the electrodes, and, similarly, a voltage difference across the electrodes produces structural deformations. The electrolytic solution comprises a mobile counterion species and an uncharged solvent. Sensing and actuation are the result of interactions between the mobile ions and the solvent, and between the solvent and the backbone polymer [56]. A comprehensive review of modeling, fabrication, and applications of IPMCs up to the early part of this decade can be found in [44, 71–73]. Recent modeling efforts include black-box linear and nonlinear models [16–18, 37] and physics-based models that employ multiphase mixture theory and Poisson–Nernst–Planck equations (see, e.g., [22, 23, 26, 41, 54, 55, 57, 61, 62, 84–86]). Advanced fabrication techniques are presented in [3, 4]. Recent applications span underwater propulsion systems [1, 9, 42, 50, 60, 88], force and displacement sensors [15, 19, 34, 70], and grippers and manipulation systems [27, 33, 45].

Despite their considerable use as sensors and actuators, there are few demonstrations of IPMC-based energy harvesting. Specifically, energy harvesting from mechanical vibrations in air is demonstrated in [20, 35, 82]. In [82], IPMCs consisting of two different electrode materials are tested and compared under cyclic bending, tension, and shear. Analytical insight into base excitation of IPMCs in air along with experimental results is reported in [20]. Longitudinal vibrations in air are modeled, tested, and compared with results from piezoelectrics in [35].

In this book chapter, we report the implementation of IPMCs as fluid energy harvesters in three different conditions exemplifying steady, oscillatory, and unsteady flows. Specifically, we summarize the integration of IPMCs in flapping flags underwater presented in [36], the base excitation of IPMC strips in water studied in [13], and the interaction between IPMCs and coherent fluid structures described in [58, 59].

9.2 IPMC Energy Conversion

As a first approximation, the IPMC sensing behavior can be described by using a lumped circuit model comprising a linear impedance in series with a voltage source that depends on the IPMC mechanical deformation. This model is derived starting from a Poisson–Nernst–Planck model of the dynamics of the counterion concentration and the electric potential within the ionomeric membrane following [13]. In this model, IPMC deformation acts as a forcing term in Poisson's equation, wherein the free charge density is described by the concentration of the mobile counterions and of the fixed charges, which varies as the IPMC deforms. The modeling approach presented therein discards convective terms in the governing equations and surface roughness of IPMC electrodes [10, 63] and focuses on small deformations producing voltage levels smaller than the thermal voltage.

Within this lumped circuit model, the linear impedance is found to be the series connection of a resistor $\mathscr{R}$ and a capacitor $\mathscr{C}$. The resistor accounts for the ion diffusion in the ionomeric membrane and it can practically be used to incorporate the resistivity of the plating electrodes. The capacitor embodies bulk and double layer effects [14] in the vicinity of the IPMC electrodes and is highly correlated to the electrode surface roughness [10, 63]. The voltage source describes IPMC sensing and is proportional to the rate of change in time t of the relative rotation $\vartheta(t)$ between the IPMC ends through the constant $\mathscr{B}/\mathscr{C}$, that is,

$$V_{\mathrm{OC}}(t) = \frac{\mathscr{B}}{\mathscr{C}} \frac{\mathrm{d}\vartheta(t)}{\mathrm{d}t} \tag{9.1}$$

Here, $\mathscr{B}$ represents an electromechanical gain, and subscript OC is used to emphasize that this voltage is the IPMC output when its terminals are left open circuited. For convenience, we refer to the rate of change of $\vartheta(t)$ as $\Omega(t)$. We further note that Eq. (9.1) implies that the overall voltage drop across the IPMC depends only on the differences between the rotations at its ends that is a measure of the average IPMC curvature. This is due to the fact that the original physics-based model assumes that the electrodes are perfect conductors.

By considering oscillations at the radian frequency ω and using a superimposed hat to identify phasors, the voltage $\hat{V}(\omega)$ across the IPMC is

$$\hat{V}(\omega) = \frac{\hat{I}(\omega)}{Y(\omega)} - \frac{\mathscr{B}}{\mathscr{C}}\hat{\Omega}(\omega) \tag{9.2}$$

where the IPMC electrical admittance $Y(\omega)$ equals $\mathrm{i}\omega\mathscr{C}/(1 + \mathrm{i}\omega\mathscr{R}\mathscr{C})$ and $I(t)$ is the current through the IPMC. Thus, the IPMC response is described by using only the parameters $\mathscr{C}, \mathscr{R}$, and $\mathscr{B}$ that are experimentally identified from a discharge experiment along with an open circuit, or short circuit, measurement by following the procedure described in [13]. We further comment that if convective terms are taken into consideration, the expression for $V_{\mathrm{OC}}(t)$ is modified to favor a dependence on the relative rotation as compared to its time derivative [11]. Practically, given the experimental identification of the electromechanical gain, the noise in experimental data and the relatively narrow range of vibrations typically considered for energy harvesting, both the approaches may be suitable.

When the IPMC electrodes are connected to an external resistive load R_l, its average power output is

$$\hat{P}(\omega) = \frac{1}{2}\frac{|\hat{V}(\omega)|^2}{R_l} = \frac{1}{2}\frac{\omega^2\mathscr{B}^2}{(\mathscr{R} + R_l)^2\omega^2\mathscr{C}^2 + 1}R_l\,|\hat{\Omega}(\omega)|^2 \tag{9.3}$$

The resistance that maximizes the power transfer is found by setting $\partial\hat{P}(\omega)/\partial R_l = 0$, that is,

$$R_l^{\mathrm{opt}}(\omega) = \frac{1}{\omega\mathscr{C}}\sqrt{1 + \omega^2\mathscr{T}^2} \tag{9.4}$$

where the time constant of the equivalent IPMC circuit is defined as $\mathscr{T} = \mathscr{R}\mathscr{C}$.

9.3 Energy Harvesting from Flapping Flags

Here, we report on an energy harvester based on a heavy and flexible flag flapping in a uniform water flow hosting an IPMC, as designed in [36] and shown in Fig. 9.1. The physics of the fluid–structure interaction and the resulting instability of the fluttering flag can be elucidated by considering the propagation of waves in a fluid-loaded elastic plate [25]. We consider the two-dimensional incompressible inviscid

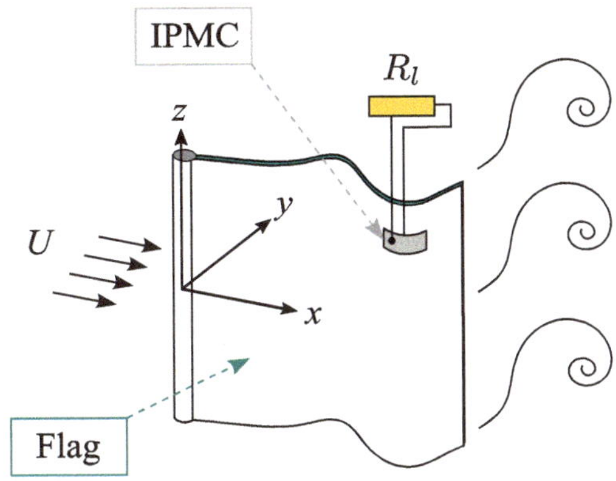

Fig. 9.1 Schematic of the flexible heavy flag flapping in a uniform water stream of velocity U, comprising the host structure and IPMC strip (Reprinted with permission from [36]. Copyright 2011, American Institute of Physics)

flow with free stream speed U and density ρ over a thin elastic plate of infinite extension along the x direction. Following [74], we neglect the near field solutions and the effect of the wake on the instability boundary [5, 80] to derive a simple closed-form expression for the flutter boundary.

We model the structure as a homogenous Kirchhoff–Love plate of length L_{flag}, width w_{flag}, bending stiffness per unit width B_{flag}, and mass per unit length m_{flag} undergoing cylindrical bending [48]. The plate is subject to a pretension T due to the fluid boundary layer and to the pressure of the surrounding fluid. The plate is coupled to the fluid problem by following [5], where the flag boundary is modeled as a bound vortex and the relative fluid velocity at the flag is set to zero. From dimensional analysis, we find that the fundamental nondimensional groups are

$$R_1 = \frac{m_{\text{flag}}}{\rho L_{\text{flag}} w_{\text{flag}}}, \qquad R_2 = \frac{B_{\text{flag}}}{\rho U^2 L_{\text{flag}}^3}, \qquad \beta = \frac{T L_{\text{flag}}^2}{B_{\text{flag}}} \tag{9.5}$$

These parameters describe the mass ratio between the flag and the surrounding fluid, the ratio between elastic and fluid forces, and the stiffening effect of the tension T with respect to the flag's inherent bending stiffness, respectively.

By analyzing the linear stability of a the fundamental propagating wave, that is, a vibration with wavelength equal to L_{flag}, we obtain a closed-form expression for the onset of flutter as a function of the three nondimensional parameters. The stability is represented in Fig. 9.2 with the stable region in the upper left side of the plot. The space (R_1, R_2, β) allows visual identification of the effect of flag mass, velocity, and tension on the stability boundary. In particular, increasing values of the tension tend to stabilize the system, while increasing values of the flag mass have the opposite effect.

The host flag is constructed following [74]: a Mylar sheet of thickness 26 μm constitutes the flexible membrane, and copper strips of dimension $102 \times 6.7 \times 0.4\,\text{mm}^3$ are glued every 4.4 mm of bare Mylar on both sides to increase the mass per unit length of the flag. The IPMC sample is integrated in the structure by

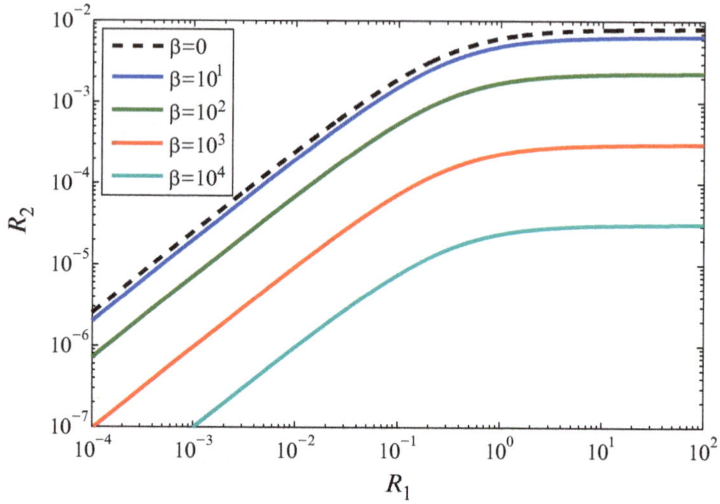

Fig. 9.2 Linear stability plot as the tension to bending stiffness ratio β is varied. The region above the boundary identifies stable oscillations of the system. In the region below the boundary the system is unstable. The *dashed line* represents the case $T = 0$ (Reprinted with permission from [36]. Copyright 2011, American Institute of Physics)

clamping it within the copper strips. The mechanical parameters of the heavy flag, that is, the mass per unit length and the bending stiffness, are estimated from its periodic structure and the physical parameters of the constituent materials yielding $B_{\text{flag}} = 2 \times 10^{-5}\,\text{N m}$ and $m_{\text{flag}} = 0.45\,\text{kg/m}$. The tension in the flag is estimated using the Blasius boundary layer solution for an infinite plate theory following [24]. From Eq. (9.5) and Fig. 9.2, we estimate that the critical flow speed of flutter instability is $U_c = 0.23\,\text{m/s}$, as what occurs beyond this speed would likely be bounded nonlinear limit-cycle oscillations.

The IPMC harvester is fabricated in-house from commercial Nafion membrane foils N117 produced by DuPont de Nemours by following the procedure developed by Oguro from Osaka National Research Institute, AIST, Japan (see also [10]). The nominal dimensions of the sample are $33 \times 11 \times 0.2\,\text{mm}^3$ and the resistance $\mathscr{R}$ and capacitance $\mathscr{C}$ are $390\,\Omega$ and $4\,\mu\text{F}$, respectively.

The device is immersed in the test section of a water tunnel of nominal dimensions $2500 \times 150 \times 150\,\text{mm}^3$ filled with tap water and the onset and development of the flutter instability is studied for increasing velocities. The water speed is measured with laser Doppler velocimetry, the post-flutter vibrations are tracked by means of image analysis and the time trace of the voltage across the load R_l is recorded with a data acquisition (DAQ) board.

A qualitative analysis of the flow before and after the onset of flapping can be garnered through flow visualization as illustrated in Fig. 9.3. In Fig. 9.3a, representing a flow speed of $U = 0.4\,\text{m/s}$, the vortex sheet aligns itself to the flow direction. In Fig. 9.3b,c, the periodic release of vorticity having alternating signs is

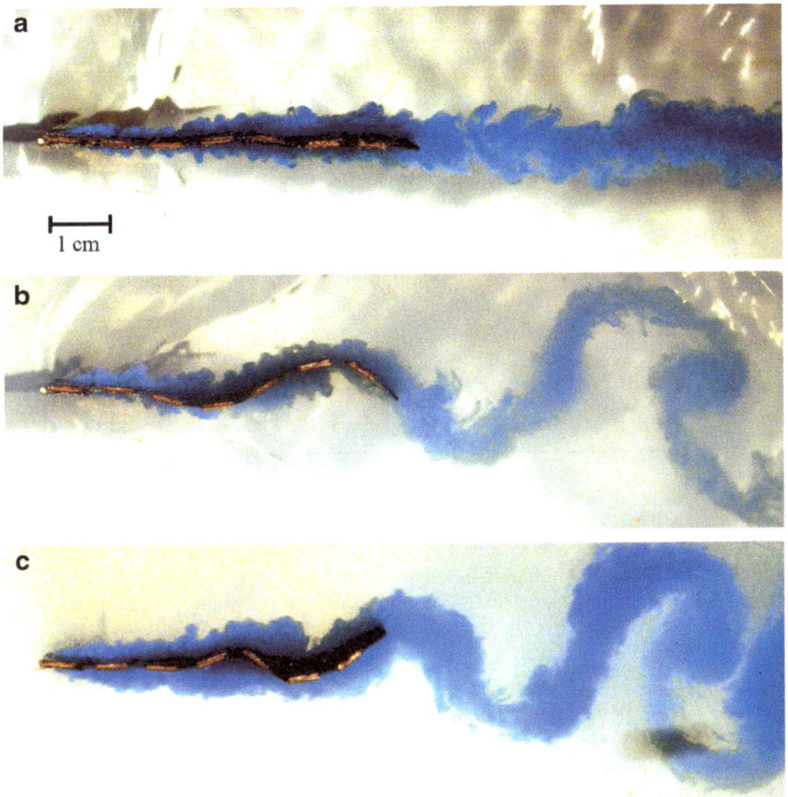

Fig. 9.3 Flow visualization of the vortex sheet originating on the boundary of the flag and released in the wake. Images are captured (**a**) before and (**b**)–(**c**) after the onset of flutter, for free stream speed values of 0.4, 0.6, and 0.8 m/s, respectively (Reprinted with permission from [36]. Copyright 2011, American Institute of Physics)

made evident by the dye lines. The free stream velocities are 0.4, 0.6 and 0.8 m/s, respectively. The shedding frequency increases with the background flow speed as can be deducted from the different numbers of folds in the wake sheets.

From the direct measurement of the IPMC voltage output for varying R_l, in Fig. 9.4 we present an estimate of the scavenged power. The maximum power conversion is obtained when the load resistance is $R_l = 10\,\mathrm{k}\Omega$, in line with the prediction of Eq. (9.4) for the considered range of vibration frequencies on the order of a few Hertz. The harvested power in the optimal conditions lies in the range 10^{-12}–10^{-10} W. This estimate is conservative since it neglects the energy content of higher harmonics that can be exploited in real applications. The power harnessed from the flow can be increased by enlarging the portion of flag surface area covered with IPMC strips. In this prototype, the IPMC strip covers approximately 2% of the area of the active flag and contributes to approximately 0.5% of its weight.

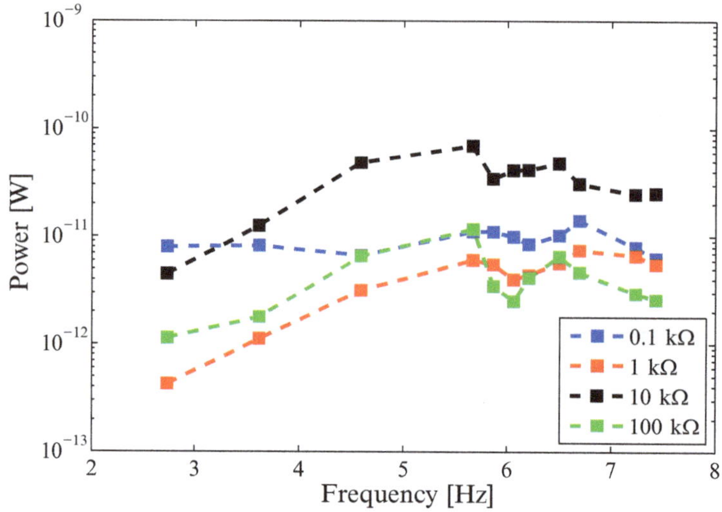

Fig. 9.4 Harvested power as a function of flapping frequency and load resistance. The estimate is performed by including harmonic components within ±0.3 Hz from the flapping frequency in the computation of the voltage amplitude (Reprinted with permission from [36]. Copyright 2011, American Institute of Physics)

9.4 Energy Harvesting from an Oscillating Fluid

Here, we consider energy harvesting from the base excitation of a cantilevered IPMC to explore the mechanics of energy scavenging from oscillatory flows. Unlike the problem considered above, in this case the IPMC is the sole structural element. Specifically, we consider a cantilevered IPMC rectangular strip vibrating in a fluid in response to harmonic base excitation of its clamped end.

The energy available for harvesting can be estimated from the analysis of a simplified model of the coupled fluid–structure problem, wherein the IPMC is modeled as a Kirchhoff–Love plate undergoing cylindrical bending and the effect of the encompassing fluid is lumped into a hydrodynamic function [8, 12, 32, 69, 83]. Such function describes both added mass and viscous damping from the surrounding fluid and can be derived by solving an auxiliary two-dimensional problem where a rigid lamina, identifying a generic beam cross section, periodically oscillates in the fluid. For small vibrations compared to the IPMC width, convective terms in the Navier–Stokes equations can be discarded in favor of an unsteady Stokes flow, which yields a hydrodynamic function that depends exclusively on the oscillatory Reynolds number [69, 83]. If oscillations become more severe, vorticity shedding and advection in the flow becomes dominant and the dependence of the hydrodynamic function on the Keulegan–Carpenter number, related to the ratio between the oscillation amplitude and the width, should be taken into consideration to elucidate nonlinear damping [8, 12, 32].

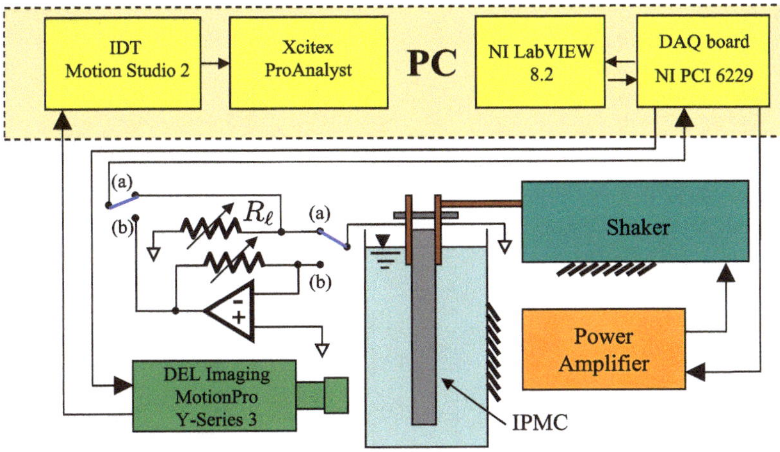

Fig. 9.5 Schematics of the experimental setup from [13]. The switch positions (**a**) and (**b**) allow for voltage, current, and power measurements in open circuit, short circuit, and loaded conditions

The IPMC sample used in this study, purchased from Environmental Robots, Inc., has nominal dimensions of $30 \times 5.8 \times 0.2\,\mathrm{mm}^3$. The mass per unit length of the sample is $3.48 \times 10^{-3}\,\mathrm{kg/m}$. The undamped in-vacuum fundamental natural frequency is 12.5 Hz and the in-vacuum loss factor is 0.14, yielding an IPMC bending stiffness per unit width of $B = 1.2 \times 10^{-4}\,\mathrm{N\,m}$. The IPMC sample is tested in a 2 l pool of deionized water. The sample is arranged in a cantilever configuration with a free vibration length of $L = 25\,\mathrm{mm}$ and the clamped end rigidly attached to a shaker to produce the base excitation. The shaker is driven with a sinusoidal voltage signal with the actuation frequency varied from 2 to 50 Hz. The clamp is electrically connected to a DAQ board to record the voltage output of the IPMC (see Fig. 9.5). The actual motion of the clamped edge of the IPMC is constantly monitored by using a laser displacement sensor. The IPMC lateral displacement is recorded through a high-speed camera and then analyzed using a commercial software.

Figure 9.6 displays experimental data on the IPMC frequency response defined by taking the base excitation as the input and the tip deflection, that is, the difference between the tip displacement and the rigid body motion, as the output. The discrepancy between the theoretical predictions from the linear theory highlights the importance of vorticity shedding in the mechanics of fluid–structure interactions. On the other hand, the model in [8] is able to reasonably capture the magnitude and the phase of the transfer function for low to moderately high frequencies. Discrepancies in the high-frequency region of the phase diagram may be attributed to noise amplification due to the indirect nature of the relative tip displacement measurements. We note that in the considered broad frequency interval, the magnitude of the transfer function varies in a rather limited range. This hints that energy harvesting may be possible in a variety of loading conditions that are typically present in underwater systems, such as wave slamming [64] and structural vibrations [52].

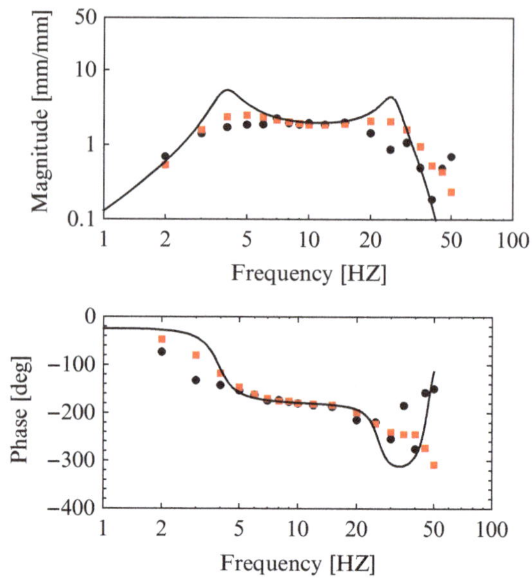

Fig. 9.6 Frequency response of the IPMC tip deflection. *Red markers* are experimental data, *solid line* is the prediction using unsteady Stokes flow from [69, 83], and *black markers* are predictions from the nonlinear hydrodynamic model in [8]

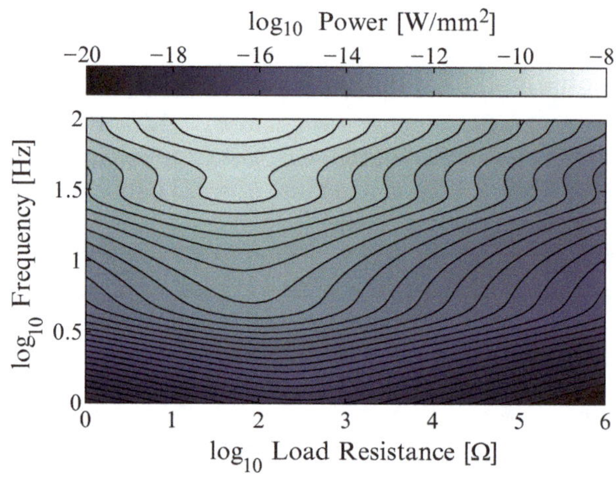

Fig. 9.7 Theoretical harvested power from IPMC underwater vibration from [13]

Two different experiments are performed to establish the feasibility of the vibrating IPMC as an energy harvester. In the first experiment, the IPMC is shunted with a resistor R_l, while in the second experiment, the short circuit current through the IPMC is measured. The electric parameters of the IPMC are $\mathscr{R} = 55.2\,\Omega$ and $\mathscr{C} = 0.51\,\mathrm{mF}$. The electromechanical coupling is $\mathscr{B} = 2.286 \times 10^{-9}\,\mathrm{A\,s^2/rad}$, which is comparable to the one for the flapping flag described above $1.85 \times 10^{-10}\,\mathrm{A\,s^2/rad}$.

Further, we determine the power harvested from the IPMC underwater vibration under loaded conditions by using Eq. (9.3) with the parameters discussed above. The influence of the vibration frequency and the electrical load on the harvested power is shown in Fig. 9.7, which is obtained by using the frequency response in Fig. 9.6.

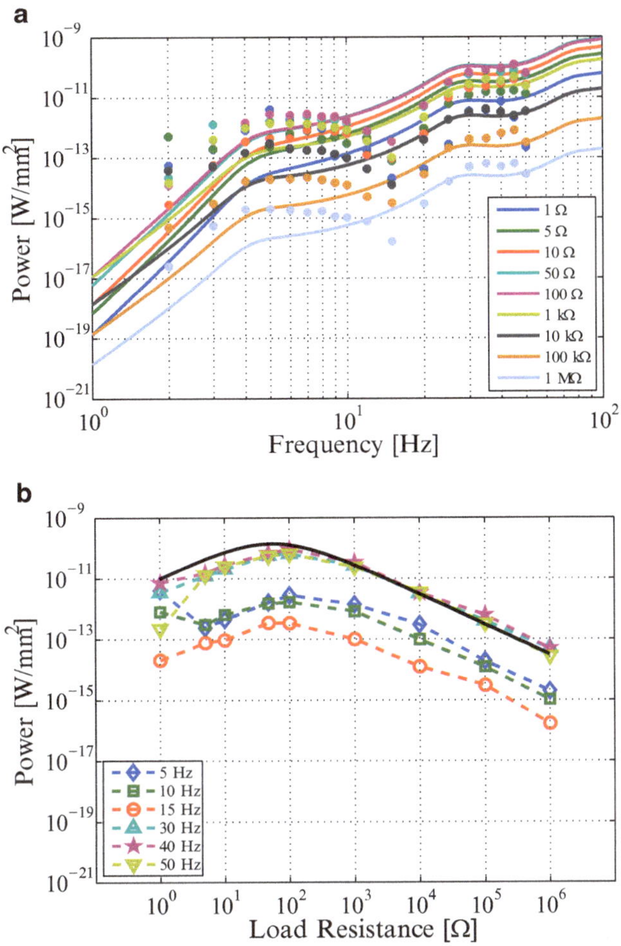

Fig. 9.8 (**a**) Comparison between the theoretical and experimental (points) power harvesting capability of the vibrating IPMC, from [13]. (**b**) Experimental data on power harvesting as a function of the shunting resistance. The *solid line* represents theoretical predictions

Note that nonlinear damping due to vorticity shedding is here approximated as a surplus of structural damping to simplify the parametric analysis. Figure 9.7 presents a contour map identifying optimal regions of operation for the harvesting device. The harvested power varies in the range 10^{-20}–10^{-8} W/mm^2 and significantly increases with increasing actuation frequency. In particular, the maxima of the harvested power occur when the actuation frequency matches the underwater resonance frequencies of the vibrating IPMC. That is, a flow oscillating at the underwater resonance frequency of the harvester will maximize the energy scavenged.

Experimental results are compared with theoretical predictions of power harvested versus input frequency in Fig. 9.8a. The curves in Fig. 9.8a show the trends of the harvested power with respect to the base excitation frequency and are parameterized with respect to the load resistance R_l. Figure 9.8b shows the dependence of the output power on the load resistance along with the existence of a maximum in correspondence to a shunting resistance in the range 10–100 Ω. The harvested power is computed using Eq. (9.3). We note that power harvested is expected to depend nonlinearly on the oscillation amplitude due to nonlinear hydrodynamic damping [8].

9.5 Energy Harvesting from Coherent Fluid Structures

Here, we consider energy harvesting from coherent fluid structures, summarizing results from [58,59]. Specifically, we study the mechanics of energy transfer from a self-propagating vortex ring in an otherwise placid fluid to an IPMC via impact. The IPMC harvester is fabricated in-house from commercial Nafion membrane foils N117 as per the flapping flag experiment. It has a total length of 36.6 mm, a nominal thickness of 200 μm, a width of $w = 14.3$ mm, its mass per unit length is $m = 5.63$ g/m, and its bending stiffness per unit width is $B = 2.83 \times 10^{-4}$ N m. The IPMC is fixed at one end by a custom clamping device with copper electrodes adhered to each arm of the clamp and immersed in a tank of dimensions $415 \times 236 \times 215$ mm^3 that is filled with tap water (see Fig. 9.10). The free vibration length is $L = 21.0$ mm and corresponds to a vibrating mass of 118 mg. In this configuration, the IPMC electrical capacitance is $\mathscr{C} = 37.8$ μF.

The vortex ring is generated by a piston submerged in water plunged through a cylinder with internal diameter of 19.1 mm. Experiments are conducted at room temperature and the water density is taken at its reference value of $\rho = 1,000$ kg/m^3. The stroke length of the piston is 75.0 mm, and the cylinder has a flat, sharp-edged exit plane. The piston is placed 165 mm below the free surface of the tank and 115 mm from the IPMC. It is situated such that the center of the vortex ring is approximately aligned with the tip of the IPMC and the direction of propagation of the ring is orthogonal to its rest position. Figure 9.9 illustrates a flow visualization experiment in which a high-speed video camera is used to record the fluid motion and the IPMC vibration. As the piston plunges into the cylinder, water is expelled and the resulting shear layer rolls up into a vortex ring. Once the plunger stops, vorticity is no longer formed and the vortex ring propagates towards the IPMC at a nearly constant speed due to self-induction. As the vortex ring approaches the IPMC, the IPMC slowly bends away from the vortex ring, and as it reaches the rest position of the IPMC, the energy transfer dramatically increases resulting in rapid and severe structural deflections.

We use time-resolved particle image velocimetry (PIV) [65] to compute the strength and trajectory of the vortex ring as it propagates towards and interacts with the IPMC. In addition, the obtained images are analyzed to study the transient vibration of the IPMC and explore its correlation with the vortex dynamics. A cross section of the vortex ring, as acquired from PIV, is seen in Fig. 9.10 along with

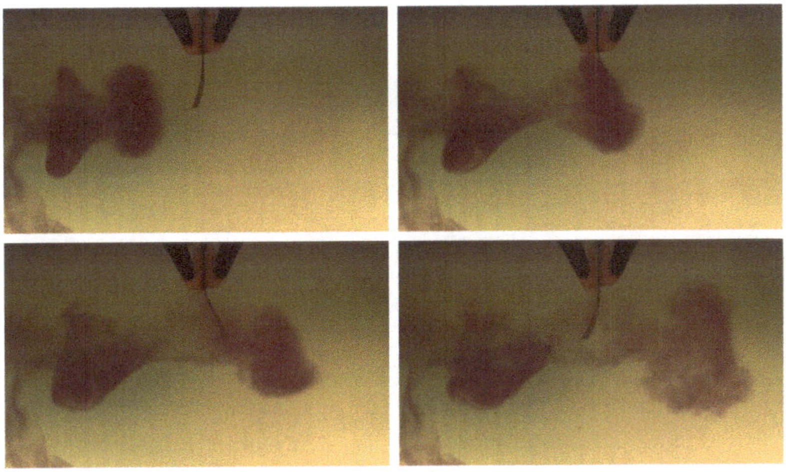

Fig. 9.9 Flow visualization images of a vortex ring impacting a cantilevered strip adapted from [59]

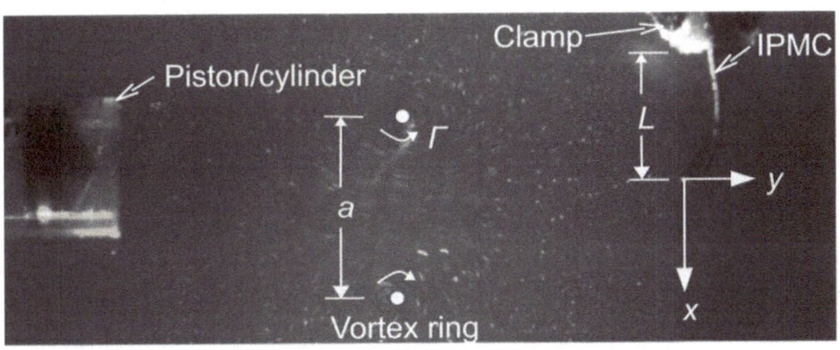

Fig. 9.10 Image of the experimental configuration with overlaid coordinate system and variable definitions (Reprinted with permission from [58]. Copyright 2011, American Institute of Physics)

pertinent nomenclature. The vortex ring circulation is denoted as Γ, its velocity is V_{vr}, and the diameter is termed a.

The circulation of the vortex is estimated by computing the line integral of the velocity field around each of the two vortices appearing in the two-dimensional slice through the center of the ring. Each experiment time is shifted so that t equals zero when the vortex ring is at $y = -2.5\,L$. The tip displacement of the IPMC, referred to as δ, is tracked in the sequence of images using a commercial software.

The physics of the interaction can be represented through three primary dimensionless parameters under the following assumptions: the fluid dynamics is modeled using potential flow theory, the vortex ring is assimilated to a vortex pair of equal and opposite circulation, and the beam is described as a Kirchhoff–Love plate undergoing cylindrical bending [59]. The mass ratio $\mu_1 = m/(\rho w L)$ relates the inertia of the beam to the inertia of the fluid and is equal to $\mu_1 = 0.019$. The second

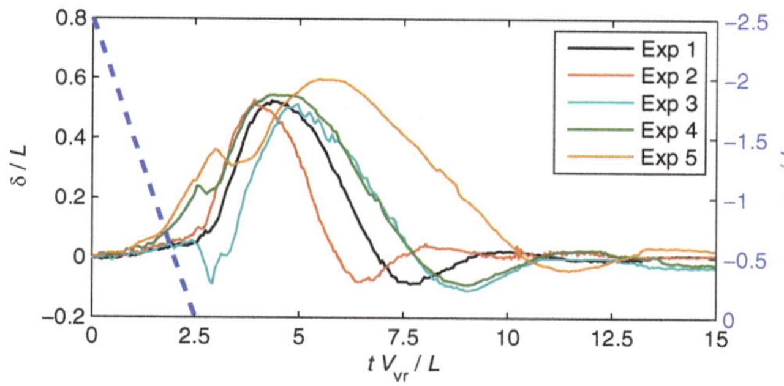

Fig. 9.11 Scaled deflection at the IPMC tip as a function of the nondimensional time tV_{vr}/L for five different experiments along with a curve indicating the vortex ring y-position on the secondary axis (Reprinted with permission from [58]. Copyright 2011, American Institute of Physics)

dimensionless parameter $\mu_2 = B/(\rho\Gamma^2 L)$ is the ratio of the restoring force in the IPMC to the fluid loading. The final parameter is a geometric parameter relating the vortex ring length scale a to the IPMC length scale L as $\mu_3 = a/L$.

Figure 9.11 displays the tip deflection as a function of time for five experiments under comparable flow conditions; the experiments are labeled as one through five (Exp 1: $\mu_2 = 0.073$, $\mu_3 = 1.3$, and $V_{vr} = 0.55\,\mathrm{m/s}$; Exp 2: $\mu_2 = 0.06$, $\mu_3 = 1.3$, and $V_{vr} = 0.46\,\mathrm{m/s}$; Exp 3: $\mu_2 = 0.03$, $\mu_3 = 1.5$, and $V_{vr} = 0.73\,\mathrm{m/s}$; Exp 4: $\mu_2 = 0.04$, $\mu_3 = 1.4$, and $V_{vr} = 0.78\,\mathrm{m/s}$; and Exp 5: $\mu_2 = 0.05$, $\mu_3 = 1.5$, and $V_{vr} = 0.80\,\mathrm{m/s}$). Included in the plot is the y-position of the vortex ring on the secondary axis y_{vr}. The maximum deflection experienced by the IPMC due to the impact is in the range $0.5 - 0.6\,L$ and the maximum deflection is reached after the vortex ring breaks down. When the vortex ring reaches the rest position of the IPMC, the deflection is typically not greater than $0.1 - 0.2\,L$ and only afterwards does the IPMC tip speed increase, suggesting that vortex breakdown plays a critical role in the energy exchange. This phenomenon is indeed not observed in the absence of viscosity, wherein secondary vortical structures are not generated [59]. After the impact takes place, the IPMC vibrates in its fundamental mode shape in a highly damped manner, ultimately returning to rest.

The energy transfer from the fluid to the solid is estimated from the energy associated with the vortex ring core and kinetic energy and the maximum strain energy in the IPMC. The former quantity is computed by using a thin core approximation with hollow core of constant volume [79], which yields 7.6, 5.5, 17.2, 15.6, and 16.8 mJ for experiments one through five, respectively. The maximum strain energy of the IPMC is estimated from the maximum tip deflection by using the fundamental mode of vibration of a cantilever beam, which yields 89, 83, 88, 96, and 120 μJ for experiments one through five, respectively. Therefore, the energy transfer computed as the ratio between the input fluid energy and the elastic energy ranges between 0.5 and 1.5%.

In contrast with the steady flow and the oscillatory experiments, the electrical response from the IPMC in this experiment is purely transient. In this case, the potential for energy harvesting is measured by estimating the electrical energy stored in the IPMC during the interaction in short circuit conditions. Thus, we find that the energy transduced into electrical form is 0.56, 0.62, 0.23, 0.56, and 1.5 nJ for experiments one through five. This indicates that the energy transfer from strain energy into electrical is in line with previously observed conversion rates on the order of 0.0003%–0.0012%.

9.6 Summary and Conclusions

In this chapter, we have presented a few implementations of IPMCs as energy harvesting devices for underwater applications. Specifically, we have described the potential use of such active materials in scavenging energy from steady and oscillatory flows and from coherent fluid structures. In this context, IPMCs offer several advantages over more commonly used active materials, including high compliance, ability of working in wet environments, and good potential for energy storage due to their large electrical capacitance. Another main advantage of these materials is that their electromechanical behavior is largely dictated by surface rather than volumetric properties, which hints at the potential for miniaturization. Nevertheless, the energy conversion rate of this material is currently inferior to more conventional active materials, such as piezoelectrics; yet, we foresee that the current efforts on IPMC fabrication will continue to improve their performance. We also note that the methodologies explored in this book chapter are not strictly confined to IPMCs as the energy harvesters nor to water as the fluid medium. Moreover, the fundamental knowledge developed in the analysis of fluid-structure interactions through direct IPMC-based strain measurement is expected to aid complementary studies on dynamic loading of marine structures.

Acknowledgements This research was supported by the National Science and Engineering Research Council, the National Science Foundation, and the Office of Naval Research under grant numbers 386282-2010, CMMI-0745753 and CMMI-0926791, and N00014-10-1-0988, respectively. The authors would also like to thank the students Matteo Aureli, Emre Basaran, Alberto Giacomello, and Chekema Prince who have contributed to the research efforts summarized in this chapter.

References

1. Abdelnour K, Mancia E, Peterson SD, Porfiri M (2009) Hydrodynamics of underwater propulsors based on ionic polymer metal composites: a numerical study. Smart Mater Struct 18(8):085,006
2. Akaydin HD, Elvin N, Andreopoulos Y (2010) Energy harvesting from highly unsteady fluid flows using piezoelectric materials. J Intell Mater Syst Struct 21(13):1263–1278

3. Akle BJ, Bennet MD, Leo DJ, Wiles KB, McGrath JE (2007) Direct assembly process: a novel fabrication technique for large strain ionic polymer transducers. J Mater Sci 42(16):7031–7041
4. Akle BJ, Leo DJ (2008) Single-walled carbon nanotubes – ionic polymer electroactive hybrid transducers. J Intell Mater Syst Struct 19(8):905–915
5. Alben S (2008) The flapping-flag instability as a nonlinear eigenvalue problem. Phys Fluid 20:104,106
6. Allen JJ, Smits AJ (2001) Energy harvesting eel. J Fluid Struct 15(3–4):629–640
7. Anton S, Sodano H (2007) A review of power harvesting using piezoelectric materials (2003–2006). Smart Mater Struct 16(3):R1–R21
8. Aureli M, Basaran ME, Porfiri M (2012) Nonlinear finite amplitude vibrations of sharp-edged beams in viscous fluids. J Sound Vib 331(7):1624–1654
9. Aureli M, Kopman V, Porfiri M (2010) Free-locomotion of underwater vehicles actuated by ionic polymer metal composites. IEEE ASME Trans Mechatron 15(4):603–614
10. Aureli M, Lin W, Porfiri M (2009) On the capacitance-boost of ionic polymer metal composites due to electroless plating: theory and experiments. J Appl Phys 105:104,911
11. Aureli M, Porfiri M (2012) Nonlinear sensing of ionic polymer metal composites. Continuum Mech Therm. (DOI 10.1007/s00161-012-0253-x)
12. Aureli M, Porfiri M (2010) Low frequency and large amplitude oscillations of cantilevers in viscous fluids. Appl Phys Lett 96(16):164,102
13. Aureli M, Prince C, Porfiri M, Peterson SD (2010) Energy harvesting from base excitation of ionic polymer metal composites in fluid environments. Smart Mater Struct 19(1):015,003
14. Bard A, Faulkner L (1980) Electrochemical methods. Wiley, New York
15. Bonomo C, Brunetto P, Fortuna L, Giannone P, Graziani S, Strazzeri S (2008) A tactile sensor for biomedical applications based on IPMCs. IEEE Sensor J 8(7–8):1486–1493
16. Bonomo C, Fortuna L, Giannone P, Graziani S (2006) A circuit to model the electrical behavior of an ionic polymer-metal composite. IEEE Trans Circ Syst 53(2):338–350
17. Bonomo C, Fortuna L, Giannone P, Graziani S, Strazzeri S (2007) A nonlinear model for ionic polymer metal composites as actuators. Smart Mater Struct 16(1):1–12
18. Bonomo C, Fortuna L, Giannone P, Graziani S, Strazzieri S (2006) A model for ionic polymer metal composites as sensors. Smart Mater Struct 15(3):749–758
19. Bonomo C, Fortuna L, Giannone P, Graziani S, Strazzieri S (2008) A resonant force sensor based on ionic polymer metal composites. Smart Mater Struct 17(1):015,014
20. Brufau-Penella J, Puig-Vidal M, Giannone P, Graziani S, Strazzeri S (2008) Characterization of the harvesting capabilities of an ionic polymer metal composite device. Smart Mater Struct 17(1):015,009
21. Bryant M, Garcia E (2011) Modeling and testing of a novel aeroelastic flutter energy harvester. J Vib Acoust 133(1):011,010
22. Chen Z, Hedgepeth D, Tan X (2009) A nonlinear, control-oriented model for ionic polymer-metal composite actuators. Smart Mater Struct 18(5):055,008
23. Chen Z, Tan X, Will A, Ziel C (2007) A dynamic model for ionic polymer-metal composite sensors. Smart Mater Struct 16(4):1477–1488
24. Connell B, Yue D (2007) Flapping dynamics of a flag in a uniform stream. J Fluid Mech 581:33–67
25. Crighton D, Oswell J (1991) Fluid loading with mean flow. I. Response of an elastic plate to localized excitation. Phil Trans Phys Sci Eng 335(1639):557–592
26. Del Bufalo G, Placidi L, Porfiri M (2008) A mixture theory framework for modeling mechanical actuation of ionic polymer metal composites. Smart Mater Struct 17(4):045,010
27. Deole U, Lumia R, Shahinpoor M, Bermudez M (2008) Design and test of IPMC artificial muscle microgripper. J Micro Nano Mechatron 4(3):1865–3936
28. Erturk A, Inman DJ (2008) A distributed parameter electromechanical model for cantilevered piezoelectric energy harvesters. Trans ASME J Vib Acoust 130(4):041,002
29. Erturk A, Inman DJ (2009) An experimentally validated bimorph cantilever model for piezoelectric energy harvesting from base excitations. Smart Mater Struct 18(2):025,009
30. Erturk A, Inman DJ (2011) Piezoelectric energy harvesting. Wiley, Chichester, West Sussex

31. Erturk A, Vieira WGR, De Marqui Jr, C, Inman DJ (2010) On the energy harvesting potential of piezoaeroelastic systems. Appl Phys Lett 96(18):184,103
32. Falcucci G, Aureli M, Ubertini S, Porfiri M (2011) Transverse harmonic oscillations of laminae in viscous fluids: a lattice boltzmann study. Phil Trans Roy Soc A Math Phys Eng Sci 369(1945):2456–2466
33. Fang BK, Ju MS, Lin CCK (2007) A new approach to develop ionic polymer-metal composites (IPMC) actuator: Fabrication and control for active catheter systems. Sensor Actuator A Phys 137(2):321–329
34. Farinholt K, Leo DJ (2004) Modelling of electromechanical charge sensing in ionic polymer transducers. Mech Mater 36(5):421–433
35. Farinholt K, Pedrazas N, Schluneker D, Burt D, Farrar C (2009) An energy harvesting comparison of piezoelectric and ionically conductive polymers. J Intell Mater Syst Struct 20(5):633–642
36. Giacomello A, Porfiri M (2011) Underwater energy harvesting from a heavy flag hosting ionic polymer metal composites. J Appl Phys 109(8):084,903
37. Gutta S, Lee JS, Trabia MB, Yim W (2009) Modeling of ionic polymer metal composite actuator dynamics using a large deflection beam model. Smart Mater Struct 18(11):115,023
38. Guyomar D, Badel A, Lefeuvre E, Richard C (2005) Toward energy harvesting using active materials and conversion improvement by nonlinear processing. IEEE Trans Ultrason Ferroelectrics Freq Contr 52(4):584–595
39. Guyomar D, Sebald G, Lefeuvre E, Khodayari A (2009) Toward heat energy harvesting using pyroelectric material. J Intell Mater Syst Struct 20(3):265–271
40. Inman D, Grisso B (2006) Towards autonomous sensing. In: Tomizuka M, Yun CB, Giurgiutiu V (eds) Proc. SPIE smart structures and materials 2006: sensors and smart structures technologies for civil, mechanical, and aerospace systems, vol. 6174. San Diego, CA, USA, pp. 61,740T:1–7
41. Jo C, Naguib HE, Kwon RH (2008) Modeling and optimization of the electromechanical behavior of an ionic polymer-metal composite. Smart Mater Struct 17(6):065,022
42. Kamamichi N, Yamakita M, Asaka K, Luo ZW (2006) A snake-like swimming robot using IPMC actuator/sensor. In: Proc. 2006 IEEE international conference on robotics and automation, pp. 1812–1817
43. Kauffman J, Lesieutre G (2009) A low-order model for the design of piezoelectric energy harvesting devices. J Intell Mater Syst Struct 20(5):495–504
44. Kim K, Shahinpoor M (2003) Ionic polymer-metal composites: II. Manufacturing techniques. Smart Mater Struct 12(1):65–79
45. Kruusmaa M, Hunt A, Punning A, Anton M, Aabloo A (2008) A linked manipulator with ion-polymer metal composite (IPMC) joints for soft- and micromanipulation. In: Proc. 2008 IEEE international conference on robotics and automation, pp. 3588–3593
46. Lefeuvre E, Badel A, Richard C, Guyomar D (2007) Energy harvesting using piezoelectric materials: case of random vibrations. J Electroceramics 19(4):349–355
47. Leonov V, Torfs T, Fiorini P, Van Hoof C (2007) Thermoelectric converters of human warmth for self-powered wireless sensor nodes. IEEE Sensor J 7(5):650–657
48. Mansfield EH (1989) The bending and stretching of plates, 2nd edn. Cambridge University Press, New York
49. Mateu L, Moll F (2005) Review of energy harvesting techniques and applications for microelectronics. In: Lopez JF, Fernandez FV, Lopez-Villegas JM, de la Rosa JM (eds) Proc. SPIE VLSI circuits and systems II, vol. 5837. Sevilla, Spain, pp. 359–373
50. Mbemmo E, Chen Z, Shatara S, Tan X (2008) Modeling of biomimetic robotic fish propelled by an ionic polymer-metal composite actuator. In: Proc. 2008 IEEE international conference on robotics and automation, pp. 689–694
51. Mitcheson P, Yeatman E, Rao G, Holmes A, Green T (2008) Energy harvesting from human and machine motion for wireless electronic devices. Proc IEEE 96(9):1457–1486
52. Murray R, Rastegar J (2009) Novel two-stage piezoelectric-based ocean wave energy harvesters for moored or unmoored buoys. In: Ahmadian M, Ghasemi-Nejhad MN (eds) Proc.

SPIE active and passive smart structures and integrated systems 2009, vol. 7288. San Diego, CA, USA, p. 72880E
53. Myers R, Vickers M, Kim H, Priya S (2007) Small scale windmill. Appl Phys Lett 90(5):054,106
54. Nemat-Nasser S (2002) Micromechanics of actuation of ionic polymer-metal composites. J Appl Phys 92(5):2899–2915
55. Nemat-Nasser S, Li J (2000) Electromechanical response of ionic polymer-metal composites. J Appl Phys 87(7):3321–3331
56. Nemat-Nasser S, Thomas C (2004) Ionomeric polymer-metal composites electroactive polymer (EAP) Actuators as artificial muscles - reality, potential and challenges, Chap. 6. SPIE Press, Bellingham, WA
57. Newbury K, Leo DJ (2003) Linear electromechanical model of ionic polymer transducers-part I: model development. J Intell Mater Syst Struct 14(6):333–342
58. Peterson SD, Porfiri M (2012) Energy exchange between a vortex ring and an ionic polymer metal composite. Appl Phys Lett 100(11):114,102
59. Peterson SD, Porfiri M (2012) Interaction of a vortex pair with a flexible plate in an ideal quiescent fluid. J Intell Mater Syst Struct 23(13):1482–1501
60. Peterson SD, Porfiri M, Rovardi A (2009) A particle image velocimetry study of vibrating ionic polymer metal composites in aqueous environments. IEEE ASME Trans Mechatron 14(4):474–483
61. Porfiri M (2008) Charge dynamics in ionic polymer metal composites. J Appl Phys 104(10):104,915
62. Porfiri M (2009) An electromechanical model for sensing and actuation of ionic polymer metal composites. Smart Mater Struct 18(1):015,016
63. Porfiri M (2009) Influence of electrode surface roughness and steric effects on the nonlinear electromechanical behavior of ionic polymer metal composites. Phys Rev E 79(4):041,503
64. Qin Z, Batra RC (2009) Local slamming impact of sandwich composite hulls. Int J Solid Struct 46(10):2011–2035
65. Raffel M, Willert C, Wereley S, Kompenhans J (2007) Particle image velocimetry: a practical guide. Springer, Berlin
66. Rastegar J, Pereira C, Nguyen H (2006) Piezoelectric-based power sources for harvesting energy from platforms with low-frequency vibration. In: White EV (ed) Proc. SPIE Smart Structures and Materials 2006: Industrial and Commercial Applications of Smart Structures Technologies, vol. 6171. San Diego, CA, USA, pp. 617,101:1–7
67. Renno J, Daqaq M, Inman D (2009) On the optimal energy harvesting from a vibration source. J Sound Vib 320(1–2):386–405
68. Roundy S (2005) On the effectiveness of vibration-based energy harvesting. J Intell Mater Syst Struct 16(10):809–823
69. Sader JE (1998) Frequency response of cantilever beams immersed in viscous fluids with applications to the atomic force microscope. J Appl Phys 84(1):64–76
70. Shahinpoor M, Bar-Cohen Y, Simpson JO, Smith J (1998) Ionic polymer-metal composites (IPMCs) as biomimetic sensors, actuators and artificial muscles-a review. Smart Mater Struct 7(6):R15–R30
71. Shahinpoor M, Kim K (2001) Ionic polymer-metal composites: I. Fundamentals. Smart Mater Struct 10(4):819–833
72. Shahinpoor M, Kim K (2004) Ionic polymer-metal composites: III. Modeling and simulation as biomimetic sensors, actuators, transducers, and artificial muscles. Smart Mater Struct 13(6):1362–1388
73. Shahinpoor M, Kim K (2005) Ionic polymer-metal composites: IV. Industrial and medical applications. Smart Mater Struct 14(1):197–214
74. Shelley M, Vandenberghe N, Zhang J (2005) Heavy flags undergo spontaneous oscillations in flowing water. Phys Rev Lett 94(9):94,302
75. Sodano H, Inman D, Park G (2004) A review of power harvesting from vibration using piezoelectric materials. Shock Vib Digest 36:197–205

76. Sodano HA, Inman DJ, Park G (2005) Comparison of piezoelectric energy harvesting devices for recharging batteries. J Intell Mater Syst Struct 16(10):799–807
77. Sodano HA, Inman DJ, Park G (2005) Generation and storage of electricity from harvesting devices. J Intell Mater Syst Struct 16(1):67–75
78. St. Clair D, Bibo A, Sennakesavababu VR, Daqaq MF, Li G (2010) A scalable concept for micro-power generation using flow-induced self-excited oscillations. Appl Phys Lett 96(14):144,103
79. Sullivan IS, Niemela JJ, Hershberger RE, Bolster D, Donnelly RJ (2008) Dynamics of thin vortex rings. J Fluid Mech 609:319–347
80. Tang L, Païdoussis M (2008) The influence of the wake on the stability of cantilevered flexible plates in axial flow. J Sound Vib 310(3):512–526
81. Taylor GW, Burns JR, Kammann SM, Powers WB, Welsh TR (2001) The energy harvesting eel: A small subsurface ocean/river power generator. IEEE J Ocean Eng 26(4):539–547
82. Tiwari R, Kim K, Kim S (2008) Ionic polymer-metal composite as energy harvesters. Smart Struct Syst 4(5):549–563
83. Tuck EO (1969) Calculation of unsteady flows due to unsteady motion of cylinders in a viscous fluid. J Eng Math 3(1):29–44
84. Wallmersperger T, Akle B, Leo D, Kroplin B (2008) Electrochemical response in ionic polymer transducers: an experimental and theoretical study. Compos Sci Tech 68(5):1173–1180
85. Wallmersperger T, Kroplin B, Gulch R (2004) Coupled chemo-electro-mechanical formulation for ionic polymer gels-numerical and experimental investigations. Mech Mater 36(5–6): 411–420
86. Wallmersperger T, Leo DJ, Kothera CS (2007) Transport modeling in ionomeric polymer transducers and its relationship to electromechanical coupling. J Appl Phys 101(2):024,912
87. Yeatman E (2009) Energy harvesting: small scale energy production from ambient sources. In: Ahmadian M, Ghasemi-Nejhad MN (eds) Proc. SPIE Active and Passive Smart Structures and Integrated Systems 2009, vol. 7288. San Diego, CA, USA, p. 728802
88. Yeom SW, Oh IK (2009) A biomimetic jellyfish robot based on ionic polymer metal composite actuators. Smart Mater Struct 18(8):085,002

Chapter 10
Flow-Induced Vibrations for Piezoelectric Energy Harvesting

Hüseyin Doğuş Akaydın, Niell Elvin, and Yiannis Andreopoulos

Abstract Fluid flow has the potential to provide significant mechanical energy input for piezoelectric harvesters. However, the efficient conversion of the bulk kinetic energy of a steady and uniform flow into time-dependent elastic energy in the piezoelectric structure remains a significant challenge. In this chapter, a comparison of two harvesters is first presented and it is shown that improved aeroelastic coupling greatly increases the performance of flow-powered harvesters. A computational framework which couples the fluidic, structural, and electrical domains is introduced and is used to simulate the "aeroelectromechanical" performance of a piezoelectric energy harvester in the wake of a bluff body. The structural response of a flexible cantilever beam to a passing vortex ring and the estimation of its aeroelastic efficiency are also presented. Finally, we discuss the potential of turbulent boundary layers for energy harvesting from turbulent boundary layers.

10.1 Introduction

The previous chapters of this book have clearly shown the potential and need of vibration energy harvesters to power small electronic devices such as wireless sensor nodes. When solar power and ambient heat energy is limited, vibration energy harvesting can provide a valuable source of additional electrical energy. There are cases however, where significant kinetic energy is available in the form of fluid flow such as in wind, ocean currents, in ventilation ducts, or in water pipes. In many of these cases, solar power, heat energy, and vibration levels might be low and fluidic energy harvesting can provide a valuable alternative power source.

H.D. Akaydın (✉) • N. Elvin • Y. Andreopoulos
Department of Mechanical Engineering, The City College of New York,
160 Convent Avenue ST233, New York, NY, 10031 USA
e-mail: dogus.akaydin@gmail.com; elvin@me.ccny.cuny.edu; andre@ccny.cuny.edu

N. Elvin and A. Erturk (eds.), *Advances in Energy Harvesting Methods*,
DOI 10.1007/978-1-4614-5705-3_10,

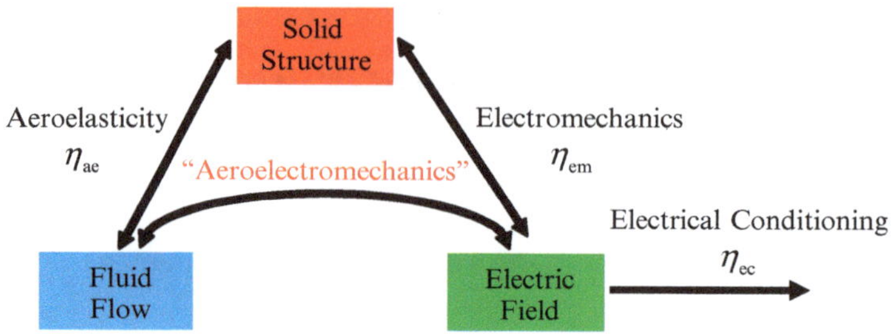

Fig. 10.1 Aeroelectromechanical interactions and conversion efficiencies in a fluidic energy harvester

Fluid flows can contain significant levels of mechanical energy in many different scales. The mechanical power carried by a cross section of a flow is proportional to the third power of the flow speed. For example, in air flowing at a speed of 1 m/s (typical for a ventilation system), the available mechanical power is about 600 mW/m^2. Air flow with a speed of 10 m/s (typical for wind power generation) carries about 600 W/m^2. Wind turbines based on electromagnetic conversion are efficient generators at large scales. However, reducing the size of a turbine to miniature scales disproportionally degrades its aerodynamic performance, increases its stall speed, lowers its power output, and in general complicates both its design and manufacturing [1]. Small scale piezoelectric converters on the other hand, can be mechanically simpler and arguably more durable than turbines and electromagnetic harvesters while directly generating higher voltages needed for AC to DC rectification. However, the bulk kinetic energy in a uniform fluid flow is useful only when it is converted to strain energy in the piezoelectric structure through oscillatory deformations. The efficiency of this "aeroelastic conversion" is the key for a successful piezoelectric energy harvester operating in a uniform fluid flow and is the focus of the present chapter.

Insertion of a piezoelectric structure in a fluid flow results in a set of mutual interactions between the flow of the fluid, structural mechanics of the harvester, and the electric field generated by the piezoelectric material. This three-way mutual interaction (so called "aeroelectromechanics") is shown in Fig. 10.1. The kinetic energy of a uniform and steady fluid flow is converted to time-dependent strain energy within the harvester through flow-induced vibrations. In this chapter, the efficiency of this conversion is called the "aeroelastic efficiency", η_{ae}. Once the time-dependent strain energy is generated in the piezoelectric structure by the fluid flow, it is converted to AC electrical energy through the direct piezoelectric effect with an associated "electromechanical efficiency", η_{em}. Finally, the AC electrical energy is conditioned through a series of electrical circuits (rectifiers, boosters, etc.) and the charge is typically stored (in devices such as rechargeable batteries, capacitors, etc.) before being supplied to the electronic device (e.g., a wireless sensor node). The efficiency of this conditioning process is termed the "electrical conditioning efficiency", η_{ec}. The total efficiency of the harvester then becomes

$$\eta_t = \eta_{ae}\eta_{em}\eta_{ec}. \tag{10.1}$$

A diagram of this efficiency breakdown is also presented in Fig. 10.1.

In this chapter, several flow-powered piezoelectric energy harvesters will be presented and compared. A computational framework to perform aeroelectromechanical simulations will be introduced. The interaction of a vortex ring with a flexible beam will be investigated from an energy harvesting perspective. Finally, the results of energy harvesting tests done by placing flexible piezoelectric beams in turbulent boundary layers will be presented.

10.2 Configuration I: Piezoelectric Beam in the Wake of a Bluff Body

The kinetic energy of a flowing fluid can be captured by placing a bluff body in a uniform and steady flow and a piezoelectric beam in its wake. At sufficiently high flow speeds, the wake of a bluff body consists of a staggered array of traveling vortices known as a "Karman Vortex Street." These vortices are periodically shed from the top and bottom of the bluff body and they rotate in alternating directions. A flexible beam placed in the wake of the bluff body may undergo relatively large oscillations due to the passing vortices [2, 3]. A snapshot of a flexible PVDF beam responding to a vortex in the wake of a circular cylinder is shown in Fig. 10.2a.

The piezoelectric harvester used in the cylinder-wake experiments consisted of a cantilevered composite beam with dimensions of 30 mm × 16 mm × 0.2 mm in a unimorph configuration with a PVDF layer thickness of $t_p = 28$ μm and a Mylar backing thickness of $t_b = 172$ μm (see [2, 3] for details). The first bending mode resonance frequency of the beam, f_1, was measured to be 48.5 Hz. The beam was placed along the centerline of the wake of a circular cylinder (30 mm in diameter and 1.2 m long) with its free tip upstream (Figs. 10.2 and 10.3).

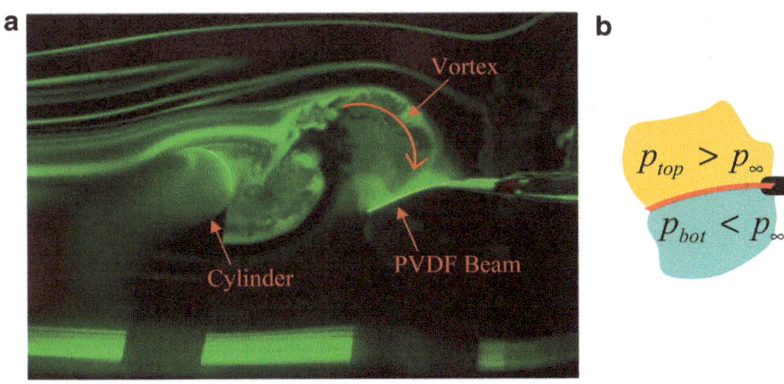

Fig. 10.2 (**a**) A snapshot of the PVDF beam placed in the wake of a circular cylinder showing streaklines and vortex shedding. The direction of the flow is from left to right. (**b**) The beam is bent due to pressure differences between the top and bottom face of the beam. The periodic change of the pressure on either face of the beam causes it to vibrate [3]

Fig. 10.3 A close-up view of the cylinder and the beam inside the wind tunnel

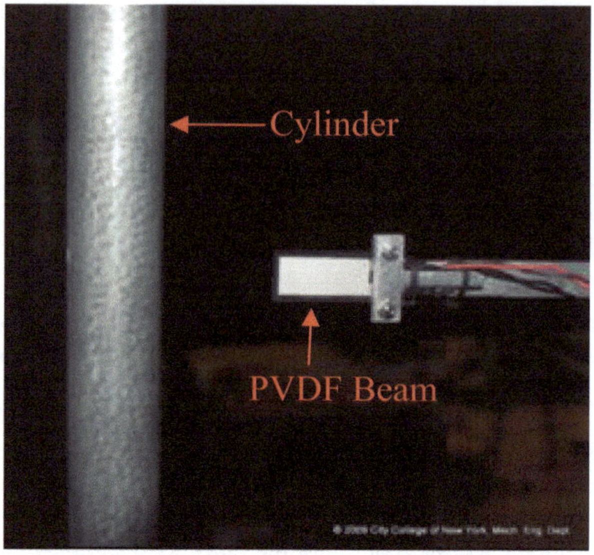

The vortex shedding frequency from the circular cylinder is given by

$$f_s = St \frac{U_\infty}{d_c}, \tag{10.2}$$

where d_c is the diameter of the cylinder and St is the Strouhal number. Strouhal number varies depending on the Reynolds number which is defined as

$$Re = \frac{\rho U_\infty d_c}{\mu}, \tag{10.3}$$

where μ is dynamic viscosity of the fluid and ρ is the fluid density [5]. For $Re < 40$, $St = 0$ since the wake is symmetric and thus, the Karman Vortex Street is not produced. The wake instabilities initiate vortex shedding at $Re \approx 40$ with $St = 0.1$. After a laminar shedding regime is passed, the wake becomes fully turbulent with $St \approx 0.2$ for $300 < Re < 20{,}000$. If Re is increased further, then St decreases slightly until $Re = 100{,}000$, at which point the cylinder's boundary layer also becomes turbulent and St increases steadily with increasing Re [5, 6]. The maximal output voltage occurs when the first natural frequency of the harvester matches frequency of the vortex shedding from the cylinder. The flow speed at harvester resonance in the first bending mode was calculated to be 7.23 m/s using Eq. (10.2). The wind tunnel tests at flow speeds between 0 and 11 m/s showed that the maximum voltage output occurred at $U_\infty = 7.13$ m/s which is in close agreement with the expected value (Fig. 10.4).

The variation of the harvested power with the location of the beam within the wake was then investigated. The beam was positioned at different locations along

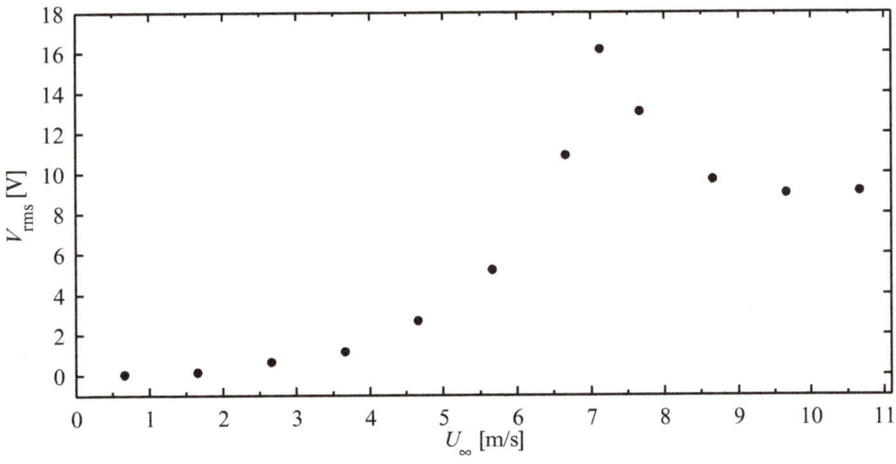

Fig. 10.4 Variation of piezoelectric voltage (RMS) with wind speed. Load resistance is $R = 10\ \mathrm{M}\Omega$

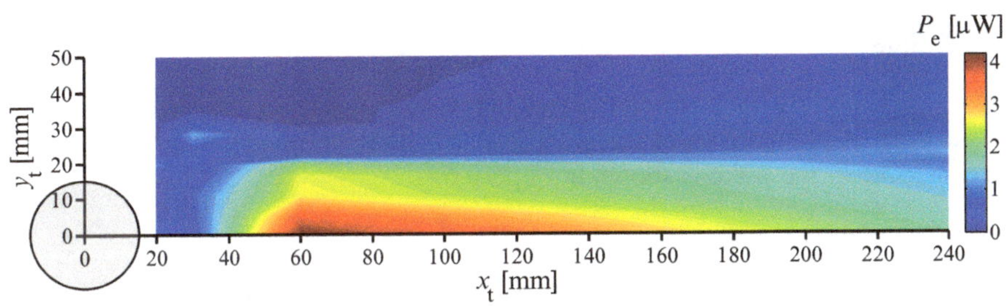

Fig. 10.5 Contours of the mean electrical power of the PVDF beam in the wake of the cylinder at $Re = 14{,}800$ with a load resistance of $R = 100\ \mathrm{k}\Omega$. The cylinder of radius 15 mm is centered at the origin as shown. x_t and y_t denote the position of the tip of the unbent beam. Only the top half of the wake is shown due to symmetry along the x-axis of the time-averaged flow

and across the wake while the wind speed was kept constant at $U_\infty = 7.13$ m/s. The piezoelectric voltage was measured over a load resistance of $R = 100\ \mathrm{k}\Omega$. The contour plot of the time-averaged electrical power shown in Fig. 10.5 depicts the "power landscape" of the configuration tested in this experiment. The maximum power was harvested when the beam was positioned along the centerline of the wake ($y_t = 0$). The harvested power is a minimum when the tip of the beam is located within the formation distance of the vortices (i.e., $d_c/2 < x_t < 1.5d_c$). However, a peak value of about 4 μW is obtained when the tip of beam is located near $x_t = 2d_c$. The power gradually decreases for increasing values of $x_t > 2d_c$. This decay is due to loss of intensity of the vortices from both viscous dissipation and growth in their sizes while moving downstream. Dimensional analysis suggests that the force acting on the beam is proportional to the square of a velocity scale, and therefore the harvested power is proportional to the cube of this velocity scale. For this

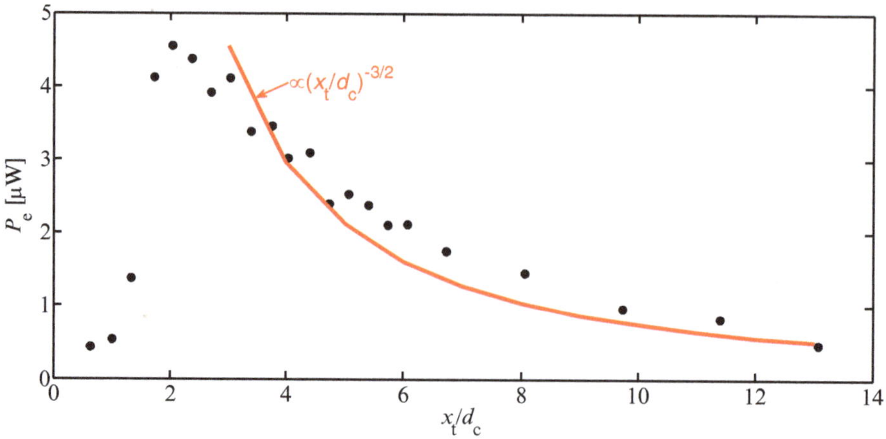

Fig. 10.6 Mean power output along the central line of the wake ($y_t = 0$). Experimental data points and decay relation proportional to $(x_t/d_c)^{-3/2}$, normalized with P_e at $(x_t/d_c) = 2$, starting at $(x_t/d_c) = 3$

particular case, the velocity scale is proportional to ΔU, which is the velocity deficit in the wake of the cylinder. Similarity arguments show that ΔU is proportional to $(x/d_c)^{-1/2}$. Therefore the power harvested in the wake is expected to be proportional to $(x/d_c)^{-3/2}$. A comparison of this relationship with experimental data taken along the centerline of the wake is shown in Fig. 10.6.

10.3 Configuration II: Piezoelectric Beam Carrying a Tip Body

In the configuration investigated in the previous section, the bluff body (i.e., circular cylinder) converted the kinetic energy of the uniform and steady fluid flow to large temporal pressure fluctuations caused by vortex shedding. These pressure fluctuations were then converted by the piezoelectric generator into electrical energy. As such, the aerodynamic forces developed on the cylinder were not transmitted to the beam directly; rather these forces were transmitted to the beam through the vortices shed off and moved downstream of the cylinder. In the configuration investigated in this section, a tip body attached to the free end of a piezoelectric beam serves to induce vortex shedding. The aerodynamic forces and moments from these shedding vortices directly cause the piezoelectric beam to vibrate (Fig. 10.7a). In other words, the advantage of this configuration is that the lift and drag forces (L and D) and the aerodynamic moments (M_a) developing on the tip body are exerted directly onto the beam rather than being captured from the wake as in the previous configuration. Therefore, it is expected that this configuration would have a higher aeroelastic efficiency than the previous configuration. The details of this study can be found in [4].

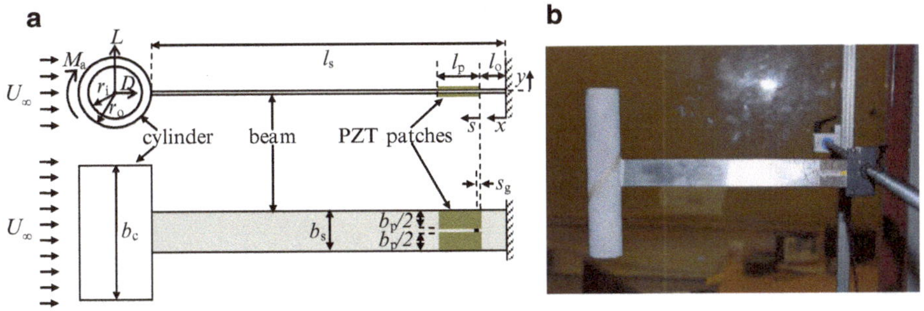

Fig. 10.7 (**a**) Top and side views of the self-excited harvester. Lift force L is perpendicular to the free-stream velocity U_∞, the drag force D is parallel to U_∞, and M_a is taken positive clockwise around the center of the cylinder. (**b**) Harvester installed in the wind tunnel. (© IOP Publishing. Reprinted with permission)

As shown in Fig. 10.7, the piezoelectric beam of this configuration consisted of an aluminum shim with four PZT-5A patches bonded near its clamped end ($l_o = 0.5$ mm) and electrically connected in series. Each patch had a length $l_p = 31.8$ mm, thickness $h_p = 0.267$ mm, and width $b_p/2 = 12.7$ mm. The hollow circular cylinder was made of paperboard, covered on both ends, and had a mass of 16 g. The outer radius of the cylinder has $r_o = 19.8$ mm and length $b_c = 203$ mm. The shim has a length $l_s = 267$ mm, width $b_s = 32.5$ mm, and thickness $h_s = 0.635$ mm. From tests on similar configurations, it was concluded that the ratios b_c/b_s and b_c/r_o should be large enough so that vortex shedding is not significantly hindered by splitter-plate and finite-length effects, and that the stiffness of the beam should be low enough to promote large-amplitude flow-induced vibrations. These ratios in the present configuration are $b_c/b_s = 6.25$, $b_c/r_o = 10.2$, and $l_s/r_o = 13.3$. A pair of strain gages was bonded near the clamped end at $s_g = 3.92$ mm for measuring strains developing on the surface of the beam during the tests.

Experimental tests at different load resistances showed that $R = 2.46$ MΩ is close to the optimal value. Using a resistor of this value, the harvester was tested in a wind tunnel at various speeds between 0 and 11 m/s. The variation of the measured average electrical power with the flow speed is shown in Fig. 10.8. A maximum electrical power of about 0.1 mW is harvested at a flow speed of 1.19 m/s. The vibration frequency at resonance was of about 3.14 Hz which is about half of what would have been expected for the case of a long, stationary cylinder using Eq. (10.2).[1] In addition, the dynamic lift force was calculated to be about 3.05 mN, which is more than six times larger than the static lift force values reported for stationary cylinders [7]. Furthermore, the tip deflection at resonance was found to be approximately $w_t = \pm 14$ mm, which is nearly 60 times larger than the tip deflection under the amount of static tip force (3.05 mN). It appears that (1) the lift forces

[1] Further experimental work is underway to determine whether the decrease in oscillation frequency is due to the motion of the cylinder or due to its finite length resulting in spiraling tip vortices emanating from both tips and directed towards the middle of the cylinder.

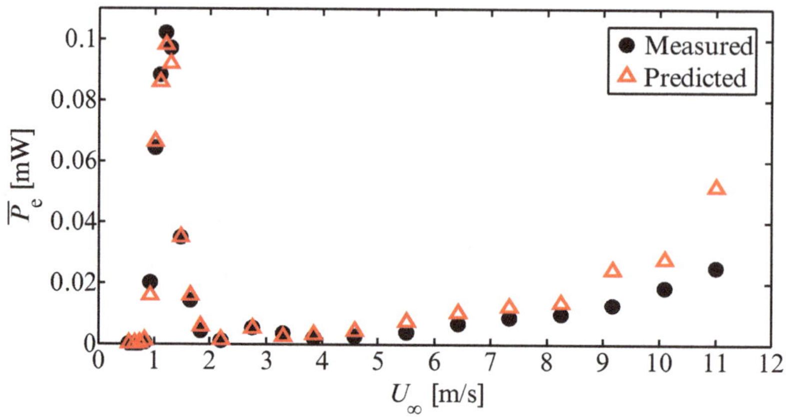

Fig. 10.8 Variation of time-averaged power depending on wind speed

increase due to cross-flow motion and (2) the amplitude of the cross-flow motion is greatly amplified near resonance. Thus the combined result of these aeroelastic effects is both increased tip deflections and consequently the electrical power output near $U_\infty = 1.19$ m/s.

In addition to the direct measurements, the voltage output was predicted using the measured strain values. Employing the strain transfer model detailed in [4], the strains developing in the piezoelectric patches were calculated based on the values measured on the surface of the shim. The strain data were then used in a distributed parameter model [8] to predict the voltage output of the patches. As seen in Fig. 10.8, power output predicted from strain measurements is in excellent agreement with the direct measurements nearly up to four times the resonance speed. When testing and optimizing energy harvesters, the use of piezoelectric materials can be costly, laborious, and restricted by the availability of specific piezoelectric materials. If the electromechanical coupling is low, the effect of harvested electrical power on the overall dynamics of the beam may be ignored. In this case, the piezoelectric components can be replaced by mechanically equivalent dummy materials when building the test models. Thus the prediction of piezoelectric power output based on strain measurements can be beneficial for evaluating a large number of designs.

10.4 Comparison of Efficiency and Power Densities

10.4.1 Efficiency and Power Densities of Configuration I

The average mechanical power available to the harvester is estimated in [3] as

$$P_f = \frac{A_f}{2}\rho U_\infty^3, \tag{10.4}$$

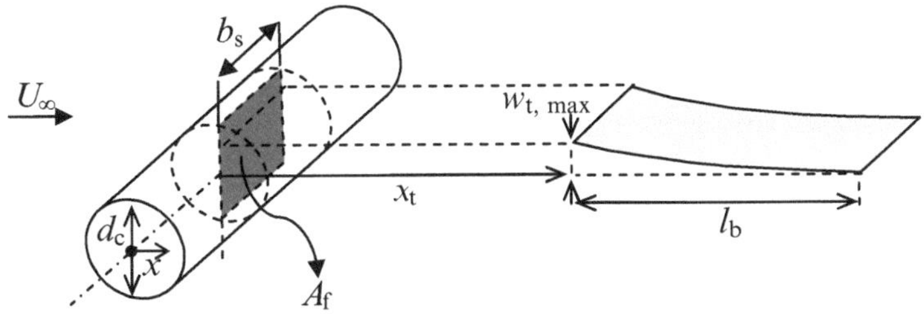

Fig. 10.9 Dimensions used for calculating the operational volume of Configuration I

where A_f is the frontal area swept by the device when vibrating in the flow. For this configuration the frontal area is defined as the product of the width of the beam with the diameter of the cylinder or twice the maximum tip deflection, whichever is greater (Fig. 10.9):

$$A_f = \max(d_c, 2w_{t,\max})\, b_s. \tag{10.5}$$

The total efficiency of the harvester at resonance conditions is found to be $\eta_t = 0.0035\%$ from

$$\eta_t = \frac{\bar{P}_e}{P_f}. \tag{10.6}$$

Using the tip force and tip velocity, the average mechanical power is found by

$$\bar{P}_m = \frac{1}{t_1 - t_0} \int_{t_0}^{t_1} F \dot{w}_t \mathrm{d}t. \tag{10.7}$$

Since tip force and tip velocity were not measured directly during the experiments, these quantities were calculated from the aeroelectromechanical simulations discussed in the next section. Using the computed values of $\bar{P}_m$ and $\bar{P}_e$ from these simulations, the electromechanical efficiency is found to be approximately $\eta_{em} = 11\%$ using

$$\eta_{em} = \frac{\bar{P}_e}{\bar{P}_m}. \tag{10.8}$$

Since the electrical conditioning is absent in this investigation, η_{ec} in Eq. (10.1) can be taken as 100%. The aeroelastic efficiency is then found to be $\eta_{ae} = 0.032\%$

$$\eta_{ae} = \frac{\eta_t}{\eta_{em}}. \tag{10.9}$$

The power density of the device is calculated based on the piezoelectric material volume and on the operational device volume (i.e., the total space occupied by the harvester when experiencing maximum deflection). The piezoelectric material volume in this unimorph beam is

$$\forall_{\text{piezo}} = t_p b_s l_b. \tag{10.10}$$

For operational device volume, two definitions are used. The first definition is based only on the operational volume of the piezoelectric beam

$$\forall_{\text{dev1}} = 2w_t b_s l_b. \tag{10.11}$$

The second definition encompasses also the upstream cylinder:

$$\forall_{\text{dev2}} = A_f \left(x_t + l_b + d_c/2\right). \tag{10.12}$$

Based on these volume definitions, the power densities are found to be $\bar{P}_e/\forall_{\text{piezo}} = 298\,\text{W/m}^3$, $\bar{P}_e/\forall_{\text{dev1}} = 408\,\text{mW/m}^3$, and $\bar{P}_e/\forall_{\text{dev2}} = 78\,\text{mW/m}^3$.

10.4.2 Efficiency and Power Densities of Configuration II

Dimensions used for calculating the operational volume of Configuration II are shown in Fig. 10.10. The frontal area of the harvester during operation is given by

$$A_f = 2\left(r_o + w_t\right) b_c. \tag{10.13}$$

Based on this definition, the flow power available for the harvester is found to be $P_f = 13.9$ mW from Eq. (10.4). Since the electrical power output at resonance is 0.1 mW, the total efficiency is $\eta_t = 0.72\%$, which is a tremendous improvement on the previous configuration. The electromechanical power is found to be $\bar{P}_m = 0.38$ mW by using the experimentally derived tip deflection and the tip force in Eq. (10.7). Finally, the electromechanical efficiency is found to be $\eta_{em} = 26\%$ from Eq. (10.8) which suggests that the aeroelastic efficiency using Eq. (10.9) is $\eta_{ae} = 2.8\%$.

The operational device volume and piezoelectric element volumes are, respectively,

$$\forall_{\text{dev}} = A_f \left(l_s + 2r_c\right) \tag{10.14}$$

and

$$\forall_{\text{piezo}} = 2 l_p b_p h_p. \tag{10.15}$$

At resonance, power per operational device volume is $\bar{P}/\forall_{\text{device}} = 23.6\ \text{mW/m}^3$ while the power per piezoelectric element volume is $\bar{P}/\forall_{\text{piezo}} = 233\,\text{W/m}^3$. Note that the corresponding numbers for the previous configuration were approximately $\bar{P}/\forall_{\text{dev2}} = 78\,\text{mW/m}^3$ and $\bar{P}/\forall_{\text{piezo}} = 298\,\text{W/m}^3$.

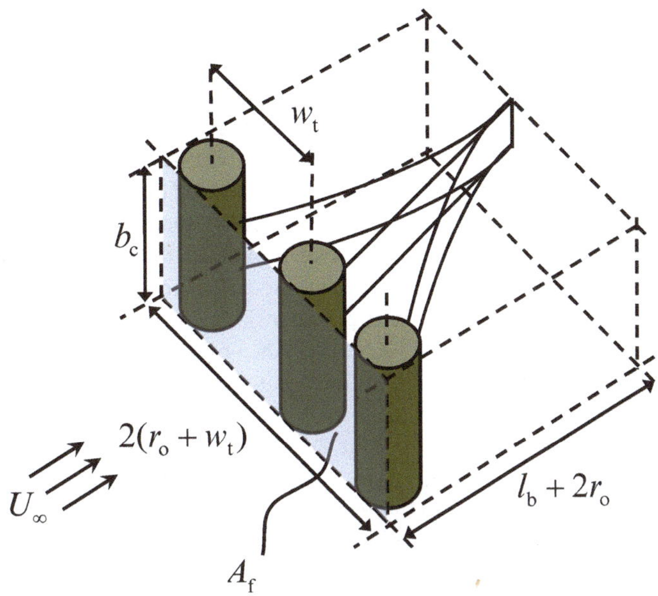

Fig. 10.10 Dimensions used to calculate the operational volume of Configuration II

Comparing the efficiencies and power densities of the two configurations, it is noted that:

1. The aeroelastic efficiency of Configuration II is nearly two orders of magnitude larger than that of Configuration I ($\eta_{ae} = 2.8\%$ as compared to $\eta_{ae} = 0.032\%$). This clearly shows the improvement in aeroelastic coupling due to the attachment of the cylinder on the tip of the harvester beam.
2. The increase in η_{em} is nearly threefold for Configuration II (26% for Configuration II compared with 11% for Configuration I). However, this increase is not only due to change of harvester configuration but also due to the change of piezoelectric material. The piezoelectric constant and relative permittivity of PZT-5A used in Configuration II are $d_{31} = 190$ pm/V and $\varepsilon_0 = 1{,}800$, respectively, whereas those of PVDF used in Configuration I are $d_{31} = 23$ pm/V and $\varepsilon_0 = 13$.
3. Power per piezoelectric material volume is comparable in both configurations although PVDF has a smaller piezoelectric coefficient. This is primarily because larger strains (>2,000 microstrain) develop in the PVDF beam of Configuration I than in the PZT patches of Configuration II. The maximum strain on the outer surface of the PZT patches is measured to be about 90 microstrain, which is far less than damage limit of 500 microstrain for PZT. The piezoelectric patches in Configuration II can thus allow for higher applied strains which would increase the voltage and power outputs.
4. Thus, although Configuration I is less efficient, it has a larger power density.

10.5 Effect of Tip Body Shape

It was shown in the previous section that self-excited vibrations of significant amplitudes can be created by adding a circular cylinder to the free end of the cantilever beam. Investigating the behavior of a circular cylinder is advantageous for comparative research purposes since the flow around circular cylinders has been extensively studied as a benchmark case; thus a large quantity of both experimental and computational data is available. In order to increase the electrical power output, the effect of tip body shapes was also studied. In this section, we present the performances of harvesters with a half cylinder (D-section) tip body and a flat plate (I-section) tip body (Fig. 10.11). The performance of these tip bodies are compared with the performance of the circular cylinder (O-section) tip body case investigated in the previous section.

Wind tunnel tests showed that both the D-section and the I-section harvesters significantly increased the harvested power (Fig. 10.12). The D-section and O-section harvesters exhibited similar resonant power output responses below $U_\infty < 1.47$ m/s. Above this speed, however, the response of the D-section harvester was very different from that of the O-section harvester investigated in the previous section. After passing through the first bending resonance, the O-section harvester resonated in the first torsional mode (at about 11.2 Hz) and then the second bending mode (at about 27.4 Hz)[2] with little tip deflection. In contrast, the D-section harvester's

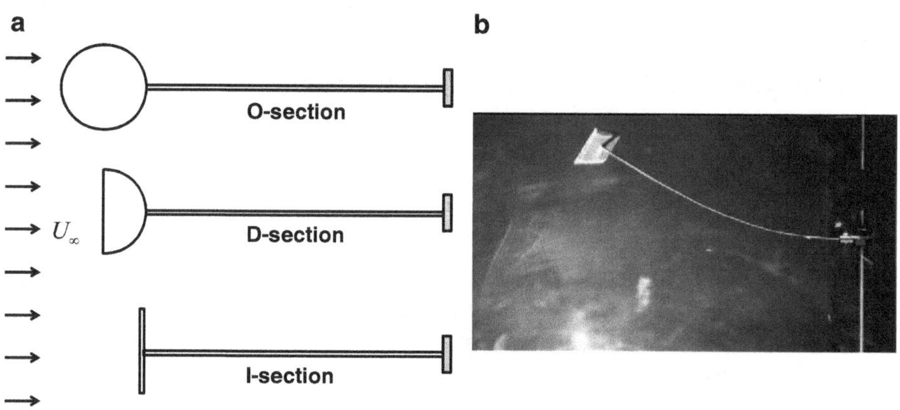

Fig. 10.11 (**a**) Tested tip body shapes. Both the "O-section" and "D-section" cylinders were 40 mm in diameter with masses of 16 g and 21 g, respectively. Two "I-section" cylinders were tested: "I-section 1" had a width of 32.5 mm and mass of 16 g; "I-section 2" had a width of 40 mm and a mass of 24 g. All cylinders had a length of $b_c = 203$ mm. (**b**) A photograph of the I-section harvester exhibiting large tip deflections

---

[2]Due to the construction and electrical connection of the harvester, the voltage outputs of the modes higher than the first bending mode were small and cannot be seen in Fig. 10.12. However, visual observation of the harvester and FFT analysis of the strain signal showed resonances in the first torsional mode and the second bending mode as detailed in [4].

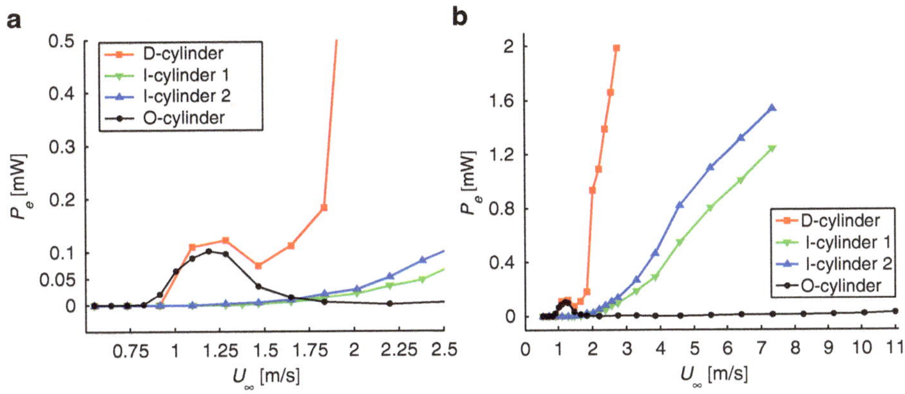

Fig. 10.12 Electrical power output using different tip body shapes

tip deflection grew larger with increasing flow speed, while its vibration frequency remained fixed at 3.2 Hz. This matching of the vortex shedding frequency to the resonant frequency of the structure with increasing flow velocity is known as "lock-in" [9] and is observed commonly in galloping instability of cylinders of noncircular cross sections [10, 11]. The I-section harvester did not exhibit a pronounced resonance as its vibration amplitude increased steadily throughout the velocity range. The increase in amplitude of oscillations for the D-section harvester was so rapid that the test had to be stopped before reaching $U_\infty = 3$ m/s to prevent damage to the piezoelectric patches. The voltage amplitudes obtained at the last data points for the D-section, I-section 1, and I-section 2 in Fig. 10.12b were approximately ±107 V, ±93 V, and ±82 V, respectively, with vibration frequencies near 3.18 Hz, 3.16 Hz, and 3.83 Hz.

At the highest flow speeds, the oscillation amplitudes of the D- and I-section harvesters exceeded 100 mm, which was more than 35% of the beam's length (Fig. 10.13b). When the tip deflections are this large, the contributions of the drag force and the aerodynamic moment to the deflection of the beam cannot be neglected. In addition, the geometric (and possibly material) nonlinearities in the structure of the beam become significant. Thus simplified models of the harvester behavior need to be used with caution, since these models generally assume that (a) the beam undergoes small deflections and (b) only lift forces act on the beam.

From a practical point of view, the D-section harvester outperformed the other shapes tested. At 1.19 m/s, the D-section harvester's total efficiency is approximately $\eta_t = 0.80\%$, which is slightly larger than that of the O-section harvester ($\eta_t = 0.75\%$) at the same speed. However, once past the resonant regime, the power output of the D-section harvester rapidly increases to 2 mW at a flow speed of 2.75 m/s. The total efficiency of the D-section harvester decreases to $\eta_t = 0.31\%$ at this speed since large tip deflections increase the operational frontal area. However, it should be noted that the efficiency of O-section harvester at 2.75 m/s is nearly zero since it produces almost no power at this speed. Because the strain data is not available for the D-section and I-section harvesters, the electromechanical efficiencies of these models are not currently known and a full breakdown of their efficiency components cannot be performed.

10.6 The Computational Framework for Aeroelectromechanical Simulations

To provide further insight into the behavior of fluidic energy harvesters, a computational framework was developed for simulating the aeroelectromechanical behavior of a piezoelectric beam in the wake of a circular cylinder. The system of equations governing the fluid flow, structural dynamics, and electrical output of a piezoelectric beam are given in Eqs. (10.16a–10.16c). Within this framework, the mutual interactions between these equations are handled in a loosely coupled manner [12], that is, each of these equations are solved in their independent domains (fluid, structural, and electrical) and appropriate information such as updated boundary conditions and electromechanical coupling is exchanged between the domains at each time step. While a monolithic approach to solve all the governing equations in a strongly coupled manner is desirable, the computational cost of such an approach for simulating realistic cases often renders a loosely coupled method more practical. Furthermore commercially available codes can be readily adopted in the loosely coupled approach.

$$\rho\left(\dot{\mathbf{V}} + \mathbf{V}\cdot\nabla\mathbf{V}\right) = -\nabla p + \mu\nabla^2\mathbf{V} \quad \text{(Navier-Stokes equations)} \tag{10.16a}$$

$$m\,\ddot{w}_{\mathrm{t}} + c\,\dot{w}_{\mathrm{t}} + k w_{\mathrm{t}} - \Theta v = F(t) \quad \text{(Actuator equation)} \tag{10.16b}$$

$$\Theta\dot{w}_{\mathrm{t}} + C_{\mathrm{p}}\dot{v} = i = -v/R \quad \text{(Sensor equation)} \tag{10.16c}$$

Incompressible flow of a Newtonian fluid is governed by the Navier–Stokes equations given in Eq. (10.16a) where p is the pressure field and $\mathbf{V}$ is the velocity vector field. If the relatively small contribution of shear forces over the surface of the beam is neglected, the normal forces due to pressure distribution are the only forces bending the piezoelectric beam. The structural dynamics of the beam is approximated by a SDOF (single degree of freedom) lumped parameter model. The governing equation of the structural dynamics is then given by Eq. (10.16b) where w_{t} is the tip deflection, m, c, and k are the lumped mass, damping coefficient, and the stiffness of the beam for the first mode respectively, Θ is the electromechanical coupling coefficient and F is the equivalent tip force found from the fluid pressure distribution over the beam [3]. Due to electromechanical coupling, the deflection of

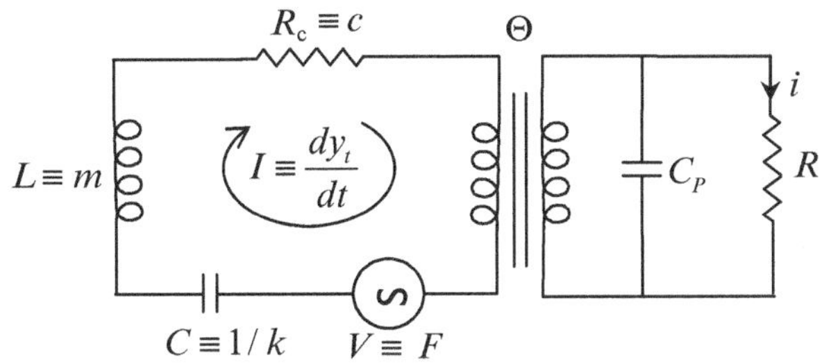

Fig. 10.13 Equivalent circuit model of the actuator and sensor equations

the harvester (with all electrode pairs connected) results in a piezoelectric voltage (v) across the generator as shown in Eq. (10.16c) where C_p is the electrical capacitance of the piezoelectric beam, R is the load resistance, and i is the current across the load resistance.

The Navier–Stokes equations were solved using *Fluent v6.2*, a commercial flow simulation software. To find the electromechanical response, the electrical equivalent of the actuator equation is coupled with the sensor equation through an ideal transformer with a conversion ratio Θ (Fig. 10.13). The solution of the coupled circuit is performed using *SpiceOpus v2.3*, an open source circuit simulator. Although a load resistor is used as the model electric circuit in this study, more complex circuits involving diodes or synchronized switch rectifiers, etc., can also be modeled within the same framework. A similar approach was successfully implemented using FEA (Finite Element Analysis) to study piezoelectric harvesters excited in multiple vibrational modes [13, 14].

The UDF (User Defined Function) facility of *Fluent* allows the exchange of data with other software during runtime. Using this ability, *SpiceOpus* was recursively called after each time step of the flow simulation. The tip displacement obtained from *SpiceOpus* was then used to find the coordinates of the other points along the beam according to the first bending mode shape. After the deformed shape of the beam was updated, the fluid mesh around the beam was updated and flow simulation was started for the next time step. This procedure is depicted in Fig. 10.14.

The actual vortex shedding mechanism behind a finite-length cylinder involves three-dimensional effects (such as shedding of vortices from the tips of the cylinder or as spanwise phase difference in vortex shedding along the cylinder). In addition, the flow in the wake is fully turbulent for $Re > 300$. An accurate simulation of these effects requires both 3D (three-dimensional) modeling of the turbulent flow and deforming mesh capabilities, which would be computationally very expensive; therefore, a 2D simulation was used to demonstrate the computational framework. The solution domain, the applied boundary conditions, and a detailed view of the mesh near the cylinder and the piezoelectric beam are shown in Fig. 10.15. A uniform flow speed of $U_\infty = 7.23$ m/s is applied at the velocity inlet while a zero gage pressure condition is imposed at the pressure outlet. A 2D pressure-based solver with double precision was used for all simulations. For the case of the circular cylinder without the beam, the k–ω model with shear stress transport (SST) option

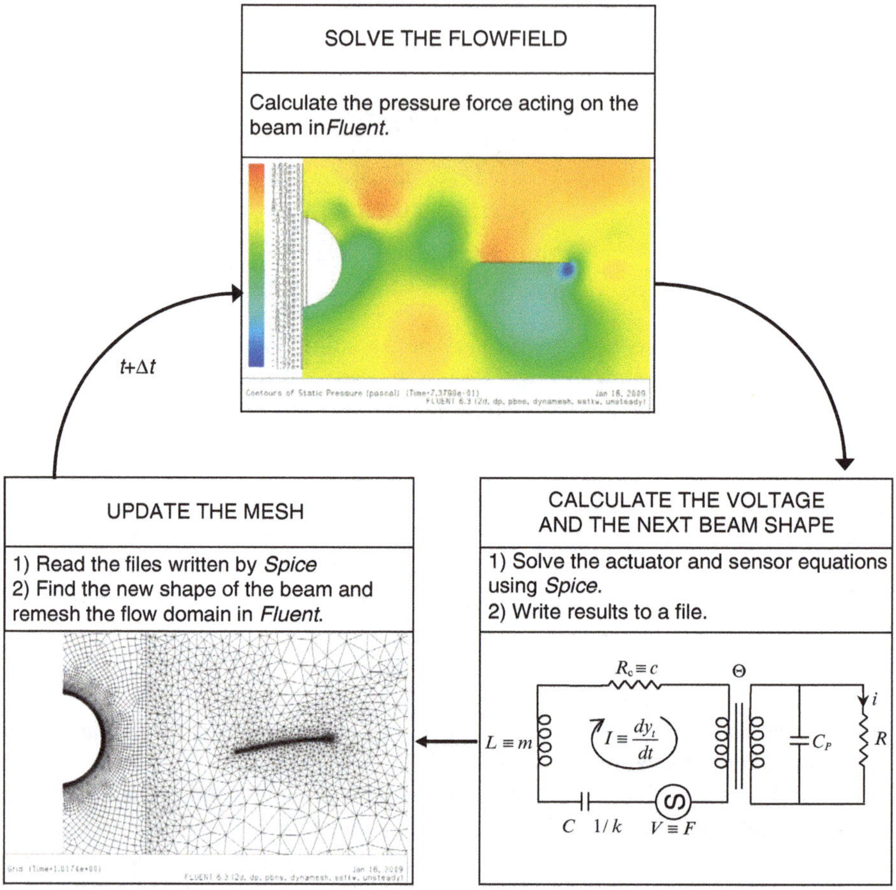

Fig. 10.14 The computational framework for aeroelectromechanical simulations

gave the best results in terms of the magnitude and frequency of lift and drag coefficients when compared to literature data [5, 15]. This turbulence model was thus used for the simulations including the beam.

The shedding frequency of the vortices from the cylinder in the absence of the beam was found to be approximately 50 Hz, which is in good agreement with the expected value of 48.5 Hz based on $St = 0.20$. The lift and drag coefficients are in reasonable agreement with those previously reported for similar flow regimes [5, 15]. When the beam is present in the wake of the cylinder, the drag force and Strouhal number is expected to decrease due to the splitter-plate effect [16, 17]. It has been shown for this flow regime that the Strouhal number decreases to 0.16 when a rigid splitter plate of a similar size and location is present in the wake [17]. Therefore, the shedding frequency is expected to be around 39 Hz instead of 48.5 Hz for a rigid beam. The shedding frequency found from the simulation is

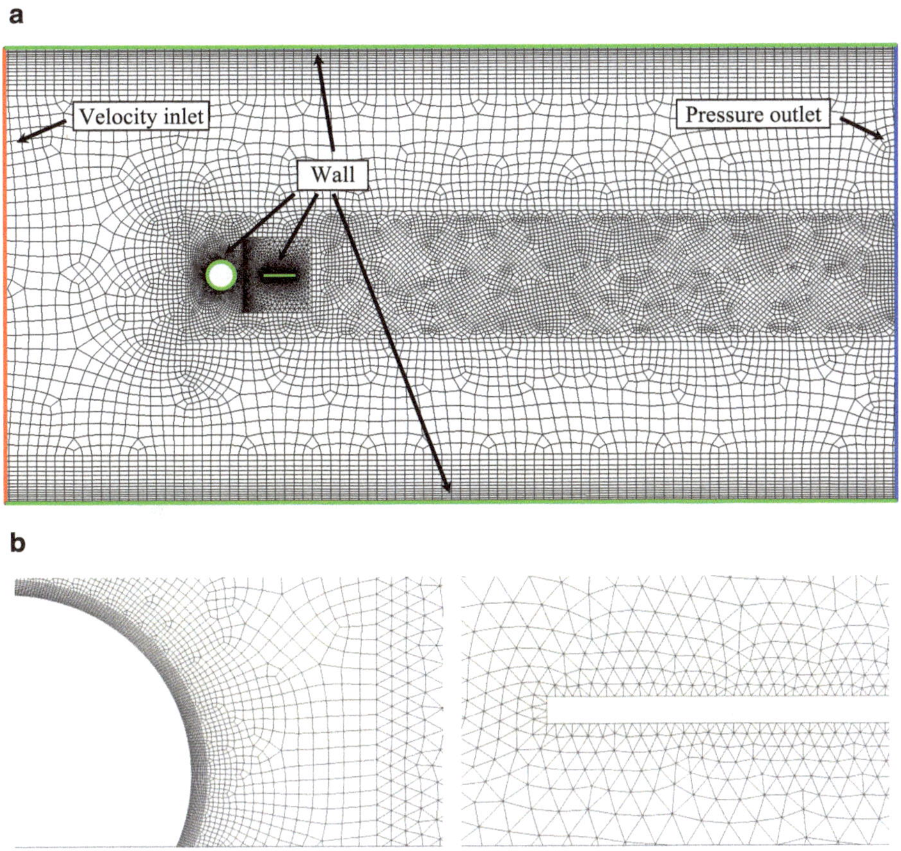

Fig. 10.15 (**a**) The solution domain and the mesh generated for the flow simulations. (**b**) Close-up views showing the boundary layers zones near the cylinder and the beam

41 Hz, which is in good agreement with the expected value. A rigid splitter plate hinged to the cylinder [18] or short and flexible ribbons attached to the cylinder [9] are known to decrease the magnitude and frequency of the aerodynamic forces due to vortex shedding. In our simulations, we observed that the vortex shedding frequency further decreased to 37 Hz when the piezoelectric beam in the wake of the cylinder was modeled to be flexible.

Figure 10.16 shows the path lines and pressure fields obtained during the simulation at an instant $t = t_1$ (left figures) and a quarter period after t_1 (right figures). In these simulations, the vortices can be identified by their low-pressure cores. At the first instant $t = t_1$, a vortex is approaching the beam from the top while another one is located directly below the beam. As the vortex approaches the beam from above, the flow impinges on the top surface of the beam and the dynamic pressure of the flow is converted to static pressure due to stagnation (Fig. 10.16a,b). At the same time, the low-pressure core of the vortex below the beam creates

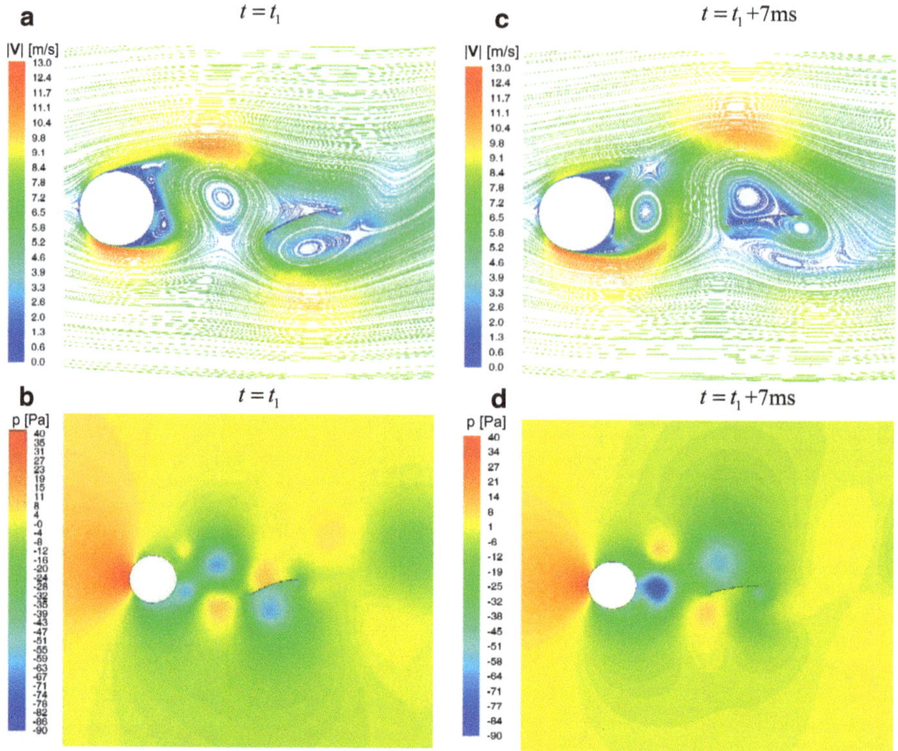

Fig. 10.16 (**a**) Pathlines (*colored* by velocity magnitude) and (**b**) Pressure contours for an instant t_1. (**c**) and (**d**) The same plots for a quarter of period (7 ms) after t_1

suction that further deflects the beam downwards. Under the combined effect of these vortices at this instant, the beam is forced downwards. As the lower vortex travels downstream, the lower pressure is replaced by a high-pressure zone while a new vortex approaching from above creates suction on the top surface of the beam (Fig. 10.16d). The combined effect at this time is to deflect the beam upwards. As this process repeats itself, the beam oscillates in response to the shed vortices.

The piezoelectric voltage data obtained from the simulation and the experiment are compared in Fig. 10.17. As shown in Fig. 10.17a,b, there is good agreement between the overall magnitude of the simulated and experimental piezoelectric voltage amplitudes. The modulation in the experimental signal is indicative of a time-dependent three-dimensionality in the vortex shedding. It is hypothesized that a phase difference exists between consecutive vortices shed along the length of the cylinder. This phase difference in vortex shedding manifests itself as a low-frequency modulation in the voltage response of the piezoelectric beam located downstream of the cylinder. Since the 3D effects are not modeled in this 2D simulation, such a modulation is not captured in Fig. 10.17b. The higher frequency

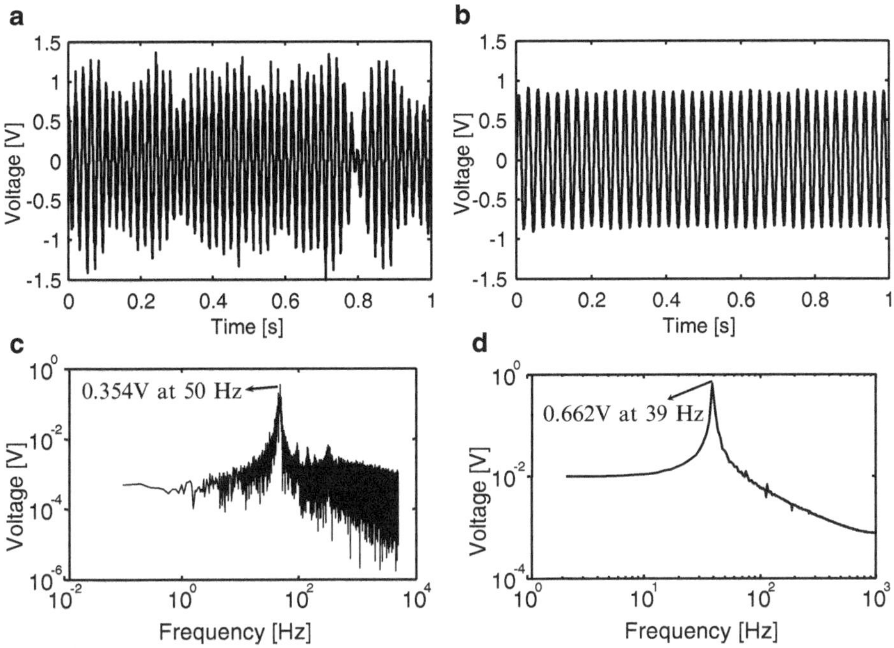

Fig. 10.17 Comparison of the simulated voltage output with experimentally measured output voltage. A one-second section of the voltage signal from (**a**) the experiment and (**b**) the simulation. Frequency content of the corresponding signals are shown in (**c**) and (**d**)

components in the experimental data (Fig. 10.17c), on the other hand, are due to secondary modes of vortex shedding and turbulent flow fluctuations. Due to the SDOF modeling of the harvester, output voltage frequencies higher than the first mode (which are seen in the experimental data in Fig. 10.17c) are not captured by the simulations (Fig. 10.17d). Finally, note that the fundamental frequency observed in the experiment is significantly higher than the simulation results (i.e., as if the splitter-plate effects are nonexistent). Since the width of the beam in the experiments was less than 2% of the span of the cylinder, the beam had virtually no upstream effect on the vortex shedding frequency. Note that in these 2D simulations, however, the width of the beam is identical to the span of the cylinder, thus the splitter-plate effect of the beam on the vortex shedding frequency is significant. Future work will focus on a complete 3D aeroelectromechanical model; better agreement between the experiments and simulations can then be expected.

Despite the differences discussed, the aeroelectromechanical simulation can be considered successful in capturing the overall behavior and electrical output of the piezoelectric beam in the wake of a cylinder. From the simulated voltage, the power is calculated to be approximately 3.7 μW as compared to the experimental measured value of 4 μW.

10.7 Interaction of a Vortex Ring with a Flexible Beam

The mechanism by which vortices deform a flexible beam in the wake of a circular cylinder was explained in the previous section. In this section, we investigate the response of a flexible beam to a single vortex ring moving with its self-induced velocity. The behavior of an IPMC (Ionic Polymer Metal Composite) beam impinged perpendicularly by a vortex ring in quiescent water was investigated recently [19, 20]. Here we study the response of a flexible Mylar beam to a passing vortex ring in a quiescent volume of air. A diagram of the test setup is shown in Fig. 10.18. In this experiments, the vortex ring was generated using a "vortex cannon," which consists of a plastic bucket with a circular opening on one side and covered with a silicone membrane on its other side. Upon tensioning and rapid release of the membrane, a vortex ring forms and travels away from the opening with a speed U_{ring}. After seeding the enclosure with atomized olive oil, a thin sheet of laser light was used to illuminate a cross section of the vortex ring as it passes over the midline of the transparent beam. The base force and the surface strain near the base were recorded together with video images using a high-speed camera (Fig. 10.19a). The images were then processed for PIV (particle image velocimetry) to generate the velocity contours of the flow field (Fig. 10.19b). The ring radius (R), core radius (r), and the strength (Γ) of the vortex ring were found using these images. The strength of the vortex in the beginning of the interaction is calculated to be approximately $\Gamma = 0.324$ m^2/s by taking a contour integral of the velocity field around the vortex.

The pressure distribution over the beam's surface due to the vortex is estimated based on the potential flow solution of a stationary vortex near a finite wall

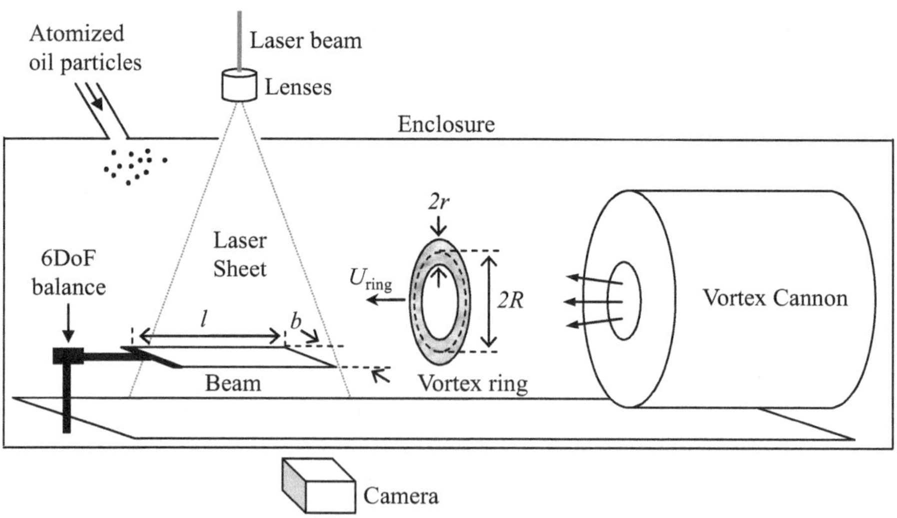

Fig. 10.18 Diagram of the experimental setup for investigating the interaction of a flexible beam with a vortex ring

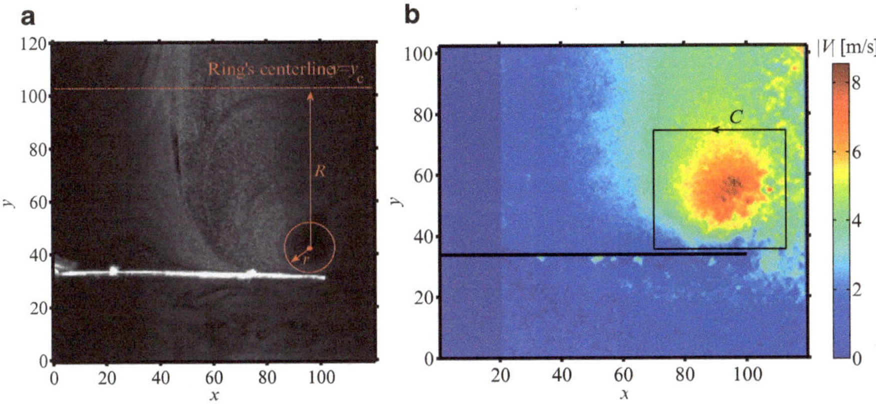

Fig. 10.19 (**a**) PIV image taken by using atomized olive oil particles. The beam is fixed on the left end. The vortex moves from the right to the left while rotating counter-clockwise. The units of x and y axes are both in mm. (**b**) The velocity contours obtained from the image. For enhanced visual contrast, the travel speed of vortex $U_{\text{ring}} = -1.8$ m/s is subtracted from the entire field. The strength of the vortex is found by taking a contour integral of the velocity field (without subtracting U_{ring}) along the contour C in the counterclockwise direction. Units of axes are in mm, velocity contours are in m/s

Fig. 10.20 The physical plane (z-plane) for a vortex near a finite rigid wall

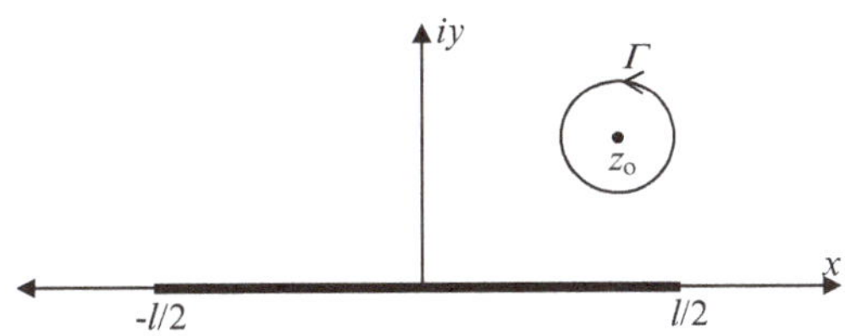

(Fig. 10.20). The complex potential of the flow field is calculated using conformal mapping and Milne-Thomson's circle theorem [21, 22]. The local flow speed on the beam is then found by taking the derivative of the complex potential with respect to the complex coordinate z. The pressure distribution on top of the beam is assumed to change quasi-steadily with the position of the vortex as it travels along the beam's length. The net pressure load over the beam is then calculated from the steady-state Bernoulli equation. Using the measured values of the vortex strength (Γ), position (z_o), and speed (U_{ring}), the numerical values of the net pressure load over the beam at consecutive time steps are calculated (Fig. 10.21a). It should be noted for this particular case that the vortex ring broke down just before reaching the middle of the beam ($z_w = 0$), therefore, the pressure loads are not calculated for $z_w < 0$ (Fig. 10.21a). The traveling load due to the vortex-induced pressure distribution was then used in the Euler–Bernoulli beam equation to calculate the tip deflection of the beam. Despite these drastic simplifications in modeling, the estimated tip deflection of the beam agrees reasonably well with the direct measurements (Fig. 10.21b).

An estimation of the aeroelastic efficiency is performed by calculating the kinetic energy supplied by the vortex and the strain energy transferred to the beam. The

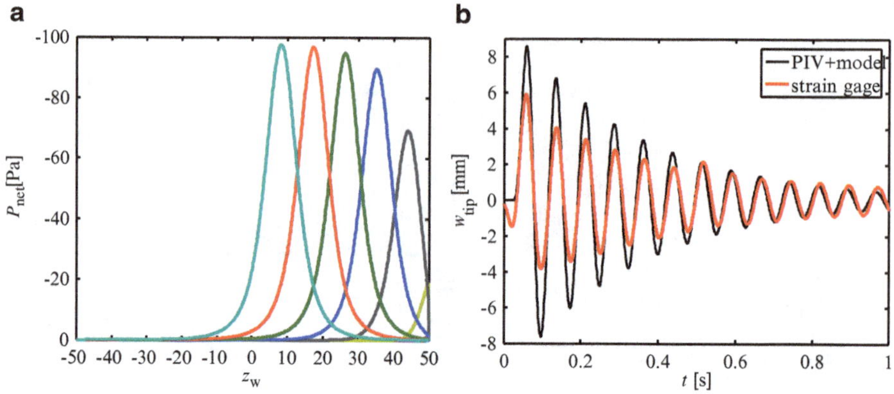

Fig. 10.21 (**a**) Pressure distribution over the beam. The time difference between each load curve is 10 ms. (**b**) The tip deflection of the beam is calculated based on a potential flow model and PIV data (PIV + model) and based on strain gage data calibrated using measured tip deflections

kinetic energy of the vortex at the earliest visible time step (Fig. 10.19a) was estimated based on corresponding PIV data (Fig. 10.19b) using two methods. In the first method, the kinetic energy density of the flow field in the PIV image was found from

$$e = \frac{1}{2}\rho\left(u^2 + v^2\right). \tag{10.17}$$

Assuming the flow field is axis-symmetric around the centerline of the vortex ring, the 3D flow field can be reconstructed by revolving the PIV plane around this centerline (Fig. 10.22). The kinetic energy of the entire vortex ring is then calculated from

$$E_v = 2\pi \int_{x=0}^{0.120} \int_{y=0}^{y_c} e\,(y_c - y)\,dy dx, \tag{10.18}$$

where $y_c = 0.103$ m is approximately the vertical coordinate of the centerline of the vortex ring and $x = 0.120$ m is the horizontal extent of the PIV frame (Fig. 10.19). In the second method, the vortex energy is calculated based on Fraenkel's second order formula [23, 24]:

$$E_v = \frac{1}{2}\rho R\Gamma^2\left[\ln\frac{8R}{r} - \frac{7}{4} + \frac{3}{16}\left(\frac{r}{R}\right)^2 \ln\frac{8R}{r}\right]. \tag{10.19}$$

The kinetic energies found by Eqs. (10.18) and (10.19) are about 13 mJ and 9 mJ respectively. The difference between the two results is attributed mainly to the rather crude determination of parameters r, R, and Γ from the PIV and to construction of a 3D flow field based on 2D images. The elastic mechanical energy transferred to the beam is found from

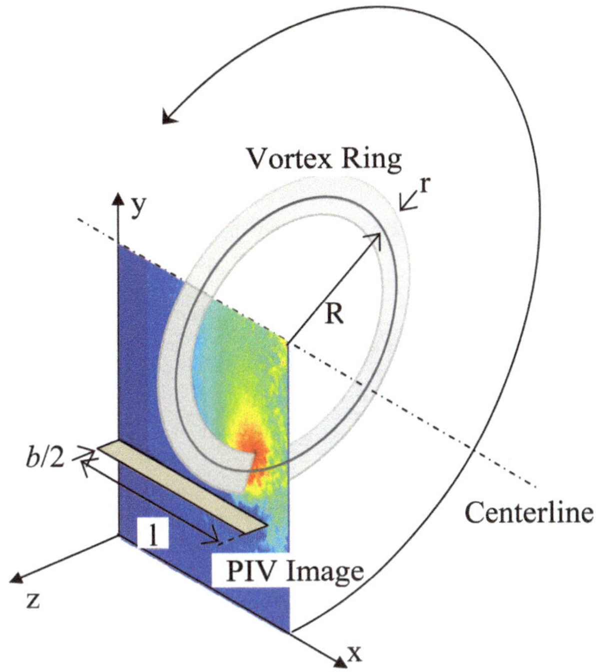

Fig. 10.22 The revolution of the planar PIV data around the centerline of the vortex ring

$$E_b = \frac{1}{2} k w_{tip}^2, \qquad (10.20)$$

where $k = 3.92$ N/m is the stiffness of the beam and $w_t = 5.9$ mm is the measured tip deflection at $t = 50$ ms. The mechanical energy transferred to the beam is then found to be $E_b = 0.068$ mJ. Note that this energy dissipates in approximately 2 s due to structural and viscous damping. Based on the kinetic energy found from Eq. (10.18), the aeroelastic efficiency of the vortex-beam interaction is calculated as

$$\eta_{ae} = \frac{E_b}{E_v} = 0.52\% \qquad (10.21)$$

10.8 Piezoelectric Energy Harvesting from Turbulent Boundary Layers

The cascading behavior of eddies in turbulent boundary layers results in temporal and spatial fluctuations of pressure over a large range of frequencies and amplitudes. Therefore the turbulent boundary layers could be a broadband, multiscale source for piezoelectric energy harvesting. We investigated the potential of turbulent boundary layers for piezoelectric energy harvesting by placing PVDF beams near the wall of a wind tunnel (Fig. 10.23). Two types of PVDF unimorph beam were used. The

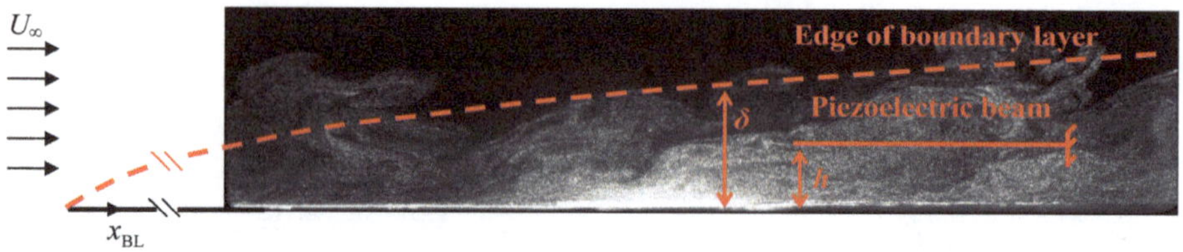

Fig. 10.23 Flow visualization image of the turbulent boundary layer

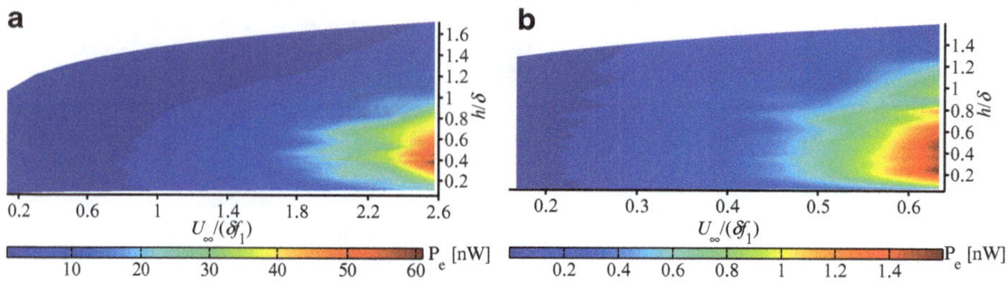

Fig. 10.24 Contour plots of the electrical power harvested in a turbulent boundary layer (**a**) for the long beam and (**b**) for the short beam

"long beam" had dimensions of 30 mm × 16 mm × 0.2 mm and the "short beam" had dimensions of 15 mm × 13 mm × 0.2 mm. The first bending mode frequencies (f_1) of the long and short beams were 48.5 Hz and 198 Hz, respectively. The measurements were performed at a distance $x_{BL} = 5.12$ m downstream of a trip wire placed near the start of the test section of the wind tunnel. The boundary layer developing over the bottom wall of this wind tunnel has been extensively characterized in [25, 26]. The thickness of the boundary layer (δ) at the position of the measurements is found from

$$\delta = c\frac{x_{BL}}{Re_{x_{BL}}^{1/5}}, \tag{10.22}$$

where the parameter $c = 0.353$ is obtained from the experimental data present in [25, 26] and Re_x is the local Reynolds number given by

$$Re_{x_{BL}} = \frac{\rho U_\infty x_{BL}}{\mu}. \tag{10.23}$$

The normalized frequency is defined as $f^* = U_\infty/(\delta f_1)$. The normalized contour plots of the generated power are shown in Fig. 10.24. The maximum electrical power harvested over a load resistance of 10 MΩ was 60 nW for the long beam and 1.6 nW for the short beam. Within the range of flow speeds tested (i.e., 0–11 m/s), the harvester power from both beams increased monotonically and reached their maximal values at $U_\infty = 11$ m/s for which $\delta = 95$ mm. However, the wall distance for the maximum power was in a range of 25–50 mm for the long beam whereas this range was 15–60 mm for the short beam. Maximum power is harvested in a region where h/δ is approximately 0.4 for both the long and the short beams; however, use of the shorter beam increases this region of maximal energy harvesting while decreasing the peak power magnitude.

10.9 Summary

In this chapter, we first demonstrated that a portion of the kinetic energy of a uniform and steady fluid flow can be converted to electrical energy by placing a cantilevered piezoelectric beam in the wake of a bluff body. It was shown that the major losses in the conversion of the flow energy to electrical energy occur during the aeroelastic conversion phase, i.e. during the creation of oscillatory motions from the uniform and steady fluid flow. Attachment of the bluff body to the piezoelectric beam was shown to increase the aeroelastic efficiency dramatically. The shape of the tip body plays a critical role in the behavior of the harvester; large-amplitude deformations and more electrical power can be obtained by changing the tip body shape from a circular cylinder to a half-circular cylinder. More investigations into the effect of tip shape need to be performed in the future to fully characterize their energy harvesting potential.

Velocity fluctuations in turbulent boundary layers can be present in a relatively large range of amplitudes and frequencies. Therefore, turbulent boundary layers seem to be favorable energy sources for broadband piezoelectric energy harvesting at various temporal and spatial scales. The wind tunnel tests presented in this chapter indicated that the harvested power in a turbulent boundary layer increased steadily with flow speed. The maximum power is harvested within a certain well-defined wall distance range inside the boundary layer. While the harvested power decreased drastically with beam length, the wall distance range over which maximal power harvesting could be obtained was greater for a shorter beam.

The study of the interaction of a flexible beam with a vortex may provide further insight in behavior of piezoelectric beam in the wake of bluff bodies as well as in turbulent boundary layers. A preliminary investigation of the case where a vortex ring passes over a flexible beam was experimentally investigated. A PIV technique to calculate the kinetic energy of a vortex was presented in this chapter and the efficiency of transmission of this fluid energy to the beam was then estimated. The development of the pressure loads over the beam due to the vortex was predicted by potential flow theory and was shown to agree reasonably well with the measured values.

Computational fluid dynamics modeling could be very beneficial for design and analysis of flow-powered energy harvesters. However, a proper fluid-structure interaction analysis can be challenging for the case of three-dimensional, turbulent flows with flow separation and large structural deformations. The inclusion of piezoelectric effects to such an aeroelastic analysis is even more complicated, as it results in an aeroelectromechanical system of equations that ideally requires a coupled treatment of the aeroelastic and electromechanical interactions. In this chapter, we introduced a computational framework to model the aeroelectromechanical behavior of a piezoelectric beam in the wake of a cylinder. The model compares well with the experimental results and it can readily be extended to more complicated cases.

References

1. Mitcheson PD, Yeatman EM, Rao GK, Holmes AS, Green TC (2008) Energy harvesting from human and machine motion for wireless electronic devices. Proc IEEE 96(9):1457–1486
2. Akaydın HD, Elvin N, Andreopoulos Y (2010) Wake of a cylinder: a paradigm for energy harvesting with piezoelectric materials. Exp Fluids 49(1):291–304
3. Akaydın HD, Elvin N, Andreopoulos Y (2010) Energy harvesting from highly unsteady fluid flows using piezoelectric materials. J Intell Mater Syst Struct 21(13):1263–1278
4. Akaydın HD, Elvin N, Andreopoulos Y (2012) Performance of a self-excited fluidic energy harvester. Smart Mater Struct 21(2):025007. doi:10.1088/0964-1726/21/2/025007
5. Panton R (1996) Incompressible flow, 2nd edn. Wiley, New York
6. Lienhard JH (1966) Synopsis of lift, drag and vortex frequency data for rigid circular cylinders. College of Engineering, Research Division Bulletin 300. Washington State University, Pullman, WA
7. Norberg C (2002) Fluctuating lift on a circular cylinder: review and new measurements. J Fluids Struct 17:57–96
8. Erturk A, Inman DJ (2008) A distributed parameter electromechanical model for cantilevered piezoelectric energy harvesters. J Vib Acoust 130:041002-1
9. Blevins RD (1977) Flow induced vibration. Litton Educational, New York
10. Alonso G, Valero E, Meseguer J (2009) An analysis on the dependence on cross section geometry of galloping stability of two-dimensional bodies having either biconvex or rhomboidal cross sections. Eur J Mech B Fluids 28(2):328–334
11. Alonso G, Meseguer J, Perez-Grande I (2005) Galloping instabilities of two dimensional triangular cross-section bodies. Exp Fluids 38:789–795
12. Kamakoti R, Shyy W (2004) Fluid–structure interaction for aeroelastic applications. Prog Aerospace Sci 40:535–558
13. Elvin NG, Elvin A (2009) A coupled finite element-circuit simulation model for analyzing piezoelectric energy generators. J Intell Mater Syst Struct 20:587–595
14. Elvin NG, Elvin A (2009) A general equivalent circuit model for piezoelectric generators. J Intell Mater Syst Struct 20:3–9
15. Dong S, Karniadakis GE, Ekmekci A, Rockwell DA (2006) Combined DNS-PIV study of the turbulent near wake. J Fluid Mech 569:185–207
16. Unal MF, Rockwell D (1988) On vortex formation from a cylinder. Part II. Control by splitter-plate interference. J Fluid Mech 190:513–529
17. Roshko A (1954) On the development of turbulent wakes from Vortex Streets. National Advisory Committee for Aeronautics Report NACA-TN-2913
18. Shukla S, Govardhan RN, Arakeri JH (2009) Flow over a cylinder with a hinged-splitter plate. J Fluids Struct 25(4):713–720
19. Peterson SD, Porfiri M (2012) Energy exchange between a vortex ring and an ionic polymer metal composite. Appl Phys Lett 100(11):114102
20. Peterson SD, Porfiri M (2012) Interaction of a vortex pair with a flexible plate in an ideal quiescent fluid. J Intell Mater Syst Struct. doi:1045389X11435995
21. Saffman PG (1993) Vortex dynamics. Cambridge University Press, Cambridge

22. Batchelor GK (2002) An introduction to fluid dynamics. Cambridge University Press, Cambridge
23. Sullivan IS, Niemela JJ, Hershberger RE, Bolster D, Donnelly RJ (2008) Dynamics of thin vortex rings. J Fluid Mech 609:319–347
24. Fraenkel LE (1972) Examples of steady vortex rings of small cross-section in an ideal fluid. J Fluid Mech 51:119–135
25. Andreopoulos J, Agui J (1996) Wall vorticity flux dynamics in a two-dimensional turbulent boundary layer. J Fluid Mech 309:45–84
26. Andreopoulos Y, Honkan A (2001) An experimental study of the dissipative and vortical motions in turbulent boundary layers. J Fluid Mech 439:131–163

Chapter 11
Airfoil-Based Linear and Nonlinear Electroaeroelastic Energy Harvesting

Carlos De Marqui Jr. and Alper Erturk

Abstract The transformation of aeroelastic vibrations into low-power electricity has received growing attention in the last few years. The goal in electroaeroelastic energy harvesting is to convert airflow energy into electricity for applications ranging from aircraft sensor systems to wireless electronic components located in high wind areas. The use of an airfoil section is a convenient approach to create instabilities and persistent aeroelastic vibrations. This chapter investigates airfoil-based electroaeroelastic energy harvesters employing piezoelectric transduction and electromagnetic induction. An airfoil with two degrees of freedom (DOF) is investigated by adding piezoelectric and electromagnetic coupling to the plunge DOF in two separate cases. The governing dimensionless electroaeroelastic equations are derived in each case for a resistive load in the electrical domain. Both linear and nonlinear electroaeroelastic methods of energy harvesting are discussed. The linear problem focuses on the response at the flutter boundary while the nonlinear configurations with free play and cubic stiffness in the pitch DOF exploit nonlinear limit-cycle oscillations. The effects of several dimensionless system parameters on the electrical power output and flutter speed are investigated. Experimental validations are presented for linear and nonlinear electroaeroelastic energy harvesting systems employing piezoelectric transduction. It is demonstrated that the free play nonlinearity (a detrimental form of nonlinearity) can be exploited to reduce the cut-in speed for persistent oscillations while the hardening stiffness (a benign form of nonlinearity) can be combined with free play to keep the response amplitude at acceptable levels.

C. De Marqui Jr. (✉)
Department of Aeronautical Engineering, Engineering School of Sao Carlos,
University of Sao Paulo, São Carlos, SP, Brazil
e-mail: demarqui@sc.usp.br

A. Erturk
G. W. Woodruff School of Mechanical Engineering, Georgia Institute of Technology,
Atlanta, GA, USA
e-mail: alper.erturk@me.gatech.edu

N. Elvin and A. Erturk (eds.), *Advances in Energy Harvesting Methods*,
DOI 10.1007/978-1-4614-5705-3_11,

11.1 Introduction

The harvesting of ambient energy toward enabling self-powered electronic systems has been investigated by numerous research groups over the last decade [1–6]. The ultimate goal in this research field is to convert sufficient ambient energy to eliminate the need for periodic, external battery charging in wireless applications. Most of the existing research on mechanical energy harvesting has focused on transforming direct vibrational input into electricity by employing the electromagnetic [7–9], electrostatic [10–12], and piezoelectric [13–15] transduction mechanisms. Another form of energy that is available for electronic devices located in high wind areas (such as sensor systems in bridges or aircraft components) is due to air flow. Wind energy has been harvested for centuries by means of large-scale windmills or wind turbines to generate orders of magnitude larger power output as compared to the power requirement of wireless electronic components. For low-power electricity generation by harvesting wind energy, one evident approach is therefore to miniaturize windmill or wind turbine configurations as done by Priya et al. [16], Myers et al. [17], Rancourt et al. [18], and Xu et al. [19]. Similar to their large-scale counterparts, miniaturized windmill and wind turbine systems require the fabrication of several components due to their rather sophisticated design. Moreover, increased cut-in speed[1] due to reduced dimensions remains an important issue, suggesting the need for more research on scalable wind energy harvesters.

As an alternative to small-scale windmills and wind turbines, researchers have recently considered directly exploiting aeroelastic vibrations for converting wind energy into electricity using simple and scalable configurations. An experimental effort of generating electricity from thin curved airfoils made of macro-fiber composite (MFC) piezoelectric structures under airflow excitation was presented by Erturk et al. [20]. For the piezoaeroelastic problem of harvesting energy from airflow excitation of a cantilevered plate with embedded piezoceramics, De Marqui et al. [21, 22] presented electroaeroelastic finite-element models coupling the electroelastic [15, 59] and aeroelastic [23–26] domains based on the vortex-lattice method [21] and the doublet-lattice method [22]. Time-domain simulations [21] were performed for a cantilevered plate with embedded piezoceramics at various airflow speeds below the linear flutter speed and at the flutter boundary. Frequency-domain simulations [22] considering resistive and resistive-inductive circuits were also presented focusing on the linear response at the flutter boundary. Bryant and Garcia [27, 28] studied the aeroelastic energy harvesting problem for a typical section by using the finite state theory of Peters et al. [29]. Erturk et al. [30] presented an experimentally validated lumped-parameter model for a wing-section (airfoil) with piezoceramics attached onto plunge stiffness members using Theodorsen's [31] unsteady aerodynamic model. Piezoelectric power generation at the flutter boundary, including the minor shift in the linear flutter speed, was

[1] The cut-in speed is the minimum wind speed at which energy can be extracted from the device.

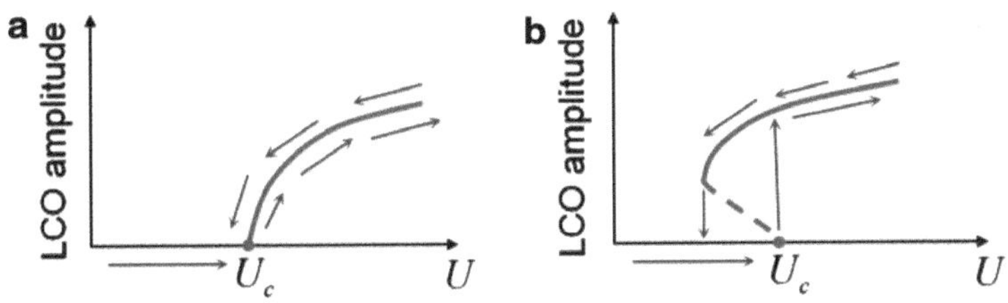

Fig. 11.1 Two types of nonlinear LCO due to (**a**) supercritical and (**b**) subcritical bifurcations with increasing airflow speed (*solid* and *dashed lines* are stable and unstable LCOs, respectively). Electroaeroelastic energy harvester configurations exhibiting subcritical bifurcations (**b**) have reduced cut-in speed [30]

discussed [30]. In addition to these recent efforts, we note that the "wingmill" concept employing aeroelastic vibrations was investigated previously for rather large-scale configurations as an alternative to conventional windmills and wind turbines [32–34].

Persistent oscillations at the linear flutter speed (neutral stability condition) is an idealized scenario for linear wind energy harvesting [21, 22, 30] and it limits the effective power generation performance to a specific airflow speed. Moreover, often nonlinearities are inherently present in aeroelastic systems. Nonlinear systems present a very rich variety of dynamic behavior such as nonlinear limit-cycle oscillations (LCOs), internal resonances, and chaotic motions (see Guckenheimer and Holmes [35], Nayfeh and Mook [36], Moon [37], and Strogatz [38]). In particular, stable aeroelastic LCO of acceptable amplitude can provide an important source of persistent electrical power over a wide range of airflow speeds, even below the linear flutter speed of the corresponding linear aeroelastic system.

In the aeroelasticity literature, many authors have investigated aerodynamic and structural nonlinearities [39–44] while the focus in this chapter is placed on structural nonlinearities. Due to their simplicity and the fundamental insight they provide, aeroelastic typical section (airfoil) models have been widely studied with concentrated (lumped) nonlinearities. Price et al. [40] investigated free play nonlinearity in the pitch DOF of a typical section and reported nonlinear LCOs below the linear flutter speed. Tang and Dowell [41] also modeled free play nonlinearity in the pitch DOF of a typical section. In addition to the concentrated structural nonlinearity, they considered aerodynamic nonlinearities. Zhao and Yang [42] examined LCOs with cubic nonlinearity modeled in the pitch DOF of an airfoil subjected to incompressible airflow. For further discussions on the subject of structural and aerodynamic nonlinearities, the reader is referred to Dowell and Tang [43] and Dowell et al. [44].

Inspired by the literature of nonlinear aeroelastic systems, researchers have recently focused on exploiting nonlinearities in electroaeroelastic energy harvesting. Erturk et al. [30] suggested that catastrophic (or "detrimental" [43, 44]) nonlinearities that are avoided in real aircraft can be employed for reducing the cut-in speed of persistent oscillations in electroaeroelastic energy harvesters. Figure 11.1a depicts the development of "good LCO" due to benign nonlinearities whereas Fig. 11.1b shows the formation of "bad LCO" due to detrimental nonlinearities [43, 44] with increasing airflow speed (U), where U_c is the linear flutter speed. "Good" and "bad" LCO result from supercritical and subcritical bifurcations (typically Hopf bifurcations [38]) respectively. What makes the form of the LCO in Fig. 11.1b detrimental to real aircraft is the possibility of attraction by the stable LCO branch *below* the linear flutter speed due to an abrupt disturbance, such as sudden maneuvering. Since the present challenge in scaling of electroaeroelastic energy harvesters is due to increased cut-in speed, the type of LCO shown in Fig. 11.1b can be proposed as a solution [30]. It is known from the literature of nonlinear aeroelasticity that the LCO associated with subcritical bifurcations often results from free play or loose joints that create bilinear stiffness. Following this argument, Sousa et al. [45] considered free play nonlinearity for the pitch stiffness and experimentally validated its effect of reducing the cut-in speed and increasing the harvested power. The case of combined nonlinearities (free play and cubic hardening stiffness) was also numerically investigated by the same authors [45]. Nonlinear plunge and pitch stiffness components (in polynomial forms) were modeled and theoretically investigated by Abdelkefi et al. [46] for piezoaeroelastic energy harvesting.

Before outlining the content of this chapter, it is worth summarizing some of the major work on airflow energy harvesting other than the efforts employing airfoil-based electroaeroelastic energy harvesting. St. Clair et al. [47] presented a design that uses a piezoelectric beam embedded within a cavity under airflow. Vortex-induced oscillations of piezoelectric cantilevers located behind bluff bodies (following the hydro-generator version introduced by Allen and Smits [48]) were investigated by Robbins et al. [49], Pobering et al. [50], and Akaydin et al. [51] through experiments and numerical simulations. Tang et al. [52] presented a rigorous analysis of the energy transfer from the fluid to the structure for self-excited vibrations due to axial flow over a cantilever. Piezoelectric energy harvesting from LCO under axial flow over a cantilever beam was discussed by Dunnmon et al. [53] recently. Kwon [54] considered a T-shaped cantilever beam that causes vortex street formation over the cantilever in response to axial flow. Very few research groups have investigated electromagnetic transduction for airflow energy harvesting although electromagnetic induction can be very effective for large displacements at low frequencies [7–9]. Zhu et al. [55] experimentally investigated an electromagnetic energy harvester that oscillates behind a bluff body while Jung et al. [56] studied electromagnetic energy harvesting from wake galloping.

This chapter presents linear and nonlinear modeling of 2-DOF electroaeroelastic typical sections for wind energy harvesting applications by using (1) piezoelectric transduction and (2) electromagnetic induction. First the conventional form of a 2-DOF linear aeroelastic typical section is reviewed along with the governing equations. Then, the electroaeroelastic forms are given for the problems of piezoelectric and electromagnetic energy harvesting in two separate configurations by coupling the respective energy harvesting interface with the plunge DOF. Free play, and combined free play-cubic hardening nonlinearities are then introduced to the pitch

DOF in the governing equations. Theoretical case studies are given with simulations using the dimensionless forms of the electroaeroelastic equations of piezoelectric and electromagnetic energy harvesting. Finally, an experimental case study [45] is presented for a typical section with piezoelectric coupling on the plunge DOF for validation of the linear and nonlinear electroaeroelastic models.

11.2 Conventional 2-DOF Linear Aeroelastic Equations

Figure 11.2 shows a schematic of the well-known linear 2-DOF typical section [23–26] including purely mechanical dissipative elements. In this schematic, the plunge displacement, h (positive downward), and the pitch angle, α (positive clockwise), are measured at the elastic axis (point P). In addition, b is the semi-chord length of the airfoil section, x_α is the dimensionless chord-wise offset of the elastic axis from the centroid (C), k_h is the stiffness per length in the plunge DOF, k_α is the stiffness per length in the pitch DOF, d_h is the damping coefficient per length in the plunge DOF, d_α is the damping coefficient per length in the pitch DOF, and U is the airflow speed.

The governing linear aeroelastic equations for the configuration shown in Fig. 11.2 are [23–26]

$$(m + m_e)\ddot{h} + mbx_\alpha\ddot{\alpha} + d_h\dot{h} + k_h h = -L \tag{11.1}$$

$$mbx_\alpha\ddot{h} + I_\alpha\ddot{\alpha} + d_\alpha\dot{\alpha} + k_\alpha\alpha = M \tag{11.2}$$

where m is the airfoil mass per length (in the span direction), m_e is the effective fixture mass (connecting the airfoil to the plunge springs) per length, M is the aerodynamic moment per length, L is the aerodynamic lift per length, and an over-dot represents differentiation with respect to time (t).

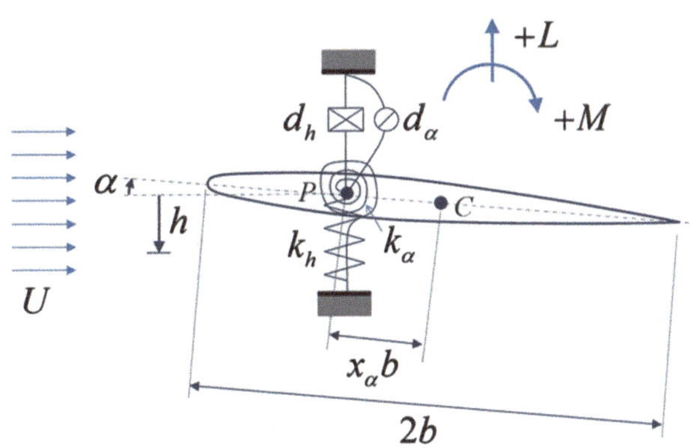

Fig. 11.2 A 2-DOF aeroelastic typical section under airflow excitation

In the following sections, the unsteady aerodynamic loads (lift and moment) due to arbitrary motions are obtained from Jones' approximation [57] of Wagner's indicial function [58], which is an approximation to the generalized Theodorsen function [23–26, 31]. The fixture mass (m_e) is defined for the case when the system slightly deviates from the ideal typical section depicted in Fig. 11.2 due to the spring mass, and other attachments in real experiments [30, 45], while it is zero in the ideal representation given by Fig. 11.2.

11.3 Linear Electroaeroelastic Energy Harvesting

11.3.1 Linear Typical Section with Piezoelectric Transduction

In this section, piezoelectric coupling is added to the plunge DOF of the typical section presented in Fig. 11.2. The schematic of an electroaeroelastic section for piezoelectric energy harvesting from airflow excitation is shown in Fig. 11.3. The piezoelectric patches are strained dynamically due to aeroelastic vibrations of the cantilever (plunge spring), yielding an oscillatory voltage output. The resultant voltage from the electrodes is connected to a resistive load in order to quantify the electrical power output. Following the derivation presented in Erturk and Inman [59] (which is a Lagrangian formulation of the piezoaeroelastic problem as an extension of the purely aeroelastic version available in textbooks on aeroelasticity [23–26]), the linear equations governing the piezoelectrically coupled typical section are obtained as

$$(m + m_e)\ddot{h} + mbx_\alpha\ddot{\alpha} + d_h\dot{h} + k_h h - \frac{\theta}{l}v = -L \tag{11.3}$$

$$mbx_\alpha\ddot{h} + I_\alpha\ddot{\alpha} + d_\alpha\dot{\alpha} + k_\alpha\alpha = M \tag{11.4}$$

$$C_p\dot{v} + \frac{v}{R_l} + \theta\dot{h} = 0 \tag{11.5}$$

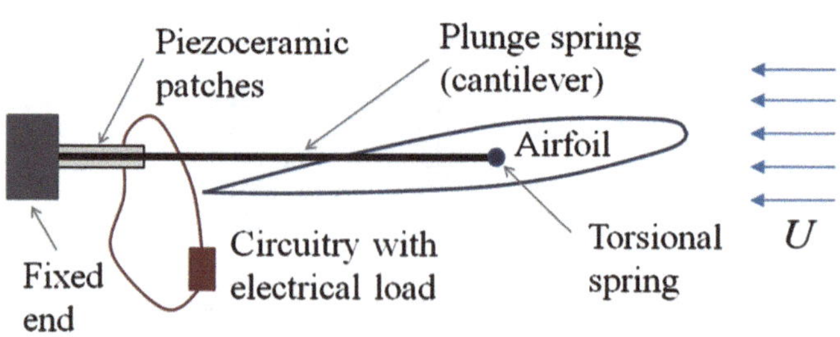

Fig. 11.3 Electroaeroelastic typical section with piezoelectric coupling on the plunge DOF and an external electrical load (piezoelectric patches are attached to the plunge spring)

where l is the span length, R_l is the electrical load resistance, v is the voltage across the resistive load, C_p is the equivalent capacitance of the piezoceramic layers, and θ is the electromechanical coupling. If there are multiple piezoelectric patches as in a bimorph configuration or due to multiple plunge cantilevers [30, 45, 59], the capacitance and electromechanical coupling terms depend on the way the patches are connected (in series or in parallel).

Equations (11.3)–(11.5) can be written in dimensionless form as

$$\beta\bar{h}'' + x_\alpha\alpha'' + \zeta_h\bar{h}' + \bar{h} - \kappa\bar{v} = -\bar{L} \tag{11.6}$$

$$x_\alpha\bar{h}'' + \bar{r}_\alpha^2\alpha'' + \zeta_\alpha\alpha' + \gamma^2\bar{r}_\alpha^2\alpha = \bar{M} \tag{11.7}$$

$$\eta\bar{v}' + \frac{\bar{v}}{\lambda_l} + \kappa\bar{h}' = 0 \tag{11.8}$$

where $\beta = (m + m_e)/m$ is a mass ratio, $\bar{h} = h/b$ is the dimensionless plunge displacement, $\zeta_h = d_h/m\omega_h$ is the plunge damping factor, $\zeta_\alpha = d_\alpha/mb^2\omega_h$ is the pitch damping factor, $\bar{r}_\alpha = r_\alpha/b$ is the dimensionless ratio of gyration, $\bar{v} = v/\tilde{v}$ (where $\tilde{v} = 1$ V is the reference voltage for normalization), $\kappa = \theta\tilde{v}/lmb\omega_h^2$ is the dimensionless electromechanical coupling, $\eta = C_p\tilde{v}^2/mb^2l\omega_h^2$ is the dimensionless equivalent capacitance, $\lambda_l = R_l mb^2 l\omega_h^3/\tilde{v}^2$ is the dimensionless load resistance, $\gamma = \omega_\alpha/\omega_h$ is the frequency ratio, $\omega_h^2 = k_h/m$ is the square of the plunge natural frequency, and $\omega_\alpha^2 = k_\alpha/I_\alpha$ is the square of the pitch natural frequency (in the uncoupled sense). The dimensionless aerodynamic loads are $\bar{L} = L/mb\omega_h^2$ and $\bar{M} = M/mb^2\omega_h^2$ while a prime ($'$) denotes differentiation with respect to the dimensionless time $\tau = \omega_h t$. The damping factors in Eqs. (11.6) and (11.7) are simply the reciprocals of the respective quality factors.

The piezoaeroelastic equations can be represented in a form similar to the state-space form proposed (for the aeroelastic problem) by Edwards et al. [60] after introducing the effect of electromechanical coupling. Therefore, the voltage output should be considered as an additional state variable. Two augmented aerodynamic states, $\mathbf{x}_a = \left\{x_1 \ x_2\right\}^t$ (where the superscript t stands for the transpose) are included in the state-space representation of the piezoaeroelastic problem. Then the state-space piezoaeroelastic equations in the matrix form are

$$\begin{bmatrix} \mathbf{I} & \mathbf{0} & \mathbf{0} & 0 \\ \mathbf{0} & \tilde{\mathbf{M}} & \mathbf{0} & 0 \\ \mathbf{0} & \mathbf{0} & \mathbf{I} & 0 \\ \mathbf{0} & \mathbf{0} & \mathbf{0} & \eta \end{bmatrix} \begin{Bmatrix} \mathbf{x}' \\ \mathbf{x}'' \\ \mathbf{x}_a{}' \\ \bar{v}' \end{Bmatrix} = \begin{bmatrix} \mathbf{0} & \mathbf{I} & \mathbf{0} & 0 \\ -\tilde{\mathbf{K}} & -\tilde{\mathbf{B}} & \mathbf{D} & \Theta_1 \\ \mathbf{E}_1 & \mathbf{E}_2 & \mathbf{F} & 0 \\ \mathbf{0} & \Theta_2 & \mathbf{0} & 1/\lambda_l \end{bmatrix} \begin{Bmatrix} \mathbf{x} \\ \mathbf{x}' \\ \mathbf{x}_a \\ \bar{v} \end{Bmatrix} \tag{11.9}$$

where $\Theta_1 = \left\{0 \ \kappa\right\}^t$, $\Theta_2 = \left\{0 \ -\kappa\right\}$, $\mathbf{x} = \left\{\alpha \ \bar{h}\right\}^t$, and $\mathbf{I}$ is the 2×2 identity matrix. The mass, stiffness, and damping related matrices in Eq. (11.9) are

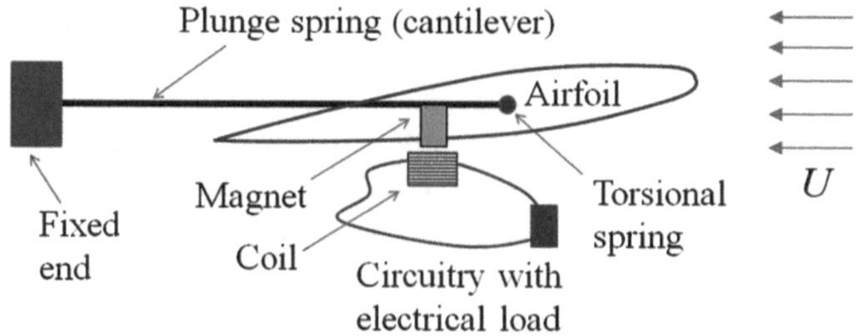

Fig. 11.4 Electroaeroelastic typical section with electromagnetic coupling on the plunge DOF and an external electrical load (magnet is attached to the plunge spring)

$$\tilde{\mathbf{M}} = \mathbf{M} - \frac{\rho b^2}{m}\mathbf{M}_{\mathrm{nc}} \tag{11.10}$$

$$\tilde{\mathbf{K}} = \mathbf{K} - \frac{\rho b^2}{m}(U/b)^2\left(\mathbf{K}_{\mathrm{nc}} + \frac{1}{2}\mathbf{R}\mathbf{S}_1\right) \tag{11.11}$$

$$\tilde{\mathbf{B}} = \mathbf{B} - \frac{\rho b^2}{m}(U/b)\left(\mathbf{B}_{\mathrm{nc}} + \frac{1}{2}\mathbf{R}\mathbf{S}_2\right) \tag{11.12}$$

Here, **M** is the structural mass matrix, **K** is the structural stiffness matrix, **B** is the structural damping matrix, ρ is the free-stream air mass density, $\mathbf{M}_{\mathrm{nc}}$, $\mathbf{K}_{\mathrm{nc}}$, and $\mathbf{B}_{\mathrm{nc}}$ are non-circulatory aerodynamic matrices related to inertia, stiffness, and damping. These matrices as well as the aerodynamic matrices **D**, $\mathbf{E}_1$, $\mathbf{E}_2$, **F**, **R**, $\mathbf{S}_1$, and $\mathbf{S}_2$ can be found in Edwards et al. [60].

Equation (11.9) can be also represented as

$$\tilde{\mathbf{x}}' = \mathbf{A}\tilde{\mathbf{x}} \tag{11.13}$$

where

$$\mathbf{A} = \begin{bmatrix} \mathbf{0} & \mathbf{I} & \mathbf{0} & \mathbf{0} \\ -\tilde{\mathbf{M}}^{-1}\tilde{\mathbf{K}} & -\tilde{\mathbf{M}}^{-1}\tilde{\mathbf{B}} & \tilde{\mathbf{M}}^{-1}\mathbf{D} & \tilde{\mathbf{M}}^{-1}\Theta_1 \\ \mathbf{E}_1 & \mathbf{E}_2 & \mathbf{F} & 0 \\ \mathbf{0} & (1/\eta)\Theta_2 & \mathbf{0} & (1/\eta)(1/\lambda_1) \end{bmatrix} \tag{11.14}$$

$$\tilde{\mathbf{x}} = \left\{\mathbf{x}\ \mathbf{x}'\ \mathbf{x}_a\ \bar{v}\right\}^{\mathrm{t}} \tag{11.15}$$

11.3.2 Linear Typical Section with Electromagnetic Induction

Electromagnetic energy harvesting from airflow excitation can be realized by employing the electroaeroelastic typical section shown in Fig. 11.4. The magnet is attached to the cantilever (plunge spring) and oscillates relative to the coil to produce electricity from electromagnetic induction in response to aeroelastic vibrations. Therefore, electromagnetic coupling is added to the plunge DOF along with a resistive load (connected to the coil) in the electrical domain of the problem. The linear electromagnetically coupled electroaeroelastic equations are then

$$(m + m_e)\ddot{h} + mbx_\alpha\ddot{\alpha} + d_h\dot{h} + k_h h - \frac{B_l}{l}I = -L \tag{11.16}$$

$$mbx_\alpha\ddot{h} + I_\alpha\ddot{\alpha} + d_\alpha\dot{\alpha} + k_\alpha\alpha = M \tag{11.17}$$

$$L_c\dot{I} + (R_c + R_l)I + B_l\dot{h} = 0 \tag{11.18}$$

where l is the span length, R_l is the electrical load resistance, R_c is the internal resistance of the coil, I is the induced electrical current, L_c is the coil inductance, and B_l is the electromagnetic coupling. The electrical equation is kept in its general form [7, 9] to account for the inherent coil inductance and resistance.

Equations (11.16)–(11.18) can be written in dimensionless form as

$$\beta\bar{h}'' + x_\alpha\alpha'' + \zeta_h\bar{h}' + \bar{h} - \chi\bar{I} = -\bar{L} \tag{11.19}$$

$$x_\alpha\bar{h}'' + \bar{r}_\alpha^2\alpha'' + \zeta_\alpha\alpha' + \gamma^2\bar{r}_\alpha^2\alpha = \bar{M} \tag{11.20}$$

$$\phi\bar{I}' + \lambda_c\bar{I} + \lambda_l\bar{I} + \chi\bar{h}' = 0 \tag{11.21}$$

where the definition of the dimensionless mass ratio (β), the dimensionless plunge displacement ($\bar{h}$), the dimensionless damping factors (ζ_h and ζ_α), the frequency ratio (γ), the dimensionless aerodynamic loads ($\bar{L}$ and $\bar{M}$), and the dimensionless ratio of gyration ($\bar{r}_\alpha$) are as defined in the previous section. In the present case, $\bar{I} = I/\tilde{I}$ is the dimensionless current (where $\tilde{I} = 1$ A is the reference current for normalization), $\chi = B_l\tilde{I}/lmb\omega_h^2$ is the dimensionless electromagnetic coupling, $\phi = L_c\tilde{I}^2/lmb^2\omega_h^2$ is the dimensionless coil inductance, $\lambda_c = R_c\tilde{I}^2/lmb^2\omega_h^3$ is the dimensionless internal resistance of the coil, and $\lambda_l = R_l\tilde{I}^2/lmb^2\omega_h^3$ is the dimensionless load resistance.

The linear equations can be represented in the state-space form by considering the dimensionless electric current ($\bar{I}$) as an additional state variable. The two augmented aerodynamic states, $\mathbf{x}_a = \{x_1 \; x_2\}^t$, are included in the following state-space representation of the problem:

$$\begin{bmatrix} \mathbf{I} & \mathbf{0} & \mathbf{0} & \mathbf{0} \\ \mathbf{0} & \tilde{\mathbf{M}} & \mathbf{0} & \mathbf{0} \\ \mathbf{0} & \mathbf{0} & \mathbf{I} & \mathbf{0} \\ \mathbf{0} & \mathbf{0} & \mathbf{0} & \phi \end{bmatrix} \begin{Bmatrix} \mathbf{x}' \\ \mathbf{x}'' \\ \mathbf{x}'_a \\ \bar{I}' \end{Bmatrix} = \begin{bmatrix} \mathbf{0} & \mathbf{I} & \mathbf{0} & \mathbf{0} \\ -\tilde{\mathbf{K}} & -\tilde{\mathbf{B}} & \tilde{\mathbf{D}} & \mathbf{X}_1 \\ \mathbf{E}_1 & \mathbf{E}_2 & \mathbf{F} & \mathbf{0} \\ \mathbf{0} & X_2 & \mathbf{0} & Z \end{bmatrix} \begin{Bmatrix} \mathbf{x} \\ \mathbf{x}' \\ \mathbf{x}_a \\ \bar{I} \end{Bmatrix} \tag{11.22}$$

where $\mathbf{X}_1 = \{0\ \chi\}^t$, $\mathbf{X}_2 = \{0\ -\chi\}$, $\mathbf{x} = \{\alpha\ \bar{h}\}^t$, $Z = (\lambda_c + \lambda_l)$, and $\mathbf{I}$ is the 2×2 identity matrix. The mass, stiffness, and damping related matrices are defined as previously given in Eqs. (11.10)–(11.12).

11.4 Nonlinear Electroaeroelastic Energy Harvesting

11.4.1 Nonlinear Typical Section with Piezoelectric Transduction

Persistent oscillations occurring at a specific wind speed (i.e., at the flutter boundary) is a limited approach in terms of robustness in device performance, although it gives useful insight into the electroaeroelastic problem. Often inherent nonlinearities are present in real aeroelastic systems, making the response bounded beyond the linear flutter boundary. Moreover, nonlinear systems offer various forms of dynamic behavior, such as large-amplitude LCOs, internal resonances, and chaotic motions [35–38]. In particular, stable aeroelastic LCOs of acceptable amplitude can provide an important source of persistent electrical power over a wide range of airflow speeds.

Concentrated nonlinearities (free play, and combined free play-cubic hardening nonlinearities) are modeled for the pitch DOF of the piezoaeroelastic generator of Sect. 11.3.1 as summarized in Fig. 11.5. The dimensionless nonlinear piezoaeroelastic equations become

$$\beta \bar{h}'' + x_\alpha \alpha'' + \zeta_h \bar{h}' + \bar{h} - \kappa \bar{v} = -\bar{L} \tag{11.23}$$

$$x_\alpha \bar{h}'' + \bar{r}_\alpha^2 \alpha'' + \zeta_\alpha \alpha' + \gamma^2 \bar{r}_\alpha^2 \alpha + f_{\text{fp}}(\alpha) + f_c(\alpha) = \bar{M} \tag{11.24}$$

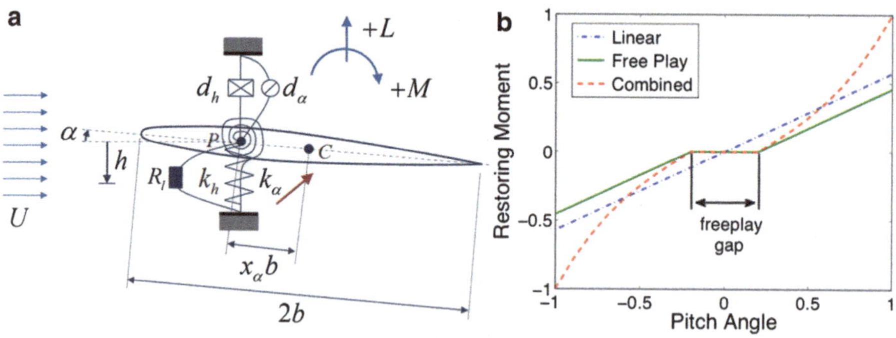

Fig. 11.5 (**a**) Electroaeroelastic typical section with a concentrated stiffness nonlinearity in the pitch DOF and (**b**) various forms of restoring moments in the pitch DOF for the linear, free play, and combined free play-cubic hardening nonlinearity configurations

$$\eta\bar{v}' + \frac{\bar{v}}{\lambda_1} + \kappa\bar{h}' = 0 \tag{11.25}$$

where the dimensionless terms of the equations are defined in Sect. 11.3.1 and the same aerodynamic representation is used as before. The nonlinear restoring moments $f_{\mathrm{fp}}(\alpha)$ and $f_{\mathrm{c}}(\alpha)$, representing free play, and combined free play-cubic hardening nonlinearities, respectively, are given by

$$f_{\mathrm{fp}}(\alpha) = \begin{cases} -\gamma^2\bar{r}_\alpha^2\alpha_{\mathrm{fp}} \\ 0 \\ \gamma^2\bar{r}_\alpha^2\alpha_{\mathrm{fp}} \end{cases} \qquad f_{\mathrm{c}}(\alpha) = \begin{cases} \sigma\gamma^2\bar{r}_\alpha^2(\alpha-\alpha_{\mathrm{fp}})^3 & \alpha > \alpha_{\mathrm{fp}} \\ 0 & -\alpha_{\mathrm{fp}} \leqslant \alpha \leqslant \alpha_{\mathrm{fp}} \\ \sigma\gamma^2\bar{r}_\alpha^2(\alpha+\alpha_{\mathrm{fp}})^3 & \alpha < -\alpha_{\mathrm{fp}} \end{cases} \tag{11.26}$$

where α_{fp} is the free play gap and $\sigma = k_{n\alpha}/k_\alpha$ is the nonlinear-to-linear stiffness ratio (where $k_{n\alpha}$ is the dimensional nonlinear spring constant). It is important to note that when $k_{n\alpha} = 0$ and $\alpha_{\mathrm{fp}} \neq 0$, the free play nonlinearity is obtained (combining the linear restoring moment and $f_{\mathrm{fp}}(\alpha)$) whereas the $k_{n\alpha} \neq 0$ and $\alpha_{\mathrm{fp}} \neq 0$ conditions yield the combined nonlinearity (combining the linear restoring moment, $f_{\mathrm{fp}}(\alpha)$, and $f_{\mathrm{c}}(\alpha)$). The linear equations are recovered when $\alpha_{\mathrm{fp}} = 0$ and $k_{n\alpha} = 0$. The restoring moments in the pitch DOF for the linear, free play, and combined free play-cubic hardening nonlinearity configurations are presented in Fig. 11.5b.

A combination of linear state-space models can be used to describe the nonlinear system in the state-space form [61], yielding the following nonlinear representation:

$$\tilde{\mathbf{x}}' = \mathbf{A}_i\tilde{\mathbf{x}} + \mathbf{a}_i + \mathbf{b}_i \tag{11.27}$$

where the state matrix and the vectors $\mathbf{a}_i$ and $\mathbf{b}_i$ change as the system reaches the free play boundaries (Fig. 11.5b). The governing first-order equations are solved using a Runge–Kutta algorithm with Henon's method [62]. Conner et al. [61] presented an adaptation of Henon's method in order to determine the switching point (or free play boundaries) in the time domain to avoid errors in numerical integration and numerical instability.

11.4.2 Nonlinear Typical Section with Electromagnetic Induction

After adding the concentrated nonlinearities (free play, and combined free play-cubic hardening nonlinearities) to the pitch DOF of the aeroelastic typical section with electromagnetic induction, the dimensionless nonlinear electroaeroelastic equations become

$$\beta\bar{h}'' + x_\alpha\alpha'' + \zeta_h\bar{h}' + \bar{h} - \chi\bar{I} = -\bar{L} \tag{11.28}$$

$$x_\alpha\bar{h}'' + \bar{r}_\alpha^2\alpha'' + \zeta_\alpha\alpha' + \gamma^2\bar{r}_\alpha^2\alpha + f_{\mathrm{fp}}(\alpha) + f_{\mathrm{c}}(\alpha) = \bar{M} \tag{11.29}$$

$$\phi\bar{I}' + \lambda_c\bar{I} + \lambda_1\bar{I} + \chi\bar{h}' = 0 \tag{11.30}$$

where the definition of the nonlinear restoring moments are given in the previous subsection. The nonlinear state-space representation is then obtained as in Eq. (11.27).

11.5 Theoretical Case Studies

This section presents two case studies using the linear electroaeroelastic models described in this chapter. The effects of several dimensionless system parameters on the electroaeroelastic behavior are investigated for two configurations: typical section with (1) piezoelectric transduction and (2) electromagnetic induction along with an external resistive load. The fixed aeroelastic system parameters used in the following simulations are based on the experimental setup by Sousa et al. [45].

11.5.1 Linear Typical Section with Piezoelectric Transduction

In the first theoretical case study, the effects of dimensionless piezoelectric coupling, load resistance, and equivalent capacitance on the dimensionless electrical power output and flutter speed of the piezoelectrically coupled typical section (Fig. 11.3) are investigated. The contours of dimensionless flutter speed, $U^* = U_c/(\omega_h b)$, versus dimensionless load resistance and electromechanical coupling are displayed in Fig. 11.6a for fixed dimensionless capacitance ($\eta = 3.66 \times 10^{-9}$). The linear flutter speed increases with increasing dimensionless coupling for all values of load resistance. Moreover, a finite optimal load that gives the largest flutter speed is obtained for each dimensionless electromechanical coupling. Figure 11.6b shows the contours of dimensionless electrical power output ($P^* = \bar{v}^2/\lambda_1$) versus dimensionless load resistance and electromechanical coupling obtained at each dimensionless flutter speed of Fig. 11.6a. The electrical power output increases with increasing dimensionless electromechanical coupling for any dimensionless load resistance. The presence of an optimal electrical load that gives the maximum power output for all values of electromechanical coupling can be observed in Fig. 11.6b. The optimal load slightly varies with changing dimensionless coupling.

Figure 11.7a displays the contours of dimensionless flutter speed versus dimensionless load resistance and capacitance for fixed electromechanical coupling ($\kappa = 5.90 \times 10^{-6}$). The linear flutter speed decreases with increasing dimensionless

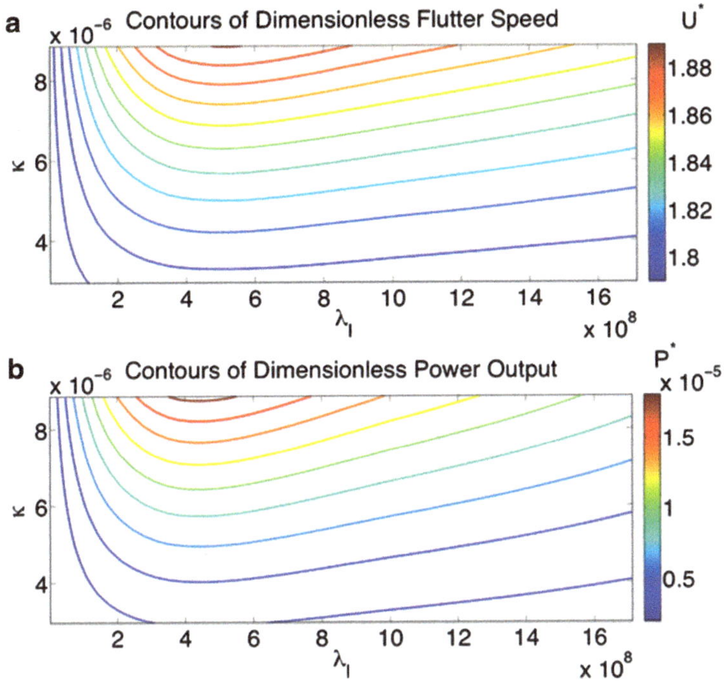

Fig. 11.6 Contours of dimensionless (**a**) flutter speed and (**b**) power output versus dimensionless load resistance and electromechanical coupling (for fixed dimensionless capacitance: $\eta = 3.66 \times 10^{-9}$)

capacitance for all values of load resistance except for very low resistance values. A finite optimal load that gives the largest flutter speed is obtained for each dimensionless capacitance. Figure 11.7b shows the contours of dimensionless electrical power output versus dimensionless load resistance and capacitance obtained at each dimensionless flutter speed of Fig. 11.7a. The electrical power output decreases with increasing dimensionless capacitance for any dimensionless load resistance except for very low resistance values. The presence of an optimal load resistance that gives the maximum power output can be observed in Fig. 11.7 for all values of dimensionless capacitance. The optimal load varies significantly with changing dimensionless capacitance.

11.5.2 Linear Typical Section with Electromagnetic Induction

Next, the effects of dimensionless electromechanical coupling, load resistance, and coil inductance on the dimensionless electrical power as well as the dimensionless linear flutter speed of the electromagnetically coupled typical section (Fig. 11.4) are

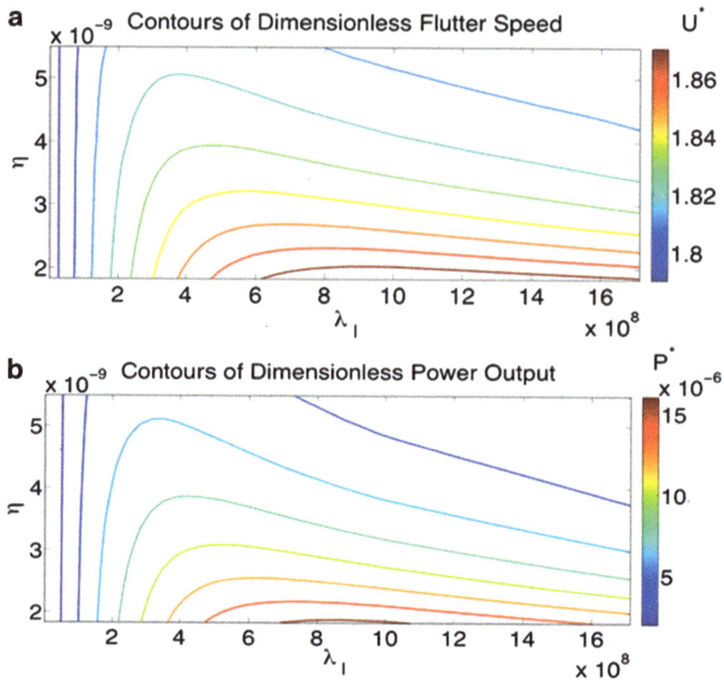

Fig. 11.7 Contours of dimensionless (**a**) flutter speed and (**b**) power output versus dimensionless load resistance and capacitance (for fixed dimensionless electromechanical coupling: $\kappa = 5.90 \times 10^{-6}$)

investigated. In addition to the basic aeroelastic setup parameters from Sousa et al. [45], the dimensionless coil resistance is taken as $\lambda_c = 0.1022$.

The contours of dimensionless flutter speed versus dimensionless load resistance and electromechanical coupling are displayed in Fig. 11.8a for fixed dimensionless coil inductance ($\phi = 0.0130$). The linear flutter speed increases with increasing dimensionless coupling for all values of load resistance. In addition, the flutter speed decreases with increasing load resistance for any dimensionless coupling. It should be noted that the short circuit stiffness is larger than the open circuit stiffness due to electromagnetic coupling in the presence of finite coil inductance [9]; this is the opposite of the piezoelectric transduction case [15, 59]. Further simulations (not shown here) reveal that the variation of the flutter speed depends on the presence of internal coil resistance. For instance, when the internal coil resistance is neglected ($\lambda_c = 0$), a finite and nonzero optimal load that gives the largest flutter speed is obtained in this case study. Figure 11.8b shows the contours of dimensionless electrical power output ($P^* = \bar{I}^2\lambda_l$) versus dimensionless load resistance and electromechanical coupling obtained at the respective flutter speed values of Fig. 11.8a. The electrical power output increases with increasing dimensionless electromechanical coupling for any dimensionless load resistance.

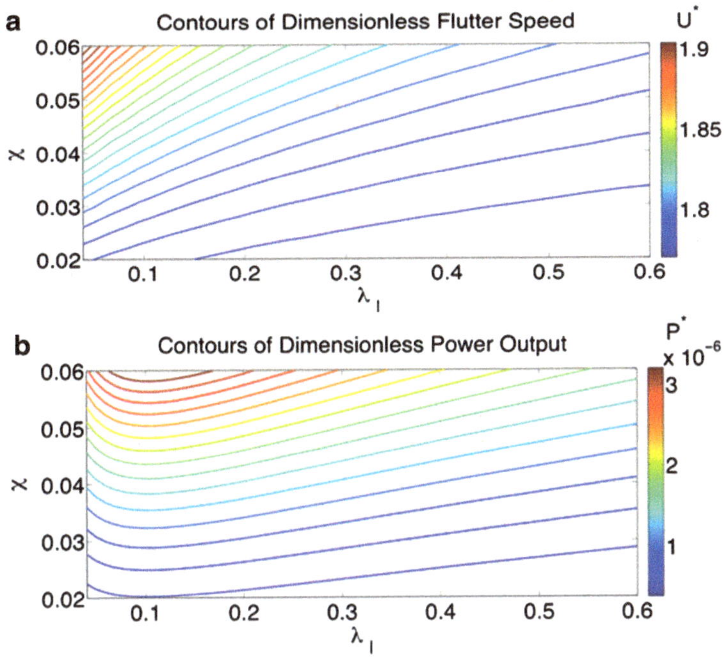

Fig. 11.8 Contours of dimensionless (**a**) flutter speed and (**b**) power output versus dimensionless load resistance and electromechanical coupling (for fixed dimensionless coil inductance: $\phi = 0.0130$)

The presence of an optimal load resistance that gives the maximum power output for all values of electromechanical coupling can be observed in Fig. 11.8b. The optimal load slightly varies with changing dimensionless coupling. The optimal value of λ_l is approximately λ_c for the range of dimensionless coupling values considered. That is, the matched resistance of the maximum power is around the internal resistance in agreement with the maximum power transfer theorem [63].

Figure 11.9a shows the contours of dimensionless flutter speed with dimensionless load resistance and coil inductance. The value of electromechanical coupling is taken as $\chi = 0.0457$. The flutter speed decreases with increasing load resistance, as discussed for Fig. 11.8a. However, the dimensionless flutter speed does not change significantly with dimensionless inductance for any value of load resistance. Figure 11.9b displays the variation of the dimensionless power output with dimensionless load resistance and coil inductance for each flutter speed of Fig. 11.9a. As in the case of flutter speed, the electrical power output is insensitive to the dimensionless inductance except for values around the optimal load resistance. As in Fig. 11.8b, there is an optimal load that gives maximum power output, which slightly varies with changing dimensionless inductance.

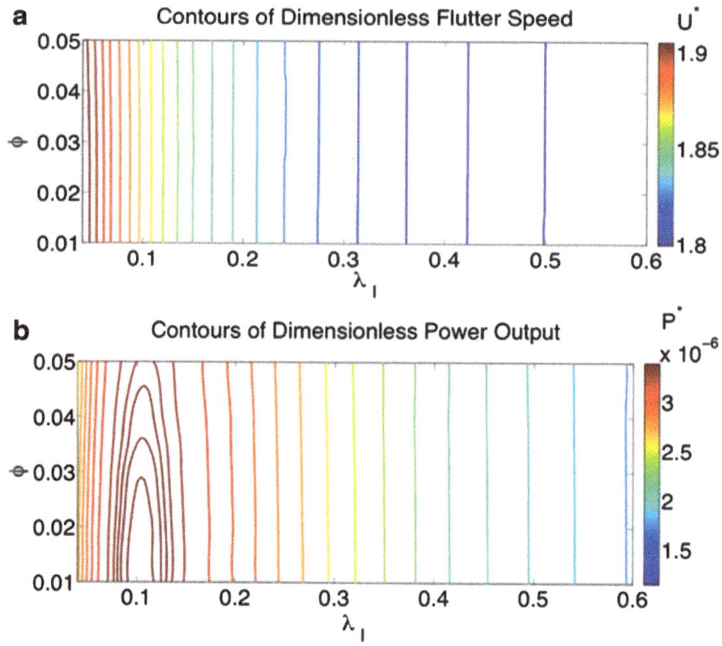

Fig. 11.9 Contours of dimensionless (**a**) flutter speed and (**b**) power output versus dimensionless load resistance and coil inductance (for fixed dimensionless electromechanical coupling: $\chi = 0.0457$)

11.6 Experimental Validations

This section presents experimental validations [45] for the linear and nonlinear electroaeroelastic models described in this chapter. In the first case study, the linear piezoaeroelastic solution is validated against the experimental piezoaeroelastic results obtained at the flutter boundary. Secondly, bilinear structural stiffness is considered in the pitch DOF through the free play nonlinearity. In the presence of free play, the torsional stiffness is zero for small airfoil rotations (due to the free play gap) and approaches the original linear torsional stiffness for relatively large rotations (Fig. 11.5b). In the third case study, a combined nonlinearity is considered in the pitch DOF. For small airfoil rotations, the torsional stiffness is zero and a stiffening behavior is considered for relatively large rotations as in Fig. 11.5b. The piezoaeroelastic behavior is investigated for a set of resistive loads ($R_l = 10^2$, 10^3, 10^4, 10^5, and 10^6 Ω) and also for a range nonlinear-to-linear pitch stiffness ratios ($\sigma = k_{n\alpha}/k_\alpha$) defined in Eq. (11.26).

Figure 11.10 shows the experimental setup used for investigating the linear and nonlinear piezoaeroelastic behavior of a typical section. It is important to note that this setup uses a readily available aeroelastic section modified with piezoelectric coupling and was not originally designed for energy harvesting; however, the results

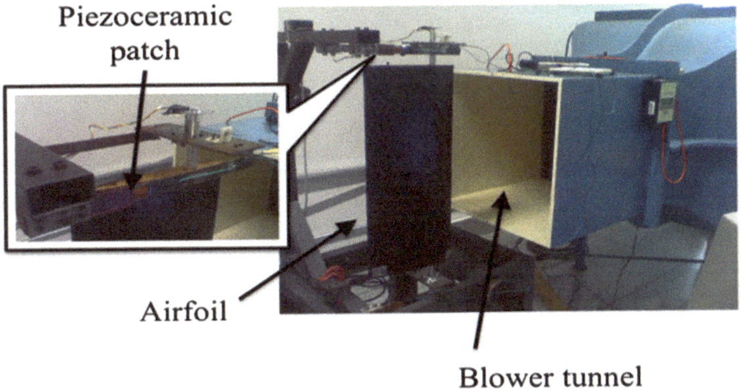

Fig. 11.10 Experimental electroaeroelastic typical section with piezoelectric coupling and a close-up view of a piezoceramic patch on the upper plunge spring

are applicable to its scaled versions. The plunge stiffness is due to four elastic beams with clamped–clamped end conditions, two of which are shown in the close-up view in Fig. 11.10. The free ends of the elastic beams are connected to the airfoil through metal plates. Therefore the experimental setup in this work slightly deviates from the ideal definition of a typical section (where springs are assumed massless), yielding the fixture mass (m_e) defined in Eq. (11.1). A shaft (or pitch axis) is mounted to the upper and the lower plates through a pair of bearings. The pitch stiffness is given by a spring wire clamped to the shaft at the elastic axis. The free end of the wire is simply supported on the top plate without a gap for the linear case and with a gap for the free play nonlinearity. Two piezoceramic patches (QP10N from Mide Technology Corporation) are attached to the root of the two plunge stiffness members (symmetrically), and their electrodes are connected in parallel to a set of external resistive loads. The manufacturer's published equivalent capacitance of $C_p = 120$ nF is used in the piezoaeroelastic model. The electromechanical coupling parameter is obtained based on distributed parameter modeling [15, 59] by considering clamped–clamped end conditions for the two beams with piezoceramic patches to yield $\theta = 1.55$ mN/V [30, 45]. The reader is referred to Sousa et al. [45] for further details of the experimental setup.

11.6.1 *Linear Piezoaeroelastic Typical Section*

In the first case study, the linear aeroelastic behavior of the piezoelectrically coupled typical section is investigated at the flutter boundary for a set of resistive loads. The real part of the relevant eigenvalue is shown in Fig. 11.11 for changing airflow speed and different resistive loads. Only the eigenvalue for the mode that becomes unstable is shown for clarity. The predicted linear short circuit ($R_l \rightarrow 0$) flutter speed is

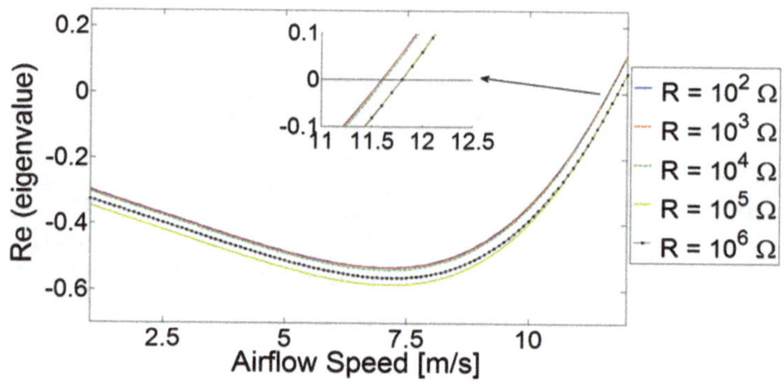

Fig. 11.11 Real part of the eigenvalue (for the mode that becomes unstable) with increasing airflow speed for a set of resistive loads along with a close-up view around the flutter boundary

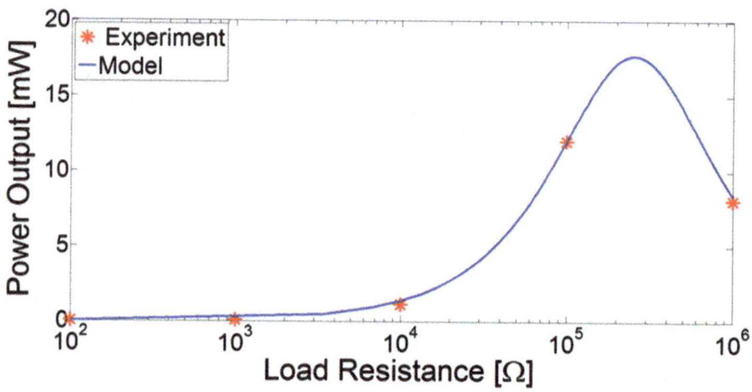

Fig. 11.12 Experimental and theoretical power outputs versus load resistance at the linear flutter boundary for the configuration with linear pitch stiffness

11.6 m/s. The experimental short circuit flutter speed is measured as 11.9 m/s. The load resistance of 100 kΩ gives the maximum experimental power output among the set of resistors considered in this chapter. The model predicts the linear flutter speed for this load as 11.8 m/s. The experimental flutter speed for the same resistive load is 12.1 m/s. Although the model slightly underestimates the experimental flutter speed, approximately the same increase (1.7%) in the linear flutter speed (with respect to short-circuit flutter speed) is observed for a resistive load of 100 kΩ.

The experimentally measured and theoretically predicted variations of power output with increasing load resistance are shown in Fig. 11.12. In order to obtain almost persistent but not divergent oscillations, i.e., to avoid post-flutter oscillations, the electrical power is experimentally measured as close as possible to the linear flutter speed of each resistive load. The model slightly overestimates the experimental power output for the set of resistive loads used in this work. The predicted power output for the load resistance of 100 kΩ is 13.8 mW and the experimental

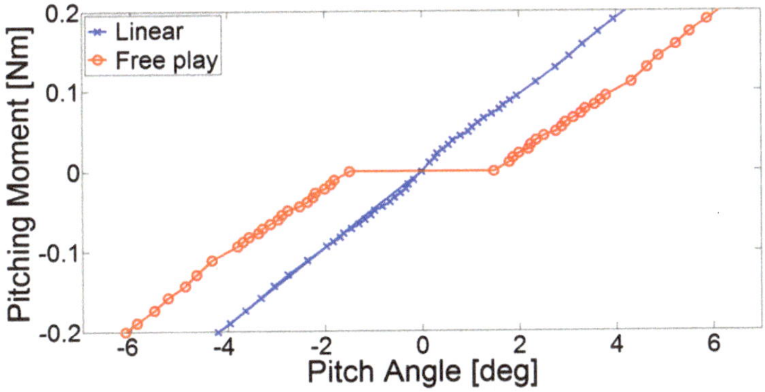

Fig. 11.13 Experimentally measured linear and bilinear (free play) pitching moments versus pitch angle

power output for the same load is measured as 12 mW. This overestimation can be attributed to the material and dissipative nonlinearities not considered in this work. The reader is referred to a series of papers by Stanton et al. [64–66] for the modeling and effects of basic nonlinearities observed in piezoelectric energy harvesting. The post-flutter response of the particular experimental setup results in dramatically large-amplitude oscillations with growing amplitude (of a divergent oscillatory nature for practical purposes) as the setup behaves quite linearly around the flutter speed. Therefore, the maximum airflow speed of effective energy harvesting without divergent oscillations is very close to the linear flutter speed. However, almost persistent oscillations occurring at a specific wind speed restricts the effective performance range of a linear electroaeroelastic energy harvester.

11.6.2 Piezoaeroelastic Typical Section with Free Play

The experimental piezoaeroelastic typical section with a free play nonlinearity is investigated now for the same set of resistors. The free play gap causing the bilinear stiffness in the pitch DOF is ±1.4° (i.e., $\alpha_{fp} = 1.4\pi/180\,\text{rad}$). The pitch stiffness outside the free play gap is given by the same stiffness as the linear case study. The experimentally measured linear pitching moment and free play pitching moment diagrams are shown in Fig. 11.13.

In the presence of free play nonlinearity, the LCO mechanism observed in the experiments leads to LCOs below the linear flutter speed for each load resistance used in this work. The lowest airflow speed to have LCO is experimentally measured as 10 m/s and the model predicts this lower bound as 10.4 m/s. It is observed that, beyond 12 m/s, the response amplitude (predicted by the present model) becomes very large for all resistive loads considered here due to the detrimental nature of free play nonlinearity. Therefore, no experimental testing was performed outside the range of 10–12 m/s.

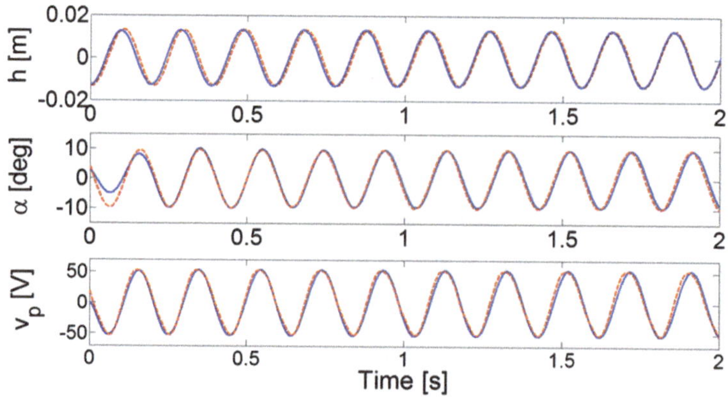

Fig. 11.14 Experimental (*dashed*) and theoretical (*solid*) nonlinear piezoaeroelastic response histories for $R_l = 100$ kΩ (experiment: 10.0 m/s, model: 10.4 m/s)

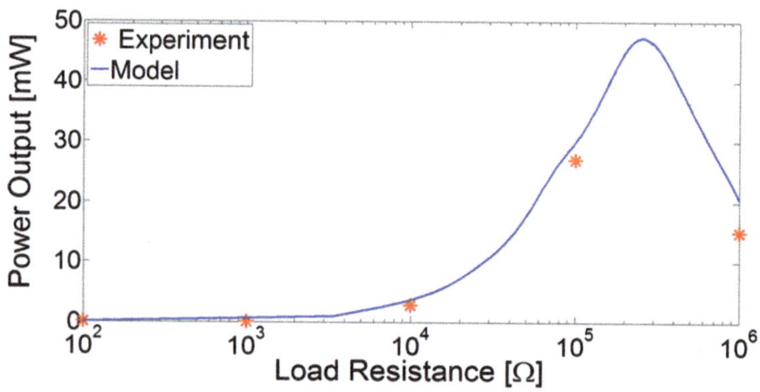

Fig. 11.15 Experimental and theoretical power outputs versus load resistance (experiment: 10.0 m/s, model: 10.4 m/s) for the configuration with a free play nonlinearity in pitch stiffness

The load resistance of 100 kΩ gives the maximum experimental power output for the set of resistors used in this case as well. The piezoaeroelastic time histories (pitch, plunge, and voltage output) for this resistive load with persistent oscillations are shown in Fig. 11.14. The model predicts the amplitudes of the pitch, plunge, and voltage response histories as well as the frequencies very accurately.

The power versus load resistance diagrams obtained from the experiments and the model are displayed in Fig. 11.15. In this figure, the experimental and the theoretical airflow speeds of LCO are 10.0 m/s and 10.4 m/s, respectively. Unlike the linear flutter boundary case (Fig. 11.12), the LCO due to free play (Fig. 11.15) is a strongly nonlinear phenomenon and the airflow speed of persistent oscillations is not affected by the resistive shunt damping effect due to piezoelectric power generation since piezoelectric coupling tends to be weak, usually on the order of mechanical damping [59]. Moreover, in this particular case the piezoelectric coupling is more

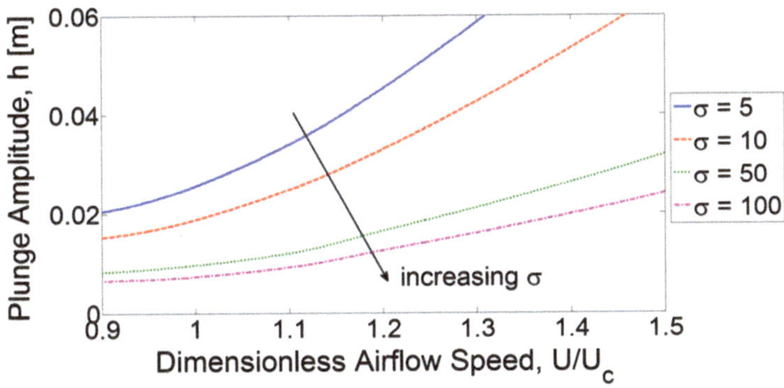

Fig. 11.16 Plunge amplitude with increasing airflow speed for four different values of the nonlinear-to-linear stiffness ratio ($R_l = 100\ \Omega$)

effective for the plunge DOF while the strong nonlinearity is in the pitch DOF. Therefore, all the experimental and theoretical data points in Fig. 11.15 are for airflow speeds of 10 m/s and 10.4 m/s, respectively. Among the set of resistors used in the experiments, the maximum experimental power output is again obtained for $R_l = 100\ \mathrm{k}\Omega$. For this electrical load, the power output is predicted by the model to be 28.6 mW, slightly overestimating the experimental value of 27.0 mW. Note that the maximum power output for this configuration with free play nonlinearity is more than twice the power output obtained in the previous case study (Fig. 11.12-linear piezoaeroelastic energy harvester excited at the flutter boundary). Moreover, the cut-in speed of LCO is reduced with the free play nonlinearity by about 16.5% (i.e., by 2.1 m/s relative to 12.1 m/s of the linear case for $R_l = 100\ \mathrm{k}\Omega$). However, the range of airflow speeds with LCO of acceptable amplitude is still relatively narrow. Therefore the response amplitude needs to be reduced to acceptable values while keeping the free play nonlinearity; this is theoretically investigated in the next section.

11.6.3 On the Exploitation of Combined Nonlinearities

Having validated the nonlinear piezoaeroelastic energy harvester model, the configuration with combined free play-hardening cubic stiffness nonlinearities (recall Fig. 11.5b) is studied in this section. The theoretical piezoaeroelastic behavior of the electromechanically coupled typical section is investigated for different values of the nonlinear-to-linear stiffness ratio (σ). The free play gap considered in this third case study is $\pm 1.4^\circ$ as in the previous section. The vibration amplitude and the electrical power output are investigated for airflow speeds ranging from 90% to 150% of the linear flutter speed of 12 m/s.

Figure 11.16 shows the plunge amplitude with dimensionless airflow speed (the ratio of airflow speed to linear flutter speed) close to short-circuit conditions

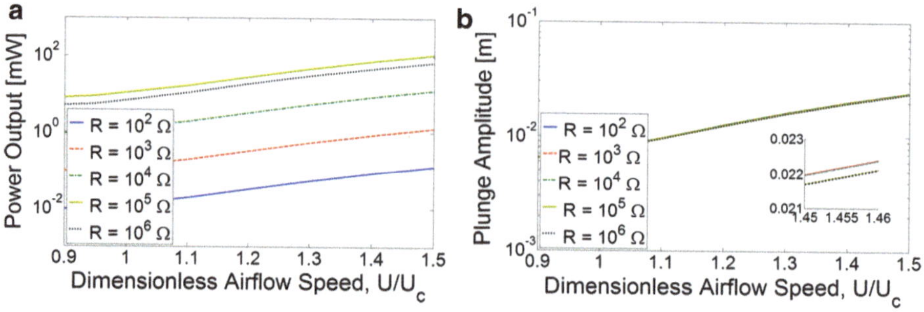

Fig. 11.17 Variations of the (**a**) peak power output and (**b**) plunge displacement amplitude with airflow speed for five different values of load resistance (fixed nonlinear-to-linear stiffness ratio: $\sigma = 100$)

($R_l = 100$ kΩ). It is observed from this figure that for a given airflow speed, the plunge amplitude is highly sensitive to the nonlinear-to-linear stiffness ratio. Note that the combination of $\sigma = 0$ ($k_{nl} = 0$) and $\alpha_{fp} \neq 0$ (not shown in the figure) corresponds to the bilinear free play nonlinearity without any hardening stiffness, and this system is prone to very large amplitude response due to the "detrimental" nature of the free play nonlinearity when employed alone. In Fig. 11.16, the amplitude of plunge displacement increases with increasing airflow speed and decreases with increasing stiffness ratio. Therefore, for the configuration with combined nonlinearities ($\sigma \neq 0$ and $\alpha_{fp} \neq 0$), the LCO response has acceptable amplitude over a wide airflow speed range. In practice, the nonlinear-to-linear stiffness ratio can be increased with additional springs exhibiting hardening nonlinearity (which is a "benign" nonlinearity in this context) to improve the range of airflow speeds with acceptable response amplitude in the presence of a free play nonlinearity.

For the case of $\sigma = 100$, the variation of electrical power output with increasing airflow speed for five different values of load resistance is shown in Fig. 11.17a. The power output increases with increasing airflow speed for any resistive load. At any airflow speed, as the value of load resistance is increased from $R_l = 100$ Ω to $R_l = 100$ kΩ, the power output increases. When the value of load resistance is further increased to $R_l = 1$ MΩ (close to open-circuit condition) the power output starts decreasing. Therefore, among the set of resistive loads considered here, $R_l = 100$ kΩ gives the maximum power output over the entire range of airflow speeds. The maximum theoretical power of 106 mW is obtained for the optimal load resistance at $U = 18$ m/s ($U/U_c = 1.5$); this value is expected to overestimate the possible experimental results due to material and dissipative nonlinearities [64–66]. Figure 11.17b shows the variation of plunge amplitude with airflow speed and $\sigma = 100$ for the set of resistive load considered in this work. The shunt damping [67] effect of power generation on the vibration amplitude is negligible for practical purposes, in agreement with the discussion given in the previous section regarding the weak effect of shunt damping on the free play nonlinearity.

11.7 Conclusions

This chapter summarized electroaeroelastic modeling and analysis of airfoil-based wind energy harvesting using either piezoelectric transduction or electromagnetic induction. The electromechanical coupling is introduced to the plunge DOF by means of piezoelectric patches or magnet-coil arrangements. The governing dimensionless electroaeroelastic equations are obtained in the presence of a resistive load in the electrical domain to quantify the power generation performance of these systems in response to airflow excitation. Both linear and nonlinear approaches of electroaeroelastic energy harvesting are examined. The linear problem focuses on the response at the flutter boundary while the nonlinear configurations exploit limit-cycle oscillations due to free play and cubic stiffness nonlinearities in the pitch DOF. The effects of several dimensionless system parameters on the electrical power output and flutter speed are investigated.

Experimental validations are presented for linear and nonlinear electroaeroelastic systems employing piezoelectric transduction. The electrical power output at the flutter speed of each resistor is successfully predicted. Energy harvesting at the linear flutter boundary restricts the performance of an electroaeroelastic energy harvester to a specific airflow speed and often inherent nonlinearities are present in these systems. Therefore, concentrated nonlinearities are introduced and modeled for the pitch DOF. Importantly, increased power output is obtained along with reduced cut-in speed due to the free play nonlinearity. Although the cut-in speed is reduced favorably, a narrow range of airflow speeds with acceptable oscillation amplitudes is observed due to the detrimental nature of the free play nonlinearity. After combining the free play nonlinearity with hardening stiffness, it is shown that the hardening stiffness helps in bringing the response amplitude to acceptable levels over a wide range of airflow speeds while the free play nonlinearity still reduces the cut-in speed of persistent oscillations. Therefore, the combination of the free play and hardening cubic stiffness nonlinearities provides the most useful scenario for electroaeroelastic energy harvesting among the methods considered in this chapter.

References

1. Paradiso JA, Starner T (2005) Energy scavenging for mobile and wireless electronics. IEEE Pervasive Comput 4:18–27
2. Beeby SP, Tudor MJ, White NM (2006) Energy harvesting vibration sources for microsystems applications. Meas Sci Technol 17:R175–R195
3. Anton SR, Sodano HA (2007) A review of power harvesting using piezoelectric materials (2003–2006). Smart Mater Struct 16:R1–R21
4. Cook-Chennault KA, Thambi N, Sastry AM (2008) Powering MEMS portable devices – a review of non-regenerative and regenerative power supply systems with emphasis on piezoelectric energy harvesting systems. Smart Mater Struct 17:043001
5. Priya S (2007) Advances in energy harvesting using low profile piezoelectric transducers. J Electroceram 19:167–184

6. Hudak NS, Amatucci GG (2008) Small-scale energy harvesting through thermoelectric, vibration, and radiofrequency power conversion. J Appl Phys 103:101301
7. Amirtharajah R, Chandrakasan AP (1998) Self-powered signal processing using vibration-based power generation. IEEE J Solid State Circuits 33:687–695
8. Glynne-Jones P, Tudor MJ, Beeby SP, White NM (2004) An electromagnetic, vibration-powered generator for intelligent sensor systems. Sens Actuators A 110:344–349
9. Elvin N, Elvin A (2011) An experimentally validated electromagnetic energy harvesters. J Sound Vib 330:2314–2324
10. Mitcheson P, Miao P, Start B, Yeatman E, Holmes A, Green T (2004) MEMS electrostatic micro-power generator for low frequency operation. Sens Actuators A 115:523–529
11. Roundy S, Wright PK, Rabaey JM (2003) A study of low level vibrations as a power source for wireless sensor nodes. Comput Commun 26:1131–1144
12. Tvedt LGW, Nguyen DS, Halvorsen E (2010) Nonlinear behavior of an electrostatic energy harvester under wide- and narrowband excitation. IEEE J Microelectromech Syst 19:305–316
13. Roundy S, Wright PK (2004) A piezoelectric vibration based generator for wireless electronics. Smart Mater Struct 13:1131–1144
14. Jeon YB, Sood R, Jeong JH, Kim S (2005) MEMS power generator with transverse mode thin film PZT. Sens Actuators A 122:16–22
15. Erturk A, Inman DJ (2009) An experimentally validated bimorph cantilever model for piezoelectric energy harvesting from base excitations. Smart Mater Struct 18:025009
16. Priya S, Chen CT, Fye D, Zahnd J (2005) Piezoelectric windmill: a novel solution to remote sensing. Jpn J Appl Phys 44:L104–L107
17. Myers R, Vickers M, Kim H, Priya S (2007) Small scale windmill. J Appl Phys 90:054106
18. Rancourt D, Tabesh A, Frechette LG (2007) Evaluation of centimeter-scale micro wind mills: aerodynamics and electromagnetic power generation. Proc PowerMEMS 2007:93–96
19. Xu FJ, Yuan FG, Hu JZ, Qiu YP (2010) Design of a miniature wind turbine for powering wireless sensors. Proc SPIE 7646:764741
20. Erturk A, Bilgen O, Fontenille M, Inman DJ (2008) Piezoelectric energy harvesting from macro-fiber composites with an application to morphing wing aircraft. In: Proceedings on the 19th international conference of adaptive structures and technologies, Monte Verità, Ascona, Switzerland, 6–9 Oct 2008
21. De Marqui C Jr, Erturk A, Inman DJ (2010) Piezoaeroelastic modeling and analysis of a generator wing with continuous and segmented electrodes. J Intell Mater Syst Struct 21:983–993
22. De Marqui C Jr, Erturk WGR, Inman DJ (2010) Modeling and analysis of piezoelectric energy harvesting from aeroelastic vibrations using the doublet-lattice method. ASME J Vib Acoust 133:011003
23. Bisplinghoff RL, Ashley H (1962) Principles of aeroelasticity. Wiley, New York
24. Fung YC (1969) Introduction to the theory of aeroelasticity. Dover, New York
25. Dowell EH, Curtiss HC Jr, Scalan RH, Sisto F (1978) A modern course in aeroelasticity. Sijthoff and Norrdhoff, Amsterdam
26. Hodges DH, Pierce GA (2002) Introduction to structural dynamics and aeroelasticity. Cambridge University Press, New York
27. Bryant M, Garcia E (2009) Development of an aeroelastic vibration power harvester. Proc SPIE 7288:728812
28. Bryant M, Garcia E (2009) Energy harvesting: a key to wireless sensor nodes. Proc SPIE 7493:74931W
29. Peters DA, Karunamoorthy S, Cao WM (1995) Finite state induced flow models. Part I: Two dimensional thin airfoil. J Aircraft 32:313–322
30. Erturk A, Vieira WGR, De Marqui C Jr, Inman DJ (2010) On the energy harvesting potential of piezoaeroelastic systems. Appl Phys Lett 96:184103
31. Theodorsen T (1935) General theory of aerodynamic instability and mechanism of flutter. Langley Memorial Aeronautical Laboratory, NACA-TR-496

32. McKinney W, DeLaurier JD (1981) The wingmill: an oscillating-wing windmill. J Energy 5:109–115
33. Ly KH, Chasteau VAL (1981) Experiments on an oscillating-wing aerofoil and application to wing-energy converters. J Energy 5:116–121
34. Jones KD, Platzer MF (1999) Oscillating-wing power generator. In: Proceedings of ASME/JSME-Joint Fluids Engineering Conference, No. 7050
35. Guckenheimer J, Holmes P (1983) Nonlinear oscillations, dynamical systems, and bifurcations of vector fields. Springer, New York
36. Nayfeh AH, Mook DT (1979) Nonlinear oscillations. Wiley, New York
37. Moon FC (1987) Chaotic vibrations. Wiley, New York
38. Strogatz SH (1994) Nonlinear dynamics and chaos. Perseus Book, Cambridge, MA
39. Lee BHL, Price SJ, Wong YS (1999) Nonlinear aeroelastic analysis of airfoils: bifurcation and chaos. Prog Aerospace Sci 35:205–334
40. Price SJ, Lee BHK, Alighanbari H (1994) Postinstability behavior of a two-dimensional airfoil with a structural nonlinearity. J Aircraft 31:1395–1401
41. Tang D, Dowell EH (2006) Flutter and limit-cycle oscillations for a wing-store model with freeplay. J Aircraft 43(2):487–503
42. Zhao LC, Yang ZC (1990) Chaotic motions of an airfoil with nonlinear stiffness in incompressible flow. J Sound Vib 138:245–254
43. Dowell EH, Tang D (2002) Nonlinear aeroelasticity and unsteady aerodynamics. AIAA J 40:1697–1707
44. Dowell EH, Edwards J, Strganac T (2003) Nonlinear aeroelasticity. AIAA J Aircraft 40: 857–874
45. Sousa VC, Anicézio MM, De Marqui C Jr, Erturk A (2011) Enhanced aeroelastic energy harvesting by exploiting combined nonlinearities: theory and experiment. Smart Mater Struct 20:094007
46. Abdelkefi A, Nayfeh AH, Hajj MR (2011) Modeling and analysis of piezoaeroelastic energy harvesters. Nonlinear Dyn 67:925–939
47. St. Clair D, Bibo D, Sennakesavababu VR, Daqaq MF, Li G (2010) A scalable concept for micropower generation using flow-induced self-excited oscillations. Appl Phys Lett 96:144103
48. Allen JJ, Smits AJ (2001) Energy harvesting eel. J Fluids Struct 15:629–640
49. Robbins WP, Morris D, Marusic I, Novak TO (2006) Wind-generated electrical energy using flexible piezoelectric materials. In: Proceedings of ASME IMECE 2006, Chicago, IL
50. Pobering S, Ebermeyer S, Schwesinger N (2009) Generation of electrical energy using short piezoelectric cantilevers in flowing media. Proc SPIE 7288:728807
51. Akaydin HD, Elvin N, Andreopoulos Y (2010) Wake of a cylinder: a paradigm for energy harvesting with piezoelectric materials. Exp Fluids 49:291–304
52. Tang L, Paidoussis M, Jiang J (2009) Cantilevered flexible plates in axial flow: energy transfer and the concept of flutter-mill. J Sound Vib 326:263–276
53. Dunnmon JA, Stanton SC, Mann BP, Dowell EH (2011) Power extraction from aeroelastic limit cycle oscillations. J Fluids Struct 27:1181–1198
54. Kwon SD (2010) A T-shaped piezoelectric cantilever for fluid energy harvesting. Appl Phys Lett 77:164102
55. Zhu D, Beeby S, Tudor J, White N, Harris N (2010) A novel miniature wind generator for wireless sensing applications. Proc IEEE Sensors 2010:1415–1418
56. Jung HJ, Lee SW (2011) The experimental validation of a new energy harvesting system based on the wake galloping phenomenon. Smart Mater Struct 20:055022
57. Jones RT (1938) Operational treatment of the non-uniform lift theory in airplane dynamics. Technical note 667. NASA, Washington, DC
58. Wagner H (1925) Über die Entstehung des dynamischen Auftriebes von Tragflügeln. Zeitschrift für Angewandte Mathematic und Mechanik 5:17–35
59. Erturk AA, Inman DJ (2011) Piezoelectric energy harvesting. Wiley, New York
60. Edwards JW, Ashley H, Breakwell JV (1979) Unsteady aerodynamic modeling for arbitrary motions. AIAA J 17(4):365–374

61. Conner MD, Virgin LN, Dowell EH (1996) Accurate numerical integration of state space models for aeroelastic systems with free play. AIAA J 34(10):2202–2205
62. Henon M (1982) On the numerical computation of Poincaré maps. Physica 5D:512–514
63. Agarwal A, Lang J (2005) Foundations of analog and digital electronic circuits. Morgan Kaufmann, San Francisco
64. Stanton SC, Erturk A, Mann BP, Inman DJ (2010) Resonant manifestation of intrinsic nonlinearity within electroelastic micropower generators. Appl Phys Lett 97:254101–254104
65. Stanton SC, Erturk A, Mann BP, Inman DJ (2010) Nonlinear piezoelectricity in electroelastic energy harvesters: modeling and experimental identification. J Appl Phys 108:074903
66. Stanton SC, Erturk A, Mann BP, Dowell EH, Inman DJ (2012) Nonlinear nonconservative behavior and modeling of piezoelectric energy harvesters including proof mass effects. J Intell Mater Syst Struct 23:183–199
67. Lesieutre GA, Ottman GK, Hofmann HF (2004) Damping as a result of piezoelectric energy harvesting. J Sound Vib 269:991–1001

Chapter 12
Acoustic Energy Harvesting Using Sonic Crystals

Liang-Yu Wu, Lien-Wen Chen, I-Ling Chang, and Chun-Chih Wang

Abstract This chapter presents the development of an acoustic energy harvester using the sonic crystal and the piezoelectric material. A point defect is created by removing a rod from a perfect sonic crystal. The point defect in the sonic crystal acts as a resonant cavity; thus, the acoustic waves at resonant frequency can be localized within the cavity of the sonic crystal. The piezoelectric material is put into the cavity to convert the ambient mechanical energy into electrical energy. The power generation from acoustic energy is based on the effect of the wave localization in the cavity and the direct piezoelectric effect of the piezoelectric material. The plane wave expansion method is employed to calculate the resonant frequency of the sonic crystal, and the finite element method is adopted to obtain the pressure and particle velocity field of the defect mode at the resonant frequency in the sonic crystal. A model for energy harvesting of the piezoelectric curved beam is also developed to predict the output voltage and power of the harvester. The larger voltage output of the piezoelectric material is associated with the larger pressure in the cavity of the sonic crystal. Two kinds of piezoelectric PVDF films (LDT4-028k and LDT2-028k) are placed inside the cavity of the sonic crystal and attached on rods. When the frequency of the incident acoustic wave is at 4.2/4.21 kHz, a maximum power is generated under a load resistance 3.9/15 kΩ for LDT4-028k/LDT2-028k PVDF film, respectively. Piezoelectric materials with higher electromechanical coupling should be selected to improve the output power. In addition, the piezoelectric beam can be designed to have the same resonant frequency with the cavity. By using properties of band gaps and wave localizations of the sonic crystal, the noise control and energy harvesting can be achieved simultaneously.

L.-Y. Wu • L.-W. Chen (✉) • I.-L. Chang • C.-C. Wang
Department of Mechanical Engineering, National Cheng Kung University,
1, Ta-Hsueh Road, Tainan, 70101, Taiwan
e-mail: liangyu99@hotmail.com; chenlw@mail.ncku.edu.tw; ilchang@mail.ncku.edu.tw; c.c.wang2011@gmail.com

N. Elvin and A. Erturk (eds.), *Advances in Energy Harvesting Methods*,
DOI 10.1007/978-1-4614-5705-3_12,

12.1 Introduction

The conversion of ambient light, thermal or mechanical energy into electrical energy is an important aspect for power generators. One of the most effective methods of implementing a power harvesting system is to use mechanical vibration to apply strain energy to the piezoelectric material. The use of piezoelectric materials to convert ambient mechanical strain energy into electrical energy is a method that is attracting a growing interest. These materials can be employed to transfer ambient motion into electric energy that may be stored and utilized by electric devices such as sensors and wireless transmitters. The electrical and mechanical behaviors of power harvesting devices have been studied using a number of approaches. Of these approaches, cantilevered beams with piezoelectric layers are the most widely used in vibration energy harvesting. Various piezoelectric energy harvesting mechanisms have been investigated by numerous researchers, and extensive discussions can be found in existing review papers [1, 2]. Research in energy harvesting involves understanding the mechanics of vibrating structures, the constitutive behavior of piezoelectric materials, and circuit theory.

One such self-powered wireless sensor using piezoelectric materials has been theoretically and experimentally demonstrated [3]. The energy harvesting system was found to provide the required energy to power circuitry that could transmit a signal containing information regarding the strain of the beam. Lu et al. [4] have given a model for the analysis of piezoelectric power generator for application in micro-electromechanical systems. The output power and voltage of the system were obtained, in which the electromechanical coupling effect was included. Jiang et al. [5] have investigated the performance of a piezoelectric bimorph in the flexural mode for scavenging ambient vibration energy. The mathematical modeling of a cantilever bimorph with a proof mass attached to its end was proposed to evaluate the physical and geometrical effects on the energy harvesting performance. The generalized Hamilton's principle for electromechanical systems was employed by Sodano et al. [6] for modeling the cantilevered piezoelectric energy harvesters which are based on the Euler–Bernoulli beam theory. They developed a mathematical model to predict the energy generated from a piezoelectric bimorph cantilever beam. Erturk and Inman [7] have demonstrated an analytical solution for the coupled problem of a piezoelectric cantilevered beam based on the Euler–Bernoulli assumptions. They explicitly obtained the coupled voltage response across a resistive electrical load and the coupled vibration response of the piezoelectric beam subjected to translational and rotational excitations. The short circuit and open circuit trends and the effect of piezoelectric coupling were also investigated extensively.

Several other systems that harvest energy have been developed or proposed [8–16]. Flowing water is one of energy sources provided by the ambient movement of a fluid which shows significant energy harvesting potential. Taylor et al. [8] developed an energy harvesting eel made of a long strip of piezoelectric polymer bimorph material. The energy harvesting eel was submerged underwater to power harvesting from flowing water. Another potential source of energy is wind current.

Priya et al. [9, 10] have designed and tested a piezoelectric windmill energy harvesting device. Oscillating piezoelectric bimorphs produce electricity as the wind flows through the windmill. Acoustic energy is another potential source for power generations. Horowitz et al. [11] and Liu et al. [12] have used an electromechanical Helmholtz resonator as an acoustic energy harvesting device. An oscillatory pressure in the cavity is generated to cause the vibration of the piezoelectric materials, and thereby the conversion of acoustic energy to electric energy. Nanogenerators driven by ultrasonic waves have been also investigated [13]. The wave bend the nanowires and continuous direct-current can thus be produced. Guigon et al. [14, 15] have investigated harvesting the mechanical energy generated by the impact of rain drops both theoretically and experimentally.

Sonic crystals are periodic composite materials, which have abundant acoustic characteristics. The propagation of acoustic or elastic waves in sonic crystals has attracted much attention over the last two decades [16–27]. The existence of acoustic band gaps in sonic crystals is of interest for applications such as acoustic filters, noise control, and transducers. A point defect is created within the sonic crystal by removing a rod from a perfect sonic crystal and can act as a resonant cavity. The transmission spectra, point defect modes, and wave localization of sonic crystals with a defect have been investigated theoretically and experimentally [19–27]. The acoustic wave can be localized in the cavity of the sonic crystal when the incident acoustic wave is at the resonant frequency of the cavity. Using this phenomenon, piezoelectric material placed into the cavity of the sonic crystal can be used to convert the acoustic energy to electric energy at the resonant frequency of the cavity.

In this chapter, a method for electric power harvesting from acoustic energy using piezoelectric materials and sonic crystals is presented. The plane wave expansion method is employed to calculate the resonant frequency and defect mode of the resonant cavity of the sonic crystal. A piezoelectric PVDF film is put into the resonant cavity of the sonic crystal and can be modeled as a curved beam. The dynamic pressure difference between the two sides of the curved beam acts as the external force to vibrate the piezoelectric curved beam, thereby, achieving acoustic energy harvesting. In the following sections, the modeling of energy harvesting from the piezoelectric curved, and the output voltage and power is calculated. The experimental voltage and power outputs are measured and compared with theoretical results.

12.2 Numerical Modeling of Sonic crystal

12.2.1 Plane Wave Expansion Method

A 2D periodic square lattice sonic crystal consisted of circular rods immersed in a fluid is shown in Fig. 12.1. In this schematic, a_0 is the lattice constant, and r_0 is the radius of rods. $\mathbf{a}_1$ and $\mathbf{a}_2$ are the primitive lattice vectors. The region of dashed

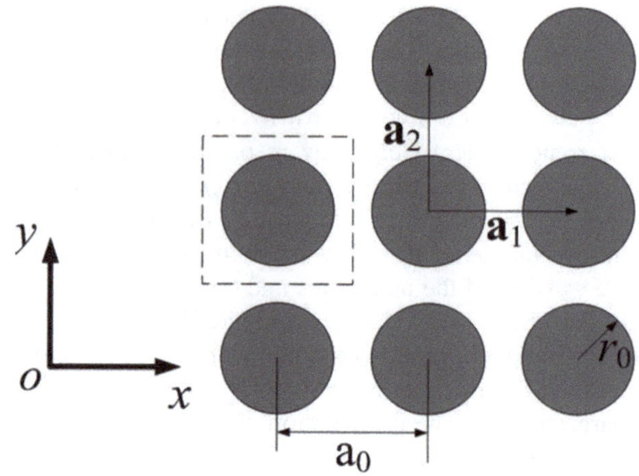

Fig. 12.1 The sonic crystal with a square lattice

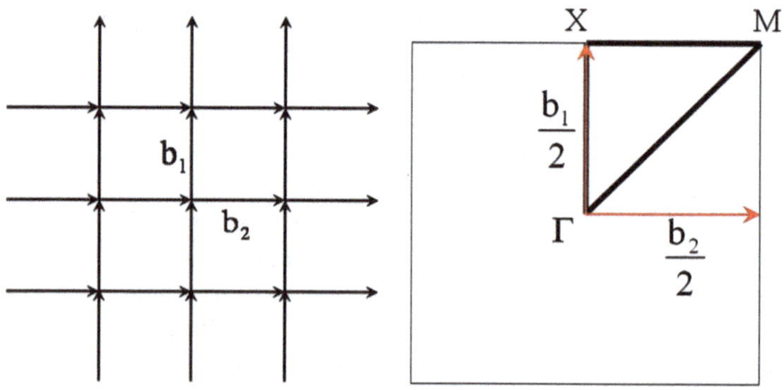

Fig. 12.2 The space of the reciprocal lattice vectors and the first Brillouin zone of the square lattice sonic crystal. The *highlighted region* represents the irreducible Brillouin zone of the square lattice

line represents a unit cell. Figure 12.2 shows the space of the reciprocal lattice vectors and the first Brillouin zone of the square lattice sonic crystal; $\mathbf{b}_1$ and $\mathbf{b}_2$ are the vectors of the reciprocal lattice space. The highlighted region represents the irreducible Brillouin zone of the square lattice. For the square lattice, $\mathbf{a}_1$ and $\mathbf{a}_2$ are expressed as $a_0(1, 0)$ and $a_0(0, 1)$, respectively, and $\mathbf{b}_1$ and $\mathbf{b}_2$ are expressed as $\frac{2\pi}{a_0}(0, 1)$ and $\frac{2\pi}{a_0}(1, 0)$, respectively.

Since the fluid does not allow the propagation of a transverse wave, only the longitudinal wave is considered. As the impedance of the solid rods is significantly larger than that of fluid, the total longitudinal waves propagating in the fluid will be almost completely reflected by the solid rods. Only a little acoustic energy can propagate in the solid rods, and the transverse waves only exist in the solid rods. Hence, the wave propagation in such a sonic crystal is predominantly in the fluid and the transverse waves can be ignored. It is a good approximation to consider the solid rods as fluid with very high stiffness and specific mass. Then, the wave equation is simplified as follows [16–18]:

$$(C_{11})^{-1}\frac{\partial^2 \boldsymbol{p}}{\partial t^2} = \nabla \cdot (\rho^{-1}\nabla \boldsymbol{p}) \tag{12.1}$$

where $\boldsymbol{p}$ is the pressure, ρ is the mass density, $C_{11} = \rho c_l^2$ is the longitudinal elastic constant, and c_l is the longitudinal speed of sound. Expanding the quantities $\rho^{-1}(\mathbf{r})$ and $C_{11}^{-1}(\mathbf{r})$ in the Fourier series as given below [17–19]:

$$\rho^{-1}(\mathbf{r}) = \sum_{\mathbf{G}} \sigma(\mathbf{G})\mathrm{e}^{\mathrm{i}\mathbf{G}\cdot\mathbf{r}}, \text{ and} \tag{12.2a}$$

$$C_{11}^{-1}(\mathbf{r}) = \sum_{\mathbf{G}} \zeta(\mathbf{G})\mathrm{e}^{\mathrm{i}\mathbf{G}\cdot\mathbf{r}} \tag{12.2b}$$

where $\mathbf{G}$ is the 2D reciprocal lattice vectors. A periodic system of cylinders (medium A) in a background of medium B is analyzed. The corresponding densities (elastic constants) are ρ_A, ρ_B (C_{11A}, C_{11B}). The Fourier coefficients can easily be determined as follows [16–18]:

$$\sigma(\mathbf{G}) = \begin{cases} \rho_A^{-1} f + \rho_B^{-1}(1-f) \equiv \overline{\rho^{-1}}, & \text{for } \mathbf{G} = 0, \text{ and} \\ (\rho_A^{-1} - \rho_B^{-1})F_G \equiv \Delta(\rho^{-1})F_G, & \text{for } \mathbf{G} \neq 0, \end{cases} \tag{12.3a}$$

and

$$\zeta(\mathbf{G}) = \begin{cases} C_{11\mathrm{A}}^{-1} f + C_{11\mathrm{B}}^{-1}(1-f) \equiv \overline{C_{11}^{-1}}, & \text{for } \mathbf{G} = 0, \text{and} \\ (C_{11\mathrm{A}}^{-1} - C_{11\mathrm{B}}^{-1})F_G \equiv \Delta(C_{11}^{-1})F_G, & \text{for } \mathbf{G} \neq 0, \end{cases} \tag{12.3b}$$

For the rods with radius r_0 in the system, the structure factor F_{G} can be obtained as [16–18];

$$F_G = 2f\frac{J_1(\mathbf{G}r_0)}{\mathbf{G}r_0}, \tag{12.4}$$

where $f = \pi r_0^2 / a_0^2$ is the filling fraction of rods with lattice constant a_0 for the square lattice, and $J_1(x)$ is the Bessel function of the first kind of order one. The pressure field $\boldsymbol{p}(\mathbf{r})$ must satisfy the Bloch theorem

$$\boldsymbol{p}(\mathbf{r}, t) = \mathrm{e}^{\mathrm{j}(\mathbf{K}\cdot\mathbf{r}-\omega t)} \sum_{\mathbf{G}} \boldsymbol{p}_{\mathbf{K}}(\boldsymbol{G})\mathrm{e}^{\mathrm{j}\mathbf{G}\cdot\mathbf{r}}, \tag{12.5}$$

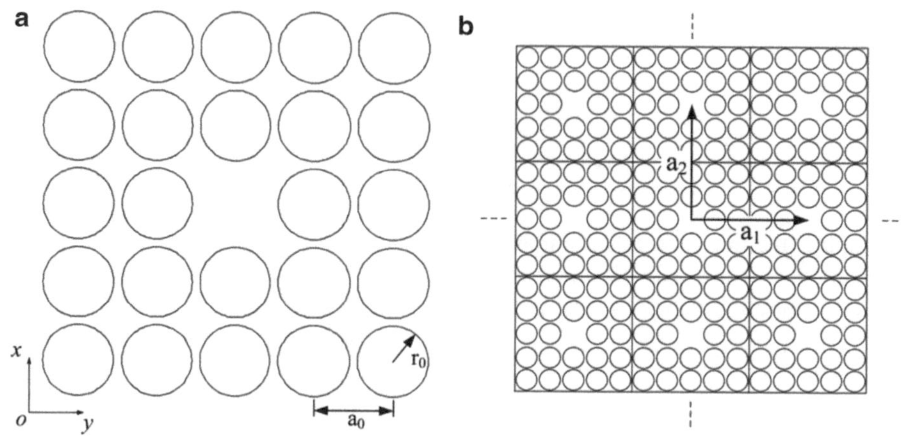

Fig. 12.3 (**a**) The 5 × 5 supercell with a point defect. (**b**) A schematic of the periodic structure composed of the supercell with a point defect, where $\mathbf{a}_1$ and $\mathbf{a}_2$ are the primitive lattice vectors

where **K** is a 2D Bloch vector. By substituting Eqs. (12.2), (12.3), and (12.5) into Eq. (12.1), the eigenvalue equation is obtained as follows [17–19]:

$$\sum_{G' \neq G} F_G(\mathbf{G}-\mathbf{G}')[\Delta(\rho^{-1})(\mathbf{K}+\mathbf{G})\cdot(\mathbf{K}+\mathbf{G}') - \Delta(C_{11}^{-1})\omega^2]p_{\mathbf{K}}(\mathbf{G}')$$

$$+ [\overline{\rho^{-1}}|\mathbf{K}+\mathbf{G}|^2 - \overline{C_{11}^{-1}}\omega^2]p_{\mathbf{K}}(\mathbf{G}') = 0. \quad (12.6)$$

where ω and $p_{\mathbf{K}}(\mathbf{G})$ are the eigenvalues and eigenvectors, respectively. The band structures are obtained by solving the eigenvalue equation, Eq. (12.6).

12.2.2 Supercell Method

The calculations of band structures are based on the plane wave expansion method. In order to analyze the sonic crystal with a point defect, the plane wave expansion calculation with a supercell method is adopted. The supercell structure consists of several unit cells. We can arrange the unit cells within the supercell to make the periodic structure with a point defect, as shown in Fig. 12.3a, and the supercell with a point defect is used as a unit cell to construct a periodic structure in the plane wave expansion calculations. A schematic of the periodic structure composed of the supercell with a point defect is shown in Fig. 12.3b, where $\mathbf{a}_1$ and $\mathbf{a}_2$ are the primitive lattice vectors. The size of the supercell should be large enough to guarantee that the coupling effect of the defect mode between neighboring supercells. The band structure of the sonic crystal with a defect can be obtained by combining the plane wave expansion method and the supercell method.

The shift property of the Fourier transform pair of $h(\mathbf{r})$ is expressed as

$$h(\mathbf{r}-r') \Leftrightarrow e^{-j\mathbf{G}\cdot r'} h_G(\mathbf{G}) \tag{12.7}$$

If a cylinder is shifted by a displacement r', we can use the shift property to obtain the Fourier coefficients of $\rho^{-1}(\mathbf{r})$ and $C_{11}^{-1}(\mathbf{r})$. If there are many cylinders in the supercell, the shift property of the Fourier transform pair of $\rho^{-1}(\mathbf{r})$ and $C_{11}{}^{-1}(\mathbf{r})$ can be expressed as

$$\sum_i \rho^{-1}(\mathbf{r}-r'_i) \Leftrightarrow \sum_i e^{-j\mathbf{G}\cdot r'_i} \rho_G^{-1}(\mathbf{G}), \text{ and} \tag{12.8}$$

$$\sum_i C_{11}^{-1}(\mathbf{r}-r'_i) \Leftrightarrow \sum_i e^{-j\mathbf{G}\cdot r'_i} C_{11G}^{-1}(\mathbf{G}) \tag{12.9}$$

where r'_i is the position vector of ith cylinder. Here, a supercell with $N\times N$ cylinders (N is odd and greater than 1) is considered, and a defect is introduced by removing a central cylinder. Figure 12.3a shows an example of the 5×5 supercell with a point defect. The Fourier coefficient of $\rho^{-1}(\mathbf{r})$ and $C_{11}{}^{-1}(\mathbf{r})$ can be easily determined, as follows [14]:

$$\sigma(\mathbf{G}) = \begin{cases} \left(N^2-1\right)\rho_A^{-1} f + \rho_B^{-1}\left[1-\left(N^2-1\right)f\right] \equiv \overline{\rho^{-1}}, & \text{for } \mathbf{G}=0, \text{ and} \\ \left(\rho_A^{-1}-\rho_B^{-1}\right) F_G \\ \times\left\{\left(\sum_{m_1=-\frac{(N-1)}{2}}^{\frac{(N-1)}{2}} \sum_{m_2=-\frac{(N-1)}{2}}^{\frac{(N-1)}{2}} e^{-j\mathbf{G}\cdot[(m_1,\,m_2)\times a_0]}\right) - e^{-j\mathbf{G}\cdot(0,\,0)}\right\} \equiv \Delta(\rho^{-1})F_G, \\ & \text{for } \mathbf{G}\neq 0, \end{cases} \tag{12.10}$$

$$\zeta(\mathbf{G}) = \begin{cases} \left(N^2-1\right) C_{11A}^{-1} f + C_{11B}^{-1}[1-(N^2-1)f] \equiv \overline{C_{11}^{-1}}, & \text{for}\mathbf{G}=0, \\ \left(C_{11A}^{-1}-C_{11B}^{-1}\right) F_G \\ \times\left\{\left(\sum_{m_1=-\frac{(N-1)}{2}}^{\frac{(N-1)}{2}} \sum_{m_2=-\frac{(N-1)}{2}}^{\frac{(N-1)}{2}} e^{-j\mathbf{G}\cdot[(m_1,\,m_2)\times a_0]}\right) - e^{-j\mathbf{G}\cdot(0,\,0)}\right\} \equiv \Delta(C_{11}^{-1})F_G, \\ & \text{for } \mathbf{G}\neq 0, \end{cases} \tag{12.11}$$

For the rods with radius r_0 in the system, the structure factor F_G can be expressed as

$F_G = 2f\frac{J_1(\mathbf{G}r_0)}{\mathbf{G}r_0}$, where $f = \frac{\pi r_0^2}{N^2 a_0^2}$ is the filling fraction of one rod in the supercell with a square lattice.

12.2.3 Velocity Fields

The pressure distribution of a defect mode can be obtained from the eigenvectors of Eq. (12.6). The pressure field can be defined as follows [28]:

$$p = -\rho \frac{\partial \mathbf{\Phi}}{\partial t} = -\mathrm{i}\rho\omega_{\mathrm{a}}\mathbf{\Phi}, \tag{12.12}$$

where $\mathbf{\Phi}$ is the velocity potential and ω_{a} is the angular frequency of the acoustic waves. Hence, one can obtain extract $\mathbf{\Phi}$ from the Eq. (12.12), as

$$\mathbf{\Phi} = -\frac{p}{\mathrm{i}\rho\omega_{\mathrm{a}}}. \tag{12.13}$$

The particle velocity field can be expressed as

$$\mathbf{v} = \nabla\mathbf{\Phi} = -\frac{\nabla p}{\mathrm{i}\rho\omega_{\mathrm{a}}}. \tag{12.14}$$

From the Eq. (12.14), the velocity of air particles is dependent upon the gradient of the pressure field and the material density. The piezoelectric beam should therefore be located in the region of high velocity so it can be vibrated by the oscillation of the air particles.

12.2.4 Numerical Results

The 2D square lattice sonic crystal, consisting of polymethyl methacrylate (PMMA) cylinders in air, are studied, where $\rho_{\mathrm{PMMA}} = 1{,}190\ \mathrm{kg/m^3}$, $\rho_{\mathrm{air}} = 1.2\ \mathrm{kg/m^3}$, $c_{\mathrm{PMMA}} = 2{,}694$ m/s, and $c_{\mathrm{air}} = 343$ m/s. The radius of each cylinder is $r_0 = 17.5$ mm. The lattice constant is $a_0 = 49$ mm, and the filling fraction of the cylinder in the unit cell, $f = \pi r_0^2 / a_0^2$, is 40%. Figure 12.4 presents the band structures of the prefect periodic structure (heavy line) and the 5×5 supercell with a point defect (light line). From the band structures of the prefect periodic structure, it can be seen that there exists one absolute band gap between 3.87 and 4.32 kHz. In this frequency region, the acoustic wave cannot propagate through the sonic crystal. However, when a point defect is created by removing a single rod from the middle of the perfect periodic structure, it is possible to find defect bands in the absolute band gap [19–27]. The inset of Fig. 12.4 shows a 5×5 perfect sonic crystal with a single rod removed from the middle. This structure can be seen as the 5×5 supercell with a point defect. The defect band can be found by calculating the band structure of 5×5 supercell with a point defect.

From the band structures of the 5×5 supercell with a point defect, one can observe that a defect band exists in the absolute band gap. The defect band is

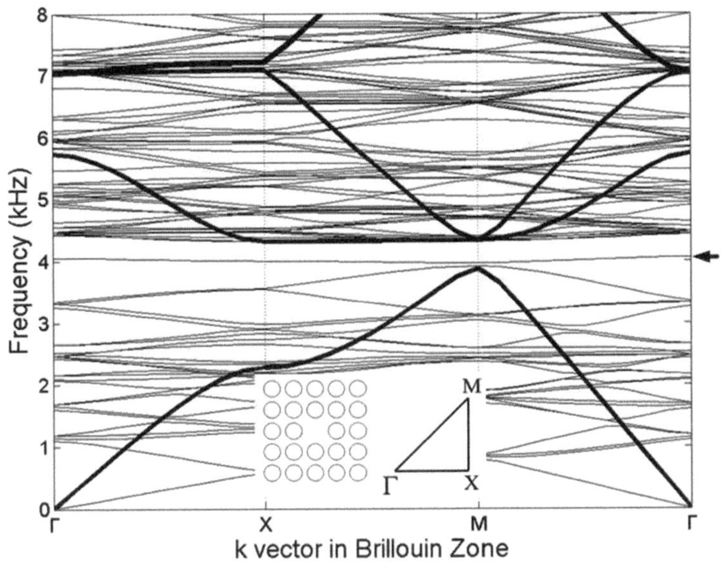

Fig. 12.4 The band structures of the prefect periodic structure (heavy lines) and the 5 × 5 supercell with a point defect (*light lines*). The *inset* shows the 5 × 5 supercell arranged in a square lattice and the irreducible Brillouin zone

indicated by an arrow in Fig. 12.4. The band structure of 5 × 5 supercell with a point defect is more complex than that of the perfect sonic crystal; the acoustic waves can propagate through the sonic crystal at the frequency of the defect band, since the defect band acts as a pass band in the band gap. Moreover, at the frequency of the defect band, which is the resonant frequency, the acoustic waves should be localized in the point defect [26, 27]. The resonant frequency of the defect band is 4.02 kHz. Figure 12.5 shows the pressure and particle velocity distributions of the defect mode at the Γ point in the reciprocal space (Fig. 12.2). It can be seen that the pressure is well localized around the defect, and the pressure has the maximum at the center of the defect. However, the minimum of particle velocity is at the center of the defect. The piezoelectric energy harvester must be placed at the position with the highest particle velocity as shown in Fig. 12.5b.

12.3 Model of Energy Harvesting System

The 31-mode of the piezoelectric materials is employed to harvest the electrical energy. A 31-mode power generator is a beam type piezoelectric energy harvester, in which a piezoelectric layer is bonded to a substructure element; the 31-mode undergoes stress through beam bending. To simplify the analysis, the generator is considered as a laminated curved beam with uniform thickness and width

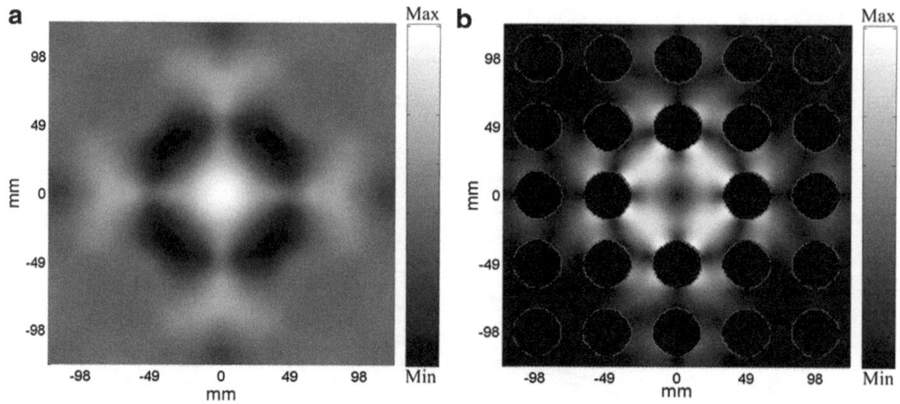

Fig. 12.5 (**a**) The pressure distribution and (**b**) the particle velocity field in the 5 × 5 sonic crystal with a point defect at the Γ point for the defect mode as shown in Fig. 12.4

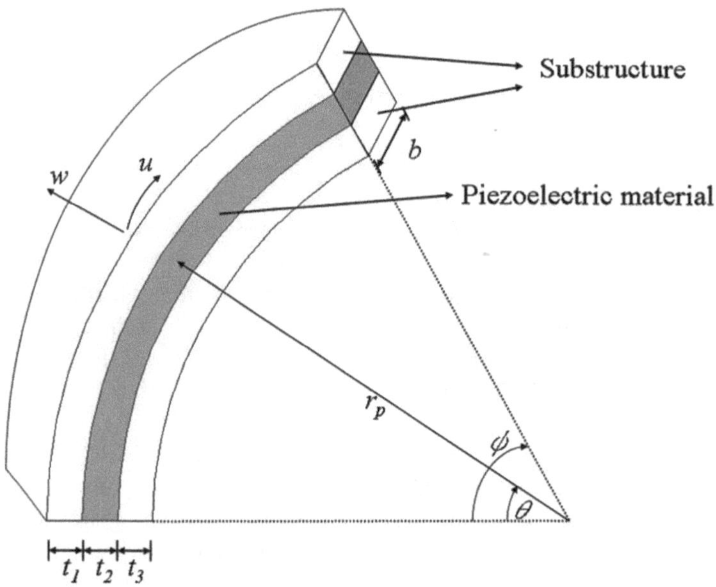

Fig. 12.6 Schematic drawing of laminated piezoelectric curved beams

in our model. Figure 12.6 shows a laminated piezoelectric curved beam. There are three layers of the laminated curved beam in this study. The center layer is the piezoelectric material. The radius of the curved beam is constant in our investigation. The mechanical strain in the circumferential direction (θ-direction), s_θ, are defined as follows [29–31]:

$$s_\theta = s^0 + z\kappa, \tag{12.15}$$

where s^0 is the extensional strain at the neutral axis, κ is the change in curvature of the neutral axis during bending and z is the distance from the neutral axis. The strain and curvature of the curved beam are expressed in terms of circumferential and radial displacement as follows [29–31]:

$$s^0 = \frac{1}{r_p}\left(\frac{\partial u}{\partial \theta} + w\right) \text{ and} \tag{12.16}$$

$$\kappa = \frac{1}{r_p^2}\left(\frac{\partial u}{\partial \theta} - \frac{\partial^2 w}{\partial \theta^2}\right), \tag{12.17}$$

where r_p, u and w are the radius of curvature, circumferential and radial displacement of the curved beam, respectively. Equation (12.15) can be rewritten as follows:

$$s_\theta = \frac{1}{r_p}\frac{\partial u}{\partial \theta} + \frac{1}{r_p}w + z\frac{1}{r_p^2}\frac{\partial u}{\partial \theta} - z\frac{\partial^2 w}{r_p^2 \partial \theta^2}. \tag{12.18}$$

Since the piezoelectric layer thickness in comparison to the length of beam is very small, the charge in the circumferential direction (direction of length) and the stress in the radial direction (direction of thickness) is assumed to be zero within the piezoelectric layer ($D_\theta = 0$ and $T_r = 0$). The electrical displacement in the radial direction within the piezoelectric layer is a function of the stress in the circumferential direction and electric field within the piezoelectric layer. The piezoelectric constitutive relation is expressed as follows:

$$D_r = d_{31}T_\theta + \varepsilon_{33}E_r, \tag{12.19}$$

where d_{31} is the piezoelectric constant in the 31 coupling direction, T_θ is the stress in circumferential direction, ε_{33} is the dielectric constant and E_r is the electric field in the radial direction within the piezoelectric layer. The electric charge can be expressed as the integral of electrical displacement on the area of the electrode surface:

$$Q = \int_A D_r \mathrm{d}A = b\int_0^{\varphi} (d_{31}Y s_\theta + \varepsilon_{33}E_r) r_p \mathrm{d}\theta, \tag{12.20}$$

where b is the width of the beam, Y is Young's modulus and φ is the angle of the curved beam. The potential difference between the upper surface and lower surface of the piezoelectric layer is denoted as v. Based on the uniform electrical field assumption, the electric field can be expressed as

$$E_r = -\frac{\partial v}{\partial z} = -\frac{v}{t_2}, \tag{12.21}$$

where t_2 is the thickness of the piezoelectric layer. It is assumed that the thickness of the layer is the same within the whole coverage area. Substituting Eqs. (12.19) and (12.21) into (12.20),

$$Q = b\int_0^{\varphi} d_{31}Y\left(\frac{1}{r_p}\frac{\partial u}{\partial\theta} + \frac{1}{r_p}w + z\frac{1}{r_p^2}\frac{\partial u}{\partial\theta} - z\frac{\partial^2 w}{r_p^2\partial\theta^2}\right)\cdot r_p\,\mathrm{d}\theta - b\int_0^{\varphi}\varepsilon_{33}\frac{v}{t_2}\cdot r_p\mathrm{d}\theta. \tag{12.22}$$

Equation (12.22) can be rewritten as

$$Q = b\cdot d_{31}Y\alpha - C_\mathrm{p}\cdot v, \tag{12.23}$$

where $\alpha = u|_0^{\varphi} + \int_0^{\varphi} w\mathrm{d}\theta + \frac{1}{r_p}\cdot z\cdot u|_0^{\varphi} - \frac{1}{r_p}\cdot z\cdot \frac{\partial w}{\partial\theta}\Big|_0^{\varphi}$, and the capacitance $C_\mathrm{p} = b\int_0^{\varphi}\frac{\varepsilon_{33}}{t_2}\cdot r_p\mathrm{d}\theta = \frac{bL\varepsilon_{33}}{t_2}$. Here, z is distance between the neutral plane and piezoelectric layer, and L is the length of the piezoelectric layer.

The current, charge and voltage are all functions of time. The frequency of these periodic functions depends on the mechanical vibration. The piezoelectric layer is connected to an external impedance. The time differential of the generated charges on the electrode surface is the current flow out to the external impedance. Since the piezoelectric layer is subjected to harmonic excitation, the electric charge can be expressed as a harmonic function. Therefore, the amplitude of the current is expressed as the frequency times the charge [4]:

$$I = \omega_\mathrm{v} Q, \tag{12.24}$$

where ω_v is the vibrating frequency of the piezoelectric layer. The relationship between current and voltage for an electrical circuit with pure load resistance R_L is

$$I = \frac{v}{R_\mathrm{L}}, \tag{12.25}$$

where the voltage and the current has the same phase. Combining Eqs. (12.23–12.25), the amplitude of the current and voltage can be expressed as

$$I = \frac{\omega_\mathrm{v} b d_{31} Y\alpha}{(1 + C_\mathrm{p}\omega_\mathrm{v} R_\mathrm{L})}, \text{ and} \tag{12.26}$$

$$v = \frac{\omega_\mathrm{v} b d_{31} Y\alpha}{(1 + C_\mathrm{p}\omega_\mathrm{v} R_\mathrm{L})}\cdot R_\mathrm{L}. \tag{12.27}$$

Since the external impedance is a pure resistance, the output voltage and the current have the same phase. The output power can be defined as

$$P = Iv = \left(\frac{\omega_v b d_{31} Y \alpha}{(1 + C_p \omega_v R_L)}\right)^2 R_L. \tag{12.28}$$

From the above equation, it is found that the output power is a function of several parameters, such as external load resistance R_L, the vibration frequency ω_v, the capacitance of piezoelectric layer C_p, as well as the material properties and dimensions of the curved beam. The value of α is determined by the radial displacement and the boundary condition of the curved beam. For weakly coupled energy harvesters, the power output reaches its maximum value when the external load resistance is

$$R_L{}^* = \frac{1}{C_p \omega_v}. \tag{12.29}$$

Therefore, the external load resistance of the maximum output power depends on the vibration frequency and the capacitance of piezoelectric layer.

12.4 Experimental Results and Discussion

12.4.1 Sonic Crystal Experiment

The experimental model used in the present study is a 2D square lattice consisting of PMMA cylinders. The 40 cm long PMMA cylinders with a radius $r_0 = 17.5$ mm are fixed on a perforated PMMA plate. Figure 12.7 presents the experimental setup. The speaker (Fostex: FF85K) is used as the sound source. An 1/8 in. microphone (Brüel and Kjær: 4138) is used as a receiver and placed in the middle of the cavity. The speaker, connected to a function generator (Motech: FG503), can either be set to a specific frequencies, or swept through a range of frequencies. Here, the range of sweeping frequencies is from 50 Hz to 20 kHz. The signals detected

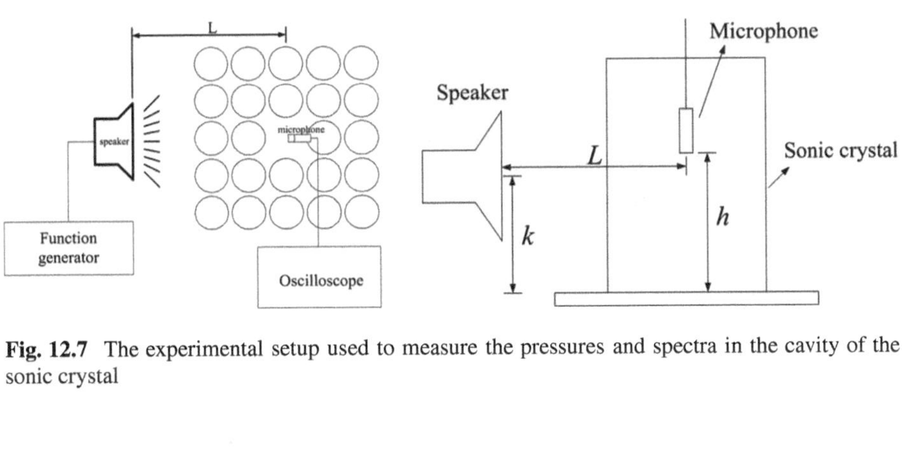

Fig. 12.7 The experimental setup used to measure the pressures and spectra in the cavity of the sonic crystal

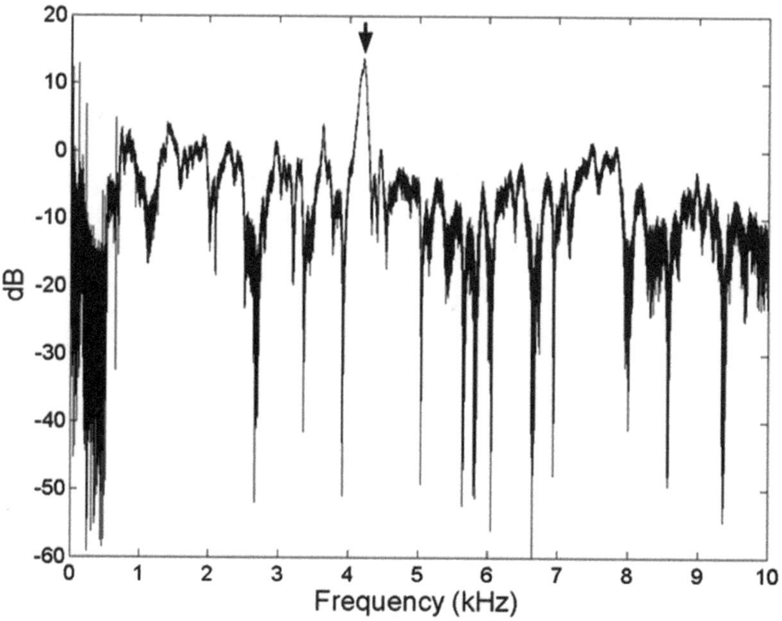

Fig. 12.8 Measured frequency spectrum in the middle of the cavity consisting of a 5×5 sonic crystal. The resonant frequency of the sonic cavity indicated by the arrow is 4.21 kHz

by the microphone are recorded by a digital sampling oscilloscope (Tektronix: TDS5032B), yielding the sound pressure in the cavity. The frequency spectrum can also be obtained by spectrum analysis. Sound-absorption sponges are utilized to enclose the sonic crystal structure and thus reduce the acoustic waves reflected from the environment. The height from the perforated PMMA plate to the center of the speaker is $k = 18$ cm. The largest pressure in the cavity at the resonant frequency is obtained when the distance between the speaker and center of the cavity is $L = 34.5$ cm and the height from the perforated PMMA plate to the microphone is $h = 10.5$ cm.

Figure 12.8 shows the measured frequency spectrum in the middle of the cavity of the 5×5 sonic crystal. The pressure in the middle of the cavity is the largest when the incident acoustic wave is at the 4.21 kHz (indicated by the arrow). This means that 4.21 kHz is the cavity resonant frequency of the sonic crystal, and the acoustic waves at this frequency can be localized in the cavity of the sonic crystal. The measured resonant frequency is close to the defect band frequency of the band structure. At the resonant frequency, the measured pressure in the cavity of the 5×5 sonic crystal is about 4.94 times larger than ones without the sonic crystal [26]. That is, the acoustic power is about 24.4 times larger than ones without the sonic crystal.

Table 12.1 Dimensions and capacitances of LDT2-028k and LDT4-028k

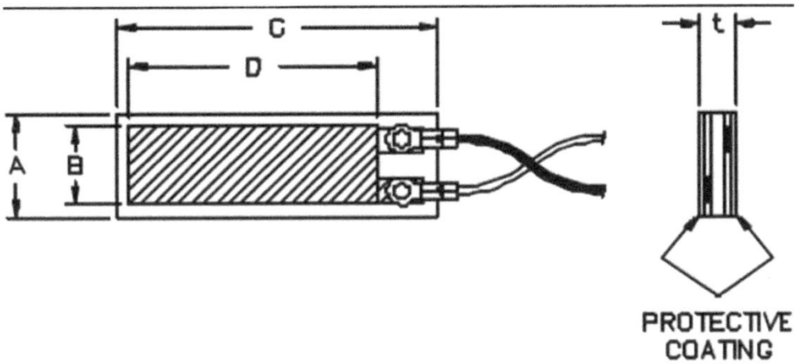

Description	LDT2-028k	LDT4-028k
A (mm)	16	22
B (mm)	12	19
C (mm)	73	171
D (mm)	62	156
t (μm)	205	205
Capacitance (nF)	2.78	11

12.4.2 *Experiment of Acoustic Energy Harvesting*

12.4.2.1 PVDF Piezoelectric Film and Experiment Setup

In this investigation, the flexible piezoelectric material, PVDF, is adopted to harvest acoustic energy. Since PVDF film has low stiffness, it is easy to vibrate by air flow. A 125μm polyester layer is laminated to a 28μm piezoelectric film element, and another 52μm polyester layer is covered on it to protect the PVDF layer. The piezoelectric polymer films are manufactured by Measurement Specialities Incorporated Inc. (MEAS). Both the LDT2-028k and LDT4-028k PVDF film's dimensions are listed on Table 12.1. It is assumed that the width of the polyester layer and PVDF layer is the same. In addition, because the laminated polyester layer, PVDF and protecting layer are polymer materials, it is also assumed that their material properties are the same. It can be observed that the neutral plane is located in the polyester. Therefore, the higher output voltage can be induced, when the PVDF film is excited in the bending mode.

Figure 12.9 shows the schematic of the experimental setup and the PVDF film in the cavity of the sonic crystal. The PVDF film connected to a load resistance can be vibrated to harvest the electric energy by the motion of air particles at the resonant frequency of the sonic crystal cavity. The speaker (Fostex: FF85K) connected with the function generator (Motech: FG503) is used as the sound source. The output voltage can be detected by the oscilloscope (Tektronix: TDS5032B) which has much higher impedance than the external load resistance R_L. The PVDF film is put in the

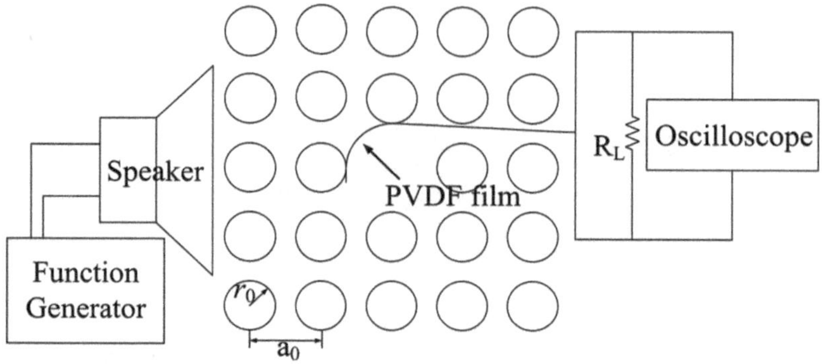

Fig. 12.9 Experimental setup for energy harvesting in the cavity of a sonic crystal

region with high air particle velocity as shown in Fig. 12.5b. The height from the perforated PMMA plate to the PVDF film is $h = 10.5$ cm, where the largest pressure is observed at the resonant frequency.

12.4.2.2 Efficiency of Energy Harvesting

The power harvesting efficiency without the sonic crystal η is the ratio of the output power to input power and can be expressed as

$$\eta = W_{\mathrm{Out}}/W_{\mathrm{In}} \tag{12.30}$$

where the input power (W_{In}) is defined as the incident acoustic power from the speaker; the output power (W_{Out}) is defined as the electric output power from the PVDF film. However, the acoustic wave can be localized in the resonant cavity of the sonic crystal, and the electric output power from the PVDF film will be affected. So, the power harvesting efficiency with the sonic crystal η' can be expressed as

$$\eta' = W'_{\mathrm{Out}}/W_{\mathrm{In}} = \alpha_{\mathrm{SC}} W_{\mathrm{Out}}/W_{\mathrm{In}} \tag{12.31}$$

where W'_{Out} is the electric output power of the PVDF film with the sonic crystal, and α_{SC} is the ratio of the power harvesting efficiency with and without the sonic crystal.

12.4.2.3 LDT4-028k

Figure 12.10 shows the photograph of the PVDF film, LDT4-028k, placed inside the cavity of the sonic crystal and attached on rods. We can see that a part of the PVDF film is outside the cavity. The frequency response of the PVDF film voltage output in the cavity of the 5×5 sonic crystal is shown in Fig. 12.11. The frequency

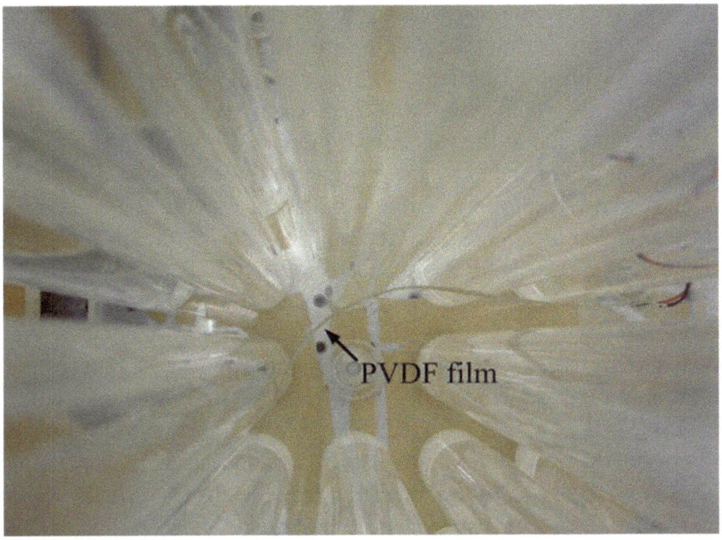

Fig. 12.10 Photographs of the PVDF film, LDT4-028k, placed inside the resonant cavity of sonic crystal

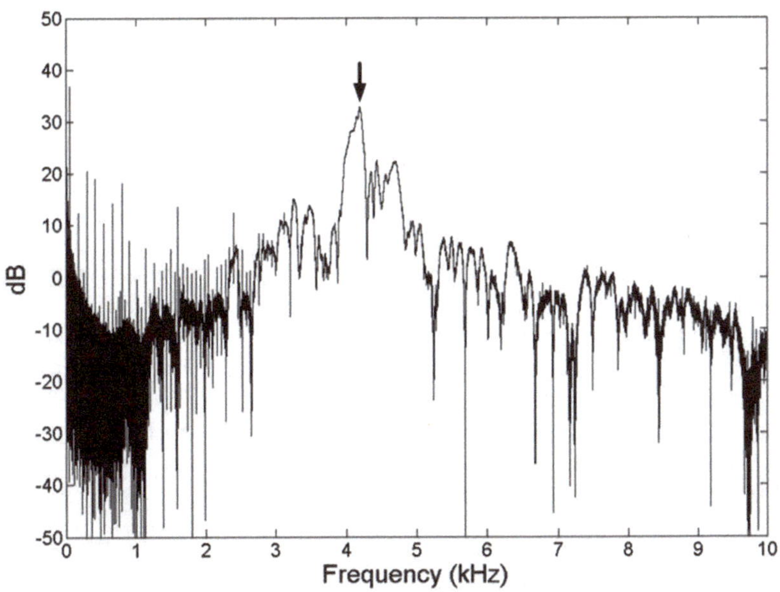

Fig. 12.11 The frequency response of the PVDF film (LDT4-028k) voltage output in the cavity of the 5×5 sonic crystal. The frequency of the peak indicated by the *arrow* is 4.2 kHz

of the peak indicated by the arrow is 4.2 kHz and it is very close to the resonant frequency of the cavity without the PVDF film. The maximum voltage output of the PVDF film occurs when the frequency of incident acoustic wave is near the resonant frequency of the cavity. This shows that the acoustic waves can be localized in the cavity of the sonic crystal at the resonant frequency, and that the voltage output of the PVDF film can be enhanced by the resonance of the cavity.

Figure 12.12 shows the output voltage with and without the sonic crystal. The PVDF film connected to a load resistance of 3.9 kΩ is vibrated by the motion of air particles and the electric energy can be harvested. The voltage variation is a sinusoidal function of time. The solid line represents the output voltage as the PVDF film is put in the cavity of 5×5 sonic crystal and the frequency of the incident acoustic wave is 4.2 kHz. The dotted line represents the output voltage of the PVDF film without the sonic crystal. At 4.2 kHz, the output voltage with the 5×5 sonic crystal is 25 times larger than that without the sonic crystal. Thus, the power harvesting efficiency with sonic crystal is 625 times larger than that without the sonic crystal ($\alpha_{SC} = 625$). This is much larger than the acoustic power enhanced by the cavity of the sonic crystal, since the acoustic wave is not only enhanced and localized but is also at the resonant mode in the cavity. Thus the PVDF film will be strongly vibrated by the acoustic wave in the cavity and hence, the larger output power can be obtained. In order to increase the output voltage of the PVDF film, the pressure in the cavity should be raised. We have previously demonstrated that the pressure in the cavity of the 7×6 sonic crystal is larger than that of the 5×5 sonic crystal [27]. Hence, we replace the original 5×5 sonic crystal with 7×6 sonic crystal, and the position of the PVDF film in the cavity does not vary. From Fig. 12.12, we can see that the PVDF film in the cavity of the 7×6 sonic crystal has higher voltage output at 4.18 kHz. The output voltage with the 7×6 sonic crystal at 4.18 kHz is 1.47 times larger than that with the 5×5 sonic crystal at 4.2 kHz.

Figure 12.13 presents the relation between the load resistances and output voltages and the relation between load resistance and output powers. The frequency of the incident acoustic wave is 4.2 kHz. The circular and triangular symbols denote the experimental results of output voltage and power, respectively. Notice that the output voltage and power increase rapidly with an increasing load resistance initially. The output voltage then increases slowly as the load resistance reaches 10 kΩ. However, the output power increases with an increasing load resistance until a critical value and then decreases. The maximum power is experimentally observed corresponding to a load resistance of 3.9 kΩ. We know that the capacitance of LDT4-028k is 11 nF and the vibrating frequency is 4.2 kHz. From Eq. (12.29), the theoretical value $R_L{}^*$ is 3.445 kΩ which is close to the experimental result of 3.9 kΩ.

12.4.2.4 LDT2-028k

We also place the PVDF film, LDT2-028k, inside the sonic crystal to harvest acoustic energy. The PVDF film is fixed on the rods and can be regarded as a curved beam with clamped–clamped boundary condition. Figure 12.14 shows the

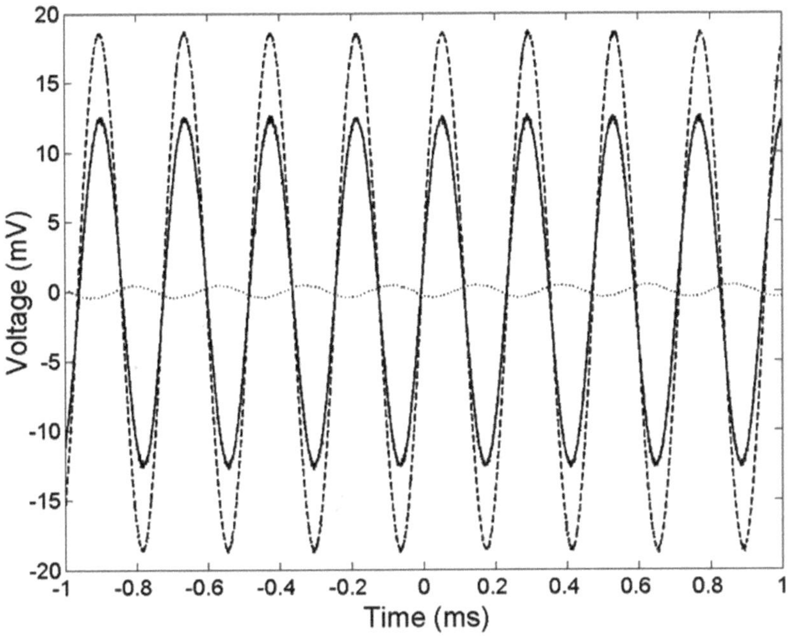

Fig. 12.12 The output voltage from the piezoelectric material in the cavity of the sonic crystal. Solid line refers to a 5 × 5 sonic crystals at 4.2 kHz. *Dotted line* represents the output voltage of the PVDF film without the sonic crystal at 4.2 kHz. And, *dashed line* refers to a 7 × 6 sonic crystals at 4.18 kHz

frequency response of the PVDF film voltage output in the cavity of the 5 × 5 sonic crystal. Similarly, the frequency of the peak indicated by the arrow is 4.21 kHz and it is very close to the resonant frequency of the cavity without PVDF film. The maximum voltage output of the PVDF film occurs as the frequency of the incident acoustic wave is near the resonant frequency of the cavity.

The pressure difference between the two sides of the PVDF film is the external force to vibrate the PVDF film. The measured pressure difference between the two sides of the PVDF film is about 7 Pa and is a sinusoidal function of time. That is, the external force applied on the PVDF film is harmonic and its frequency is 4.21 kHz. We assume that the external force is uniform distribution along the beam. The finite element software, *COMSOL Multiphysics*® [32], is employed to calculate the forced vibration of the curved beam. The dimensions and material properties of laminated piezoelectric curved beams are listed in Table 12.2. For the LDT2-028k, the width of polyester layer is 0.016 m, but the width of the piezoelectric element is 0.012 m. The width of polyester layer is adopted in the vibration calculation, and the width of piezoelectric element is employed in the energy harvesting calculation. From Eq. (12.28), we know that a larger value of α corresponds to a larger generated power. For the clamped–clamped boundary condition, α can be express as $\int_0^{\varphi} w d\theta$. Thus, increasing the radial vibration amplitude can improve the output power. The output

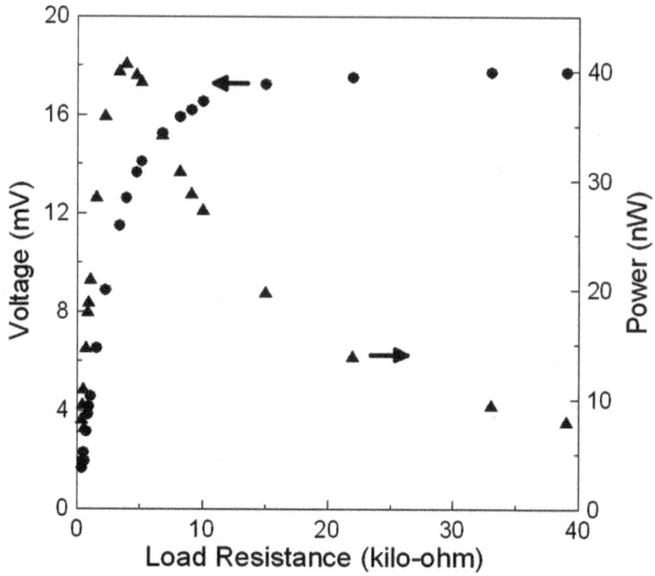

Fig. 12.13 Measured output voltage and power delivered to the load versus load resistance. The *circular* and *triangular symbols* denote the experimental results of output voltage and power, respectively

voltage and power can be obtained using Eqs. (12.27) and (12.28). Figures 12.15 and 12.16 show the relationships between the load resistances and output voltages and the relationships between load resistance and output powers, respectively. The frequency of the incident acoustic wave is 4.21 kHz. The solid lines represent theoretical prediction using Eqs. (12.27) and (12.28) and the diamond symbols denote the experimental results. The experimental results are similar for the LDT4-028k case. Initially, the output voltage and power increase rapidly with an increasing load resistance; when the load resistance reaches 30 kΩ, the output voltage increases slowly with increasing load resistance. However, the power increases gradually with an increasing load resistance until a critical value is reached and then decreases. The differences between predicted and measured values increase as the load resistances increase. The experimentally observed maximum power at a load resistance of 15 kΩ. The capacitance of the LDT2-028k film is 2.78 nF and the vibrating frequency is 4.21 kHz. From Eq. (12.29), the theoretical value R_L^* is 13.6 kΩ. The experimental result is close to the theoretical result.

We know that the displacement of the curved beam is dependent on the radius of the curved beam. The value of α as shown in Eq. (12.23) is dependent on the radius of the curved beam strongly. The curvature of the curved beam is not uniform in the experiment, and the radius of the curved beam varies along the length. In the calculation, the whole curved beam is assumed to have the uniform curvature and the radius of the curved beam is estimated from the experimental

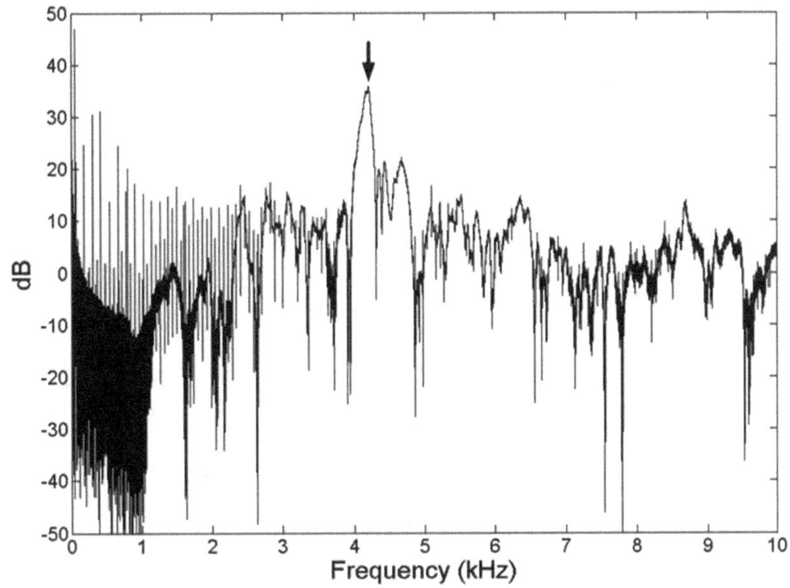

Fig. 12.14 The frequency response of the PVDF film (LDT2-028k) voltage output in the cavity of the 5 × 5 sonic crystal. The frequency of the peak indicated by the *arrow* is 4.21 kHz

Table 12.2 Dimensions and material properties of the laminated piezoelectric curved beams

Properties	Value
Radius of curvature r_p	0.06 m
Width of beam of polyester layer	0.016 m
Width of beam of piezoelectric element b	0.012 m
Thickness of polyester layer t_1	125 μm
Thickness of piezoelectric layer t_2	28 μm
Thickness of polyester layer t_3	52 μm
Angle of curved beam ψ	π/3
Young's modulus Y	3.5 GPa
Piezoelectric constant in the 31 coupling direction d_{31}	23×10^{-12} N/m V

model. Therefore, the radius of the curved beam employed in the calculation is one of the reasons for the difference between the experimental and theoretical results. Moreover, in order to simplify the electromechanical coupling problem, we employ a semi-piezoelectric constitutive equation as shown in Eq. (12.19) to solve the energy harvesting system. The piezoelectric layer's circumferential strains induced by electric fields are not taken into account. As load resistances are small, output voltages are also small and the circumferential strains induced by the electric field can be ignored. However, a larger load resistance corresponds to a higher output voltage. When the output voltage is large enough, the circumferential strains

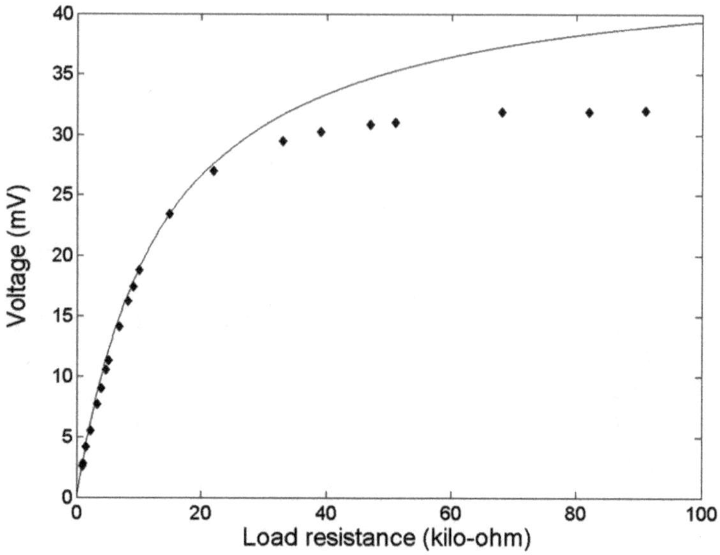

Fig. 12.15 Measured output voltage delivered to the load versus load resistance, compared to theory. *Solid line* and the *diamond symbols* refer to theoretical and experimental results of output voltage, respectively

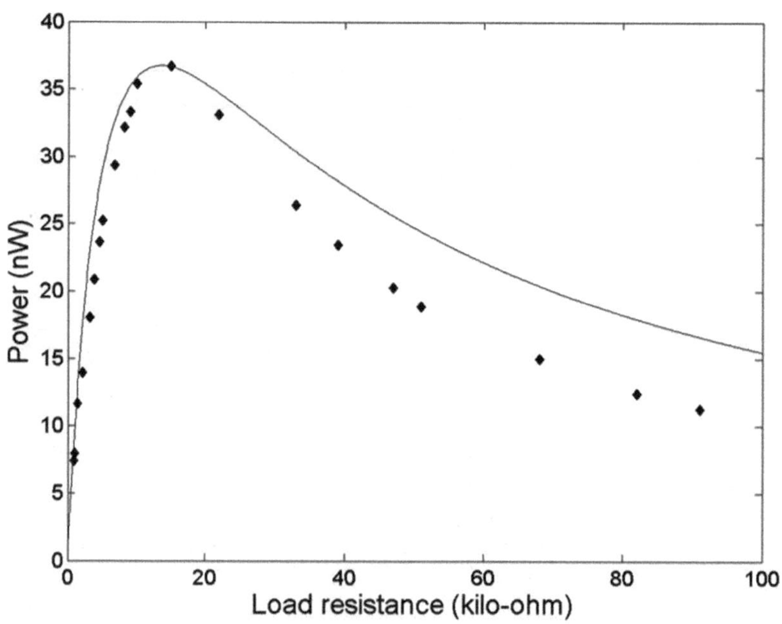

Fig. 12.16 Measured output power delivered to the load versus load resistance, compared to theory. *Solid line* and the *diamond symbols* refer to theoretical and experimental results of output power, respectively

induced by the electric field must be considered. We think that this is why the theoretical predicted values deviate from measured value for load resistances greater than 20 kΩ. In general, the experimental output voltage and power are in agreement with the calculated ones.

12.5 Conclusion

An acoustic energy harvester based on the sonic crystal and piezoelectric material is presented in this chapter. A model for energy harvesting of the piezoelectric curved beam is also developed to predict the output voltage and power of the harvester. Since sonic crystals have the characteristic of manipulating wave propagation and thus both the noise control and energy harvesting can be achieved simultaneously. The plane wave expansion method is employed to calculate the resonant frequency of the sonic crystal and obtain the pressure and particle velocity field of the defect mode at the resonant frequency. The acoustic waves at the resonant frequency can be localized in the cavity of the sonic crystal. The piezoelectric curved beam is placed at the position with the highest air particle velocity of vibrates by oscillations of these air particles. The piezoelectric material then converts the acoustic energy to electric energy at the sonic crystal's resonant frequency. The larger voltage output of the piezoelectric material is associated with the larger pressure in the cavity of the sonic crystal. The external load resistance of the maximum output power is dependent on the vibration frequency and the capacitance of piezoelectric layer. The output power is dependent on the input level of the acoustic source. To improve the output power, piezoelectric materials with higher electromechanical coupling should be selected. In addition, the piezoelectric beam can be designed to have a resonant frequency which is the same as the resonant cavity and incident wave. Thus larger a power output should be reached. We also can design a new cavity to obtain the larger localizing efficiency.

In this chapter, our energy harvesting design only can work at a monochromatic frequency; a multi-frequency or broadband design should be developed and explored. Since the damping of the piezoelectric beam affects the vibrating amplitude, resonant frequency and harvestable energy, it is important to consider the damping effect in further work. Further studies of the acoustic energy generator that harvests environmental acoustic energy are needed before the environmental acoustic energy generator can be fully realized.

References

1. Sodano HA, Inman DJ, Park G (2004) A review of power harvesting from vibration using piezoelectric materials. Shock Vib Dig 36(3):197–205
2. Anton SR, Sodano HA (2007) A review of power harvesting using piezoelectric materials (2003–2006). Smart Mater Struct 16(3):R01–R21

3. Elvin NG, Elvin AA, Spector M (2001) A self-powered mechanical strain energy sensor. Smart Mater Struct 10(2):293–299
4. Lu F, Lee HP, Lim SP (2004) Modeling and analysis of micro piezoelectric power generators for micro-electromechanical-systems applications. Smart Mater Struct 13(1):57–63
5. Jiang S, Li X, Guo S, Hu Y, Yang J, Jiang Q (2005) Performance of piezoelectric bimorph for scavenging vibration energy. Smart Mater Struct 14(4):769–774
6. Sodano HA, Park G, Inman DJ (2004) Estimation of electric charge output for piezoelectric energy harvesting. Strain 40(2):49–58
7. Erturk A, Inman DJ (2008) A distributed parameter electromechanical model for cantilevered piezoelectric energy harvesters. J Vib Acoust 130:041002
8. Taylor GW, Burns JR, Kammann SM, Powers WB, Welsh TR (2001) The energy harvesting eel: a small subsurface ocean/river power generator. IEEE J Ocean Eng 26(4):539–547
9. Priya S, Chen CT, Fye D, Zahnd J (2005) Piezoelectric windmill: a novel solution to remote sensing. Jpn J Appl Phys 44:L104–L107
10. Priya S (2005) Modeling of electric energy harvesting using piezoelectric windmill. Appl Phys Lett 87:184101
11. Horowitz SB, Sheplak M, Cattafesta LN III, Nishida T (2006) A MEMS acoustic energy harvester. J Micromech Microeng 16(9):S174–S181
12. Liu F, Phipps A, Horowitz S, Ngo K, Cattafesta L, Nishida T, Sheplak M (2008) Acoustic energy harvesting using an electromechanical Helmholtz resonator. J Acoust Soc Am 123(4):1983–1990
13. Wang X, Song J, Liu J, Wang ZL (2007) Direct-current nanogenerator driven by ultrasonic waves. Science 316:102–105
14. Guigon R, Chaillout JJ, Jager T, Despesse G (2008) Harvesting raindrop energy: theory. Smart Mater Struct 17(1):015038
15. Guigon R, Chaillout JJ, Jager T, Despesse G (2008) Harvesting raindrop energy: experimental study. Smart Mater Struct 17(1):015039
16. Kushwaha MS (1996) Classical band structure of periodic elastic composites. Int J Mod Phys B 10:977–1094
17. Kushwaha MS, Halevi P (1996) Giant acoustic stop bands in two-dimensional periodic arrays of liquid cylinders. Appl Phys Lett 69:31–33
18. Kushwaha MS, Djafari-Rouhani B (1998) Giant sonic stop bands in two-dimensional periodic system of fluids. J Appl Phys 84:4677–4683
19. Sigalas MM (1997) Elastic wave band gaps and defect states in two-dimensional composites. J Acoust Soc Am 101:1256–1261
20. Sigalas MM (1998) Defect stated of acoustic waves in a two-dimensional lattice of solid cylinders. J Appl Phys 84:3026–3030
21. Wu F, Hou Z, Liu Z, Liu Y (2001) Point defect states in two-dimensional phononic crystals. Phys Lett A 292:198–202
22. Khelif A, Choujaa A, Djafari-Rouhani B, Wilm M, Ballandras S, Laude V (2003) Trapping and guiding of acoustic waves by defect modes in a full-band-gap ultrasonic crystal. Phys Rev B 68:214301
23. Wu F, Liu Z, Liu Y (2004) Splitting and tuning characteristics of the point defect modes in two-dimensional phononic crystals. Phys Rev E 69:066609
24. Zhang X, Dan H, Wu F, Liu Z (2008) Point defect states in 2D acoustic band gap materials consisting of solid cylinders in viscous liquid. J Phys D Appl Phys 41:155110
25. Zhao YC, Yuan LB (2009) Characteristics of multi-point defect modes in 2D phononic crystals. J Phys D Appl Phys 42:015403
26. Wu LY, Chen LW, Liu CM (2009) Acoustic pressure in cavity of variously sized two-dimensional sonic crystals with various filling fractions. Phys Lett A 373:1189–1195
27. Wu LY, Chen LW, Liu CM (2009) Experimental investigation of the acoustic pressure in cavity of a two-dimensional sonic crystal. Physica B 404:1766–1770
28. Kinsler LE, Frey AR, Coppens AB, Sanders JV (1999) Fundamentals of acoustics, 4th edn. Wiley, New York

29. Washington G (1996) Smart aperture antennas. Smart Mater Struct 5:801–805
30. Moskalik AJ, Brei D (1997) Quasi-static behavior of individual C-block piezoelectric actuators. J Intell Mater Syst Struct 8:571–587
31. Susanto K (2009) Vibration analysis of piezoelectric laminated slightly curved beams using distributed transfer function method. Int J Solids Struct 46:1564–1573
32. COMSOL 3.5a The COMSOL group. Stockholm, Sweden

Part V
Advances in Electronics

Chapter 13
Power Conditioning Techniques for Energy Harvesting

S.G. Burrow, P.D. Mitcheson, and B.H. Stark

Abstract Over the last 10 years the most dramatic progress in energy harvesting has concerned the power conditioning sub-system. Whilst the mechanical components and transducers have seen incremental improvements, often related to modelling and understanding behaviour, the power conditioning systems have undergone revolution: from simple peak rectifiers to complex topologies and control, with several examples integrated onto ICs. In this chapter the basic interactions of the dynamic mechanical and electrical system are described, and it is suggested that analysing the power factor at the input to the power conditioning system can be a useful tool in understanding behaviour. The benefits of electrical tuning are then described: this is an important topic, which the authors believe will be a key direction of future research. The chapter also gives an overview of published circuits for electromagnetic, piezoelectric and electrostatic transduction mechanisms.

13.1 Introduction

Rectification, filtering and voltage regulation are the minimum requirements to make the AC electrical output from a vibration energy harvester useable in the majority of applications. The drive to maximise the usable power from energy harvesting sources has seen researchers optimise these sub-systems through the adoption of more complex implementations, such as synchronous rectifiers, switched-mode converters and peak power trackers. In recent published work,

---

S.G. Burrow (✉) • B.H. Stark
University of Bristol, Faculty of Engineering, Queen's Building, University Walk, Clifton BS8 1TR, UK
e-mail: Stephen.Burrow@bristol.ac.uk; bernard@bristol.ac.uk; stark@bristol.ac.uk

P.D. Mitcheson
Imperial College London, Department of Electrical and Electronic Engineering, Exhibition Road, South Kensington, SW7 2AZ, London
e-mail: paul.mitcheson@imperial.ac.uk

N. Elvin and A. Erturk (eds.), *Advances in Energy Harvesting Methods*,
DOI 10.1007/978-1-4614-5705-3_13,

researchers have gone further and considered the overall power harvesting system, mechanical and electrical, as a coupled dynamic system, enabling frequency tuning and compensation for transducer output impedance [1, 2].

Optimising a vibration harvester system is a multi-physics problem and whilst these complex and highly non-linear coupled electro-mechanical systems can be modelled numerically, such simulations often provide little insight into design. For those with an electrical background, the best starting point is often to exploit electrical analogies to express the mechanical system as electrical components and thus the problem can be tackled with the techniques of analogue circuit analysis [3]. Conversely mechanical engineers may prefer combining analytical approximations to the sub-system behaviour. Either way, due regard must be paid to the non-idealities in each domain, which can be significant.

In this chapter the basic considerations for describing the harvester as a power source are laid out and the commonly used peak rectifier circuit described, illustrating the complexities of cross-domain analysis. Then the chapter considers a range of electrical power conditioning sub-systems currently reported for energy harvesting.

13.2 System Interactions

13.2.1 Loading the Harvester

Along with the source vibration, the input impedance of the power conditioning subsystem determines the operating point of the mechanical harvester system. This input impedance will be synthesised by the actions of the passive and active components of the power conditioning sub-system, in many cases resulting in non-linear or discontinuous impedance. Additionally, the application of feedback control allows negative impedances (negative resistance, negative inductance/capacitance) to be synthesised.

The tools needed to study the interaction of harvester and power conditioning sub-systems follow from classical power electrical analysis. By considering the real power flow between harvester and power conditioning system, described by peak power operation, and the reactive power flow, characterised by distortion and displacement power factor components, the behaviour of any harvesting system can be determined.

13.2.2 Real Power Flow: Peak Power Operation

The goal of the harvesting system is typically to produce the maximum useable output power, and since small-scale harvesters are not thermally limited, this typically means operating around some peak power point. The special case of a source with resistive output impedance and a resistive load is well known as the peak

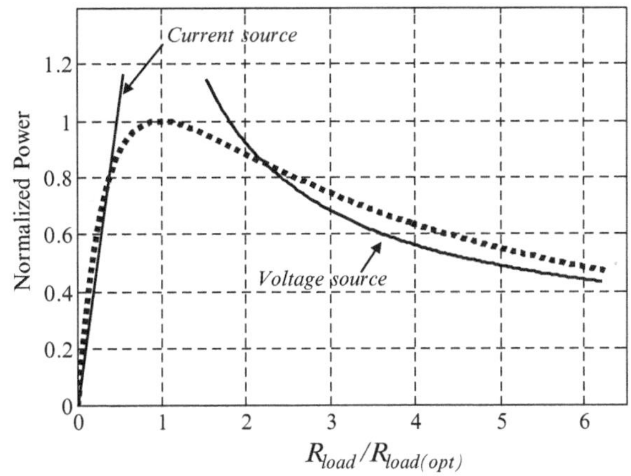

Fig. 13.1 General peak power curve with characteristics of current and voltage sources superimposed

power transfer theorem, but the general characteristic—that peak power is achieved with some value of real load impedance between zero and infinity—is applicable to all power conditioning circuits. Even where the energy harvester has a complex output impedance, e.g. when excited away from mechanical resonance or due to transducer parasitic impedance, or where the load is non-linear, e.g. a peak rectifier, the characteristic described in Fig. 13.1 will apply.

This characteristic takes on a particular importance when considering how to control power flow from the harvester, particularly for voltage regulation schemes. Many power conditioning circuits use feedback control to match input power to the power drawn by the load: a voltage regulation feedback loop will adjust converter input impedance (thus input power) to maintain a fixed output voltage against changes in load current. However, when operating a system in the region of the peak power point, the relation between impedance and power is not monotonic and such feedback would be unstable. A similar problem can occur in switched-mode converters with input filters, and thus designers will typically attempt to ensure the output impedance of the filter is at least 4 times less than the input impedance of the converter [4]. Applying this work-around to energy harvesters would result in operation significantly below maximum peak power.

Since it is difficult to use the synthesised input impedance of the converter to regulate harvester output power (thus system output voltage), a better approach may be to use a peak power tracking algorithm to maintain the power condition system input impedance at the optimum value for peak power, or indeed set the converter to a fixed input impedance as in [5], followed by an additional dissipative voltage regulation stage.

13.2.3 Reactive Power

In AC systems the concept of reactive power is used to describe the energy re-circulating between source and load each cycle; this is contrasted with real power, which describes the energy dissipated each cycle. Power factor relates real and reactive power for a particular load. Two mechanisms produce reactive power, the first is energy storage elements in the load, e.g. capacitive or inductive reactance, introducing a phase shift between current and voltage waveforms. In this case, the resulting power factor is referred to as 'displacement power factor'. The second occurs when frequency components are present in either current or voltage waveforms of the load that are not present in the other; the resulting power factor is described as 'distortion power factor'. Distortion power factor can result from the introduction of harmonics due to non-linearity or switching frequency components unrelated to the fundamental. Both can be seen in energy harvesting applications, but whilst it may be beneficial to introduce displacement power factor, having obvious positive benefits for frequency tuning (described later) or to 'resonate out' transducer parasitic impedance terms, distortion power factor is normally a negative consequence of switching circuits, tolerated since it permits efficient conversion.

Many power conditioning circuits will introduce both displacement and distortion components, and since distortion components can sometimes mask displacement components it may be helpful to perform frequency analysis on load waveforms to reveal the true behaviour of power conditioning sub-systems.

13.3 Rectification

13.3.1 Harvester/Peak Rectifier Behaviour

The ubiquitous 'peak rectifier' (diodes followed by a reservoir capacitor, C_r) is an arrangement widely used as the first power conditioning stage in energy harvesting applications, and yet, despite their apparently simple nature, the analysis of such these systems is complex. Analytical approaches applied to even the basic arrangement shown in Fig. 13.2a yield transcendental equations as both input and output voltages of the rectifier vary with time, making the conduction period difficult to define; this is illustrated in Sen [7] for the case of a low impedance source.

Additionally the rectifier of a vibration energy harvesting system is required to load the harvester near the peak power point, implying an apparent load impedance of similar magnitude to the source impedance. This operating condition is not commonly encountered although there are parallels in the analysis of vacuum diodes by Waidelich [6]. Waveform sketches depicting a peak rectifier fed from a low-impedance source and a source with impedance of similar magnitude to the load are shown in Fig. 13.2b, c. Both waveforms clearly illustrate the significant distortion components introduced by the rectifier, introducing reactive power flow.

Further adding to the complexity of analysis is the fact that these harmonic-rich waveforms created by the non-linear diode response prevent reduction of the harvester mechanical equivalent circuit to a Thevenin (or Norton) source: the full equivalent circuit, as illustrated in Fig. 13.3, should be considered. The most convenient way to analyse this system is with numerical techniques.

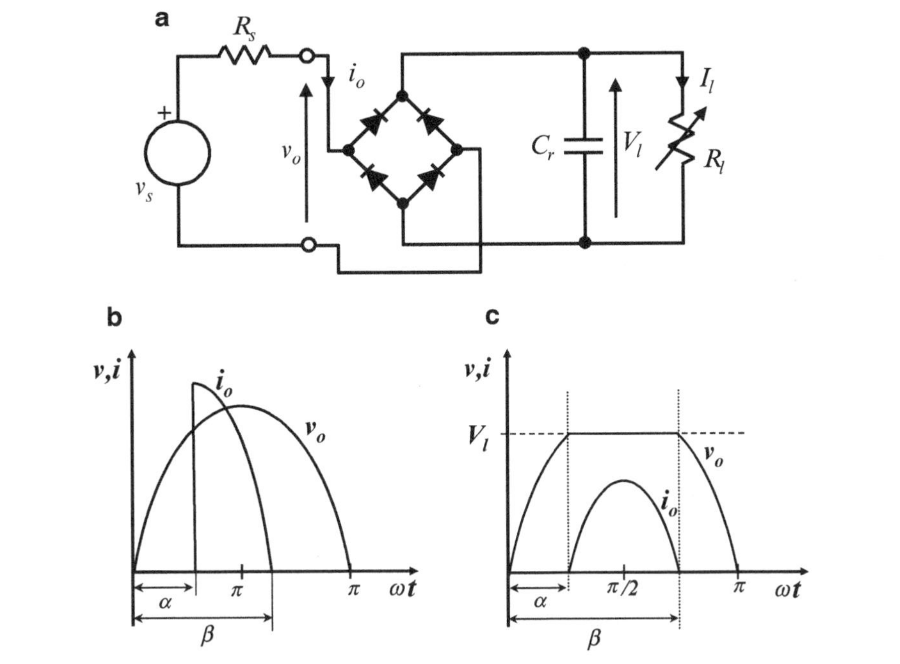

Fig. 13.2 Peak rectifier and associated waveforms for differing source impedances. (**a**) Voltage source feeding a peak rectifier. (**b**) Source impedance << Load impedance. (**c**) Source impedance ≈ Load impedance [6]

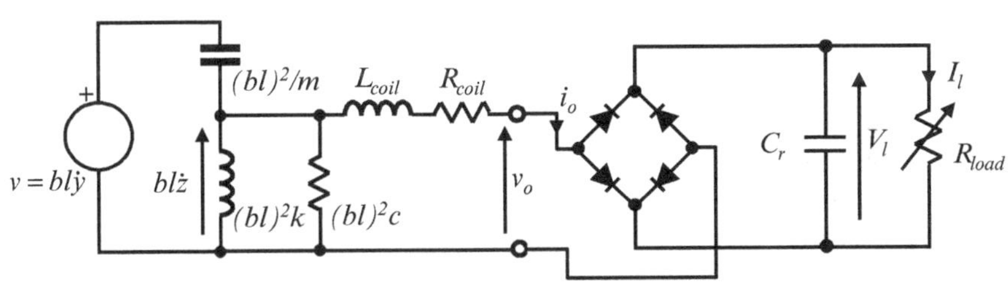

Fig. 13.3 Equivalent circuit of an electromagnetic harvester and peak rectifier

To illustrate the behaviour of the peak rectifier, Fig. 13.4 [8] plots results of a numerical simulation of average load power against load resistance for a harvester connected to three differing loads: (1) variable load resistance only (no rectifier); (2) bridge rectifier and variable load resistance (i.e. no capacitor, C_r); and (3) bridge rectifier with variable load resistance and reservoir capacitor, C_r (where C_r is large so that the ripple component V_l is negligible).

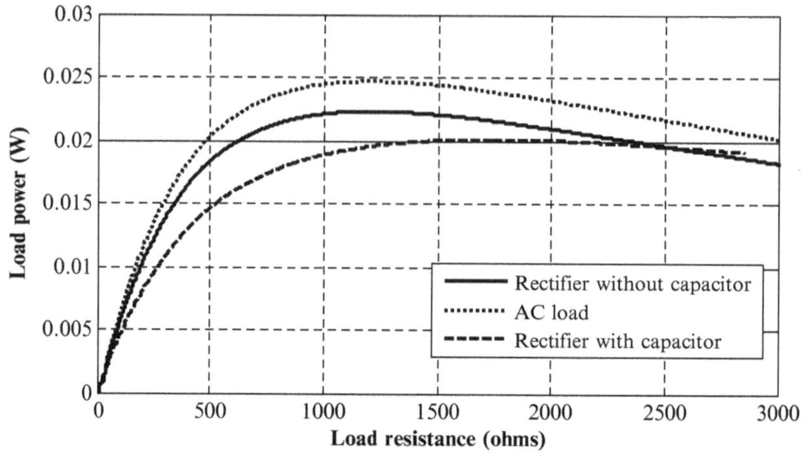

Fig. 13.4 Numerical simulation of harvester with (1) ac resistive load, (2) feeding a resistive load through a rectifier (i.e. C_r omitted) and (3) feeding a peak rectifier, $C_r = 2{,}000\ \mu F$ [8]

The first case with resistance only load acts as a benchmark, and indicates that for this harvester peak power is delivered into a resistive load of around 1,250 Ω. The addition of the rectifier without a reservoir capacitor reduces the magnitude of the power available for a given load resistance, but the response is of similar form with peak power delivered into 1,250 Ω. The reason for the reduction of power compared to the simple resistive load is losses, primarily conduction losses due to forward voltage drop, in the rectifier diodes.

The addition of the reservoir capacitor after the rectifier changes both the load at which maximum power is achieved and value of maximum power. The increase in load for maximum power occurs since the action of the capacitor, C_r, results in a higher average voltage appearing across the output of the rectifier, compared to the input. Thus for a given power the load resistance on the DC side must be greater than a notional load connected on the AC side. The reduction in maximum power when using a rectifier and reservoir capacitor is due to in part to diode losses and also to conduction losses associated with the distortion components of the load current.

By equating the rectifier output voltage, V_l, and load resistance at various load power levels, it can be shown that rectifier output voltage is a proxy for the impedance seen by the harvester with $V_l = 0$ appearing as a short circuit and $V_l = v_{o(peak)}$ appearing as an open circuit. It then follows that it is possible to control the power output of the harvester by controlling the rectifier output voltage. This approach has been used by several authors including Ottman et al. [9] who used a boost converter to control the voltage across a peak rectifier circuit and thus the loading on a piezoelectric harvester.

13.3.2 Diode Losses: Forward Drop and Reverse Leakage

When designing passive rectifiers for low-power vibration harvesting applications, two important loss mechanisms, forward voltage drop and reverse leakage, need to be considered.

The diode forward voltage drop, which leads to power dissipation in the diode during conduction, causes a reduction in rectifier efficiency related to the ratio of diode voltage drop to rectifier input voltage. Thus losses due to diode voltage drop become significant at lower harvester output voltages. Harvesters employing electromagnetic transduction typically have low output voltages, hence Schottky junction diodes, with lower drop compared to pn junction diodes, are often favoured as the rectifying elements in these applications.

Harvesters using piezoelectric transduction typically feature higher output voltages and, at the low-output power levels of many harvesters, reverse leakage can become a significant source of power loss. Reverse leakage, where a small current flows whilst the diode is reverse biased, increases with applied voltage and is greater for Schottky devices than silicon, hence, conversely, pn junction diodes are often preferred with peizoelectric transducer.

13.3.3 Synchronous Rectifiers

Within the field of power electronics a well-established practice is the use of active switching elements over passive diodes. With carefully chosen device characteristics, significantly lower conduction losses can be achieved. The reduction in losses must be balanced against the power consumed by the control/driver electronics for the active switch.

Figure 13.5 shows a circuit implementing a half-wave rectifier from discrete components [8]. This circuit illustrates the concept of an active switching device, in this case a MOSFET, commutated by drive circuitry measuring the potential across the device. Figure 13.6 shows experimental results comparing active rectifiers with passive devices taken from [8]. The effect of the quiescent power consumption of the active synchronous circuit reducing efficiency at low-power levels can clearly be seen.

13.4 Switched-Mode Power Conditioning Circuits

The widely used peak rectifier circuit is a simple way to implement the first stage of an energy harvesting power conditioning system, but it is unable to provide any voltage level conversion and it presents a non-linear load to the harvester, introducing distortion components to the load current. Linear voltage regulators may be added after the peak rectifier to provide conditioning but these result in

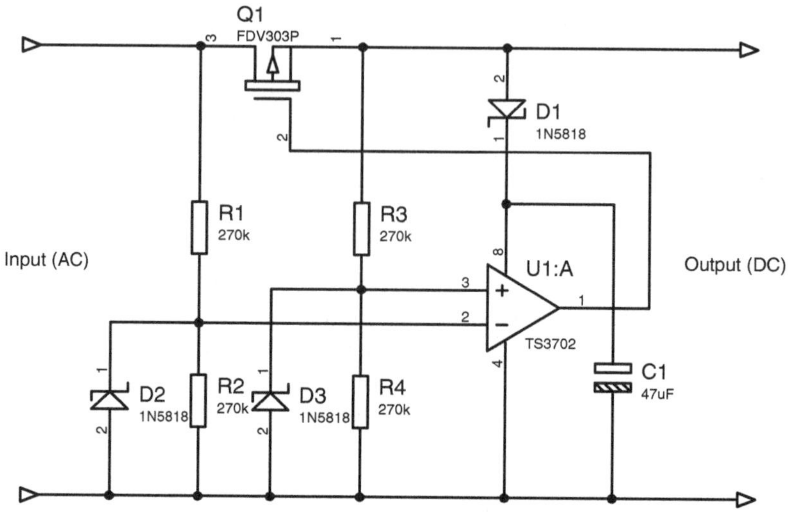

Fig. 13.5 Half-wave synchronous rectifier circuit realised from discrete components [8]

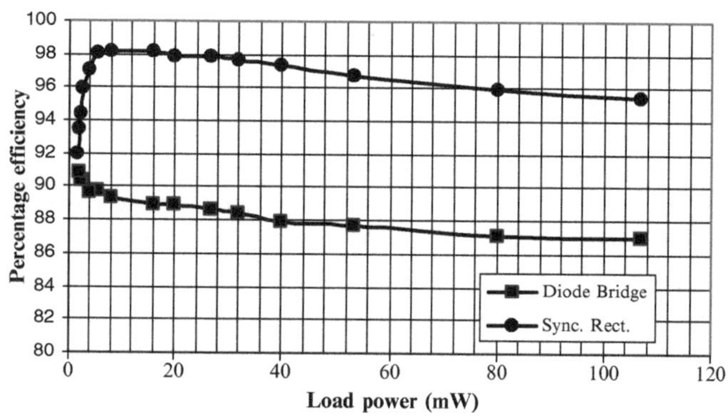

Fig. 13.6 Comparison of passive and synchronous rectifier efficiencies, measured at 50 Hz input frequency with constant output voltage of 4 VDC [8]

highly inefficient power conversion and can only regulate to a value below the input. To facilitate efficient voltage conversion and provide rectification with minimal distortion components, we need to introduce switched-mode power converters.

Switched-mode circuits act as true 'power converters' transforming power at the input voltage level to an equivalent power at the desired output voltage. The high frequency switching action, from where this class of circuits derive their name, allows high conversion efficiencies at the cost of additional circuit complexity and quiescent power demand, for example to power gate-drive circuits. Another benefit is their ability to synthesise a linear input impedance, which can be readily controlled; this can be varied to track the harvester's maximum power point under varying source conditions or to adapt to changes in the load demand.

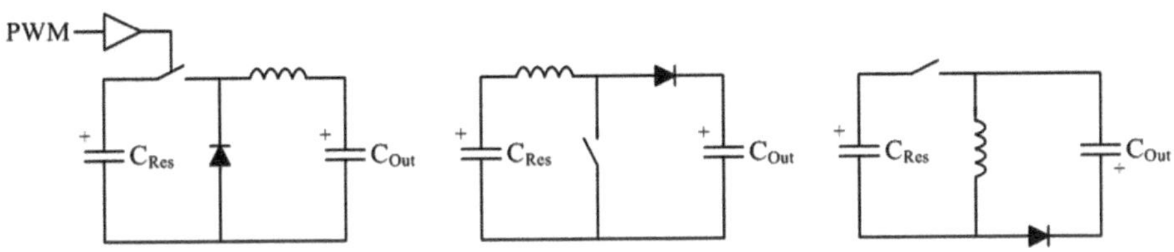

Fig. 13.7 From left to right: Buck, boost and inverting buck-boost converters

Many topologies exist; however, it is useful to consider two differing basic architectures for energy harvesting applications. In the first architecture a rectifier performs AC–DC conversion and a subsequent DC–DC switched mode converter provides voltage level conversion (this may employ a peak rectifier as in [9], with accompanying distortion components in the harvester output current, or appear as a resistance emulator as in [5], behaving similar to the rectifier without capacitor from Fig. 13.4). Alternatively switched mode circuit topologies that can provide direct AC–DC conversions (including voltage level shifting) can be used.

13.4.1 DC–DC Inductive Converters

The three most frequently used switched-inductor converter topologies are named with reference to their voltage gain, and shown in Fig. 13.7: The step-down (or buck) converter [10], step-up (or boost) converter [11], and step-up-down (or buck-boost) converter [12].

Power conversion efficiencies of 80–90% are regularly reported. The quiescent demand of the control and generation of the pulse width modulated (PWM) gate drive signals determines the lower limit on the power range, e.g. a quiescent power consumption of around 2 μW is reported in [13], resulting in high efficiencies at least down to 100 μW. Low-voltage inputs are additional strength of switched inductor circuits. The boost converter proposed in [11] achieves an efficiency of 60–70% at input voltages in the range of 50–250 mV at 10–100 μW power levels when connected to a 600 mV supply to drive the switches. The minimum input voltage reported is 20 mV with the control circuit consuming only 1 μW of quiescent power.

In contrast to higher power systems that usually operate under continuous inductor current with a certain ripple, switched-mode converters for energy harvesters are usually operated in Discontinuous Current Mode (DCM), where the switch is off long enough per switching cycle for the inductor current to fall to zero. This can reduce switching losses [14], because of the lower switching frequencies and the elimination of reverse displacement current in the diode [15] or high-side MOSFET [11]. In addition, pulse frequency modulation techniques (PFM) have been shown to incur switching losses that scale with the output power levels, thus improving the efficiency at low power [11].

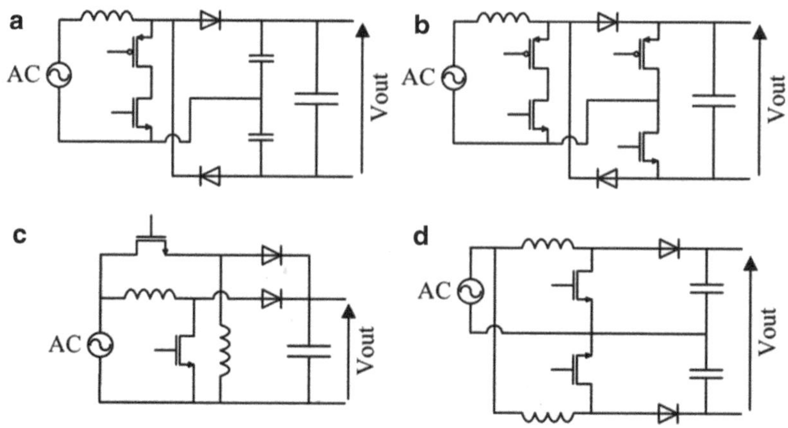

Fig. 13.8 Direct AC–DC converter topologies: (**a**) single inductor with split capacitor, (**b**) single inductor with secondary side switches [20], (**c**) combined boost–buck-boost [21] and (**d**) dual-boost converter [14]

13.4.2 DC–DC Switched-Capacitor Converters

Switched capacitor converters transfer charge between capacitors via switches without the use of inductors, facilitating integration of the converter. DC–DC switched-capacitor converters, or charge pumps, use actively controlled switches to reconfigure capacitor networks, thereby pumping charge up to higher potentials. This is opposed to switched-capacitor circuits that use passive diodes and therefore require an AC input, as for example voltage multipliers. Milliwatt-level, regulated output DC–DC charge pumps have been reported, with efficiencies of 85–90% [16] and quiescent current consumption of around 120 μA (at an input voltage of 1.8 V) [17], depending on the switching frequency. Self-contained cold-start from low-input voltages of around 200 mV is possible, as shown in [18], where a combination of inductive and capacitive voltage boosting results in an efficiency of 30–35% at 100–200 μW output power at 1.2 V from a 200 mV input.

13.4.3 Direct Switch-Mode AC–DC Converters

Direct AC–DC converters are switch-mode converters that require no separate rectification stage. These topologies usually contain one or more switched inductors. They can operate at very low input voltages and offer more effective current and voltage regulation than switched-capacitor converter topologies.

Salmon offers a topological overview and comparison of switch-mode boost rectifier topologies in [19]. Figure 13.8 shows four example converter topologies. In this figure, circuits (a) and (b) are single-inductor topologies proposed in [20], where (a) is based on split capacitors and (b) relies on additional MOSFET switches, referred to as secondary side switches. Both topologies achieve conversion efficiency in the range of 60% at an input power level of around 100 mW; however, the control circuit for the split-capacitor architecture is simpler due to switching devices and no added requirements for sensing the polarity of the input voltage.

Circuit (c) is a combination of a boost and a buck-boost converter [21], and (d) is a combination of two boost converters [14]. Both concepts rely on alternate operation of the converters during the negative and the positive input voltage cycles. The reported efficiencies for the boost–buck-boost topology is 61% when connected to a 100 mW source, whilst the dual-boost converter achieves a power conversion efficiency of 44% based on simulation results with an input power of 49 mW.

13.5 Electrical-Domain Frequency Tuning

So far the ability of power conditioning circuits to provide rectification and voltage step-up/step-down in order to interface the AC harvester to a DC load or storage device has been described, along with their ability to control the damping force seen by the harvester through a synthesised input resistance. However, some power conditioning circuit architectures are additionally able to synthesise reactive load impedance and thus introduce displacement power factor, previously described. Loading an energy harvester with a reactive load is particularly useful in two circumstances. Firstly, where the frequency of the vibration source exciting the harvester varies over time, and secondly where the output impedance of the harvester has imaginary components even when excited at mechanical resonance. In both cases the reactive component of the load impedance allows the overall system dynamics to be altered, maximising power generation for a particular condition.

Modifying overall electromechanical system dynamics by the introduction of *passive* reactive load components has been widely reported in literature dealing with vibration isolators/absorbers [22, 23] and has application to energy harvesting where the goal is to compensate for reactive components of the harvester output impedance at fixed excitation frequencies. Modifying the system dynamics with active electronics synthesising a reactive load presents great opportunities to adaptively and automatically re-tune a harvesting system.

The theoretical basis for electrical tuning of an energy harvester together with experimental verification of static load points is described in [1]. The phenomenon can be understood by looking at the equivalent circuit of an energy harvester, including the mechanical system, transducer and load circuit, shown in Fig. 13.9.

The mechanical system of the harvester is shown on the primary side of the transformer and the electrical part of the harvester (including the inductance and series resistance of the transduction mechanism) is shown on the secondary. It can be seen that the application of reactive electrical components create a fourth order

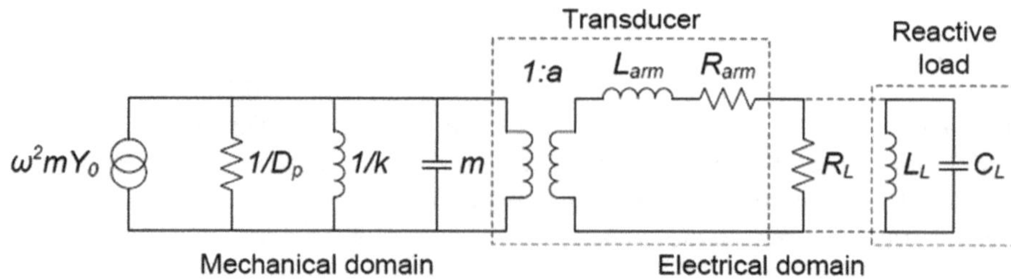

Fig. 13.9 Electrical equivalent circuit of an electromagnetic harvester with both resistive and reactive load shown [2]

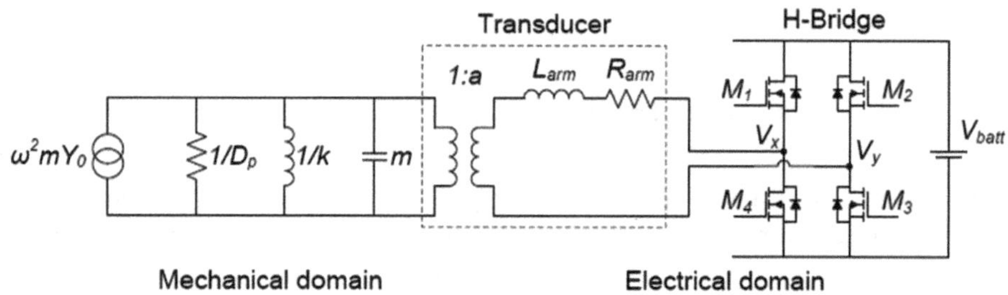

Fig. 13.10 Electromagnetic harvester with power electronic interface to a battery [2]

resonant system (in practice L_L and C_L are resolved to a single reactive element) and modifying them allows the resonant frequencies of the system to be altered.

It is possible to use several topologies of switched mode converter to synthesise load resistance, including the bi-directional boost rectifier illustrated in Fig. 13.10. This circuit provides AC–DC conversion in a single stage, as well as providing the possibility of emulating a variable reactive load.

A prototype of this system has been constructed using a large inertial energy harvester designed for harvesting energy from a rocking boat [2]. The circuit was constructed using discrete components and the control of the switches was accomplished using a PIC microcontroller. The output power from the system when different reactive loads are being emulated is shown in Fig. 13.11 [24] as a function of driving frequency.

As can be seen in Fig. 13.11, the resonant frequency of the generator has been successfully modified by the addition of load reactances. With the load emulating a positive capacitance, the resonant frequency of the system drops and as can be seen, the power generated at low frequencies is increased over the condition the emulated reactance is zero. In contrast, when the interface emulates a negative capacitance, the power output at high frequencies is increased. This interface also has the advantage of rectifying the generated energy and storing it in a battery.

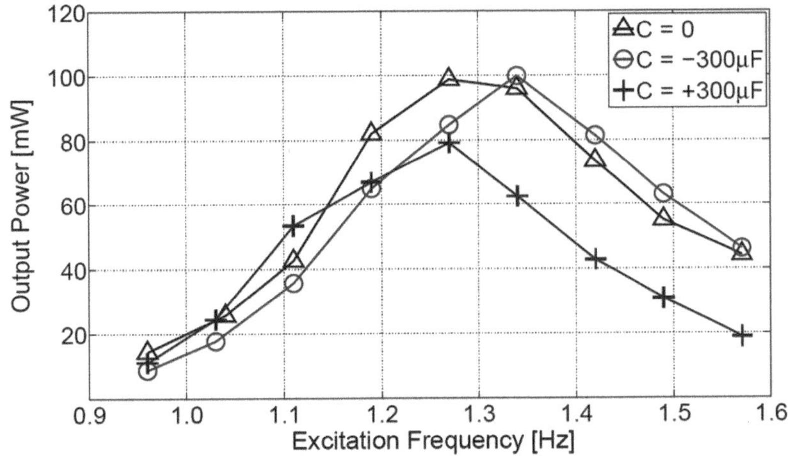

Fig. 13.11 Output power of the harvester when the interface circuit emulates different reactive loads [24]

13.6 Interface Circuits for Piezoelectric Harvesters

The simple equivalent circuit model for a piezoelectric transducer is a current source in parallel with a capacitance. The source current is proportional to the relative velocity of the piezoelectric material, and the shunt capacitance is the clamped capacitance of the piezoelectric beam. An important effect of this shunt capacitance is to limit the maximum damping that can be achieved, irrespective of the external load resistance (external damping) applied. Under many circumstances this results in an inability of a piezo transducer to operate a resonant vibration harvester at the maximum limit of power generation. Referring to Fig. 13.1, this implies the harvesting system is always operating in the under damped, voltage source region. In some applications the damping seen by the mechanical system from the electrical side is so small and the effects of backward coupling can be ignored altogether.

Where the internal capacitance of the piezo element forces the system to operate in the voltage source region, increasing the electrical damping will always increase output power: however, it must be noted that this power is much less than could be achieved by the resonant system if optimally damped. Limitations caused by internal capacitance become less significant as the frequency increases; however, the low frequencies of many machine vibrations or human motion present challenges for piezo transducers. Simply using more piezo material does not help the problem since the increase in coupling coefficient is accompanied by an increase in shunt capacitance.

At first glance the obvious approach to mitigate for device capacitance is to introduce load inductance, similar to the technique described in the previous section; however, quick calculation will satisfy the reader that a physical inductor would

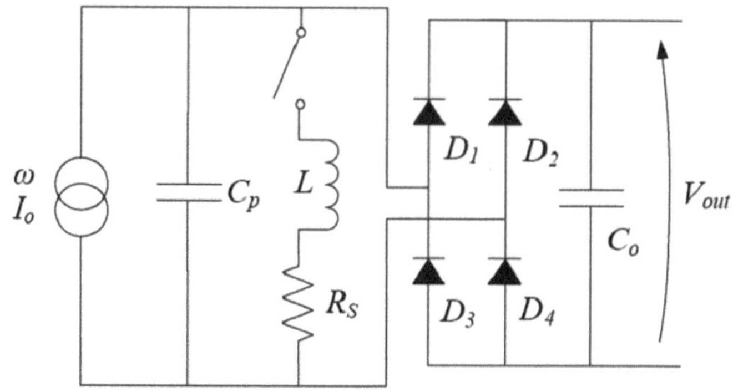

Fig. 13.12 Original SSHI circuit with DC output [25]

be prohibitively large and synthesising a linear inductance with an electronic converter would result in high terminal voltages and a low-power factor, incurring additional losses. To avoid these drawbacks, several researchers have developed non-linear electrical switching techniques, which introduce reactive load current (providing 'tuning' to compensate for device capacitance) with minimal additional components. The first implementation of an interface circuit which increased the effective coupling in this way was demonstrated in [25] by Guyomar et al. The technique, termed *synchronous switched harvesting on inductor* (SSHI) by the authors, involves flipping the polarity of the developed charge on the piezoelectric material twice per cycle when the cantilever reaches its maximum displacement. The circuit that achieves this is shown in Fig. 13.12. Frequency analysis of the load current and voltage waveforms would show that the fundamental components are displaced relative to one another, mimicking the action of physical load inductance.

When the piezoelectric cantilever beam reaches maximum deflection in either direction, the charge on C_p is flipped through the switch and inductor arrangement. This has the effect of increasing the maximum voltages present on the capacitor and ensuring that the force on the cantilever is always in a direction which opposes the relative motion of the cantilever tip, increasing the potential for the transducer to do mechanical work and thus increases the amount of energy converted. The piezoelectric output is then rectified and smoothed using a standard bridge rectifier and capacitor.

Recently a new technique has been demonstrated which, for a given constraint on circuit volume, outperforms the SSHI family of techniques. This new method is known as *single-supply pre-biasing* [26] and, rather than flipping the charge on the piezoelectric capacitance at the extremes of the motion, charge is extracted and re-injected with opposite polarity at the extremes of motion, using the circuit of Fig. 13.13.

The operation of the circuit is as follows: Switches S_1 and S_4 always operate as a pair, and S_2 and S_3 also operate as a pair. When the piezoelectric material reaches its maximum deflection, one of the pairs of switches is activated, discharging the energy on the piezoelectric capacitance into the DC link capacitor through the

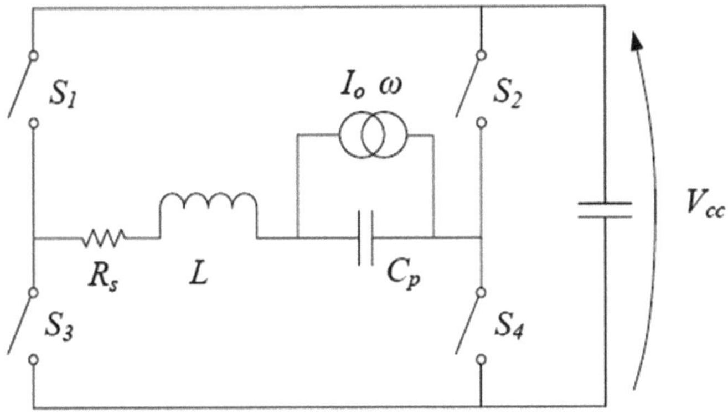

Fig. 13.13 Single supply pre-biasing circuit for a piezoelectric harvester [27]

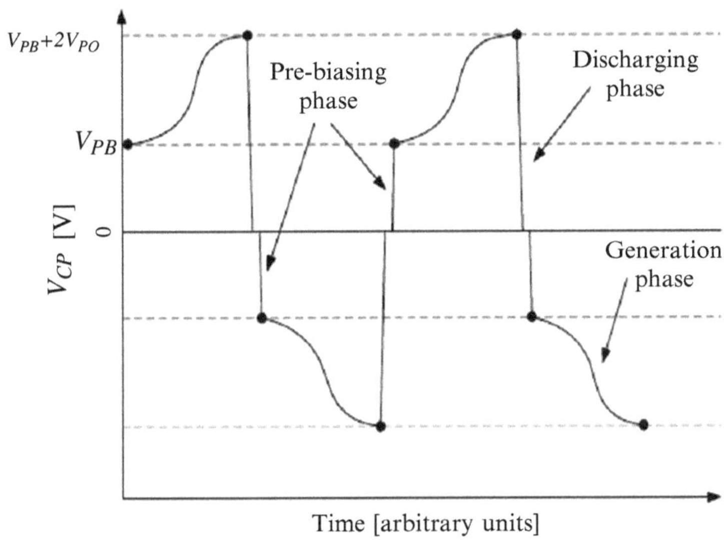

Fig. 13.14 Piezoelectric pre-biasing voltage waveforms [27]

series inductor. This corresponds to the discharging phase shown in Fig. 13.14. This resonant discharge pulse happens very quickly compared to the mechanical excitation frequency of the system. As soon as this discharge phase is complete and the voltage on the piezoelectric capacitor has reached zero, the opposite switch pair activates and injects some charge onto the piezoelectric capacitor, of opposite polarity to what had just been generated (the pre-biasing phase in Fig. 13.14). This increases the force with which the transducer is able to oppose the relative motion between mass and base, thus increasing the electrical damping. The piezoelectric material then moves to its opposite extreme of position, increasing the voltage (see the generation phase in Fig. 13.14) and the process repeats.

The ultimate limit of the maximum power extraction by this circuit from a piezoelectric material has been shown in [27] to be

$$P = \frac{4Q}{\pi^2} \frac{{I_0}^2}{\omega C_p}$$

where Q is the quality factor of the charge and discharge path, ω is the angular frequency of generator mechanical excitation, C_p is the clamped capacitance of the piezoelectric material and I_0 is the magnitude of the current produced by the bending of the piezoelectric material. This is a factor of $8Q/\pi$ greater than the maximum power that can be extracted through a bridge rectifier and a factor of two greater than the original SSHI technique. Rectification and energy storage are built-in features of the single supply pre-biasing technique. As with any system that uses actively controlled switches, there is some control overhead power and so for any of the piezoelectric charge modification techniques to generate more net power, the power required by the control circuit should be minimised.

13.7 Interface Circuits for Electrostatic Harvesters

Electrostatic energy harvesters use a variation in capacitance between two charged electrodes to convert energy from a mechanical to electrical form, as work can be done against the attractive force between the two capacitor plates. Normally, this is achieved by physical motion of the electrodes relative to each other [28–30] although examples of structures with moving dielectrics have been reported [31]. There are two common ways to operate such transducers: constant voltage and constant charge [32]. These terms refer to the constraint placed on the electrodes during the part of the cycle when the capacitance decreases, which is the part of the cycle when electrical energy is generated.

The basic requirements for the interface circuit to operate an electrostatic harvester are therefore:

- Pre-charge the electrodes whilst they are at their maximum capacitance.
- Keep the electrodes connected to a constant voltage source, or open circuit during electrode separation, depending on the chosen mode of generator operation.
- Extract the energy into a storage element.

The simplest power electronic topologies that are capable of performing these actions are shown in Fig. 13.15 for constant charge and constant voltage modes.

Figure 13.15a shows the simplest interface circuit for a constant charge mode transducer, highlighting the unavoidable parasitic components present in the MOSFET switches. The variable capacitor, C_{var}, is primed from the battery, V_{supply}, whilst at maximum capacitance, with MOSFETs M_1 and M_2 acting as the switches in a synchronous boost converter, transferring energy from the battery to C_{var}.

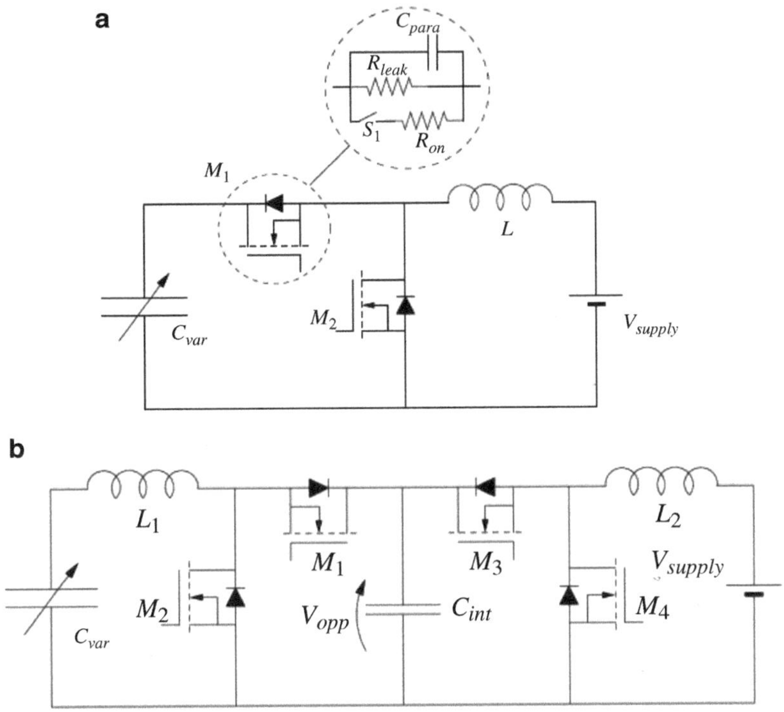

Fig. 13.15 Interface circuit topologies for electrostatic harvesters [33]. (**a**) Constant charge mode interface. (**b**) Constant voltage mode interface

The capacitance of the electrodes then reduces when both MOSFETs are switched off, increasing the voltage on C_{var}. The energy is then transferred from the capacitor to the battery by the action of the MOSFETs, this time acting as a buck converter and transferring power in the opposite direction to the pre-charge phase.

Figure 13.15b shows the interface circuit for operating the generator in constant voltage mode. An intermediate high voltage source (represented by the capacitor C_{int}) is used as the high voltage source, which keeps C_{var} at constant voltage during the energy conversion phase. C_{int} is pre-charged before harvester operation by the boost action *M3 and M4.* At the start of the generation cycle, *M1* and *M2* act as a buck converter to pre-charge the variable capacitor through *L1*. When C_{var} has reached the same value as V_{int}, *M1* is held on and *M2* is held off, clamping the variable capacitor voltage to the same potential as V_{int}. The capacitance of the transducer reduces, pushing charge from the variable capacitor into the intermediate capacitance at near constant voltage. As the intermediate voltage is generally far in excess of the low voltages used to drive a low-power electronic load circuit, *M3* and *M4* transfer the generated energy through *L2* in a buck operation into the low-voltage battery.

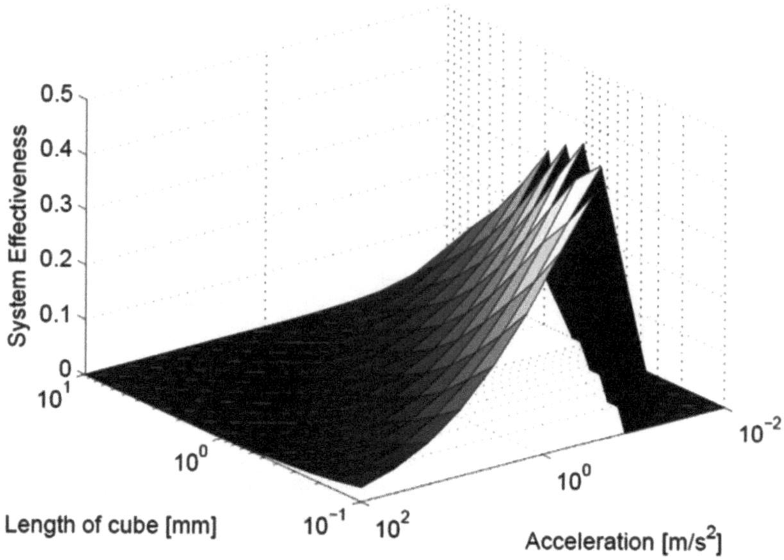

Fig. 13.16 Effectiveness of a constant charge harvester operated at 1 kHz [33]

The basic requirement for the electrostatic harvester to operate optimally is that the charge (in constant charge mode) or the voltage (in constant voltage mode) is set so that the optimal damping force (which allows maximum power to be coupled into the generator) is achieved. Under typical harvester acceleration inputs and for typical dimensions, the required voltages can easily reach hundreds of volts, as is the case also in MEMS electrostatic actuators. Whilst the interface circuits in Fig. 13.15 are conceptually simple, and if operated at power levels of hundreds of mW or above could readily achieve high efficiency, the need to operate them at very low power with high voltage blocking capability and small quantities of transducer charge presents a significant challenge [34].

In the constant charge case, it is important that when the capacitance of the transducer reduces, there is a minimal amount of charge sharing or charge leakage through M_1 (Fig. 13.15); otherwise constant charge mode operation is not achieved, the charge on the transducer falls below that which achieves the optimal damping force and the energy generated is reduced. Clearly, reducing the cross-sectional area of M_1 will reduce charge sharing effects, but a trade-off is that conduction losses in the circuit increase and thus an optimal device area exists. Recently, it has been shown that the maximum effectiveness of a constant charge energy harvester connected to an interface circuit, with optimised semiconductor devices (in terms of blocking voltages and cross-sectional area), is relatively poor for most regions of operation, with a small region where the system performance is acceptable [33]. The effectiveness of the system, as a function of generator size and input acceleration, is shown in Fig. 13.16. As can be seen, there is a narrow operating envelope where the

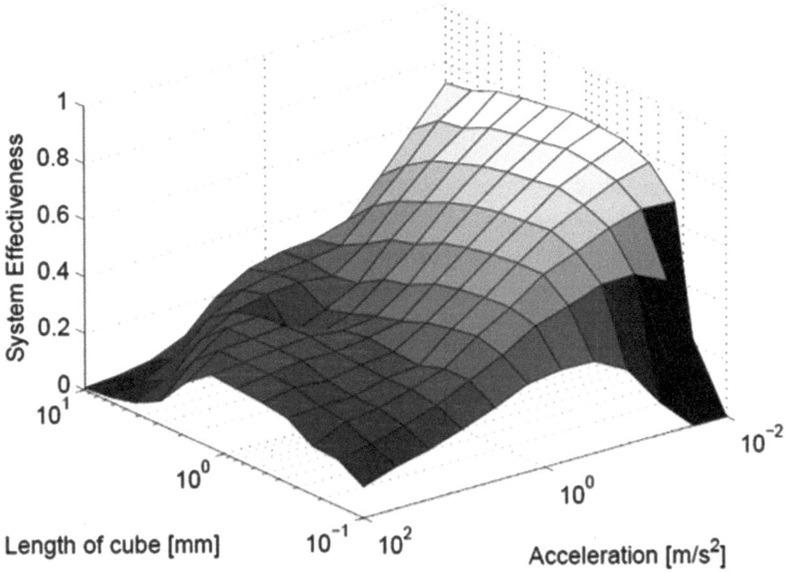

Fig. 13.17 Maximum effectiveness of constant voltage harvester operated at 1 kHz [33]

constant charge mode of operation can achieve an effectiveness of over 20%. The same optimisation has been done also on the constant voltage mode generator and the result is shown in Fig. 13.17.

As can be seen, the constant voltage mode is capable of operating at significantly better effectiveness than the constant charge mode of operation and is thus the preferred option. This is due to the fact that when operated at constant voltage, the transducer force does not diminish due to leakage currents in the system, and so an optimal electrical damping can be maintained throughout the generation stroke. It should however be noted that the constant voltage mode of operation requires the use of additional switches and the use of an electret [35], which increase complexity. To date, many examples of both types of harvester operations exist; however, most of the more recent work has concentrated on constant voltage devices using electrets.

13.8 Conclusion

All of the circuits described in this chapter reinforce the idea of a useful energy harvester as a coupled dynamic electromechanical system. Applying too simplistic an approximation, in either domain, results in design performance well below that which might be achieved. Compounding this is the fact that skills in the two emergent research topics in energy harvesting today, namely structural dynamics and low-power power-electronics design, are rarely found in the same research grouping. When these skill sets can be successfully combined, energy harvesting systems can fulfil their potential as viable power sources and exciting possibilities of autonomous, adaptive systems open up.

References

1. Cammarano A, Burrow SG, Barton DAW, Carrella A, Clare LR (2010) Tuning a resonant energy harvester using a generalized electrical load. J Smart Mater Struct 19:055003. doi:10.1088/0964-1726/19/5/055003
2. Wong KH, Toh TT, Mitcheson PD, Holmes AS, Burrow SG (2012) Tuning the resonant frequency and damping of an energy harvester using power electronics. IEEE Trans Circuits Syst II Express Briefs 59(1). doi:10.1109/TCSII.2011.2173966
3. Gardino P, Brennan MJ (2002) On the origins and development of mobility and impedance methods in structural dynamics. J Sound Vib 249(3):557–573
4. Middlebrook RD (1978) Design techniques for preventing input-filter oscillations in switched-mode regulators. In: Powercon proceedings, pp A3.1–A3.16
5. Burrow SG, Clare LR (2009) Open-loop power conditioning for vibration energy harvesting. Electron Lett 45(19):999–1000
6. Waidelich DL (1941) Diode rectifying circuits with capacitance filters. Trans AIEE 61:1161
7. Sen PC (2008) Power electronics. Tata McGraw-Hill, New Delhi. ISBN -13:978-0-07-462400-5
8. Clare LR, Burrow SG (2008) Power conditioning for energy harvesting. Proc SPIE 6928:69280A
9. Ottman GK, Hofmann HF, Bhatt AC, Lesieutre GA (2002) Adaptive piezoelectric energy harvesting circuit for wireless remote power supply. IEEE Trans Power Electron 17(5): 669–676
10. Tabesh A, Fréchette LG (2010) A low-power stand-alone adaptive circuit for harvesting energy from a piezoelectric micropower generator. IEEE Trans Industr Electron 57:840–849
11. Carlson EJ, Strunz K, Otis BP (2010) A 20 mV input boost converter with efficient digital control for thermoelectric energy harvesting. IEEE J Solid State Circuits 45:741–750
12. Lefeuvre E, Audigier D, Richard C, Guyomar D (2007) Buck-boost converter for sensor-less power optimization of piezoelectric energy harvester. IEEE Trans Power Electron 22: 2018–2025
13. Ramadass YK, Chandrakasan AP (2010) An efficient piezoelectric energy harvesting interface circuit using a bias-flip rectifier and shared inductor. IEEE J Solid State Circuits 45:189–204
14. Mitcheson PD, Green TC, Yeatman EM (2007) Power processing circuits for electromagnetic, electrostatic and piezoelectric inertial energy scavengers. Microsyst Technol 13:1629–1635
15. Dwari S, Dayal R, Parsa L (2008) A Novel direct AC/DC converter for efficient low voltage energy harvesting. 34th Annual IEEE Conference of Industrial Electronics (IECON), pp 484–488
16. Lee H, Mok PKT (2007) An SC voltage doubler with pseudo-continuous output regulation using a three-stage switchable opamp. IEEE J Solid State Circuits 42:1216–1229
17. Lam YH, Ki WH, Tsui CY (2006) An integrated 1.8 V to 3.3 V regulated voltage doubler using active diodes and dual-loop voltage follower for switch-capacitive load. Symposium on VLSI circuits digest of technical papers, pp 85–86
18. Richelli A, Colalongo L, Tonoli S, Kovacs-Vajna ZM (2009) A 0.2−1.2 V DC/DC boost converter for power harvesting applications. IEEE Trans Power Electron 24:1541–1546
19. Salmon JC (1993) Circuit topologies for single-phase voltage-doubler boost rectifiers. IEEE Trans Power Electron 8:521–529

20. Dwari S, Parsa L (2008) Efficient low voltage direct AC/DC converters for self-powered wireless sensor nodes and mobile electronics. 30th International Telecommunications Energy Conference (INTELEC), pp 1–7
21. Dwari S, Parsa L (2010) An efficient AC–DC step-up converter for low-voltage energy harvesting. IEEE Trans Power Electron 25:2188–2199
22. Kim J, Ryu YH, Choi SB (2000) New shunting parameter tuning method for piezoelectric damping based on measured electrical impedance. Smart Mater Struct 9:868–877
23. Davis CL, Lesieutre GA (2000) An actively tuned solid-state vibration absorber using capacitive shunting of piezoelectric stiffness. J Sound Vib 232:601–617
24. Toh TT et al (2011) Electronic resonant frequency tuning of a marine energy harvester. Proceedings of PowerMEMS, Seoul, Korea
25. Guyomar D, Badel A, Lefeuvre E, Richard C (2005) Toward energy harvesting using active materials and conversion improvement by nonlinear processing. IEEE Trans Ultrason Ferroelectr Freq Control 52:584–595
26. Dicken J, Mitcheson PD, Stoianov I, Yeatman EM (2012) Power-extraction circuits for piezoelectric energy harvesters in miniature and low-power applications. IEEE Trans Power Electron 27:4514–4529
27. Dicken J, Mitcheson P, Elliot A, Yeatman E (2011) Single-supply pre-biasing circuit for low-amplitude energy harvesting applications. Proceedings of PowerMEMS 2011, Soeul, Korea
28. Meninger S et al (2001) Vibration-to-electric energy conversion. IEEE Trans VLSI Syst 9: 64–76
29. Halvorsen E et al (2009) An electrostatic energy harvester with electret bias. Solid-state sensors, actuators and microsystems conference. TRANSDUCERS 2009. International, pp 1381–1384
30. Mitcheson PD et al (2004) MEMS electrostatic micropower generator for low frequency operation. Sens Actuators A Phys 115:523–529
31. Choi D et al (2010) Electrostatic energy harvester for low-frequency vibration by human physical motions using liquid. Proceedings of PowerMEMS, Belgium, pp 119–122
32. Mitcheson PD, Sterken T, He C, Kiziroglou M, Yeatman EM, Puers R (2008) Electrostatic microgenerators. Meas Control 41:114–119
33. Mitcheson PD, Green TC (2012) Maximum effectiveness of electrostatic energy harvesters when coupled to interface circuits. IEEE Trans Circuits Syst 1 Regul Pap 1–14. ISSN: 1549-8328
34. Stark BH, Mitcheson PD, Miao P, Green TC, Yeatman EM, Holmes AS (2006) Converter circuit design, semiconductor device selection and analysis of parasitics for micropower electrostatic generators. IEEE Trans Power Electron 21:27–37
35. Suzuki Y (2011) Recent progress in MEMS electret generator for energy harvesting. IEEJ Trans Electr Electron Eng 6:101–111

Chapter 14
Asynchronous Event-Based Self-Powering, Computation, and Data Logging

Shantanu Chakrabartty

Abstract Asynchronous self-powering refers to an energy scavenging approach where energy for sensing, computation, and nonvolatile storage is harvested directly from the signal being sensed. The approach eliminates the need for energy regulation modules, energy storage, analog-to-digital converters, microcontrollers, and random-access memory, all of which are commonly used in traditional energy scavenging sensors. In this chapter, we describe the fundamental principles of asynchronous self-powering by considering a case study of a sensor designed for structural health monitoring (SHM) applications. In this regard, we describe how the device physics governing the operation of nonvolatile analog memory could be combined with the physics of piezoelectric and electrostatic transducers such that the resulting circuits can operate at fundamental limits of self-powering. For the sake of completeness, we describe an architecture of a system-on-chip that uses ambient strain variations to asynchronously self-power and compute signal-level and signal-velocity statistics.

14.1 Introduction

Energy scavenging sensors operate by harvesting energy from the ambient environment and are attractive for long-term monitoring applications where the usage of batteries is considered to be impractical. Applications of such sensors include structural health monitoring (SHM) in civil engineering [1, 2] and implant health monitoring in biomedical systems [3–5]. Typically, energy scavenging sensors can be classified into two categories based on their powering mechanisms as shown in Fig. 14.1. The first category shown in Fig. 14.1a operates by scavenging energy

S. Chakrabartty (✉)
Michigan State University, East Lansing, MI, USA
e-mail: shantanu@egr.msu.edu

N. Elvin and A. Erturk (eds.), *Advances in Energy Harvesting Methods*,
DOI 10.1007/978-1-4614-5705-3_14,

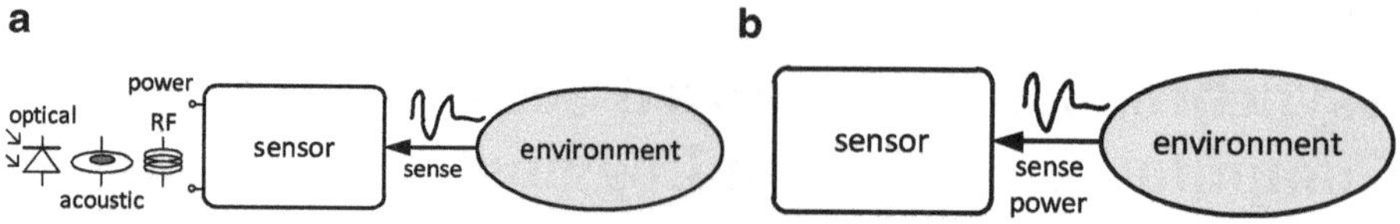

Fig. 14.1 Classification of energy scavenging sensors: (**a**) auxiliary powered sensor; (**b**) self-powered sensor

from ambient power sources (e.g., thermal, vibration, or solar) [6–8] or from a remote source (e.g., radio frequency (RF) or acoustic beacon) [9]. The scavenged energy (with or without buffering with an energy storage element) is then used to power the sensing, computation, storage, and communication modules of the sensor. In comparison, the second category of energy scavenging sensors operates using self-powering (shown in Fig. 14.1b) where the working energy is scavenged directly from the signal being sensed [3, 10]. For example, a piezoelectric element could be used for sensing ambient mechanical strain, and the same strain signal could be used to power the computation and storage operations. The main advantage of self-powering is that signal events of interest can be continuously monitored without relying on any additional auxiliary sources of power.

One possible approach for designing self-powered sensors is to use conventional power conditioning and energy buffering techniques, where a rechargeable energy storage device (e.g., rechargeable batteries or super-capacitors) is periodically refreshed using the energy scavenged from the sensed signal. This would require modularizing the operation of the sensor into different modes. The modes are illustrated in Fig. 14.2 which plots the energy content of the sensed signal with respect to time (not drawn to scale). Each of the time-epochs t_1–t_4 represents the onset of an operational mode: (a) a *refresh mode*, when the energy is transferred from the transducer to the storage; (b) an *operational mode*, when the energy is used for sensing, computation and storage; (c) a *sleep mode*, when the energy is used to power only a wake-up circuit (i.e., when the sensed signal is not present); and (d) a *shutdown mode*, when the energy is used to save the necessary sensor state variables in nonvolatile memory and deactivates all the sensor modules. This modular approach of power management, computation. and storage is amenable for ultralow power digital signal processor architectures [11, 12] and has been the hallmark of traditional self-powered and energy scavenging systems.

However, when the signal being sensed is bursty, infrequent, low bandwidth, or contains low energy (e.g., strain signals in SHM applications where energy content is less than 1 μW and the maximum frequency 10 Hz [13]), conventional methods of voltage/current multiplication [14] and voltage/power regulation cannot be directly applied. In such cases, the necessary computation has to be performed during every signal cycle, and the harvested energy cannot be accumulated. This is illustrated in Fig. 14.3, where after the sensor starts up, data processing and storage occur in conjunction with power scavenging and without any energy buffering. Therefore, computation has to be performed during every signal cycle, and all the variables have to be continuously stored on a nonvolatile storage device. This method of

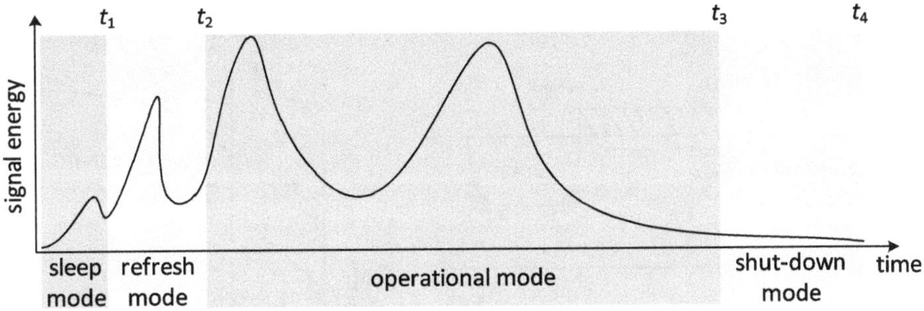

Fig. 14.2 Different functional modes in synchronous self-powering

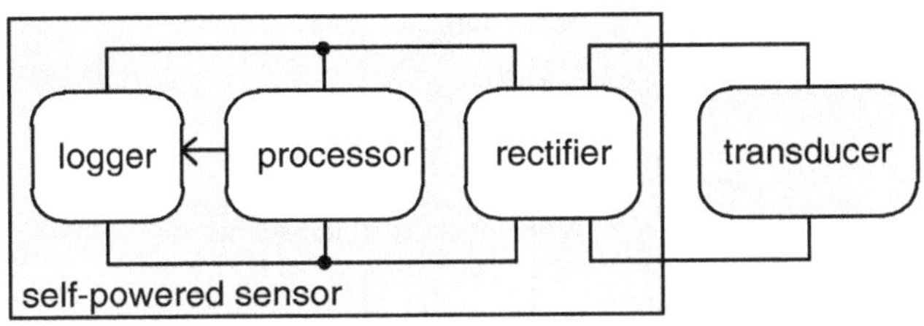

Fig. 14.3 Single-cycle computation and storage in an asynchronous self-powered sensor

powering, sensing, and computing is referred to as "asynchronous" self-powering and requires novel operating principles.

The material presented in this chapter is based on our previous and current research work in the area of self-powering [3, 15, 16] which is described in an integrated manner in the subsequent sections.

14.2 Principles of Asynchronous Self-Powering

The two key requirements for asynchronous powering are:

1. *Nonvolatile computation:* All the internal state and computational variables (intermediate and final) have to be continuously stored on a nonvolatile memory to account for infrequent availability of power (i.e., blackouts).
2. *Minimal power/voltage regulation:* Harvested energy cannot be accumulated or multiplied using conventional methods of voltage/current multiplication or regulation [14]. As a result, conventional data-acquisition technique using a microcontroller is not feasible. Therefore, all necessary computation has to be performed in the "analog" domain, during every signal cycle.

One of the technologies which satisfies both the above requirements is a self-powering approach reported in [16] that exploits operational primitives inherent in the interface physics between a piezoelectric transducer and floating-gate (FG)

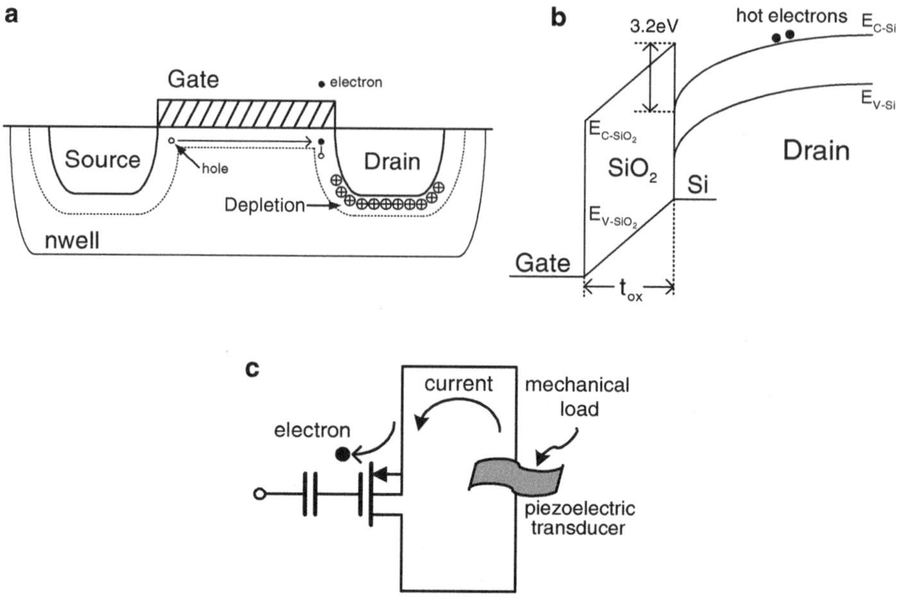

Fig. 14.4 (**a**): Illustration of IHEI process in a pMOS floating-gate transistor; (**b**): illustration of IHEI using an energy band diagram; (**c**): illustration of the concept of piezoelectricity driven IHEI

transistors. The principle referred to as piezoelectricity-driven impact-ionized hot-electron injection (p-IHEI) obviates the use of energy buffering or energy regulation while facilitating continuous storage of computational variables on a nonvolatile memory.

14.2.1 *Impact-Ionized Hot-Electron Injection in Floating-Gate Transistors*

A floating-gate transistor is a metal-oxide-semiconductor field effect transistor (MOSFET) whose polysilicon gate is completely surrounded by an insulator, which in a standard CMOS fabrication process is silicon dioxide (SiO_2) [17]. Because the gate is surrounded by a high-quality insulator, any electrical charges injected onto this gate are retained for a long time (>8 years) [17, 18]. This makes floating-gate transistors attractive for designing nonvolatile CMOS memory [19]. Figure 14.4a shows the cross section of a p-channel floating-gate MOS transistor which is used to illustrate how electrons can be injected onto the floating gate using the mechanism of impact-ionized hot-electron injection (IHEI). Note that an n-channel floating-gate MOSFET could also be used, except that the electron injection mechanism and the corresponding circuit models are slightly different from its p-channel counterpart.

IHEI in the pMOS transistor occurs when a high electric field is formed at the drain-to-channel depletion region. Due to this high electric field, the holes, which are the primary carriers in pMOS transistors, gain sufficient energy to dislodge electrons by impact ionization (see Fig. 14.4a). The released hot electrons accelerate towards the channel region and gain kinetic energy in this process. When the kinetic energy exceeds the silicon and silicon dioxide (>3.2 eV) energy barrier and if the momentum vector is correctly oriented towards the Si-SiO_2 barrier, the electrons are successfully injected into the oxide. The injection process is also shown using an energy band diagram in Fig. 14.4b. As electrons are injected into the oxide and onto the floating gate, its floating-gate potential decreases.

One of the disadvantages of using IHEI as a computational medium is that it requires a large voltage for operation. For example in a 0.5-μm CMOS process, a drain-to-source voltage greater than 4.1 V is required to start IHEI in a pMOS transistor. Fortunately, commonly available piezoelectric transducers like ceramic or polymer piezoelectric materials are capable of generating large voltages (>10 V), though with limited current driving capability (<1 μA). The limited current driving capability is not a problem for IHEI since it has been shown that when the pMOS transistor is biased in weak inversion (subthreshold), the injection efficiency (ratio of injection current and source/drain current) is practically constant for different values of source current [20]. This implies that electron injection can operate at ultra low current level which is ideal for self-powered sensing applications. Therefore, pMOS floating-gate transistors, when coupled with piezoelectric transducers, could be used for self-powered monitoring of mechanical events such as counting the number of loading cycles. The principle of operation of the piezo-driven usage monitor is shown in Fig. 14.4c, where a piezoelectric transducer converts mechanical energy into electrical energy which is then used to inject electrons onto the floating gate. The total number of electrons on the floating gate is therefore indicative of the number of mechanical events.

However, IHEI is a positive feedback process. As more electrons are injected into the floating gate, its potential decreases which in turn increases the drain current through the pMOS transistor. Increase in the drain current increases the probability of impact ionization, thus increases the hot-electron injection current. If left uncontrolled, IHEI will lead to the breakdown of the transistor. Therefore, the current through the transistor needs to be carefully controlled in order to perform any useful and long-term computation. Next we derive a mathematical model of a floating gate injector which is driven by a constant current source and is also powered by a piezoelectric transducer.

14.2.2 *Model of a Constant Current Floating-Gate Injector*

A circuit model of a constant current floating-gate injector is shown in Fig. 14.5. It consists of a pMOS floating-gate transistor whose source terminal V_s is driven by a constant current I_s through a triggering switch $SW1$.

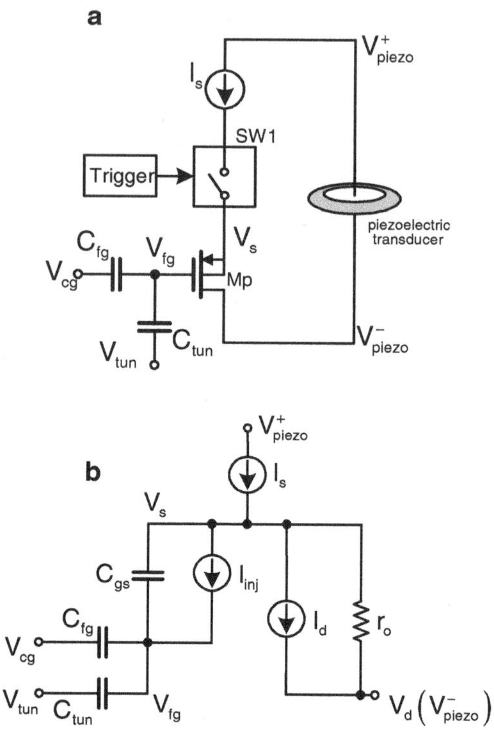

Fig. 14.5 (**a**): Schematic of a floating-gate injector; and (**b**): its equivalent circuit model

The current source in Fig. 14.5 is powered by the signal being monitored, which for this example could be the voltage generated by a piezoelectric transducer. The floating-gate voltage, denoted by V_{fg}, is controlled by the control gate voltage V_{cg} and tunneling voltage V_{tun} through capacitive coupling. The respective coupling capacitors are denoted by C_{fg} and C_{tun}. For the control gate capacitance C_{fg}, the respective plates of the capacitor are formed by polysilicon layers, whereas the tunneling capacitor is implemented using a moscap. The tunneling node V_{tun} is used for removing electrons from the floating gate and in this device is also used for equalizing any residual charges found on the floating gate post-fabrication. The schematic in Fig. 14.5 also consists of a triggering switch $SW1$ which is used to selectively turn ON and OFF the current flow through the floating-gate transistor. For the analysis presented in this section, both V_{cg} and V_{tun} are assumed to be constant, and the source voltage V_s has been assumed to be properly initialized to a predetermined value V_{s0}. Also, the trigger switch will be considered to be always ON. Under these conditions, the current source drives the source node such that it creates a high enough electric field at the drain-to-channel region to trigger the onset of the injection process. As hot electrons are injected onto the floating-gate node, the potential V_{fg} decreases resulting in the decrease of potential V_s. To understand the dynamics of this simplified circuit, an empirical model for the injection is combined with an empirical model of the pMOS transistor. An equivalent circuit of this model is shown in Fig. 14.5b where I_s denotes the source current, I_d is the drain current, I_{inj} is the injection current, r_o is the drain-to-source impedance, $V_{s,d}$ is the source and drain voltages, C_{fg} is the floating-gate capacitance, C_{tun} is the tunneling capacitance, and C_{gs} is the gate-to-source (bulk) capacitance. It is important to note that the values of the currents, r_o and C_{gs}, are dependent on the voltages and currents and should not be confused with a small signal model. We will use a simple injection current model [20] for this analysis which is given by

$$I_{inj} = \beta I_s \exp\left((V_s - V_d)/V_{inj}\right), \tag{14.1}$$

where β and V_{inj} are injection parameters which are a function of the transistor size and the process parameters. The current source I_s in Fig. 14.5b also ensures that the floating-gate transistor is biased in weak inversion. For the source-to-drain voltage $V_{ds} > 200$ mV, the current I_s can be expressed as [21]

$$I_s = I_0 \exp\left(\frac{-V_{fg}}{nU_T}\right) \exp\left(\frac{V_s}{U_T}\right), \tag{14.2}$$

where I_0 is the characteristic current, V_{fg} and V_s are the floating-gate voltage and source voltage respectively, n is the slope factor [21], and U_T is the thermal voltage (26 mV at 300 K). Integrating the models described by Eqs. (14.1) and (14.2) into the equivalent circuit model in Fig. 14.5b and solving the resulting differential equation, the expression of V_s is obtained as follows using the equivalent circuit in Fig. 14.5b. If the floating gate is properly initialized, the charge on the floating node Q_g is given by

$$Q_g = C_{fg}(V_g - V_c) + C_{tun}(V_g - V_{tun}) + C_{gs}(V_g - V_s). \tag{14.3}$$

The injection current I_{inj} modifies the floating-gate charge according to

$$\frac{dQ_g}{dt} = I_{inj} = I_s \exp\left(\frac{V_s}{V_{inj}}\right) \tag{14.4}$$

which is connected to Eq. (14.3) by

$$\frac{dQ_g}{dt} = C_t \frac{dV_g}{dt} - C_{gs}\frac{dV_s}{dt} \tag{14.5}$$

where $C_t = C_{fg} + C_{tun} + C_{gs}$. Also applying current conservation at the node V_s leads to

$$I_s = I_{inj} + I_d + \frac{V_s}{r_o} + C_{gs}\frac{dV_s}{dt} - C_{gs}\frac{dV_{fg}}{dt}, \tag{14.6}$$

where the drain current I_d of the transistor M_P in weak inversion is given by

$$I_d = I_0 \exp\left(\frac{-V_{fg}}{nU_T}\right) \exp\left(\frac{V_s}{U_T}\right). \tag{14.7}$$

Equations (14.3)–(14.7) form a set of coupled differential equations whose closed form solution is difficult to obtain. We will therefore assume that $I_s \approx I_d$ for the following derivation which is reasonable since the I_{inj} and the current charging the bulk/source capacitance C_{gs} is small compared to the source current I_s. Since the source current I_s is constant, the expression of the floating-gate voltage with the other parameters can be obtained from (14.2) as

$$V_{fg} = nV_s - nU_T \ln\left(\frac{I_s}{I_0}\right). \tag{14.8}$$

Equation (14.8) when combined with the differential equation (14.5) and (14.4) leads to

$$I_{inj} = -(nC_t - C_{gs})\frac{dV_{fg}}{dt}. \tag{14.9}$$

Equating (14.9) with (14.1), the following equation is obtained:

$$\beta I_s e^{V_s/V_{inj}} = -C_t\frac{dV_{fg}}{dt} = -C_t\frac{dV_s}{dt}, \tag{14.10}$$

using Eq. (14.8). The above equation can be simplified as

$$\frac{dV_s}{dt} = -K_1 \exp(K_2 V_s),$$

with

$$K_1 = \frac{\kappa\beta I_s}{C_t},\ K_2 = \frac{1}{V_{inj}}.$$

Solving this first-order differential equation leads to the desired relationship:

$$V_s(t) = -\frac{1}{K_2}\ln(K_1K_2t + \exp(-K_2V_{s0})), \tag{14.11}$$

$$V_s(t) = -\frac{1}{K_2}\ln(K_1K_2t + \exp(-K_2V_{s0})), \tag{14.12}$$

with the values of K_1 and K_2 given by

$$K_1 = \frac{\beta I_s}{nC_t - C_{gs}},\ K_2 = \frac{1}{V_{inj}},$$

where $C_t = C_{fg} + C_{tun} + C_{gs}$ denotes the total capacitance at the floating node.

Derivation of the expression and any underlying assumptions are described in detail in Appendix I. V_{s0} in Eq. (14.12) is the initial source voltage, and t represents the total duration for which the triggering switch (injection) is enabled. The plot of $V_s(t)$ as predicted by Eq. (14.12) is shown in Fig. 14.6 which also shows the measured results obtained from a prototype fabricated in a 0.5-μm CMOS process. The results show that the mathematical model is in close agreement with the measured results. In particular, it can be seen in Fig. 14.6 that the response of the injector consists of two distinct regions of operation. The first region is the linear

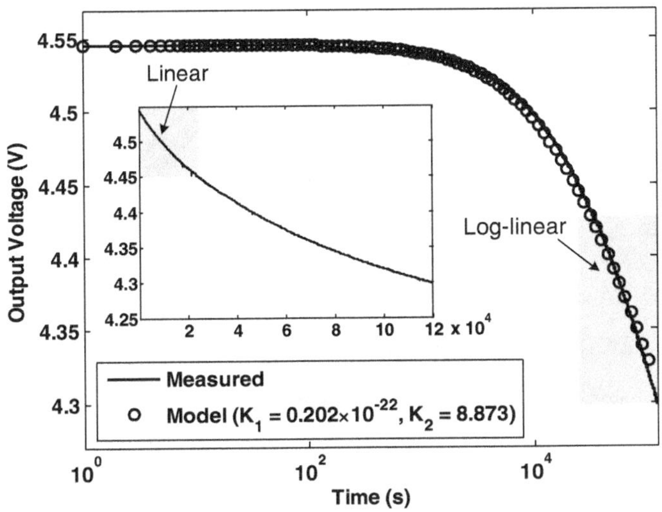

Fig. 14.6 Theoretical and measured response of the floating-gate injector plotted on a logarithmic scale and on a linear scale (inset)

region (see inset in Fig. 14.6) occurs under the condition $t \ll (\exp(-K_2 V_{s0}) / K_1 K_2$ for which the Eq. (14.12) can be simplified as

$$V_s(t) = V_{s0} - K_1 \exp(K_2 V_{s0})\, t. \tag{14.13}$$

In deriving (14.13) we have used the approximation $\ln(1 + x) \approx x$. Since the output of the injector $V_s(t)$ is a linear function of the injection duration t, the linear region is useful for monitoring short-term events (with a cumulative monitoring period less than 100 s). However, for long-term monitoring, the second region of operation called the "log-linear" region is of importance.

Under the condition $t \gg \exp(-K_2 V_{s0}) / K_1 K_2$, Eq. (14.12) can be simplified to

$$V_s(t) = -\frac{1}{K_2} \ln(K_1 K_2 t), \tag{14.14}$$

which shows that the voltage is a logarithmic function of the injection duration. The response is illustrated in Fig. 14.6 using both measured data and empirical models, where it is shown to be valid for large durations (t > 10^3 s). In fact, the log-linear model is valid even beyond 10^5 s, where the injection currents are as low as one electron per second. This can be readily verified from the measured response in Fig. 14.6, where the change in voltage observed on the floating-gate node (with capacitance of 100 fF) over a duration of 10^4 s is 20 mV. Another interesting result that can be seen from Eq. (14.14) is that the effect of V_{s0} can be neglected when t is sufficiently large and V_s is only dependent on the two constants, K_1 and K_2. The slope of the log-linear response is therefore completely determined by $1/K_2$, while K_1 only introduces an offset. This offset captures all the artifacts arising due to biasing conditions, ambient temperature, and fabrication parameters.

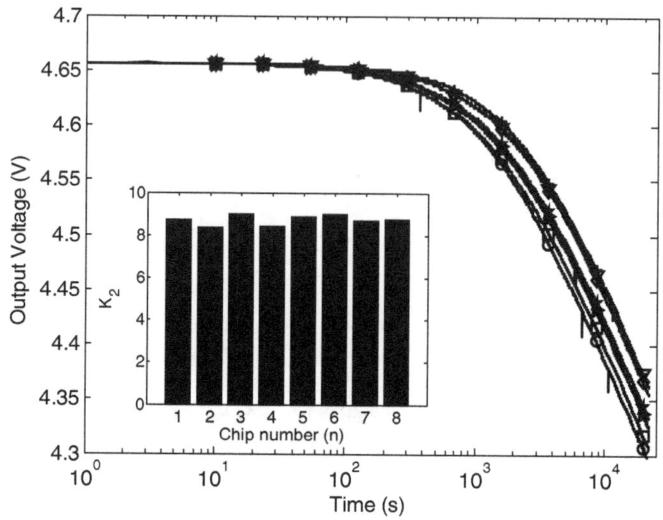

Fig. 14.7 Injector responses measured using 8 prototypes fabricated in the same and different runs

Thus, Eq. (14.14) also provides a model for compensating for these artifacts using a simple differential offset cancelation technique. Equation (14.14) can be written in its differential form as

$$\Delta V_s(\Delta t) = \frac{1}{K_2} \ln \left(\frac{t_0}{t_0 + \Delta t} \right), \tag{14.15}$$

where t_0 denotes a reference injection time with respect to which the differential time interval Δt is measured. It can be readily seen from Eq. (14.15) that the differential operation is independent of the parameter K_1. However, for Eq. (14.15) to be useful, the robustness of the parameter K_2 still needs to be addressed. Figure 14.7 shows the responses obtained from eight injectors, three of which were measured from different prototypes fabricated in the same run where five of which were measured using prototypes fabricated in different runs. For these measurements, the mismatch in the parameter K_2 was calculated to be 4.3%. The results demonstrate that the response of the injector is robust to fabricated related mismatch. We have also calibrated the temperature sensitivity of the parameter K_2 which was measured to be 0.01 V^{-1}/°C.

14.3 Event-Based Analog Self-Powered Sensor/Processor

One of several applications where p-IHEI circuits could be used is for computation of level-crossing statistics. Level-crossing statistics is the number of occurrence of the event when an attribute of the signal exceeds a predefined threshold. For

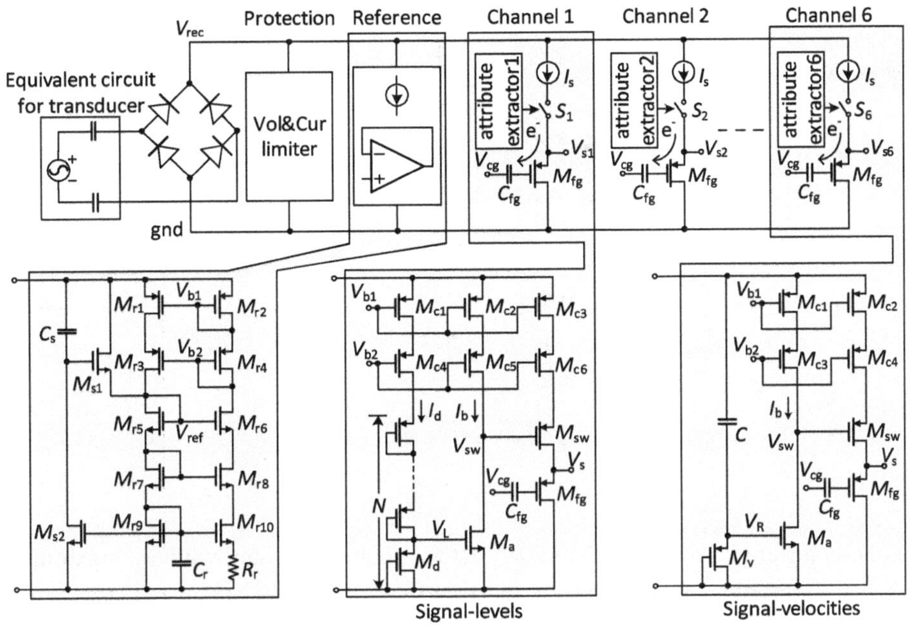

Fig. 14.8 Architecture of the event-based analog processor which multi functional p-IHEI channels (insets) show the circuit-level schematics of the reference, signal-level and signal-velocity monitoring modules

example, strain level-crossings have been shown to be important for predicting the onset of mechanical fatigue in structures of SHM applications [22]. Also, when the sensor signal admits certain statistical properties (e.g., Gaussian statistics), it has been shown [23] that level-crossing measurements are sufficient to estimate complex signal properties (e.g. signal spectra and autocorrelation). Figure 14.8 shows the architecture of the p-IHEI-based self-powered processor that senses, computes, and stores multi functional level-crossing statistics of the sensor signal. This functionality is achieved by using different attribute extraction modules that control the triggering switches S_1–S_6 for each of the channels 1–6.

14.3.1 Signal-Level Monitoring Function

Channels 1–3 in Fig. 14.8 have been designed to compute and store the number of events when the transducer signal-level exceeds different predefined thresholds. A cascoded voltage/current reference circuit shown in Fig. 14.8 (inset) generates the current source required to drive each of the p-IHEI channels and bias voltages for the control gate V_{cg}. The attribute extraction circuit which detects signal-level events is shown in Fig. 14.8 (inset) for one of the channels. A current-starved series of

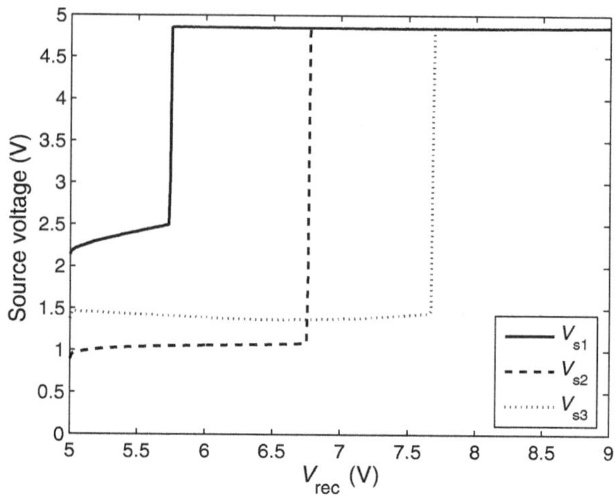

Fig. 14.9 Measurement results showing V_{s1-3} when the input voltage is varied

diode-connected pMOS transistors M_d act as a voltage divider, with the maximum current compliance determined by the drain current through M_{c1} and M_{c4}. This avoids loading the transducer when the rectified voltage V_{rec} becomes large. The output V_L of the divider circuit is given by

$$V_L \approx \frac{1}{N} \cdot V_{rec}, \tag{14.16}$$

and voltage V_L serves as an input to a common-source amplifier stage which controls the gate voltage of the p-IHEI switch V_{sw} according to

$$V_{sw} = \begin{cases} V_{rec} & \text{if } V_{rec} < N \cdot \left[V_{TH} + n \cdot U_T \cdot \ln\left(\frac{L}{W} \cdot \frac{I_b}{I_0}\right)\right] \\ 0 & \text{otherwise.} \end{cases} \tag{14.17}$$

The transition voltage for V_{sw} therefore depends on the threshold voltage V_{TH}, the slope factor n, the thermal voltage U_T, the characteristic current I_0, the aspect ratio W/L, and the biasing current I_b for M_a. It can be seen from Eq. (14.17) that different signal-level events can be defined by inserting different numbers of pMOS diodes M_d. Measured results from channels 1–3 verify the generation of the events as shown in Fig. 14.9. Channels 1–3 are designed to detect events when the transducer voltage level exceed 5.75, 6.78, and 7.69 V, which are typical values for a 1 cm^3 piezoelectric transducer (PZT) under open-load conditions. Note, a minimum of 4.2 V is required to trigger hot-electron injection on floating-gate transistors fabricated in a 0.5-μm CMOS process.

14.3.2 Signal-Velocity Monitoring

Channels 4–6 in Fig. 14.8 have been designed to count the events when the signal-velocity exceeds a predefined threshold. Statistics of signal velocity or the rate of change in the signal levels is important for detecting transient events like impact. The circuit implementation is shown in Fig. 14.8 (inset) and uses a different attribute extraction circuit compared to signal-level monitoring circuits. The attribute extraction circuit consists of a nonlinear high-pass filter (formed by the capacitor and a pMOS diode) followed by a common-source amplifier. Under DC operating conditions, the pMOS diode is biased in the linear region, and its input impedance is determined mainly by its aspect ratio. Therefore, filters with ultra low cutoff frequencies (less than 20Hz) can be implemented without using large on-chip capacitors. When the input voltage V_{rec} increases from 0 to V_{H} at $t = 0$ (during power-up), V_{R} follows V_{H} due to capacitive coupling. Assuming V_{H} is large enough, V_{sw} is pulled LOW which triggers hot-electron injection on M_{fg}. Subsequently, C discharges through M_{v} which occurs in two phases. During the first phase, V_{R} is above the threshold voltage (V_{TH}) of M_{v} and therefore quickly drops from V_{H} towards V_{TH}. In the second phase, the capacitor discharges due to the subthreshold current and can be expressed as [21]

$$I_{\mathrm{dis}} = -C \cdot \frac{\partial V_{\mathrm{R}}}{\partial t} = S \cdot I_0 \cdot \exp\left(\frac{V_{\mathrm{R}}}{U_{\mathrm{T}}}\right), \tag{14.18}$$

where S is the aspect ratio of M_{v}, I_0 is the characteristic current, and U_{T} is the thermal voltage. Considering the reference current I_{b} and assuming that the subthreshold slope and the characteristic current I_0 of transistors M_{v} and M_{a} are similar, the solution to Eq. (14.18) is expressed as

$$\Delta t = \frac{C \cdot U_{\mathrm{T}}}{S \cdot I_{\mathrm{b}}} \cdot \exp\left(-\frac{V_{\mathrm{TH}}}{U_{\mathrm{T}}}\right), \tag{14.19}$$

where Δt is the total duration during which V_{R} transitions from V_{TH} to the switching threshold of the common-source amplifier. Since the discharging time during the first stage can be neglected compared to a much longer discharging time during the second or the subthreshold stage, Eq. (14.19) is a reasonable approximation of the total discharging delay or the duration for which the hot-electron injection is ON.

Figure 14.10a shows measured responses $V_{\mathrm{sw}4-6}$ for three different channels (with different aspect ratios S) when V_{rec} is switched from 0 to 6 V. The aspect ratios of $M_{\mathrm{v}4-6}$ were designed to be in proportion of 1:10:20 for which the Δt_{4-6} were measured to be 140, 15, and 8 ms, respectively, as shown in Fig. 14.10a (offset by the startup time for the reference circuit which was measured to be 30 ms). The ratio of Δt_{4-6} is in close agreement with the reciprocal of the size ratio as predicted by Eq. (14.19). Figure 14.10b,c show the measured responses of two channels when

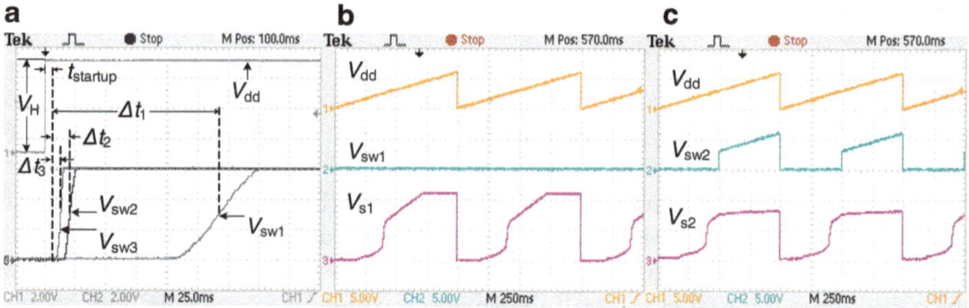

Fig. 14.10 Measurement results when a step input is applied to the signal-velocity circuit: (**a**) step responses of V_{sw4-6}; (**b**) response when the signal velocity exceeds the triggering threshold; (**c**) response when the signal velocity is below the triggering threshold

signals with different velocities were applied. A periodic ramp signal was applied to V_{rec} which was increased from 0 to 6 V at a rate of 6 V/s. As shown in Fig. 14.10b, the fourth channel (set to a lower cutoff frequency) is always enabled because V_{sw4} is always LOW. The fifth channel (set to a higher cutoff frequency) is never enabled as shown in Fig. 14.10c. Thus, the circuit in Fig. 14.8 can compute level-crossing statistics of signal velocities.

14.4 System Implementation and Evaluation

Even though the scavenged energy used by the asynchronous p-IHEI-based self-powering is sufficient for computation and storage, it is not sufficient for wireless interrogation and communication, remote configuration, and data transfer. Therefore, a system-level implementation of a p-IHEI processor has to integrate a second operational mode (i.e., the interrogation mode) where the sensor IC harvests operation energy from the incident radio frequency (RF) signal and performs sensor data digitization, framing, telemetry, and high-voltage floating-gate configuration/programming. The implementation of a complete p-IHEI system-on-chip (SOC) has been reported in [15] where the asynchronous self-powered p-IHEI-based analog processor circuits have been integrated with a 13.56MHz radio-frequency (RF) programming interface.

The SOC has been fabricated in a 0.5-μm standard CMOS process with a die area of 1.6×1.6 mm and mounted on a custom sensor board. The reader and sensor boards used for evaluation are shown in Fig. 14.11 and were designed for the task of infrastructure monitoring. The size of the sensor board was designed to be embedded within a structural stub used routinely in concrete pavements. The reader board hosts the Texas Instruments TRF7960 chipset [24], peripheral components, a matching network, and a stripline printed circuit board (PCB) antenna coil. Under matched conditions, the output power transmitted from the

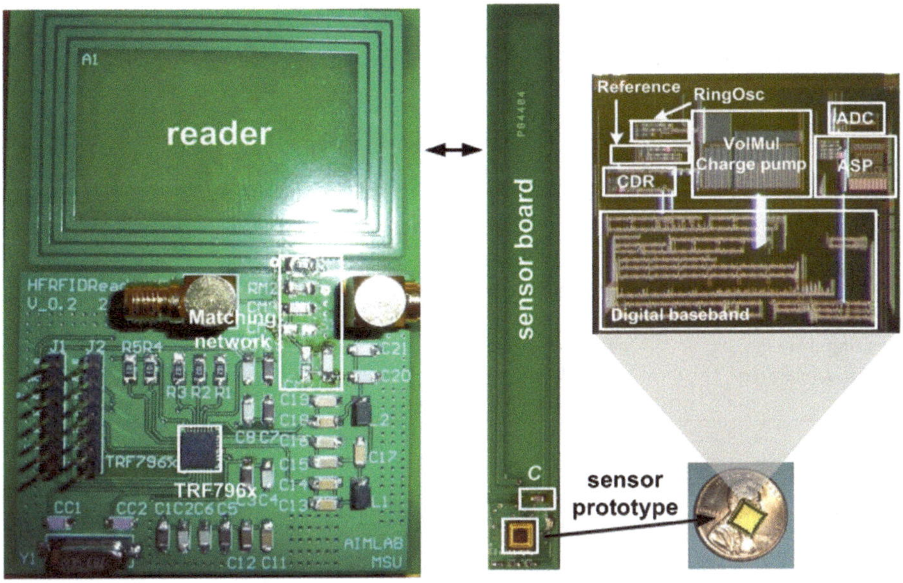

Fig. 14.11 Complete 13.56 MHz reader and sensor system and the micrograph of the sensor IC prototype

reader is 200 mW (23 dBm) when referred to a 50 Ω load at 5 V voltage supply. Details of RF modules and antenna design have been omitted for the sake of brevity. The 3-dB bandwidth of the transceiver was measured to be around 2 MHz. Since the communication protocol between the reader and sensor is customized, the protocol processing block in the TRF7960 chipset was disabled, and a field-programmable gate array (FPGA) board (SPARTAN-3) was used as a digital transceiver. The micrograph of the sensor prototype is also shown in Fig. 14.11, and the main specifications of the IC are summarized in Table 14.1.

14.4.1 Sensor Measurement Results

The first set of experiments were conducted to evaluate the performance of the analog level-crossing processor. A periodic pulse signal with $T_{\text{cycle}} = 1$ s was applied to the FG injector channels as the rectified voltage. In three groups of measurements, the magnitude of the pulse signal was programmed to different levels, and the corresponding injection responses were recorded for the source voltages during 3,000 transducer loading cycles. Before each experiment, all channels were initialized to around 3.1 V with the control gate voltage shorted to ground which made V_s exceed 4 V during the injection. From Fig. 14.9, the trigger level for each channel has been measured to be 5.75, 6.78, and 7.69 V, respectively.

Table 14.1 Main specifications for proposed sensor IC

Fabrication process	0.5-μm standard CMOS	
Die size	1.6×1.6 mm	
Sensor input voltage range	5–9 V	
Level-crossing algorithm	3-channel signal-level monitoring	6–8V
	3-channel signal-velocity monitoring	3–12 V/s
Power (monitoring mode)	< 90 nA	(540 nW@6V)
Power (interrogating mode)	Analog processor & ADC	15 μW@6V
	Digital-baseband	7.6 μW@2V
	Digital-base band (charge pump activated)	71 μW@2V
RF carrier frequency	13.56 MHz	
Reading distance	40 mm	
Wireless telemetry	Uplink	40 kHz PIE
	Downlink	150 kHz Manchester

As shown in Fig. 14.12a, the magnitude of the pulse signal was programmed to 6 V, and thus only channel 1 was triggered by hot-electron injection, while the other two channels remained unchanged. Likewise, channels 1 and 2 were triggered for injection when $V_{rec} = 7$V (shown in Fig. 14.12b), and all three channels were triggered for $V_{rec} = 8$ V (shown in Fig. 14.12c).

The total current consumption for the 3-channel array for the level-crossing processor was also measured which is shown in Fig. 14.12d. When V_{rec} is below 5.75 V, only the reference circuit is activated. Then at each trigger level, one more FG injector begins conducting, and additional current starts to flow which is 18 nA in the design. Note that the exponential trend in the measured current arises from the breakdown of the startup circuit in the reference (M_{s1} in Fig. 14.8), which can be easily fixed with a different startup topology. However, even with the exponential trend, the total current consumption of the 3-channel FG injector array is still below 100 nA which is more than sufficient to be driven by many piezoelectric transducer.

The second set of experiments were conducted to evaluate the performance for strain-velocity detection. A period ramp signal was first applied as V_{rec} which was increased from 0 to 6 V during $t = T$. After conducting multiple tests on the FG injector array in Fig. 14.8, it was found that the time required for channel 4 to start injecting was $T \leq 10$ s; the time required for channel 5 to start injecting was $T \leq 1$ s and for channel 6, $T \leq 0.5$ s, which is in the similar ratio as the time constants for each channel given by Eq. (14.19).

Then, a periodic ramp signal was applied to emulate the typical signal velocities observed in infrastructure monitoring applications: V_{rec} was increased from 0 to 6 V (in T seconds) and then reduced to 0 with a cumulative time period of $T_{cycle} = 2.5$ s. The corresponding measured results are presented in Fig. 14.13. In Fig. 14.13a, T was programmed to 2 s (denoted as 3 V/s), and only channel 4 was triggered for injection. In Fig. 14.13b, T was changed to 1 s (denoted as 6 V/s), and both channels 4 and 5 were triggered. All channels started to inject when T was programmed to 0.5 s (denoted as 12 V/s) as shown in Fig. 14.13c.

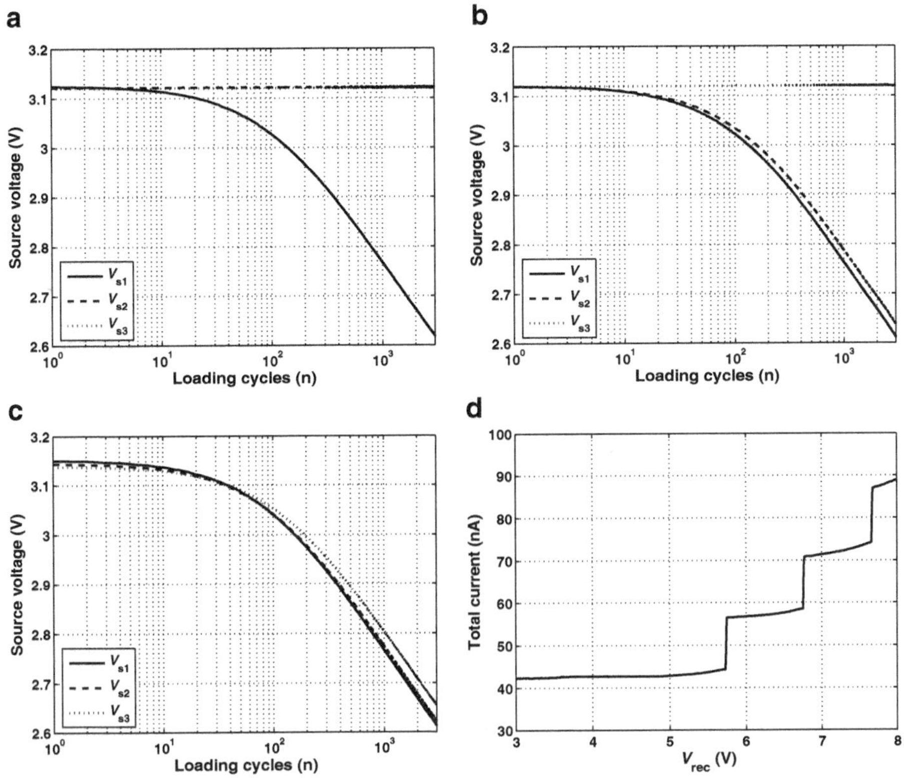

Fig. 14.12 Measured results during signal-level monitoring: (**a**) 6 V pulse; (**b**) 7 V pulse; (**c**) 8 V pulse; (**d**) total current consumption

The power consumption of the IC was measured for these experiments and are presented in Fig. 14.13d. In the case of 3 V/s, the total current consumption can be attributed to the reference circuit along with channel 4. In the case of 6 V/s, two of the channels are triggered when V_{rec} is sufficiently high and the peak current increased by 18 nA. Likewise, the peak current consumption increased to about 90 nA when the signal rate was 12 V/s and when all three channels were triggered. The measurement results showed that the total current consumption is always below 100 nA which is sufficient to be powered directly by the transducer.

14.5 Interfacing with Piezoelectric Transducers

In this section, we present an example where the signal-level monitoring modules were interfaced with a piezoelectric transducer for sensing and computing strain-level statistics. The direct piezoelectric effect is the ability of certain crystalline

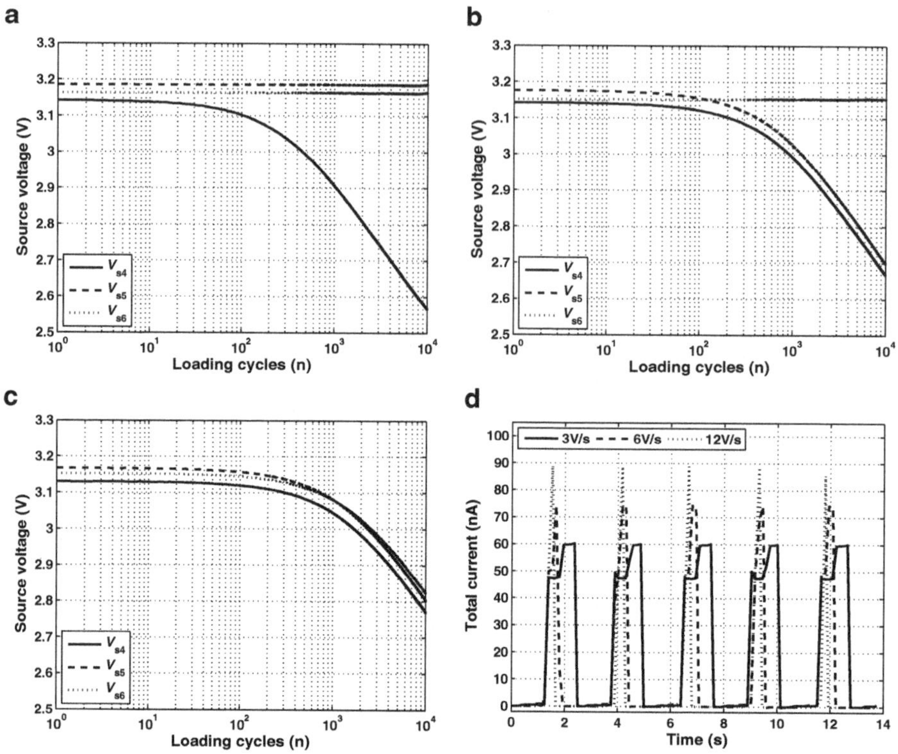

Fig. 14.13 Measured results during signal-velocity monitoring: (**a**) 3 V/s; (**b**) 6 V/s; (**c**) 12 V/s; (**d**) total current consumption

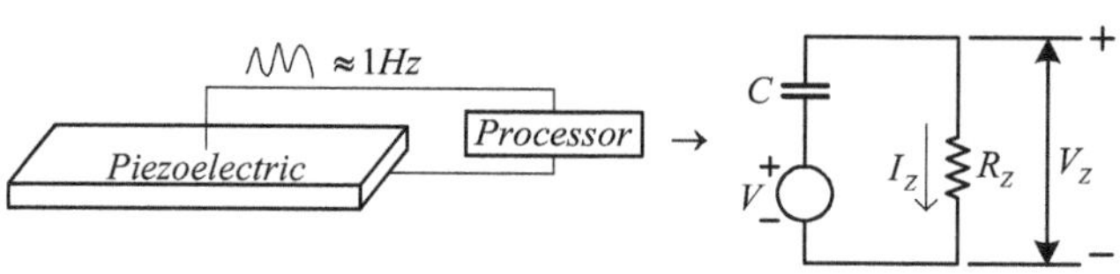

Fig. 14.14 Equivalent circuit model for the piezoelectric transducer at sub-resonance frequencies with the processor as a load

materials to generate electric charge from an applied mechanical stress. For a piezoelectric material with dimensions $L \times b \times h$ polled through its thickness, the open source voltage (V) generated across the material for an applied mechanical force (F) along its length is given by

$$V = \frac{F g_{31}}{b} = S Y^{\mathrm{E}} h g_{31} = \frac{S Y^{\mathrm{E}} d_{31} h}{\epsilon} \tag{14.20}$$

where g_{31} and d_{31} are piezoelectric constants, S is the applied mechanical strain, Y^{E} is the short-circuit elastic modulus, and ϵ is the electrical permittivity. Another important property of piezoelectric materials is its intrinsic capacitance given by

$$C = \epsilon \frac{LW}{h}. \tag{14.21}$$

Typically, the infrasonic loading is an order of magnitude smaller than the transducer resonant frequency. As a result, the piezoelectric transducer can be modeled by a quasi-static electrical circuit that consists of an open AC voltage source (computed according to Eq. (14.20)) coupled to a series capacitance given by Eq. (14.21).

As described earlier, piezoelectric materials can generate large voltage signals (>10 V) but exhibit limited current driving capability (<1 μA). This attribute makes the transducer ideal for operating floating-gate injectors because the injection efficiency (injection current/source current) is invariant of the transistor bias current (could operate at pico-ampere current levels). However, a piezoelectric transducer also acts as an AC-coupled voltage source, and the frequency of loading for representative structures (e.g., bridges or biomechanical implants) is relatively low (<1 Hz). This frequency is typically an order of magnitude lower than the resonant frequency of the transducer implying limited power harvesting capability. To understand the limitations imposed by the low-frequency operation, consider a simplified equivalent model of the prototype processor interfacing with a piezoelectric transducer as shown in Fig. 14.14 [25, 26]. The transducer has been modeled using an AC voltage source connected to a decoupling capacitor, C. The processor has been modeled as a simple resistive load R_{Z}.

For a harmonic mechanical loading of the piezoelectric transducer at a frequency of f Hz (sufficiently below the resonant frequency), the magnitude of the voltage across the load is found to be

$$V_{\mathrm{Z}}(f) = \frac{2\pi f R_{\mathrm{Z}} C V}{\left(1 + 4\pi^2 f^2 R_{\mathrm{Z}}^2 C^2\right)^{1/2}}. \tag{14.22}$$

The power delivered to the load (processor) is given by $P_{\mathrm{Z}} = V_{\mathrm{Z}}^2(f)/R_{\mathrm{Z}}$ and can be optimized with respect to R_{Z}. The optimal value of R_{Z} is given by

$$R_{\mathrm{Z}} = \frac{1}{2\pi f C}. \tag{14.23}$$

For a loading frequency of 1 Hz and for a typical capacitance (10 nF) of a polyvinylidene fluoride (PVDF)-type piezoelectric material the optimal impedance of the processor is determined to be 15 MΩ. Thus, for a 5 V input voltage, this loading condition is equivalent to a maximum current of 300 nA. Meanwhile, the total current drawn by the fabricated prototype has been measured to be 160 nA at 6.7 V which is clearly less than the current drawn under the optimal loading condition. Thus, the proposed level-crossing processor is ideal for self-powered sensing using piezoelectric transducers.

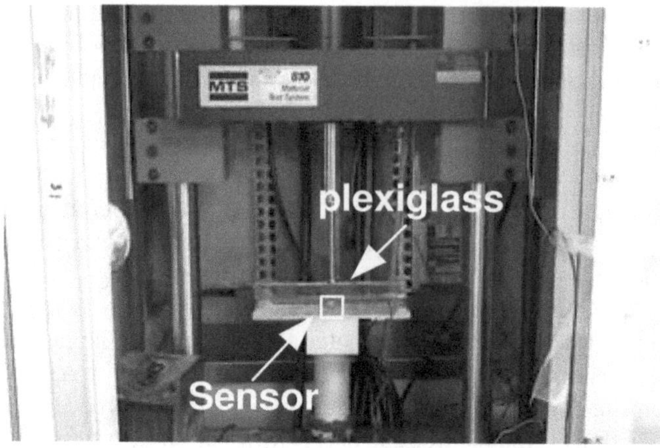

Fig. 14.15 MTS setup used for real-time evaluation of the analog processor with a PVDF transducer

For our experiment, the integrated sensor (piezoelectric transducer and analog processor) was attached to a plexiglass beam, and the setup was mounted on a mechanical testing system (MTS) as shown in Fig. 14.15. The MTS machine was then programmed to generate two distinct strain levels of 2,100 $\mu\varepsilon$ and 2,500 $\mu\varepsilon$, respectively.[1] The mechanical loading was cyclically applied to the plexiglass beam. Figure 14.16a shows the measured results when 2,100 $\mu\varepsilon$ was applied and only the first channel was shown to record a change in output voltage. When loading cycles corresponding to 2,500 $\mu\varepsilon$ was applied, both channels 1 and 2 recorded the changes in voltage, while the channel 3 voltage remained unchanged (shown in Fig. 14.16b). Although no calibration was performed in advance, the offset in measured output voltage could be used to determine the count of loading cycles with different strain levels.

14.6 Discussions and Conclusions

We have described an asynchronous, event-based self-powering technique where the energy for sensing, computation, and nonvolatile memory storage is harvested directly from the signal being sensed. The sensor exploits a log-linear response of the hot-electron injection process on a floating-gate transistor when biased in weak inversion. The array of floating-gate injectors could be configured to respond to different levels of the input signal attributes (signal level and signal velocity)

[1] 1 $\mu\varepsilon$ (called micro-strain) refers to a deformation of 10^{-6} m for the dimension of the structure being 1 m.

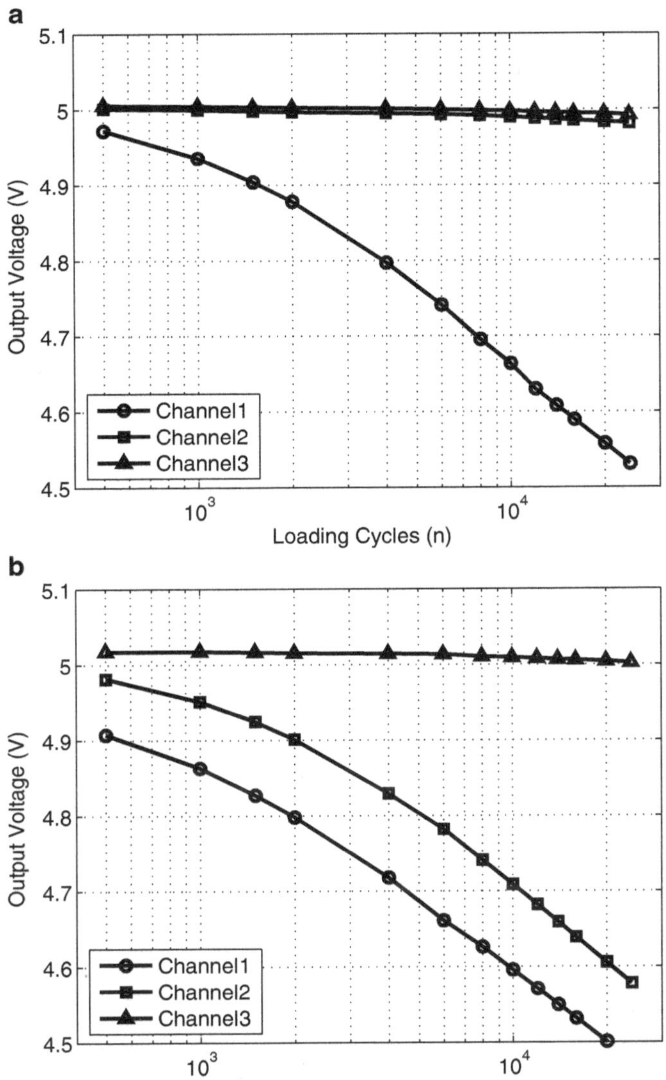

Fig. 14.16 Voltage responses measured when the prototype is interfaced with a PVDF transducer and subjected to controlled cyclic strain levels with magnitude (**a**) 2,100 $\mu\varepsilon$ and (**b**) 2,500 $\mu\varepsilon$

which could then be used for long-term and embedded SHM applications. It was shown that the self-powered design is suitable for integration with electro-capacitive transducers (e.g., piezoelectric transducers) that can generate open-load voltages greater than 5V and driving currents less than 200 nA.

Note that even though the power dissipation of the self-powered unit is measured to be less than 800 nW which makes it suitable for self-powered usage monitoring at a harmonic loading of 1 Hz (which is typical of most mechanical loading), there are applications which require monitoring at frequencies than 0.1 Hz. These applications include earthquake monitoring or monitoring strain cycles due to daily temperature variations. In such cases, the required power dissipation is less than 30 nW [13]. We believe that the current design can achieve this requirement by starving the injection currents using a higher value of current reference resistance R or using a linear hot-electron injection technique reported in our previous work [27].

Another important consideration in IHEI-based processor design is its long-term reliability. Most high-voltage failure mechanisms in a CMOS process are attributed to (a) avalanche breakdown and (b) oxide breakdown. In avalanche breakdown, the impact ionization leads to a positive feedback process that culminates with the failure of the transistor. In this work, the avalanche process is carefully controlled by starving the source current of the transistor. As a result, the injector is a negative feedback circuit where the number of electrons injected into the oxide is significantly limited. In fact we have operated the injector continuously for more than 12 months without observing any failure. The second failure mechanism is due to the oxide breakdown where repeated application of high electric field creates electron-traps, finally leading to its breakdown. For the 0.5-μm CMOS process voltages greater than 15 V are required for quantum mechanical tunneling and for creating stress-related artifacts. The voltage range is beyond the operating range of the processor and can be limited by over-voltage protection diodes. Also, the integrated prototypes consist of electro static discharge diodes which avoid failure due to unwanted voltage/current spikes generated by the piezoelectric transducer.

Acknowledgements This research was supported in part by a research grant from the National Science Foundation (NSF), CMMI: 0700632, CAREER: 0954752, AIR:1127606, and by a contract from the Federal Highway Administration (FHWA), contract no: DTFH61-08-C-00015.

References

1. Lynch JP, Loh KJ (2006) A summary review of wireless sensors and sensor networks for structural health monitoring. Shock Vib Digest 38(2):91–128
2. Sazonov E, Li H, Curry D, Pillay P (2009) Self-powered sensors for monitoring of highway bridges. IEEE Sensor J 9(11):1422–1429
3. Lajnef N, Elvin N, Chakrabartty S (2008) A piezo-powered floating-gate sensor array for long-term fatigue monitoring in biomechanical implants. IEEE Trans Biomed Circ Syst 2(3):164–172
4. Chen H, Liu M, Hao W, Chen Y, Jia C, Zhang C, Wang Z (2009) Low-power circuits for the bidirectional wireless monitoring system of the orthopedic implants. IEEE Trans Biomed Circ Syst 3(6):437–443
5. Platt SR, Farritor S, Garvin K, Haider H (2005) The use of piezoelectric ceramics for electric power generation within orthopedic implants. IEEE/ASME Trans Mechatron 10(4):455–461

6. Ramadass YK, Chandrakasan AP (2011) A battery-less thermoelectric energy harvesting interface circuit with 35mV startup voltage. IEEE J Solid State Circ 46(1):333–341
7. Amirtharajah R, Chandrakasan AP (1998) Self-powered signal processing using vibration-based power generation. IEEE J Solid State Circ 33(5):687–695
8. Alippi C, Galperti C (2008) An adaptive system for optimal solar energy harvesting in wireless sensor network nodes. IEEE Trans Circ Syst I Regular Papers 55(6):1742–1750
9. Yoo J, Yan L, Lee S, Kim Y, Yoo H-J (2010) A 5.2mW self-configured wearable body sensor network controller and a 12μW wirelessly powered sensor for a continuous health monitoring system. IEEE J Solid State Circ 45(1):178–188
10. Elvin N, Elvin A, Choi DH (2003) A self-powered damage detection sensor. J Strain Anal Eng Design 38(2):115–125
11. Hanson S, Mingoo S, Yu-Shiang L, Yoong FZ, Daeyeon K, Yoonmyung L, Liu N, Sylvester D, Blaauw D (2009) A low-voltage processor for sensing applications with picowatt standby mode. IEEE J Solid State Circ 44(4):1145–1155
12. Amirtharajah R, Chandrakasan A (2004) A micropower programmable dsp using approximate signal processing based on distributed arithmetic. IEEE J Solid State Circ 39(2):337–347
13. Elvin N, Lajnef N, Elvin A (2006) Feasibility of structural monitoring with vibration powered sensors. Smart Mater Struct 15(4):977–986
14. Nakamoto H, Yamazaki D, Yamamoto T, Kurata H, Yamada S, Mukaida K, Ninomiya T, Ohkawa T, Masui S, Gotoh K (2006) A passive UHF RFID tag LSI with 36.6% efficiency CMOS-only rectifier and current-mode demodulator in 0.35μm FeRAM technology. In: Proceedings of IEEE International Solid-State Circuits Conference-Digest of Technical Papers, Feb 2006, pp. 1201
15. Huang C, Chakrabartty S (2012) An asynchronous analog self-powered sensor-data-logger with a 13.56MHz RF programming interface. IEEE J Solid State Circ DOI:10.1109/JSSC.2011.2172159
16. Huang C, Lajnef N, Chakrabartty S (2010) Calibration and characterization of self-powered floating-gate usage monitor with single electron per second operational limit. IEEE Trans Circ Syst I Regular Papers 57(3):556–567
17. Dorio C, Hasler P, Minch B, Mead CA (1996) A single-transistor silicon synapse. IEEE Trans Electron Dev 43(11):1972–1980
18. Ozalevli E, Hasler PE (2008) Tunable highly linear floating-gate CMOS resistor using common-mode linearization technique. IEEE Trans Circ Syst I Regular Papers 55(4):999–1010
19. Chakrabartty S, Cauwenberghs G (2007) Sub-microwatt analog VLSI trainable pattern classifier. IEEE J Solid State Circ 42(5):1169–1179
20. Hasler P (1997) Foundations of learning in analog VLSI. Ph.D. dissertation, Department of Computation and Neural Systems, California Institute of Technology, Pasadena, CA
21. Vittoz E, Fellrath J (1977) CMOS analog integrated circuits based on weak inversion operation. IEEE J Solid State Circ 12(3):224–231
22. Suresh S (1998) Fatigue of materials, 2nd edn. Cambridge solid state science series. Cambridge University Press, Cambridge. ISBN-10: 0521578477
23. Brown J Jr (1967) Generalized form of prices theorem and its converse. IEEE Trans Inform Theor 13(1):27–30
24. TRF7960 Datasheet, Multi-standard fully integrated 13.56-MHz RFID AFE and data framing reader system, Texas Instrument, http://focus.ti.com/docs/prod/folders/print/trf7960.html
25. Ferrari M, Ferrari V, Marioli D, Taroni A (2006) Modeling, fabrication and performance measurements of a piezoelectric energy converter for power harvesting in autonomous microsystems. Instrum Meas 55:2096–2101
26. Mateu L, Moll F (2007) System-level simulation of a self-powered sensor with piezoelectric energy harvesting. Sensor Tech Appl, 399–404
27. Huang C, Sarkar P, Chakrabartty S (2011) rail-to-rail hot-electron injection programming of floating-gate voltage bias generators at a resolution of 13 bits. IEEE J Solid State Circ 46(11). DOI:10.1109/JSSC.2011.2167390

Chapter 15
Vibration-Based Energy-Harvesting Integrated Circuits

Gabriel Alfonso Rincón-Mora

Abstract Wireless microsensors that monitor and detect activity in factories, farms, military camps, vehicles, hospitals, and the human body can save money, energy, and lives. Unfortunately, miniaturized batteries exhaust easily, so deploying these tiny devices outside a few niche markets is difficult. Luckily, harnessing ambient energy, especially of the kinetic kind, offers hope because environmental motion is often abundant and consistent. The challenge is that tiny transducers convert only a small fraction of the available energy into the electrical domain, and the circuits that transfer and condition power dissipate some if not most of that energy. As a result, increasing the power drawn from the transducer and reducing the energy losses in the system are of paramount importance. This chapter therefore begins by evaluating how circuits transfer and consume power. Then, because electrostatic and piezoelectric transducers normally derive considerably more power per unit volume than their electromagnetic counterparts, the chapter focuses on how the former two technologies harness ambient kinetic energy. The ensuing discussion details how transducers, integrated circuits, and sample prototype integrated-circuit (IC) implementations of these circuits draw energy from ambient vibration to generate power with which to continually replenish a small on-board battery.

15.1 Introduction

Wireless microsensors can enjoy popularity in a wide variety of applications like biomedical implants and tire-pressure monitoring systems because they offer in-situ, real-time, nonintrusive processing capabilities. Miniaturized platforms, however,

---

G.A. Rincón-Mora (✉)
School of Electrical & Computer Engineering, Georgia Institute of Technology,
777 Atlantic Drive, Atlanta, GA, 30332-0250, USA
e-mail: Rincon-Mora@gatech.edu

N. Elvin and A. Erturk (eds.), *Advances in Energy Harvesting Methods*,
DOI 10.1007/978-1-4614-5705-3_15,

limit the energy that onboard batteries can store, so lifetimes are short. In this regard, ambient energy is an attractive alternative to batteries because harnessing energy from light, heat, radio-frequency (RF) radiation, and motion can continuously replenish an otherwise easily exhaustible reservoir.

Of these ambient sources, solar light produces the highest power at roughly 10–15 mW/cm^2. Unfortunately, artificial lighting, which for many applications is a more practical source, produces drastically lower power levels: one to two orders of magnitude (i.e., 10× to 100×) lower. Harnessing thermal energy is viable, but microscale dimensions constrain temperature gradients below 5–10 °C, the fundamental mechanism from which thermopiles draw power. Mobile electronic devices today radiate plenty of RF energy, but power drops with distance to impractically low levels (i.e., in the nW/cm^3 to μW/cm^3 range). Harvesting kinetic energy may not compete with solar power, but in contrast to artificial lighting, thermal, and RF sources, consistent vibrations that generate moderate power levels are typical for a vast range of applications. More on harvesting energy can be found in [1, 2, 3] and the references they cite.

15.2 Harvesting Kinetic Energy in Vibrations

Although operating conditions ultimately dictate which kinetic energy-harvesting method is optimal, piezoelectric transducers are relatively mature and produce comparatively higher power than their counterparts. On-chip piezoelectric devices, however, remain the subject of ongoing research. Electrostatic harvesters offer an edge in this regard because microelectromechanical systems (MEMS) technologies can more easily integrate variable, parallel-plate capacitors on chip. Electromagnetic transducers, on the other hand, not only generate lower power but also are more difficult to miniaturize.

Irrespective of the means, these transducers harness ambient kinetic energy by converting mechanical energy in vibrations (e.g., E_{ME} in Fig. 15.1) into the electrical domain (as E_{EE}). More specifically, piezoelectric bimorph strips produce charge when bent, parallel-plate capacitors when their plates separate, and coils when moved across a magnetic field. To harness and use the electrical energy (i.e., E_{EE})

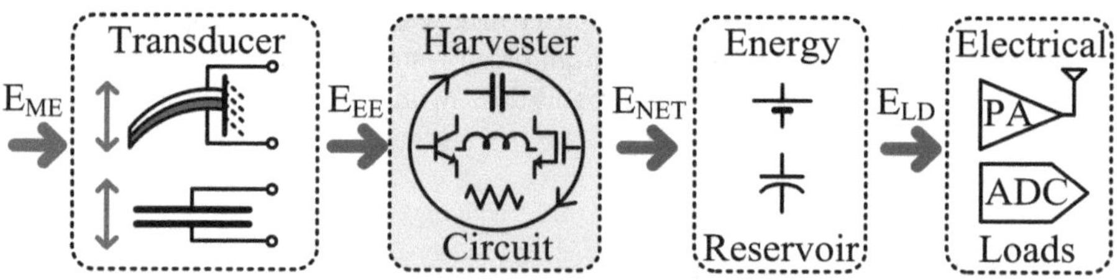

Fig. 15.1 Harvesting electrical energy from ambient kinetic energy

that these devices generate, harvester circuits condition transducers to draw and deliver a net energy gain E_{NET} to intermediate reservoirs that can supply power to electrical loads on demand.

In practice, miniaturized harvesters do not supply energy to the load directly (as E_{LD}) because the mechanical input (i.e., E_{ME}) is often unpredictable and, therefore, an unreliable source of power for both steady-state and time-varying loads. Moreover, the act of transferring energy requires work; in other words, microelectronic circuits dissipate power. As a result, electrical losses in the harvester can reduce transduced energy E_{EE} to the extent that the system can no longer generate a net positive gain E_{NET}. More on harvesting kinetic energy can be found in [4, 5] and the references they cite.

15.3 Power Conditioners

To produce a net energy gain E_{NET} from the miniscule amounts of electrical energy E_{EE} that tiny transducers produce, harvesters must dissipate little power. In fact, because the basic objective is to transfer energy, the overriding measure of success is how much output power P_O the harvesting circuit nets from E_{NET} of the available input power P_{IN} it receives from E_{EE}. Said differently, the power lost (P_L) across the circuit in relation to P_{IN} determines the efficacy of the transfer, which is why electrical–electrical efficiency η_{EE} is an important metric:

$$\eta_{EE} \equiv \frac{P_O}{P_{IN}} = \frac{P_{IN} - P_L}{P_{IN}} = 1 - \frac{P_L}{P_{IN}} \tag{15.1}$$

For instance, if a transducer supplies 100 μW and the system dissipates 25 μW, the reservoir receives 75 μW with an efficiency of 75%.

Efficiency η_{EE} varies with P_{IN} because losses do not change at the same rate as P_{IN} does. Quiescent losses in P_L, for example, usually become a smaller fraction of P_{IN} when mechanical stimulation (i.e., P_{IN}) is high, which means P_L/P_{IN} and, therefore, η_{EE} vary with P_{IN}. Unfortunately, because microsystems produce little power, almost any loss can be substantial, especially when vibration levels are low. For this reason, achieving a good η_{EE} at low P_{IN} is probably one of the most difficult challenges to tackle in a miniaturized harvester.

15.3.1 Linear Switch

Perhaps the simplest means of directing energy from one point to another is through a linearly conducting switch, as Fig. 15.2 illustrates with p-channel metal–oxide–semiconductor field-effect transistor (PMOSFET) M_{PS}. In this scheme, a controller modulates the conductance of M_{PS} linearly to match the time-varying sourcing

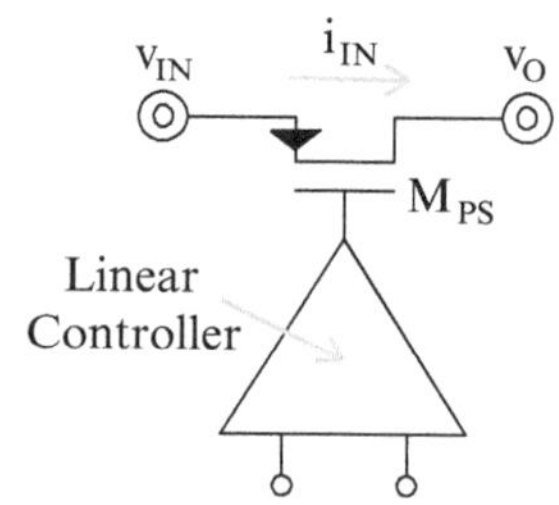

Fig. 15.2 Linear switch

capabilities of the transducer. Because only a few transistors ultimately implement this function, reaction time and quiescent power P_Q are usually low. Plus, the circuit generates little noise because M_{PS} does not switch periodically between fully on and off states.

Unfortunately, a linear switch suffers from limited voltage flexibility and efficiency. For source current i_{IN} to flow to the output, for example, input voltage v_{IN} must exceed output voltage v_O. Also, the transducer and reservoir alone set the average voltage across the switch, which means the power lost in the switch (P_{SW}) rises linearly with input current:

$$P_{SW} = i_{IN}(V_{IN} - V_O) \propto i_{IN(AVG)} \tag{15.2}$$

Although all points mentioned warrant attention, this last one is critical because efficiency never rises above output-to-input voltage ratio V_O/V_{IN}, which is not a user-defined parameter:

$$\eta_L \equiv \frac{P_O}{P_{IN}} = \frac{P_{IN} - P_{LOSS}}{P_{IN}} = \frac{i_{IN}v_{IN} - P_{SW} - P_Q}{i_{IN}v_{IN}} = \frac{v_{O(AVG)}}{v_{IN(AVG)}} - \frac{P_Q}{P_{IN}} < \frac{V_O}{V_{IN}} \tag{15.3}$$

15.3.2 Switched Capacitors

Introducing one or more intermediate steps into the energy-transfer process decouples the needs of the input from those of the output, adding flexibility to the conditioning function. Intermediate capacitors can accomplish this by receiving and releasing energy periodically from v_{IN} to v_O in alternate half cycles of a switching period. This way, charging several "flying" capacitors in parallel from v_{IN} in one phase and de-energizing them in series to v_O in the other phase, as Fig. 15.3 shows, "boosts" v_{IN} to a v_O that is higher than v_{IN}, which is otherwise impossible with a linear switch. Conversely, just to show how flexible the scheme can be, energizing capacitors in series and connecting them in parallel with v_O in the alternate half cycle "bucks" v_{IN}.

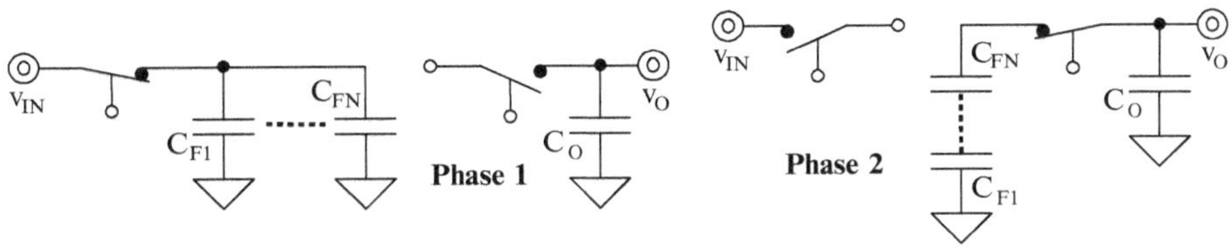

Fig. 15.3 Alternating switching phases of a boosting "charge pump"

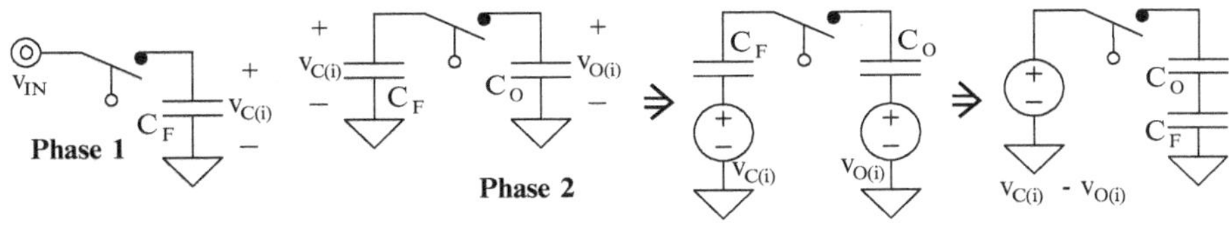

Fig. 15.4 Equivalent switch-loss models across alternating phases

Because the voltages across conducting switches are dynamic and decrease with time, switched capacitors do not dissipate the static and often limiting conduction power that a v_{IN}–v_O voltage drop across a linear switch does. Still, the switches consume power when conducting current with decreasing, but nonetheless finite, voltages across them. The initial voltage across the switch that connects v_{IN} to the flying capacitors in Phase 1 of Fig. 15.3, for example, is the voltage the flying capacitors drooped in Phase 2. To quantify the loss this voltage drop incurs, consider that, while v_{IN} in Phase 1 of Fig. 15.4 loses energy E_{IN} to charge equivalent flying capacitance C_F from $v_{C(i)}$ to v_{IN}:

$$E_{IN} = Q_C V_{IN} = C_F \left(V_{IN} - V_{C(i)}\right) V_{IN}, \tag{15.4}$$

C_F receives

$$E_C = E_{C(FIN)} - E_{C(INI)} = 0.5 C_F v_{IN}{}^2 - 0.5 C_F v_{C(i)}{}^2, \tag{15.5}$$

so the switch dissipates the difference as

$$E_{SW} = E_{IN} - E_C = 0.5 C_F \left(V_{IN}{}^2 - 2V_{IN} V_{C(i)} + V_{C(i)}{}^2\right) = 0.5 C_F V_{SW(i)}{}^2, \tag{15.6}$$

where $\Delta v_{SW(i)}$ is the initial voltage across the switch: $v_{IN} - v_{C(i)}$. Similarly, the fully charged flying capacitors and v_O also present a voltage difference across their connecting switch at the beginning of Phase 2. Because connecting capacitors in parallel is equivalent to charging them in series from a source whose voltage is the initial voltage across the switch, as the switch-loss model of Phase 2 in Fig. 15.4 depicts, the switch consumes

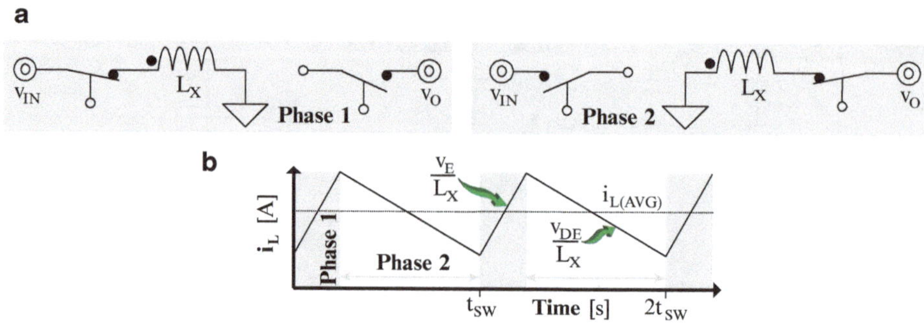

Fig. 15.5 (**a**) Switching phases and (**b**) inductor-current waveform of a switched-inductor converter

$$E_{SW} = 0.5(C_F \oplus C_O)(V_{C(i)} - V_{O(i)})^2 = 0.5\,(C_F \oplus C_O)\,V_{SW(i)}^2 \propto i_{IN(AVG)}^2 \tag{15.7}$$

of the energy that was initially stored in C_F as $0.5C_F v_{C(i)}^2$ at the end of Phase 1. In other words, switched-capacitor circuits generally and necessarily consume $0.5C_{EQ}\Delta v_{SW(i)}^2$ energy per cycle across their interconnecting switches.

Although this fundamental switching loss is by no means negligible, it decreases with reductions in source power P_{IN} because all the capacitors droop less (so $\Delta v_{SW(i)}$ is smaller) with less current, or equivalently, E_{SW} decreases quadratically with reductions in P_{IN}. In practice, however, v_{IN} also supplies the energy that switching noise and leakage currents draw from the capacitors, the quiescent power that energizes the controller, and the power parasitic gate and base capacitors require to charge and discharge every time they switch.

15.3.3 Switched Inductor

Just as charge pumps employ capacitors to transfer energy, magnetic-based switching converters use inductors to draw, temporarily cache, and release charge. They achieve this by energizing and de-energizing inductors in alternating cycles. Accordingly, L_X draws current from an input source v_{IN} by connecting L_X's other terminal to a lower voltage, like ground in Phase 1 of Fig. 15.5a. During this time, inductor current i_L rises almost linearly, as the waveform of Fig. 15.5b illustrates, because inductor voltage v_L, which equals $L_X di_L/dt$, is positive and nearly constant across the switching period t_{SW} at, in this case, $v_{IN} - 0$, or more generally, at energizing voltage v_E. Reversing the polarity of v_L to, say, $-v_O$, as shown in Phase 2 of Fig. 15.5a, causes L_X to release the energy it received in Phase 1. Because the circuit either regulates the voltage across a load or charges a battery with a well-defined, low-ripple, slow-changing voltage, v_O is usually also nearly constant, and in this case, also L_X's de-energizing voltage v_{DE}. As a result, like Fig. 15.5b shows, i_L generally rises linearly at v_E/L_X when v_{IN} energizes L_X (across Phase 1) and falls linearly at v_{DE}/L_X when L_X delivers energy to v_O (through Phase 2).

A switched-inductor conditioner embodies a few idiosyncrasies worth mentioning. First, L_X conducts dc current, so i_L's ripple Δi_L rides on a steady-state offset $i_{L(AVG)}$ or I_L that can also be zero. Conducting current continuously like this places the inductor in continuous-conduction mode (CCM), and prompting L_X to stop conducting current momentarily, which is equivalent to i_L reaching zero and remaining there for a fraction of the switching period, amounts to operating in discontinuous-conduction mode (DCM). What sets $i_{L(AVG)}$ is either the power needs of a load, or in the case of harvesting chargers, the sourcing capabilities of its driving transducer.

Another feature of a switched-inductor circuit is that it can buck or boost v_{IN} because L_X can release energy to any output voltage, as long as energizing and de-energizing voltages v_E and v_{DE} remain positive and negative, respectively. Permanently attaching L_X's input terminal to v_{IN}, for example, is a special embodiment of the general form shown in Fig. 15.5a that requires v_O to exceed v_{IN} for L_X to release its energy. Likewise, connecting L_X's output terminal to v_O directly demands that v_{IN} surpass v_O for L_X to energize. These two special cases implement the well-known boost and buck configurations discussed in the literature. One last peculiarity to note is that inductors, unlike capacitors, only receive or deliver charge, not store it statically over time. This is not really a problem in conditioners because the overriding aim of inductors is to transfer energy, unlike rechargeable batteries and large capacitors whose purpose is to store energy.

Interestingly, while capacitors hold the initial voltage across a switch by sourcing or sinking whatever instantaneous currents are necessary, an inductor maintains its current steady by instantly swinging its voltage until it finds a suitable supply or load for the charge it carries. In other words, after a connecting switch opens, the inductor swings its terminal voltage instantaneously until it matches the voltage of an accommodating source or sink. The series resistance R_{SW} that each interconnecting switch exhibits therefore drops root-mean-square voltage $i_{L(RMS)}R_{SW}$ and dissipates conduction power

$$P_{SW} = i_{L(RMS)}{}^2 R_{SW} \propto i_{IN(AVG)}{}^2. \tag{15.8}$$

Because R_{SW} is normally low (e.g., below 1 Ω), P_{SW} is also low, which is why switched-inductor converters enjoy so much popularity.

These conduction losses may be low, but not zero. Plus, like with switched capacitors, P_{SW} decreases quadratically with reductions in input power P_{IN}. v_{IN} also supplies the quiescent power that energizes the feedback controller and the energy that parasitic gate and base capacitors require to charge and discharge every time they switch. Notice that, since P_{IN} in microsystems is low and P_{SW} decreases quadratically with P_{IN}, second-order quiescent and gate/base-drive losses often are as significant as, and in some cases higher than, first-order switch losses P_{SW}.

15.3.4 Conclusions

Aside from functionality, power losses are arguably one of the most important parameters to consider when selecting a conditioner. In this respect, while a linear switch drops v_{IN} to v_O continuously, irrespective of input power P_{IN}, the terminal voltages of switches in a charge pump and a magnetic-based converter decrease with P_{IN}. In other words, switched conditioners dissipate less conduction power than linear switches when P_{IN} is low, because unlike in the latter case where P_{SW} decreases linearly with P_{IN}, P_{SW} decreases quadratically with P_{IN}. Flying capacitors, however, typically require more than four switches to reconfigure a network to bridge v_{IN} to v_O, so the conclusion is not unequivocal. A switched inductor, on the other hand, normally needs less than four switches to energize and de-energize an inductor. As a result, switched-inductor converters designed for miniaturized applications generally tend to dissipate less conduction power than charge pumps and linear switches.

When input power is considerably low, conduction power drops to the point where quiescent and gate/base-drive losses become increasingly important, which is when a linear switch might gain an edge. The facts that P_{IN} is not consistently low and v_{IN} is not always above v_O in harvesting applications, however, favor the switched inductor. Ultimately, the fundamental challenge of deriving kinetic energy from tiny electrostatic and piezoelectric transducers and delivering it to a load or a battery is to consume only a small fraction of the power drawn. This is the reason why operating the inductor discontinuously (i.e., in DCM) and switching it infrequently (i.e., at a low frequency) are as important as reducing quiescent power. More on power conditioners can be found in [6–9] and the references they cite.

15.4 Electrostatic Harvesters

An electrostatic harvester harnesses energy from the work vibrations exert against the electrostatic force of a motion-sensitive, parallel-plate variable capacitor C_{VAR}. The capacitor is the transducing agent because it converts mechanical energy into the electrical domain. More particularly, as motion separates C_{VAR}'s plates, capacitance decreases and, in response, either C_{VAR}'s voltage v_C rises or C_{VAR}'s charge q_C drops. In other words, C_{VAR}'s stored energy increases (with v_C) or C_{VAR} supplies current ($\Delta q_C/\Delta t$). Either way, the magnitude and range of the transducer's capacitance ultimately limit the extent to which it draws energy from vibrations. More on electrostatic harvesters can be found in [10–16] and the references they cite.

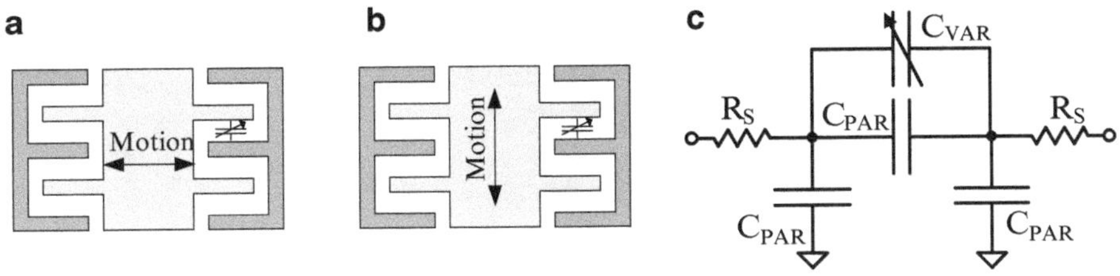

Fig. 15.6 (**a**, **b**) Sample MEMS variable capacitors and (**c**) equivalent electrical model

15.4.1 Transducer

A motion-sensitive variable capacitor is normally a pair of parallel plates whose overlap area or separation distance changes with motion. Figure 15.6a, b illustrates examples of how interleaved, spring-loaded, comb-like structures manufactured with microelectromechanical systems (MEMS) technologies can move laterally across a two-dimensional plane in response to vibrations. In one case, sliding sideways alters the overlap area between adjacent fingers and, in the other, up–down motion changes the separation distance of interleaved fingers. Fabricated MEMS varactors of this sort can feature static capacitances between 50 and 400 pF that can vary anywhere between 0.2× and 4× their original value. In practice, each parallel plate introduces not only series parasitic resistances to C_{VAR}, as the model of Fig. 15.6c shows, but also parasitic capacitances to the underlying substrate. Unfortunately, series resistors consume Ohmic power and shunt capacitors drain charge from C_{VAR}.

15.4.2 Energy-Harvesting Process

Fundamentally, C_{VAR} draws energy from the work that motion exerts to oppose an existing electrostatic force between its plates. Therefore, the first step in the energy-harvesting process is to establish that force by depositing an initial charge into C_{VAR}. This way, when disconnected, v_C rises when motion lowers C_{VAR} because q_C, which is constant, is $C_{VAR}v_C$. Here, v_C's squared rise in C_{VAR}'s energy (i.e., $0.5C_{VAR}{v_C}^2$) offsets C_{VAR}'s linear fall to produce a net gain in energy. Alternatively, when clamped to pre-charge voltage V_{PC}, C_{VAR}'s charge $C_{VAR}V_{PC}$ drops when C_{VAR} falls, which means C_{VAR} sources current. In other words, to harness ambient energy E_{HARV}, electrostatic harvesters must first invest energy E_{PC} to pre-charge C_{VAR} at C_{MAX}, as Fig. 15.7 shows, and as C_{VAR} falls to C_{MIN} across the harvesting phase, C_{VAR} harnesses energy from vibrations. After delivering and storing E_{HARV} elsewhere in the system, the circuit can recover what remains of E_{PC} in C_{VAR} as E_{REC} before C_{VAR} otherwise uses E_{REC} during the reset phase to help vibrations pull C_{VAR}'s plates together, which is how C_{VAR} resets to C_{MAX}.

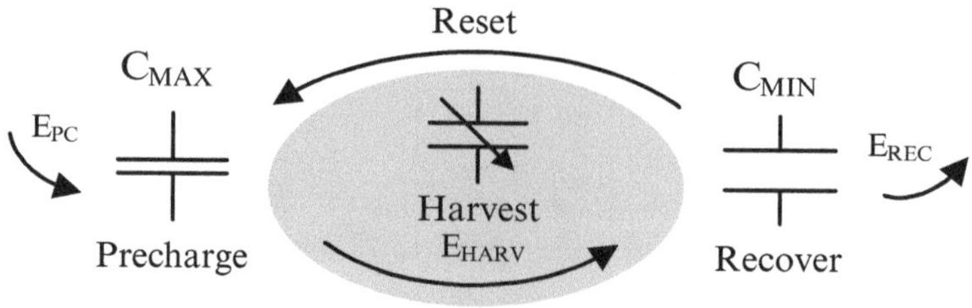

Fig. 15.7 Electrostatic harvesting process: Pre-charge, Harvest, Recover, and Reset

The damping force present in C_{VAR} when it decreases to C_{MIN} determines how much energy C_{VAR} extracts from the motion of its plates. As such, raising v_C—which is a manifestation of this force—as high as possible produces the most gain. In more explicit terms, output energy per cycle E_O is highest when keeping v_C at the highest possible level throughout the entire harvesting period. When constraining C_{VAR} to $V_{C(MAX)}$ this way, harvested energy E_{HARV} together with what remains in C_{VAR} at C_{MIN} (as E_{REC}) surpasses the investment needed to charge C_{VAR} at C_{MAX} to $V_{C(MAX)}$ (with E_{PC}) to produce a net gain in E_O:

$$E_{PC} = 0.5C_{MAX}\,V_{C(MAX)}^2, \tag{15.9}$$

$$E_{HARV} = q_C V_{C(MAX)} = \left(\Delta C_{VAR} V_{C(MAX)}\right) V_{C(MAX)} = (C_{MAX} - C_{MIN})\,V_{C(MAX)}^2 \tag{15.10}$$

and

$$E_{REC} = 0.5C_{MIN}\,V_{C(MAX)}^2, \tag{15.11}$$

so

$$E_O = E_{HARV} + E_{REC} - E_{LOSS} = 0.5(C_{MAX} - C_{MIN})V_{C(MAX)}^2 - E_{LOSS}, \tag{15.12}$$

where E_{LOSS} refers to conduction, gate/base-drive, and quiescent losses in the system. What this means is that fixing C_{VAR}'s voltage to the maximum allowable voltage the circuit can tolerate produces more power from motion than constraining its charge.

15.4.3 Voltage-Constrained Harvesters

While constraining C_{VAR} to a battery V_{BAT} (e.g., 2.7–4.2-V lithium-ion cell) is convenient, it is not necessarily optimal because V_{BAT} is seldom near the maximum

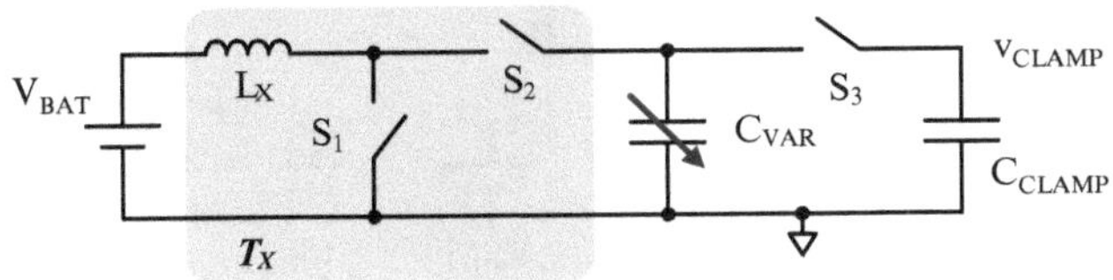

Fig. 15.8 Circuit that clamps C_{VAR} to v_{CLAMP}

voltage that the transducer or circuit can sustain. At the cost of printed-circuit-board (PCB) real estate, a large off-chip capacitor C_{CLAMP} can, instead, clamp C_{VAR} at a higher voltage to produce more power. Permanently connecting C_{CLAMP} to C_{VAR}, however, forces the system to completely discharge and again pre-charge C_{CLAMP} together with C_{VAR}, the transfer losses of which are significant at an elevated voltage.

A diode D_{CLAMP} between C_{CLAMP} and C_{VAR} avoids having to discharge and pre-charge C_{CLAMP} every cycle by connecting C_{CLAMP} to C_{VAR} asynchronously only when v_{VAR} is close to v_{CLAMP}. For example, the harvester can pre-charge C_{VAR} to V_{BAT} and allow kinetic energy in motion raise (in charge-constrained fashion) v_{VAR} until D_{CLAMP} and C_{CLAMP} clamp C_{VAR} to v_{CLAMP}. The drawbacks here are that D_{CLAMP} consumes power and C_{VAR} does not harvest at an elevated voltage for the *entire* harvesting period. Figure 15.8 presents an alternate strategy where switched-inductor conditioning block T_X pre-charges C_{VAR} all the way to v_{CLAMP} before synchronous switch S_3 connects C_{VAR} to C_{CLAMP}. This way, C_{VAR} harvests close to v_{CLAMP} through the entire harvesting phase. Note that the system pre-charges C_{CLAMP} only once (during startup) to a voltage that is slightly below the chip's breakdown level.

Operationally, closing S_1 energizes inductor L_X from V_{BAT} and subsequently opening S_1 and closing S_2 de-energize L_X into C_{VAR} to pre-charge C_{VAR} at C_{MAX} to v_{CLAMP}. Once done, with both S_1 and S_2 open, S_3 connects C_{VAR} to C_{CLAMP} and C_{VAR} falls to C_{MIN} to harvest energy at v_{CLAMP}. Because C_{CLAMP} is substantially higher than C_{VAR}, v_{CLAMP} is nearly constant through the harvesting period. When C_{VAR} reaches C_{MIN}, S_2 and S_3 close to energize L_X from C_{CLAMP} and C_{VAR} until C_{CLAMP} discharges to its pre-charged state, at which point S_3 opens to allow S_2 to discharge C_{VAR} further. S_2 then opens and S_1 closes to drain L_X into V_{BAT}. After this, vibrations (and what little remnant energy remains in C_{VAR}) push C_{VAR}'s plates closer together to raise C_{VAR} to C_{MAX}, which is when the cycle repeats. Notice that, while C_{VAR} cycles once every several milliseconds, energy transfers require only microseconds, so transfers are practically instantaneous.

Energy Transfers: Since transferring less energy incurs less conduction losses, reducing the energy that L_X transfers is important, as is reducing the number of transfers. In the case of Fig. 15.8, while V_{BAT} invests E_{PC}, recovers E_{REC}, and receives E_{HARV}, C_{CLAMP} both receives and sources E_{HARV}, as Fig. 15.9a shows. On the other hand, investing E_{PC} from what C_{CLAMP} gains with E_{HARV}, as in Fig. 15.9b, reduces the energy that L_X transfers from C_{CLAMP} to V_{BAT} to $E_{HARV} - E_{PC} + E_{REC}$,

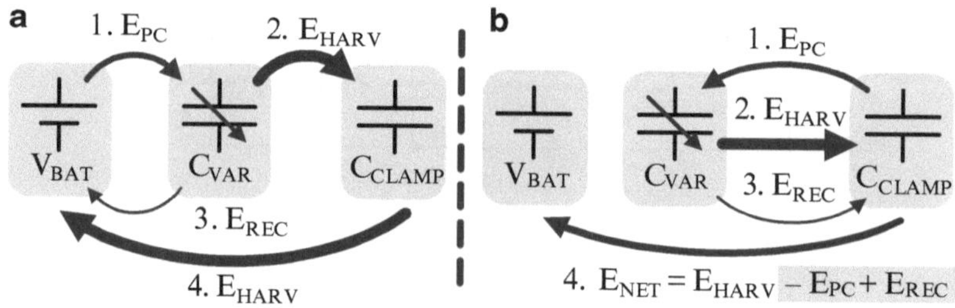

Fig. 15.9 (**a**) Battery- and (**b**) C_{CLAMP}-derived pre-charge investment strategies

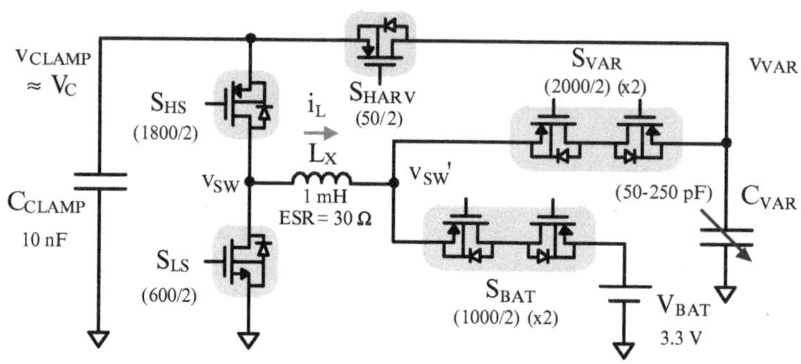

Fig. 15.10 Harvesting power stage (with MOSFET dimensions in μm)

which refers to net gain E_{NET}, while keeping all other transfers at equivalent levels. This means that reinvesting some of E_{HARV} to pre-charge C_{VAR} saves energy (i.e., reduces losses) and, as a result, yields more gain. What is more, since E_{REC} is normally low (because C_{MIN} is low), transfer losses can overwhelm E_{REC} to the extent that eliminating the recovery phase can also produce higher gains.

Circuit Implementation: The switching network of Fig. 15.10 realizes the low-loss sequence described in Fig. 15.9b. Here, with S_{VAR} closed, v_{CLAMP} falls slightly 1–1.5 μs after C_{VAR} peaks to C_{MAX}, as Fig. 15.11 illustrates, because S_{HS} energizes L_X and C_{VAR} from C_{CLAMP}. S_{LS} then drains L_X into C_{VAR} between 1.5 and 2 μs to finish pre-charging C_{VAR} to v_{CLAMP}. After that, with S_{VAR}, S_{HS}, and S_{LS} open, S_{HARV} closes across roughly 8.5 ms to steer C_{VAR}'s E_{HARV} into C_{CLAMP} until C_{VAR} falls to C_{MIN}, which raises v_{CLAMP} above its pre-cycle state. At C_{MIN}, S_{VAR} closes to drain C_{VAR}'s E_{REC} into C_{CLAMP} by energizing L_X with S_{LS} and de-energizing L_X with S_{HS}, which is why v_{VAR} falls to zero between 8.414 and 8.415 ms. The sequence continues by disengaging all switches so that vibrations can raise C_{VAR} back to C_{MAX} through the 8.5 ms of the reset phase. Finally, just before the next pre-charge phase, between 17.047 and 17.048 ms, v_{CLAMP} falls back to its pre-cycle value because S_{BAT} closes and S_{LS} and S_{HS} switch to transfer the net energy gained E_{NET} in C_{CLAMP} to V_{BAT}.

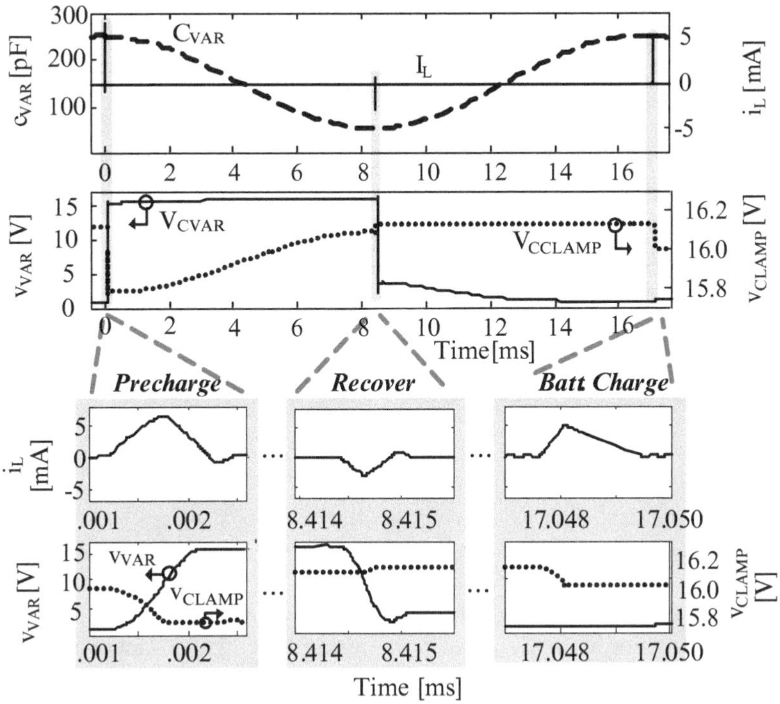

Fig. 15.11 Sample time-domain waveforms

15.4.4 Synchronization and Control

One of the fundamental functions of the controller is to synchronize the harvester circuit to C_{VAR}'s changing states. In other words, the controller must transition the circuit across its operational modes at particular times across C_{VAR}'s cycle. To be more precise, it must pre-charge C_{VAR} at C_{MAX} and recover E_{REC} from C_{VAR} at C_{MIN}, and start the harvesting phase immediately after Pre-charge and the reset phase after Harvest. The other basic function is to determine how much energy the inductor should receive and supply when engaged to transfer energy.

Monitor C_{VAR}: The onset and conclusion of every phase ultimately hinge on when C_{VAR} reaches C_{MAX} and C_{MIN}. Fortunately, floating C_{VAR} in the reset phase, when capacitance rises to C_{MAX}, ensures q_C is constant and equal to $v_C C_{VAR}$, so C_{VAR}'s v_C falls as C_{VAR} rises in charge-constrained fashion. Sensing, as a result, when v_C reaches its minimum voltage indicates when C_{VAR} peaks. For this purpose, comparator $CP_{PCH\text{-}DET}$ in Fig. 15.12 trips when v_C begins to rise above its delayed counterpart v_{CDLY} to prompt the system to start pre-charging C_{VAR}. Since this comparator only senses when C_{VAR} peaks to C_{MAX}, $CP_{PCH\text{-}DET}$ should not operate through C_{VAR}'s entire cycle, because it would otherwise consume more

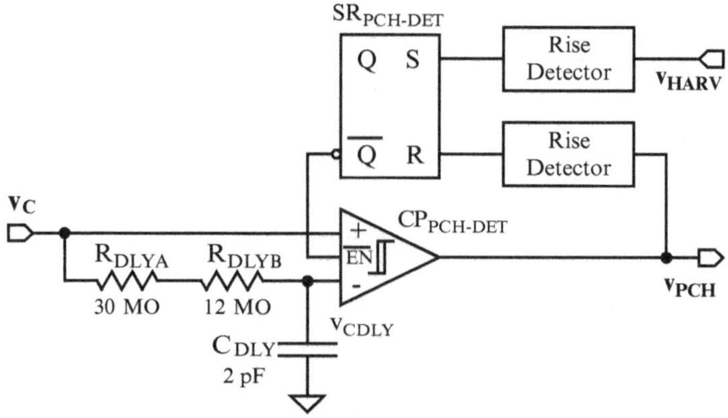

Fig. 15.12 Duty-cycled (C_{MAX} and $v_{C(MAX)}$) peak-sensing comparator

power than necessary. This is why end-of-harvest signal v_{HARV} and $CP_{PCH\text{-}DET}$'s own output (v_{PCH} in Fig. 15.12) enable and disable (i.e., duty-cycle) the comparator by setting and resetting a Set–Reset (SR) latch.

Since C_{VAR} does not float in the harvesting phase of voltage-constrained systems, v_C does not transition with C_{VAR}, as in the reset phase, which means v_C no longer monitors C_{VAR} during Harvest. However, because v_C is practically constant through this period, q_C in C_{VAR} decreases as long as C_{VAR} falls. As a result, the harvesting current that q_C generates (in i_{HARV}) drops to zero when C_{VAR} stops falling at C_{MIN}. Therefore, sensing when the voltage across a series resistor (R_S) in the path of i_{HARV} falls to zero indicates when C_{VAR} reaches C_{MIN}. To this end, a comparator (CP_{END}) with millivolts of offset whose input terminals connect across R_S trips when $i_{HARV}R_S$ drops within millivolts of zero to prompt the system to end the harvesting phase. Again, as with C_{MAX}, CP_{END} need only operate during the harvesting period, so an end-of-pre-charge signal and CP_{END}'s own output can duty-cycle CP_{END} to save energy. Note that R_S can be the resistance of a connecting switch when engaged.

Control Energy Transfer L_X: The controller must also determine how long L_X should energize to pre-charge C_{VAR} to its target v_{CLAMP} and to drain C_{CLAMP} into V_{BAT}. Charging C_{VAR} as close to v_{CLAMP} as possible is important to minimize the Ohmic power that the switch between C_{VAR} and C_{CLAMP} dissipates when engaged. The system must, as a result, include tuning features. Manually adjusting L_X's energizing time τ_E is possible when the energy transferred is consistent across cycles. Operating conditions in a real-life application, however, may change with time, as will the performance of the components in the system as they age, which means automatic tuning may be necessary.

The system should energize L_X for a nominal τ_E and disconnect L_X when it no longer has energy. Sensing when L_X's current i_L falls to zero via a sense resistor is one way of determining when to stop de-energizing L_X. Alternatively, a diode or its equivalent can disconnect L_X asynchronously when i_L attempts to reverse direction, which is when L_X no longer has energy.

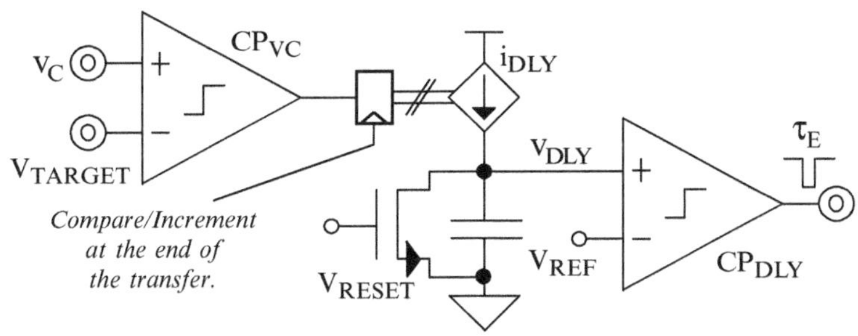

Fig. 15.13 Circuit that tunes energizing time τ_E from cycle to cycle with a feedback loop

A way of tuning the transfer process to ensure L_X does not draw more energy than necessary is to compare the voltage of the receiving or sourcing capacitor to its target value at the end of the transfer and lengthen or shorten τ_E for the ensuing cycle as needed. The circuit in Fig. 15.13, for example, implements a feedback loop that automatically adjusts τ_E across cycles so τ_E remains near its optimal setting to charge or discharge a capacitor to its target V_{TARGET}. In this particular embodiment, i_{DLY} establishes τ_E by setting how fast v_{DLY} rises to V_{REF} and comparator CP_{VC} increments i_{DLY} up or down at the end of every cycle until v_C at the end of the transfer is within millivolts of V_{TARGET}.

15.4.5 Battery-Constrained Example

Figure 15.14 illustrates a prototyped embodiment of a battery-constrained electrostatic energy-harvesting charger, which is a simplified implementation of the one in Fig. 15.10. The fundamental aim here was to remove the clamping capacitor C_{CLAMP} and use the battery V_{BAT} in its place. That way, V_{BAT} constrains C_{VAR} while at the same time receiving its charge through the entire harvesting phase. The benefits of eliminating C_{CLAMP} are less off-chip components (i.e., less printed-circuit-board area), less CMOS switches (i.e., less silicon area), lower peak voltage (i.e., use components with a lower breakdown voltage), and less energy transfers (i.e., lower circuit complexity and losses). The fundamental trade-off is lower output power because C_{VAR} now draws energy from motion with a lower electrostatic force across its plates. A side effect of this is that the energy left in C_{VAR} at the end of the harvesting period (i.e., E_{REC} at C_{MIN}) is minimal. As such, because power losses through the system are on the same order as the power generated, the system also skips the recovery process.

With these simplifications, pre-charge comparator $CP_{P\text{-}STRT}$ and delay resistor–capacitor R_D–C_D in Fig. 15.14 (from Fig. 15.12) start the process by prompting M_{PE} to energize L_X when C_{VAR} peaks to C_{MAX}. After that, M_{ND} de-energizes L_X into C_{VAR} to pre-charge C_{VAR} to V_{BAT}, which corresponds to every time v_C rises to 3.6 V

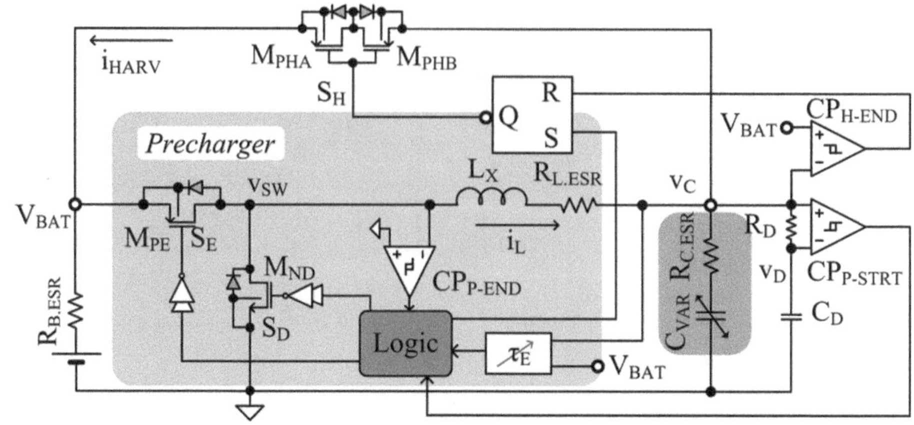

Fig. 15.14 Battery-constrained electrostatic energy-harvesting charger

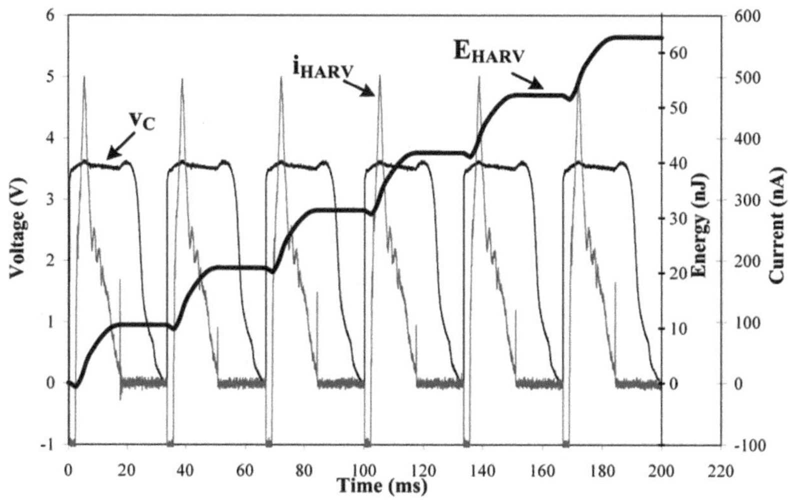

Fig. 15.15 Experimental waveforms at 30 Hz when C_{VAR} ranges between 152 and 977 pF and L_X is 10 μH

in Fig. 15.15. Harvesting switch M_{PHB} then shorts to both constrain C_{VAR} at V_{BAT} and receive its charge in i_{HARV} (while v_C remains at 3.6 V in Fig. 15.15). When i_{HARV} is close to zero (i.e., when C_{VAR} nears C_{MIN}), the voltage across M_{PHB} is within millivolts of zero, so comparator $CP_{H\text{-}END}$ trips to disconnect M_{PHB}. Vibrations then raise C_{VAR} to C_{MAX} through the reset phase, which is when v_C falls to near zero in Fig. 15.15. $CP_{P\text{-}STRT}$ senses when C_{VAR} reaches C_{MAX} to restart the entire cycle.

Note that the drop in harvested energy E_{HARV} at the beginning of every cycle in Fig. 15.15 corresponds to the energy V_{BAT} invests to pre-charge C_{VAR}. E_{HARV}, however, rises more than it drops across every cycle, which means the battery

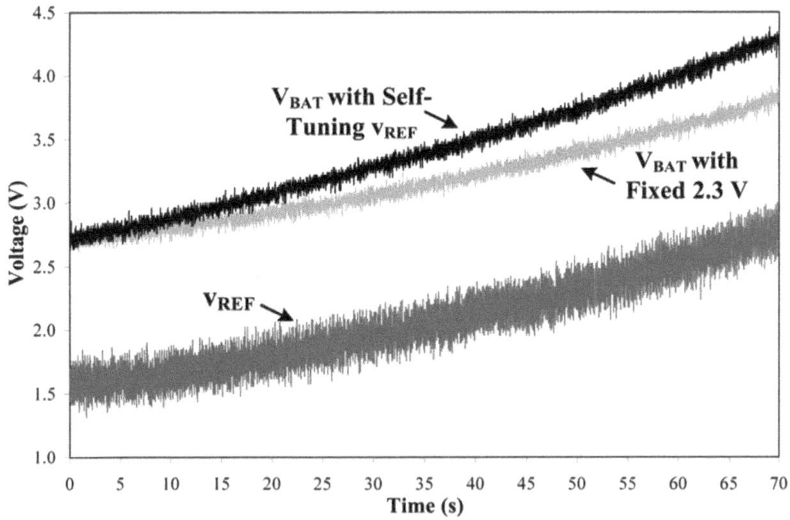

Fig. 15.16 Charging profile of a 1-μF capacitor with and without a tuning loop

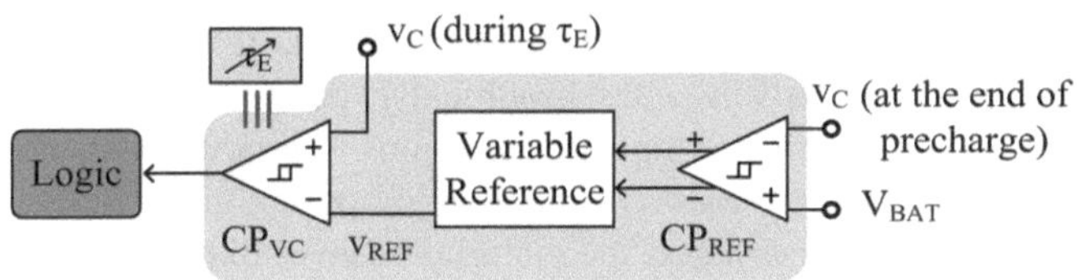

Fig. 15.17 Feedback loop that automatically tunes the inductor's energizing time τ_E

ultimately charges, as Fig. 15.16 demonstrates. Of course, including the tuning-loop concept of Fig. 15.13 as implemented in Fig. 15.17 charges V_{BAT} even faster, because the system dissipates less unnecessary losses. In this case, CP_{VC} stops energizing the inductor during Pre-charge when C_{VAR}'s voltage v_C reaches v_{REF}, and CP_{REF} then compares v_C and V_{BAT} at the end of Pre-charge to determine in which direction to increment τ_E via v_{REF}. Continually adjusting the inductor's energizing time this way ensures C_{VAR}'s pre-charge voltage tracks V_{BAT} as V_{BAT} charges across time and discharges when connected to a load.

15.5 Piezoelectric Harvesters

A piezoelectric harvester extracts kinetic energy from the work ambient vibration exerts on a piezoelectric material when mechanically straining it. While affixed to a stationary base, like Fig. 15.1 shows, a piezoelectric strip generates ac charge in response to mechanical displacements. In this fashion, the device converts mechanical

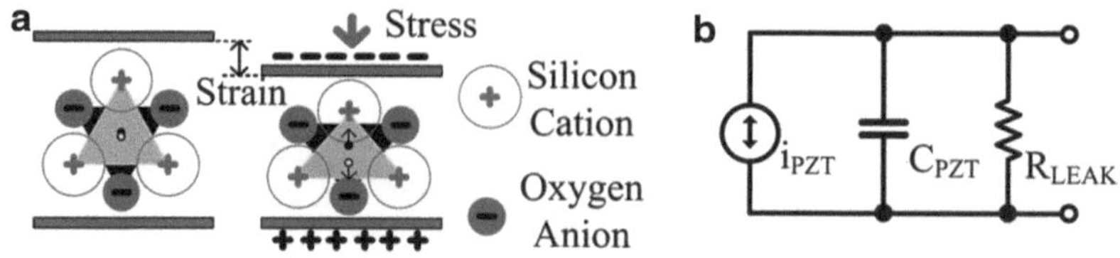

Fig. 15.18 (**a**) Ionic charges in piezoelectric crystals and (**b**) equivalent circuit model

energy into the electrical domain. The role of the harvesting circuit is to therefore condition and steer this charge into a temporary or long-term reservoir from which an electronic system can derive power to perform its prescribed functions. More on piezoelectric harvesters can be found in [17–19] and the references they cite.

15.5.1 Transducer

When vibrations strain a piezoelectric material, the stress rearranges the internal structure of the lattice and, in doing so, offsets the charge balance of the crystal. As a result, as Fig. 15.18a illustrates graphically, positive and negative charges (i.e., cations and anions) shift in opposite directions to produce a surface potential. The potential and its associated currents change continuously with variations in mechanical deformation. In essence, the material behaves like an alternating current source (i.e., i_{PZT} in Fig. 15.18b) that charges and discharges the capacitance across the surfaces of the material (i.e., C_{PZT}). In practice, the transducer leaks some of the charge it generates (through R_{LEAK}), but not much.

While typical piezoelectric transducers, which are relatively mature and reliable technologies, can produce high peak voltages, their miniaturized counterparts cannot because the amount of charge generated is proportional to strain and surface area, both of which decrease considerably with reductions in volume. Furthermore, typical environments supply mostly "weak" levels of vibrations, from which energy (and voltage) is so low that harvesters might not be able to operate.

15.5.2 Bridge Harvesters

To rectify and channel the ac power that a piezoelectric transducer generates into an intermediate storage device like a battery or capacitor, harvesters often employ full-wave bridge rectifiers. These circuits, as Fig. 15.19 shows, steer charge to the output only when i_{PZT} charges C_{PZT} above the barrier voltage that two conducting diodes (i.e., $2V_D$) and an output capacitor C_{RECT} (i.e., V_{RECT}) produce. Since harvesting more energy amounts to channeling more of i_{PZT} into C_{RECT} (not C_{PZT}), C_{RECT}

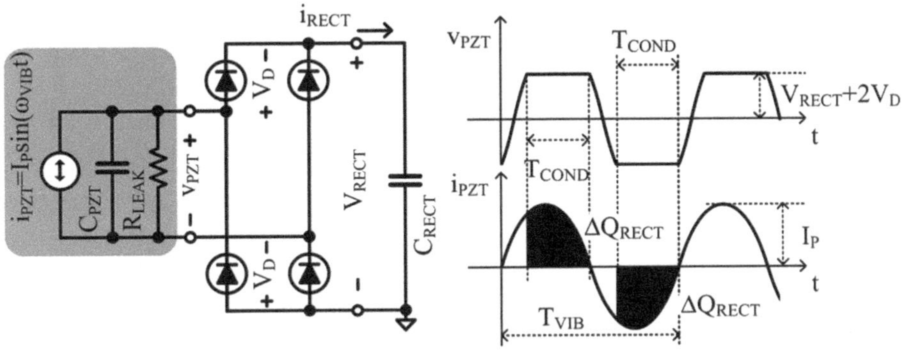

Fig. 15.19 Full-wave bridge rectifier and relevant transducer waveforms

is by design substantially larger than C_{PZT}, which means C_{RECT} presents a lower impedance path for i_{PZT} to flow than C_{PZT} does. Accordingly, in one vibration cycle T_{VIB}, V_{RECT} remains practically unchanged and, after i_{PZT} charges C_{PZT} to the aggregate sum of $2V_D$ and V_{RECT}, v_{PZT} clamps and i_{PZT} flows into C_{RECT} through conduction time T_{COND}.

Because leakage in piezoelectric transducers is typically negligible (i.e., R_{LEAK} is large) and C_{PZT} is a reactive component (which does not consume power), most of the energy i_{PZT} carries through the diodes reaches the output, except for the power the diodes dissipate. Rectifier efficiency η_{RECT}, which refers to the ratio of rectified output energy (per cycle) E_{RECT} to the input energy of the rectifier (per cycle) E_{EE}, therefore deteriorates (i.e., falls) with increasing diode voltage V_D:

$$\eta_{RECT} = \frac{E_{RECT}}{E_{EE}} = \frac{2\int_{T_{COND}} i_{RECT} V_{RECT} \mathrm{d}t}{2\int_{T_{COND}} i_{PZT}\left(V_{RECT} + 2V_D\right) \mathrm{d}t} \approx \frac{V_{RECT}}{V_{RECT} + 2V_D}. \tag{15.13}$$

Replacing the diodes with low-threshold (from Fig. 15.20a), threshold-shifted (from Fig. 15.20b), or synchronized (from Fig. 15.20c) MOS switches increases η_{RECT} because MOSFETs reduce V_D from 0.5 to 0.7 V to millivolts. Alternatively, replacing two of the four diodes in Fig. 15.20d with cross-coupled PMOS transistors that derive gate drive from opposing peak voltages, like Fig. 15.20e shows, reduces the overhead from $V_{RECT} + 2V_D$ to $V_{RECT} + V_D + V_{SD}$.

Still, a fundamental limitation with rectifiers is that v_{PZT} must exceed V_{RECT} (which is a manifestation of the energy already stored in C_{RECT}) to extract energy from the transducer. In other words, even if V_D were zero, the rectifier ceases to harness energy when v_{PZT}'s open-circuit peak voltage V_P falls below V_{RECT}, which can easily happen under weak vibration levels. Plus, low-threshold transistors leak power, threshold-shifted and synchronized MOSFETs dissipate quiescent power, and synchronized switches require time to respond.

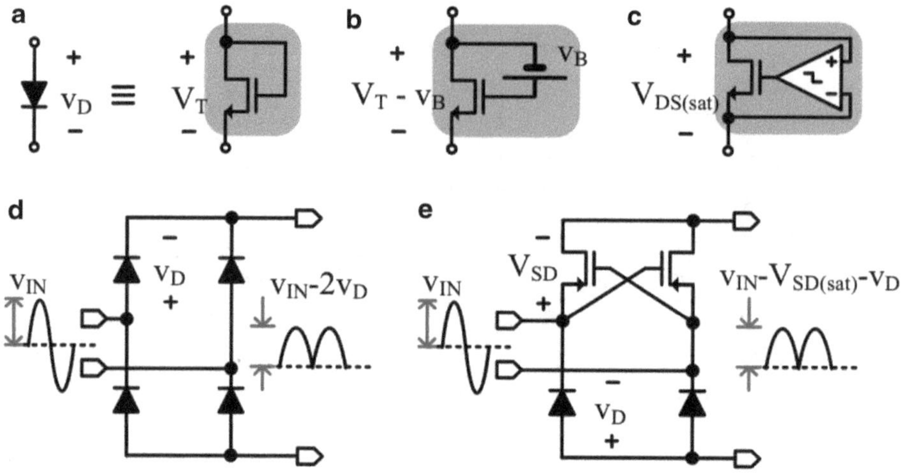

Fig. 15.20 (**a**) Diode-connected, (**b**) V_T-shifted, and (**c**) synchronized switches and (**d**) diode-based and (**e**) diode–cross-coupled-MOS rectifiers

One way of reducing the input threshold of a bridge is by conditioning C_{RECT} so V_{RECT} can always remain below v_{PZT}'s peak voltage. This way, the system can harness some of the energy C_{PZT} receives under weak vibrations that the transducer would have otherwise transferred back into the mechanical domain. To quantify how much energy E_{RECT} reaches C_{RECT} with respect to V_{RECT}, first consider that, without the rectifier, i_{PZT} would charge C_{PZT} from $-V_P$ to V_P with Q_{PZT}' or $C_{PZT}2V_P$. When the bridge loads the transducer, because a portion of Q_{PZT} charges C_{PZT} from $-(V_{RECT}+2V_D)$ to $(V_{RECT}+2V_D)$, as shown in Fig. 15.19, the rectifier steers a fraction of Q_{PZT}' as Q_{RECT}:

$$\Delta Q_{RECT} = \Delta Q_{PZT}' - 2\left(V_{RECT}+2V_D\right)C_{PZT} = 2V_P C_{PZT} - 2\left(V_{RECT}+2V_D\right)C_{PZT}, \tag{15.14}$$

into C_{RECT} (at V_{RECT}) for each half cycle. Therefore, E_{RECT} grows linearly with V_P and becomes a second-order function of V_{RECT}:

$$E_{RECT} = 2\left(V_{RECT}\Delta Q_{RECT}\right) = 4V_{RECT}C_{PZT}\left[V_P - \left(V_{RECT}+2V_D\right)\right]. \tag{15.15}$$

As a result, tuning V_{RECT} to $0.5(V_P-2V_D)$ allows a bridge to produce the maximum possible energy per cycle:

$$E_{RECT(MAX)} = C_{PZT}\left(V_P - 2V_D\right)^2 \leq C_{PZT}V_P^2, \tag{15.16}$$

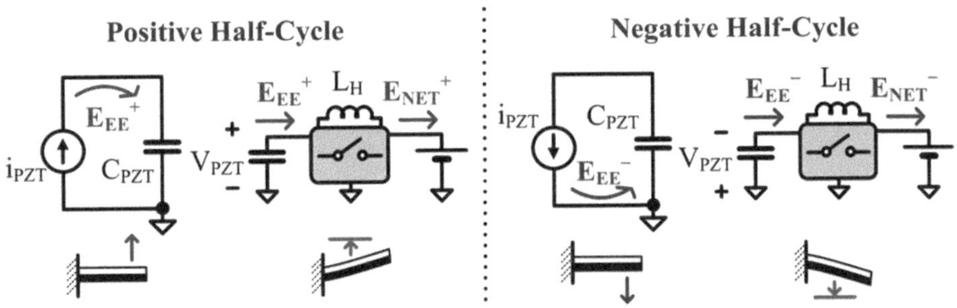

Fig. 15.21 Energy-harvesting cycle for a switched-inductor piezoelectric harvesting system

which with ideal diodes (where V_D is zero) rises to $C_{PZT}V_P{}^2$. Although this performance represents an improvement, monitoring v_{PZT} and adjusting V_{RECT} "on the fly" can easily consume more energy than the system can harness under weak vibration levels, even if V_D were zero. In other words, the power that the additional circuitry requires imposes a lower bound on v_{PZT}'s V_P below which the bridge cannot produce a net energy gain.

15.5.3 Switched-Inductor Harvesters

Switched-inductor systems circumvent the fundamental input threshold limitation that bridges impose by draining all of C_{PZT}'s energy into an intermediate inductor L_H. Since L_H empties C_{PZT} irrespective of how small v_{PZT}'s V_P is, there is no threshold above which v_{PZT} must rise to extract energy from C_{PZT}. Without this basic threshold, a system can harness kinetic energy from very weak vibrations. This way, input power need only exceed the parasitic power losses that the electronics dissipate to generate output power.

Energy-Transfer Sequence: An inductor-based system of this sort, as Fig. 15.21 illustrates, waits for vibrations to charge C_{PZT} to V_P across each half cycle before de-energizing C_{PZT} into L_H and then L_H into a local battery V_{BAT}. Since the vibration cycle (T_{VIB}) is usually in milliseconds and transferring energy from C_{PZT} to L_H and L_H to V_{BAT} requires only a few microseconds, C_{PZT}'s v_{PZT} remains practically unchanged at V_P through C_{PZT}–L_H and L_H–V_{BAT} transfers. As long as vibrations persist, the cycle repeats after the harvester charges V_{BAT} from negative v_{PZT} voltages.

Considering L_H harnesses C_{PZT}'s energy at V_P at the end of the positive and negative half cycles $E_{C(PEAK)}{}^+$ and $E_{C(PEAK)}{}^-$:

$$E_{L.HARV} = E_{C(PEAK)}{}^+ + E_{C(PEAK)}{}^- = 2E_{C(PEAK)} = 2\left[0.5C_{PZT}\,(2V_P)^2\right]$$
$$= 4C_{PZT}\,V_P{}^2, \qquad (15.17)$$

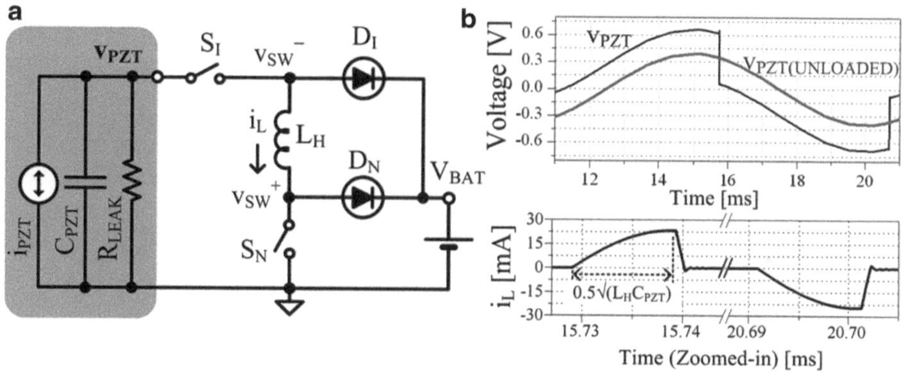

Fig. 15.22 (**a**) Switched-inductor circuit and (**b**) simulated waveforms

and the circuit consumes conduction, gate-drive, and control-circuit power as E_{LOSS}, V_{BAT} ultimately charges with

$$E_{NET} = E_{L.HARV} - E_{LOSSES} = 4C_{PZT}\,V_P{}^2 - E_{LOSS}. \tag{15.18}$$

In other words, the system can harness up to $4C_{PZT}V_P{}^2$ across each vibration period, which is four times more energy than ideal bridges can (at $C_{PZT}V_P{}^2$) under optimal operating conditions, when V_D is 0 V and V_{RECT} is $0.5V_P$. The primary reason for this improvement over the bridge is that the inductor fully exhausts C_{PZT} to 0 V every half cycle. Said differently, while a diode bridge allows vibrations to charge and discharge C_{PZT} by $2V_P$ from $-V_P$ to $+V_P$ and back to $-V_P$, v_{PZT} in switched-inductor harvesters rise and fall an equivalent amount from zero, that is, from zero to $2V_P$ and again from 0 to $-2V_P$. As a result, the energy a bridge-based harvester collects in C_{PZT} through each half cycle peaks to $0.5C_{PZT}V_P{}^2$ while that of its switched-inductor counterpart peaks to $0.5C_{PZT}(2V_P)^2$, which means inductors can draw four times as much energy from vibrations than bridges can.

Circuit Implementation: Figure 15.22 shows an implementation of the switched-inductor network described above. First, through the positive half cycle, switch S_I decouples the power stage from the transducer until v_{PZT} reaches its positive peak. The system then discharges C_{PZT} into L_H by closing S_I and S_N. Since LC resonance drives this energy transfer, L_H energizes fully after one quarter of L_H–C_{PZT}'s resonance period, after $0.25T_{LC}$ or $0.5\pi(L_HC_{PZT})^{0.5}$. After this, S_N opens and inductor current i_L charges the parasitic capacitance at switching node $v_{SW}{}^+$ quickly until diode-switch D_N forward biases and depletes L_H into V_{BAT}. Similarly, S_N decouples the circuit from the transducer during the negative half cycle until v_{PZT} reaches its negative peak. S_I and S_N then discharge C_{PZT} into L_H for one quarter of L_H–C_{PZT}'s resonance period. Afterwards, S_I opens and i_L flows into $v_{SW}{}^-$ until D_I forward biases and conducts i_L into V_{BAT}.

Notice that, without the harvesting circuit, the unloaded transducer charges C_{PZT} to lower voltages $v_{PZT(UNLOADED)}$, which demonstrates the conditioning effect that a harvester has on its transducer. As mentioned earlier, the circuit can harness energy from small v_{PZT} voltages because L_H energizes as soon as v_{PZT} is nonzero. Plus, L_H, like a current source, raises the voltage of the switching nodes until the diode switches clamp it to V_{BAT}. In contrast, a bridge-based harvester charges C_{PZT} to lower v_{PZT} voltages, harnesses energy only when v_{PZT} rises above its rectified output V_{RECT}, and requires a dc–dc converter to convey energy from V_{RECT} to V_{BAT}, the combined effect of which is less output power with respect to its switched-inductor counterpart.

15.5.4 Synchronization and Control

Every half cycle of the switched-inductor network discussed above includes two phases: harvest and transfer. First, through the harvesting phase, the system should wait until v_{PZT} peaks, when C_{PZT} has the most energy. At that point, in the transfer phase, the harvester should drain C_{PZT} into L_H and then L_H into V_{BAT}. The network again harvests in the opposite half cycle until v_{PZT} peaks in the opposite direction, at which point L_H again transfers energy from C_{PZT} to V_{BAT}. To achieve all this, the controller must (1) detect when v_{PZT} peaks and (2) energize and de-energize L_H long enough to drain C_{PZT} and deposit its energy into V_{BAT}.

Monitor v_{PZT}: One way of monitoring when v_{PZT} peaks is by comparing v_{PZT} with its delayed counterpart, as done by the comparator shown in Fig. 15.12. Here, as v_C rises through the positive half cycle, v_C is above its delayed counterpart v_{CDLY}. When v_C stops increasing and begins to fall, just past v_C's peak, v_C drops below v_{CDLY}, tripping the comparator that prompts the system to discharge C_{PZT} into L_H. Similarly, v_C is below v_{CDLY} through the negative half cycle until v_C ceases to fall and begins to rise, at which point v_C surpasses v_{CDLY} to signal the system to again drain C_{PZT} into L_H.

Control Energy Transfer $L_{(x)}$: The key to controlling L_H is timing. Once v_{PZT}'s peak-sensing comparator starts L_H's energizing period, after v_{PZT} peaks in the positive or negative half cycle, the system should wait for C_{PZT} to discharge fully into L_H, but no longer because C_{PZT} would otherwise pull energy back from L_H. A comparator can sense when switching nodes $v_{SW}{}^{+}$ and $v_{SW}{}^{-}$ in Fig. 15.22 approach zero, when C_{PZT} is almost empty, to stop energizing L_H. For this, the comparator should respond within a small fraction of L_H–C_{PZT}'s quarter resonance period, which is how long energy transfers fully between C_{PZT} and L_H. Alternatively, if the frequency of vibrations is predictable and consistent, the designer can set and tune this time manually with the time constant of an adjustable resistor–capacitor network. The advantage to trimming the delay manually is less power because a fast-responding comparator requires considerably higher quiescent current than a slower one.

L_H's de-energizing period is simpler to control. Once the aforementioned comparator or tunable delay stops energizing L_H, L_H's current i_L charges $v_{SW}{}^{+}$ and $v_{SW}{}^{-}$

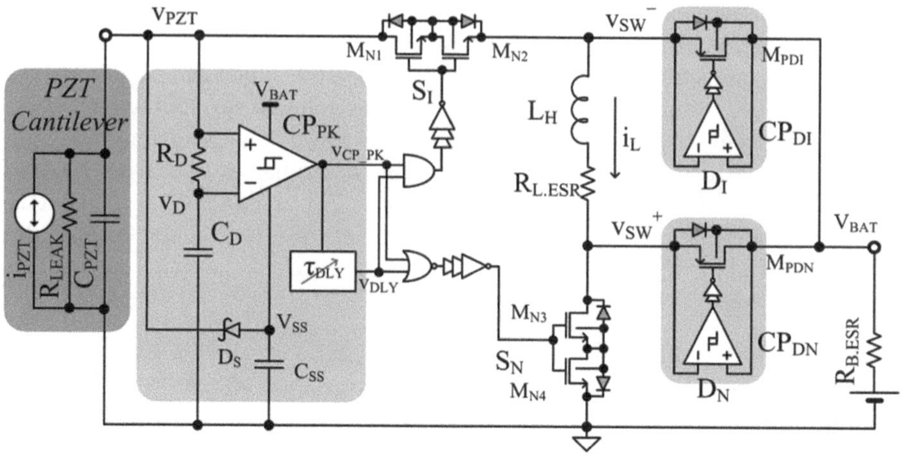

Fig. 15.23 Switched-inductor piezoelectric energy-harvesting charger

at the end of the positive and negative half cycles, respectively, until the diodes D_N and D_I forward bias and steer i_L into V_{BAT}. Once L_H exhausts its energy, the diodes keep reverse current from flowing back into L_H. In other words, the diode switches open automatically and end the de-energizing period asynchronously, without the aid of a controller.

15.5.5 Switched-Inductor Example

Figure 15.23 shows a switched-inductor embodiment of the piezoelectric harvester described above. Here, n-channel MOSFETs M_{N1}–M_{N2} and M_{N3}–M_{N4} implement switches S_I and S_N in Fig. 15.22, which energize L_H from C_{PZT}. The purpose of using series transistors in an isolated p-type well is to eliminate the undesirable conduction path the body diode would otherwise establish when using only one n-channel or p-channel FET while v_{PZT} swings, for example, above v_{SW}^- (and V_{BAT}) and below ground. This way, back-to-back body diodes block reverse current because the p-well potential tracks the lower of the two terminal voltages present across the off-state switch. This happens because the diode attached to the most negative potential always discharges the p well to that voltage when the switch is off; otherwise, the FETs short the well to the switch's terminal voltages, so the circuit always biases the well. Two series p-type transistors would also block reverse current, but fully engaging the PFETs when v_{PZT} is below ground requires a gate-drive voltage that is well below v_{PZT}'s negative peak. In contrast, V_{BAT} is sufficiently high to engage the NFETs when v_{PZT} falls below ground and v_{PZT}'s negative peak is sufficiently low to disengage them when v_{PZT} rises.

Comparators CP_{DN} and CP_{DI} along with the p-channel transistors that they drive (i.e., M_{PDN} and M_{PDI}) implement D_N and D_I in Fig. 15.22. The motivation for using M_{PDN} and M_{PDI} is to reduce the voltage (and power) dropped across D_N and D_I when they conduct. To this end, CP_{DN} and CP_{DI} close M_{PDN} and M_{PDI} when L_H's i_L raises v_{SW}^+ and v_{SW}^- above V_{BAT}. Because the comparators do not respond instantaneously, however, M_{PDN} and M_{PDI}'s body diodes first engage, but only momentarily, until the comparators react and close their corresponding FETs. The comparators should therefore respond quickly, to not only limit the power lost in the diodes but also keep V_{BAT} from discharging into L_H through M_{PDN} and M_{PDI} when the diodes open. Note that connecting M_{PDN} or M_{PDI} early when S_N or S_I is still on shorts and discharges V_{BAT} to ground or v_{PZT}, so ensuring there is a dead time between these adjacent switches is important.

CP_{PK}, R_D, and C_D compare v_{PZT} against its delayed counterpart to detect when v_{PZT} peaks and signals the system to drain C_{PZT} into L_H. Because v_{PZT} drops below ground, CP_{PK}'s allowable input common-mode range (ICMR) must extend below ground to include v_{PZT}'s negative peak voltage. For that, CP_{PK} powers from V_{BAT} to the negative supply voltage that Schottky diode D_S and C_{SS} establish with V_{SS}, which they set by sensing and holding v_{PZT}'s lowest voltage. Note that reducing power in CP_{PK} is crucial because CP_{PK} must detect peaks (i.e., operate) through the entire vibration period. Fortunately, the cycle is long, so designers can trade speed for power by, for example, operating in subthreshold. When implemented, CP_{PK} included an offset and a delay that caused it to trip about 1 ms later than expected at 5 and 9 ms in Fig. 15.24 instead of 4 and 8 ms. As a result, because v_{PZT} was not at its peak, the system extracted less energy than it could have.

With delay τ_{DLY} tuned to approximately a quarter of C_{PZT}–L_H's resonance period $0.25T_{LC}$ or $0.5\pi(L_H C_{PZT})^{0.5}$, τ_{DLY} ends the C_{PZT}–L_H energy-transfer period. This way, with *priori* knowledge of L_H and C_{PZT}, the system estimates (rather than senses) when C_{PZT} fully depletes, thereby sacrificing accuracy for the sake of lower circuit complexity and lower power. Delay τ_{DLY} in the implemented system of Fig. 15.23 was τ_{LE} at about 10 µs, from 4,979 to 4,990 ms and 8,877 to 8,888 ms in Fig. 15.24. L_H then drained asynchronously through D_N and D_I in roughly 1 µs (i.e., τ_{LDE}) at 4,990 and 8,888 ms to source charge into V_{BAT} with i_{BAT}. With vibration levels that normally produce 0.65 and 0.9 V peak–peak voltages across an unloaded $44 \times 13 \times 0.4\text{-mm}^3$ piezoelectric cantilever at 100 Hz, the prototyped system charged 1.2- and 1.5-mAh lithium-ion batteries by about 200 and 40 mV, respectively, in 300 minutes, as Fig. 15.25 shows. This means the network not only harnessed more energy than it dissipated but also drew energy from weak vibration levels.

15.6 Conclusions

Ultimately, the objective of the system is to output power, not simply harness it from a transducer, because losses in the system can easily overwhelm any gains. In this respect, a switched-inductor converter can be efficient, but not completely lossless in practice. The controller, for example, which generally consists of fast comparators,

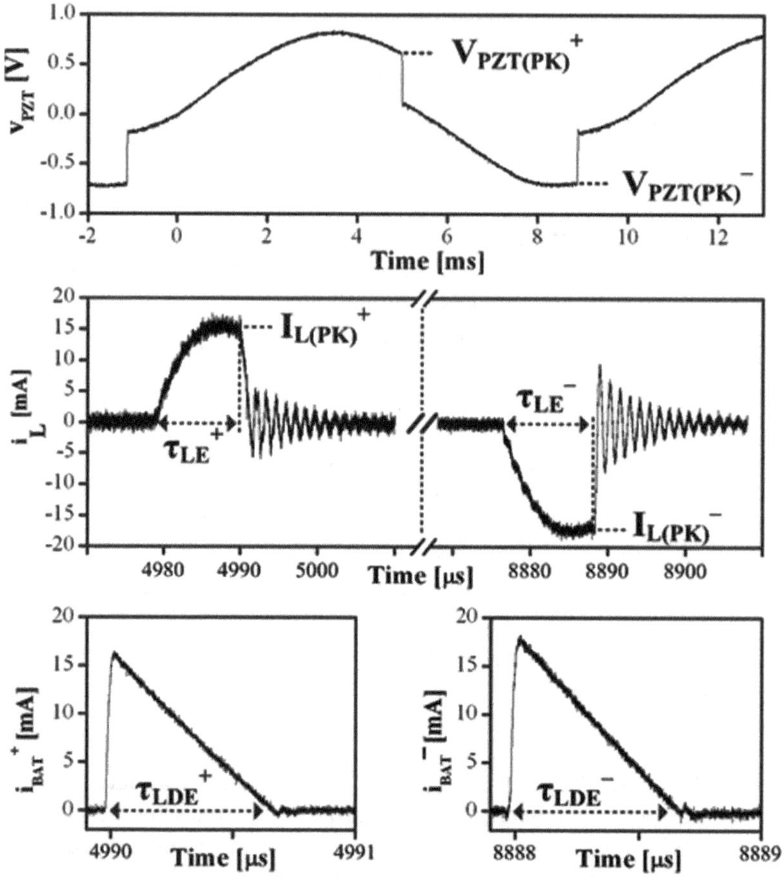

Fig. 15.24 Experimental waveforms of C_{PZT}'s v_{PZT}, L_H's i_L, and V_{BAT}'s i_{BAT}

slow comparators, logic, and bias blocks, requires power to function. Comparators that monitor inductor current, like $CP_{P\text{-}END}$ in the electrostatic case of Fig. 15.14 and CP_{DI} and CP_{DN} in the piezoelectric counterpart of Fig. 15.23, must respond quickly because the energy-transfer process that they control occurs in microseconds of the millisecond period. Luckily, they operate only for a small fraction of the period, so the strong-inversion power they consume amounts to little energy per cycle. Comparators that monitor the transducer, like $CP_{P\text{-}START}$ and $CP_{H\text{-}END}$ for C_{VAR} in Fig. 15.14 and CP_{PK} for v_{PZT} in Fig. 15.23, need not respond in microseconds because they monitor millisecond events, so they need less power. Unfortunately, they, like bias circuits, run through the entire vibration cycle, which means they can dissipate considerable energy, if not biased in subthreshold (with nanoamps). Logic also consumes energy, but not as much the comparators because the frequency of vibrations is typically low between 50 and 300 Hz and transistors are generally small, so they require little energy to drive.

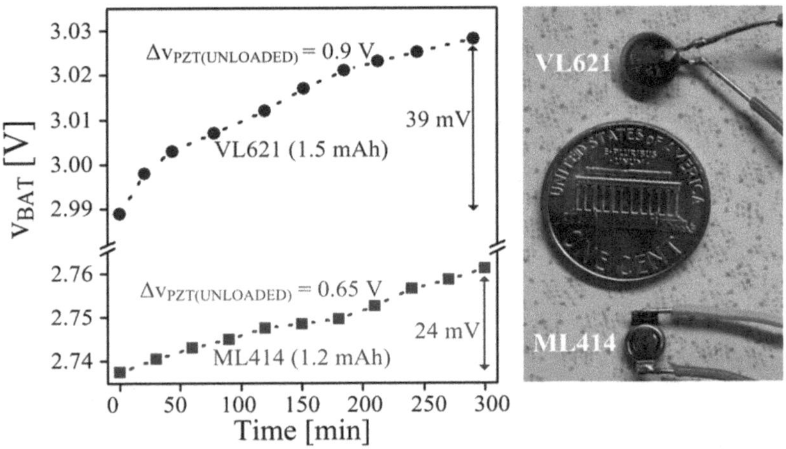

Fig. 15.25 Charging profile of 1.2- and 1.5-mAh Li-ion batteries

More fundamentally, the switches and parasitic resistors in the current-conduction path of the network dissipate power, as do the circuits that drive charge to and from the switches' gates. Since conduction losses increase with channel resistance and drive losses with gate capacitance, smaller transistors dissipate more conduction power and less drive energy than larger ones, and vice versa. Accordingly, using minimum channel lengths and balancing transistor widths in the power stage to minimize overall losses in the system are important guidelines to consider. Another design trade-off is the number of times the inductor energizes and de-energizes in one cycle. Transferring energy multiple times per cycle, for example, increases the number of switching events (and associated switching losses) and decreases i_L's peak magnitude (which reduces conduction power). In other words, sizing and duty-cycling the constituent components of the system for low power dissipation are as critical as arranging them into an optimal configuration that draws the most energy from the transducing source. More on reducing losses in energy-harvesting systems can be found in [15, 19–21] and the references they cite.

References

1. Torres EO, Rincón-Mora GA (2005) Harvesting ambient energy. Electronic Engineering Times, Aug 29, 2005
2. Torres EO, Rincón-Mora GA (2006) Harvesting energy into lithium-ion batteries. Planet Analog, Feb 14, 2006
3. Torres EO, Rincón-Mora GA (2008) Energy-harvesting system-in-package (SiP) microsystem. ASCE J Energy Eng 134(4):121–129
4. Kwon D, Rincón-Mora GA, Torres E (2010) Harvesting kinetic energy with switched-inductor DC-DC converters. IEEE International Symposium on Circuits and Systems, Paris, France, 30 May–2 Jun 2010

5. Kwon D, Rincón-Mora GA, Torres EO (2011) Harvesting ambient kinetic energy with switched-inductor converters. IEEE Trans Circuits Syst I 58(7):1551–1560
6. Rincón-Mora GA (2009) Analog IC design with low-dropout regulators. McGraw-Hill, New York (400 pages)
7. Rincón-Mora GA (2009) Power IC design. Raleigh, Lulu (264 pages)
8. Kim S, Rincón-Mora GA (2009) Achieving high efficiency under micro-watt loads with switching buck DC–DC converters. J Low Power Electron 5(2):229–240
9. Gildersleeve M, Forghani-zadeh HP, Rincón-Mora GA (2002) A comprehensive power analysis and a highly efficient, mode-hopping DC-DC converter. IEEE Asia-Pacific Conference on ASICs, Taipei, Taiwan, pp 153–156
10. Peterson K, Rincón-Mora GA (2012) High-damping energy-harvesting electrostatic CMOS charger. IEEE International symposium on circuits and systems, Seoul, Korea, 20–23 May 2012
11. Torres EO, Rincón-Mora GA (2005) Long lasting, self-sustaining, and energy-harvesting system-in-package (SiP) sensor solution. International conference on energy, environment, and disasters, Session A-2, ID 368, Charlotte, NC, July 2005, pp 1–33
12. Torres EO, Rincón-Mora GA (2006) Electrostatic energy harvester and Li-Ion charger for micro-scale applications. IEEE Midwest symposium on circuits and systems, San Juan, Puerto Rico, 6–9 Aug 2006, pp 65–69
13. Torres EO, Rincón-Mora GA (2009) Energy budget and high-gain strategies for voltage-constrained electrostatic harvesters. IEEE International symposium on circuits and systems, Taipei, Taiwan, 24–27 May 2009, pp 1101–1104
14. Torres EO, Rincón-Mora GA (2009) Electrostatic energy-harvesting and battery-charging CMOS system prototype. IEEE Trans Circuits Syst I 56(9):1938–1948
15. Torres EO, Rincón-Mora GA (2010) A 0.7 um BiCMOS electrostatic energy-harvesting system IC. IEEE J Solid State Circuits 45(2):483–496
16. Torres EO, Rincón-Mora GA (2010) Self-tuning electrostatic energy-harvester IC. IEEE Trans Circuits Syst II 57(10):808–812
17. Kwon D, Rincón-Mora GA (2009) A rectifier-free piezoelectric energy harvester circuit. IEEE International symposium on circuits and systems, Taipei, Taiwan, 24–27 May 2009, pp 1085–1088
18. Kwon D, Rincón-Mora GA (2010) A single-inductor AC-DC piezoelectric energy-harvester/battery-charger IC converting $+-$(0.35 to 1.2 V) to (2.7 to 4.5 V). IEEE International solid-state circuits conference, San Francisco, CA, pp 494–495
19. Kwon D, Rincón-Mora GA (2010) A 2-μm BiCMOS rectifier-free AC-DC piezoelectric energy harvester-charger IC. IEEE Trans Biomed Circuits Syst 4(6):400–409
20. Rincón-Mora GA, Torres E (2006) Energy harvesting: a battle against power losses. *Planet Analog*, Oct 8, 2006
21. Kwon D, Rincón-Mora GA (2010) A 2-μm BiCMOS rectifier-free AC-DC piezoelectric energy harvester-charger IC. IEEE Trans Biomed Circuits Syst 4(6):400–409, Invited

Part VI
Materials Development and MEMS Fabrication

Chapter 16
Stretching the Capabilities of Energy Harvesting: Electroactive Polymers Based on Dielectric Elastomers

Roy D. Kornbluh, Ron Pelrine, Harsha Prahlad, Annjoe Wong-Foy, Brian McCoy, Susan Kim, Joseph Eckerle, and Tom Low

Abstract Dielectric elastomer actuators are "stretchable capacitors" that can offer muscle-like strain and force response to an applied voltage. As generators, dielectric elastomers offer the promise of energy harvesting with few moving parts. Power can be produced simply by stretching and contracting a relatively low-cost rubbery material. This simplicity, combined with demonstrated high energy density and high efficiency, suggests that dielectric elastomers are promising for a wide range of energy-harvesting applications. Indeed, dielectric elastomers have been demonstrated to harvest energy from human walking, ocean waves, flowing water, blowing wind, pushing buttons, and heat engines. While the technology is promising and advances are being made, there are challenges that must be addressed if dielectric elastomers are to be a successful and economically viable energy-harvesting technology. These challenges include developing materials and packaging that sustain a long lifetime over a range of environmental conditions, designing the devices that stretch the elastomer material uniformly, and system issues such as practical and efficient energy-harvesting circuits.

This chapter was adapted from "Dielectric elastomers: Stretching the capabilities of energy harvesting," by Roy D. Kornbluh, Ron Pelrine, Harsha Prahlad, Annjoe Wong-Foy, Brian McCoy, Susan Kim, Joseph Eckerle and Tom Low, in MRS Bulletin, Volume 37 (March 2012), pp. 246–253. Reprinted with the permission of Cambridge University Press.

R.D. Kornbluh (✉) • R. Pelrine • H. Prahlad • A. Wong-Foy • B. McCoy
• S. Kim • J. Eckerle • T. Low
SRI International, 333 Ravenswood Avenue, Menlo Park, CA, 94025, USA
e-mail: roy.kornbluh@sri.com; ron.pelrine@sri.com; harsha.prahlad@sri.com; annjoe.wong-foy@sri.com; brian.mccoy@sri.com; susan.kim@sri.com; joseph.eckerle@sri.com; thomas.low@sri.com

N. Elvin and A. Erturk (eds.), *Advances in Energy Harvesting Methods*,
DOI 10.1007/978-1-4614-5705-3_16,

16.1 Introduction

Imagine the ability to generate electric power simply by stretching and relaxing a low-cost rubbery material. Such is the promise of electroactive polymers, and in particular, the type of electroactive polymer known as the "dielectric elastomer." The term "electroactive polymers" (EAPs) typically refers to materials that can deform in response to the application of an electrical stimulus (although other mechanical responses are possible too). The many types of EAPs can be divided into two categories: ionic, where mass transport from a flowing charge causes the deformation, and electronic, where a voltage-induced electric field creates the deformation. Bar-Cohen [1] and Carpi [2] provide good reviews of EAPs. EAPs offer unique properties compared with more conventional transducer technologies, such as those based on rigid materials such as piezoelectrics, magnetostrictives, or electromagnetics. Because of the relatively soft polymer composition and large strain, response can be similar to natural muscle; such EAPs are often called "artificial muscles" and find applications in biologically inspired robots, as well as human prosthetic or orthotic devices.

Dielectric elastomers, a type of electronic EAP, have shown great promise for a variety of applications. Dielectric elastomer transducers are composed of deformable polymer films that respond to an electric field applied across their thickness. In a sense, they are stretchable capacitors. When acting as an actuator, dielectric elastomers are capable of large strains (in some cases >100%) [2], with the relatively fast response and high efficiency associated with electric-field-activated materials [3]. A number of materials, including relatively inexpensive, commercially available ones such as natural rubber, silicone rubbers, and acrylic elastomers, can be used for the component materials of dielectric elastomers. The elastomers can be quite soft, suggesting their potential for a variety of applications involving human interaction or unusual mechanical load-matching requirements. The simple structure, wide availability, and unique properties of dielectric elastomers have allowed researchers to explore their use in a wide variety of actuator applications. Brochu and Pei [4] and Carpi et al. [5] include surveys of state-of-the-art dielectric elastomers in their reviews. Figure 16.1 shows an example of how dielectric elastomers can be incorporated into a muscle-like actuator. While the promise of muscle-like actuation technology has not yet been completely fulfilled, dielectric elastomers are more than laboratory curiosities and are emerging on the commercial market. Artificial Muscle, Inc. (Sunnyvale, CA, USA) is now providing actuators that are incorporated into a handheld gaming console to provide enhanced tactile feedback.

Although first reported in 2001 [6], the use of dielectric elastomers as electrical power generators has been less widespread. Only in the past 4 years has research in their use as generators increased dramatically, as evidenced by an increase in publications on this topic (some of which are cited herein). Interest in new approaches to power generation is not surprising, given the interest in developing clean and renewable sources of energy, as well as more convenient ways to recharge

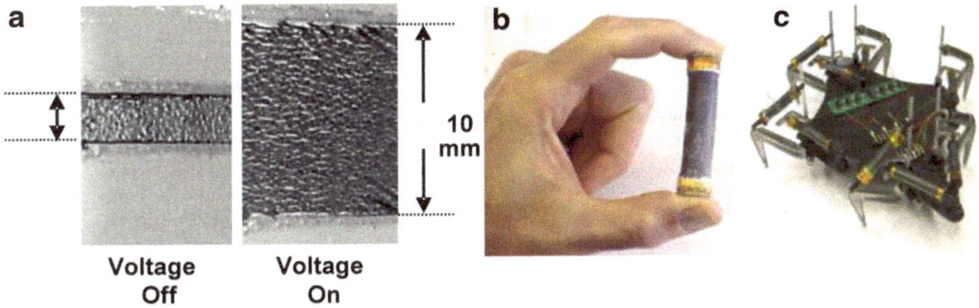

Fig. 16.1 Examples of dielectric elastomer actuation: acrylic elastomer material undergoing 300% linear strain (**a**), material incorporated into a linear roll-type "artificial muscle" actuator (**b**), and artificial muscle actuators incorporated into a biologically inspired six-legged robot (**c**) (*Source*: SRI International)

the batteries of the ever-growing number of power-hungry mobile electronic devices. This chapter considers applications that address both needs as it presents the promises and challenges of dielectric elastomer energy harvesting. First, the basic technology of dielectric elastomers and their use in electric power generation is presented. Next the use of this technology for power generation is discussed. Specific examples of a variety of dielectric elastomer generators (shoe-mounted generators, ocean wave harvesters, and a new type of fuel-burning engine-generator system) are presented. Finally, the challenges to the adoption of this technology for power generation are discussed.

Much of the information in this chapter is derived from the authors' own experiences developing dielectric elastomer energy-harvesting systems for applications, including power-generating boots and buoys that harvest the power of ocean waves. Additional information on such generators may be found in Ashley [7], Chiba et al. [8], and Prahlad et al. [9].

16.2 Background on Dielectric Elastomer Power Generation

This section provides general information on the use of dielectric elastomers as generators. The basic principles of operation and the governing equations are introduced below, followed by discussions of material and transducer configuration issues.

16.2.1 Principles of Operation

The basic operational element of a dielectric elastomer generator, shown in Fig. 16.2, is a film of an elastically deformable, insulating polymer that is coated

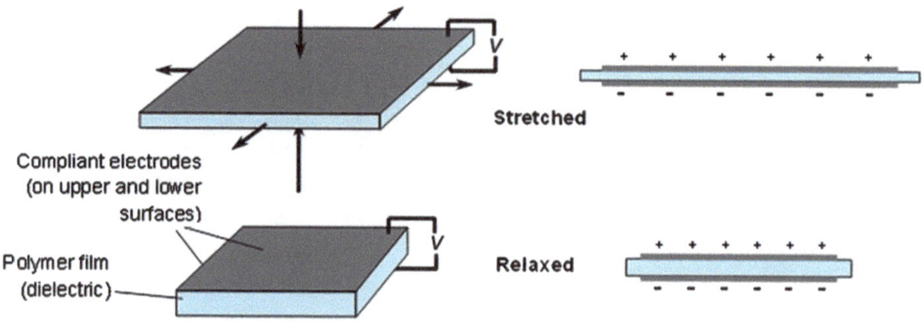

Fig. 16.2 Basic operational element of a dielectric elastomer generator: perspective view (*left*) and edge view (*right*). (*Source*: SRI International)

on each side with a compliant electrode. In generator mode, dielectric elastomers convert the mechanical work of stretching the polymer film into electrical energy. To achieve this conversion, it is necessary to add electrical charge to the surface of the polymer film while it is in a stretched state, allowing the elastic forces on the film to relax the film to a state of lower stretch. When the film relaxes, it shrinks in area and increases in thickness. If most of the charge on the film is conserved, then both geometric effects tend to increase the electrical energy on the film, since like charges on each electrode are forced together while unlike charges on the opposite electrodes are pulled apart. This increase in energy may be many times greater than that required to initially place the charge on the film.

The maximum amount of energy that can be converted using a given amount of film depends on the material properties. Several different material properties come into play, including the maximum strain that can be imposed before mechanical failure, the maximum electric field that can be supported before electrical breakdown, and the need to maintain elastic restoring forces.

Writing simple equations for the amount of energy that can be extracted is not easy, due to the highly nonlinear elastic behavior and the complex interactions with the energy source. If we assume that a given amount of stretch can be imposed on the film, it is easier to see how a "stretchable capacitor" generator functions and how certain material and operational parameters affect the amount of energy generated.

While we cannot immediately discern the maximum amount of energy that can be produced from a given volume of material by this simplification, we can determine the energy output for certain operational cycles. There are four basic steps in the simplest operational cycles (1) the film is stretched by tensile forces to its maximum stretch state, (2) a voltage or charge is applied to the film, (3) the film relaxes from its internal elastic energy, and (4) charge is removed from the film to return it to its initial state. Three common operational cycles are constant charge, constant voltage, and constant field. The names of these cycles refer to what occurs during step 3.

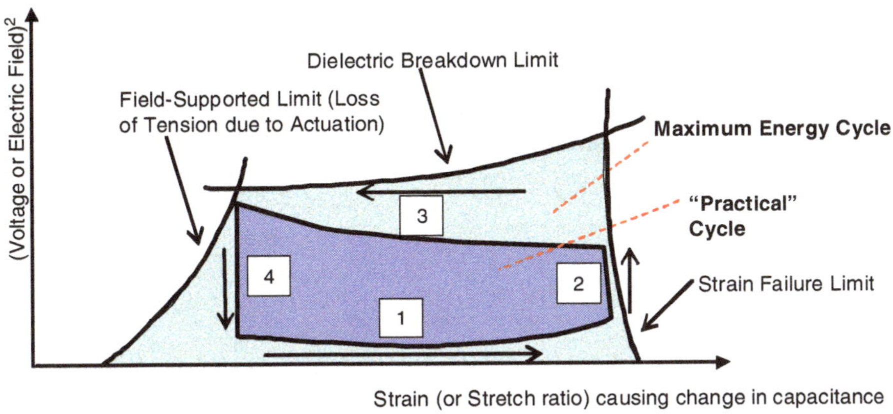

Fig. 16.3 Graphic representation of performance limits and energy cycles. Key to steps indicated in the figure: (1) the film is stretched by tensile forces to its maximum stretch state; (2) a voltage or charge is applied to the film; (3) the film relaxes from its internal elastic energy; and (4) charge is removed from the film to return it to its initial state

Figure 16.3 illustrates an energy-harvesting cycle. Note that the cycles must all be contained within the operational boundaries defined by the material limitations. The horizontal axis is a variable that represents the geometric change in the film, which is related to the change in capacitance. The vertical axis is the square of the voltage or electric field across the film. By choosing the correct variables, the energy that can be extracted for each cycle (not including losses) is proportional to the area enclosed by the cycle curve (e.g., capacitance versus the voltage squared).

The net amount of energy per unit volume of film that can be extracted for the constant charge (u_Q), constant voltage (u_V), and constant field (u_E) cycles, respectively, are [10]

$$u_Q = \frac{1}{2}\varepsilon_p {E_{\max}}^2 \left[\frac{(\gamma^2 - 1)}{\gamma^2}\right], \tag{16.1}$$

$$u_V = \frac{1}{2}\varepsilon_p {E_{\max}}^2 \left[\frac{(\gamma^2 - 1)}{\gamma^2}\right], \tag{16.2}$$

$$u_E = \varepsilon_p {E_{\max}}^2 \ln(\gamma), \tag{16.3}$$

where $E_{\max}$ is the maximum field that is applied during the cycle, ε_p is the permittivity of the film, and γ is the area stretch ratio (stretched area/unstretched area). These equations can show the benefits of one cycle compared with another for different stretch conditions, if we select materials based on the maximum field level. It is possible to implement cycles that can exceed these energy outputs by more closely approaching the material performance limits or including lower losses. Electrical losses result from resistive losses in the electrodes and leakage losses across the film, as well as additional losses in the harvesting circuit and any storage or transmission systems. Graf et al. [11] were able to model these losses by making simplifying assumptions, such as constant polymer material conductivity and electrode resistance. Mechanical losses include viscoelastic losses in the polymer and electrodes, as well as those in any mechanical transmission system that couples to the external driving loads.

16.3 Materials

The performance of a material for dielectric elastomer generators depends on a combination of electrical and mechanical properties. From the simplified analysis of energy harvesting presented previously, it can be seen that it is generally desirable to have a material that has high dielectric breakdown strength and high permittivity (dielectric constant). To minimize losses, it is desirable to select a material with low leakage and other dielectric losses. The importance of leakage depends on the frequency of operation. A vibrational energy-harvesting system might operate at more than 100 Hz, while an ocean wave power-harvesting system might operate at less than 0.1 Hz. On the mechanical side, it is generally desirable to have a material that can sustain large stretch ratios. It is also desirable to minimize viscoelastic losses as well as creep and stress relaxation effects.

The question of selecting the best material stiffness is more complex. At first glance, it would seem best to choose a material with a low stiffness so that smaller forces are needed to produce the desired polymer stretch, allowing a simpler generator structure and fewer mechanical losses. However, softer materials may experience a loss of tension in the film in the field-supported region of operation at a lower electric field. Further, many soft materials would be more prone to pull-in failure (a mode of electrical failure caused by opposing electrodes attracting each other with a force greater than the opposing force offered by the elasticity of the dielectric material that separates them) due to mechanical instabilities resulting from film defects or thinner film regions (a source of dielectric failure in softer insulating films) [12].

To date, the most common candidate materials considered for dielectric elastomer generators, the same as for dielectric elastomer actuators, have been those based on commercial formulations of acrylics and silicones [13]. These materials have a favorable combination of high dielectric breakdown strength, high elongation, and relatively low mechanical and electrical losses. While some materials can withstand electric fields of more than 200 MV/m, maximum operating electric fields for practical devices are typically 100 MV/m or less (depending on lifetime [14]). Other materials under development by researchers include styrene ethylene butadiene styrene and acrylonitrile rubbers, as well as polyurethane-based polymers [2]. Recognizing the advantages of high permittivity in achieving greater energy density (as is evident from Eqs. (16.1), (16.2) and (16.3), for example), many researchers have experimented with adding particulates to elastomers to increase the permittivity (summarized nicely in Brochu and Pei [4]). Recently, Kofod et al. [15] have shown that certain nanoparticles can increase the dielectric constant of the elastomers without adversely affecting the breakdown strength or leakage.

It is important to note that the best choice of material may not be the one with the greatest energy density; there are also economic considerations. The effect of economic factors is more critical in large-scale energy harvesting (such as ocean wave power, as discussed later). Koh et al. [16] have rigorously modeled the electromechanics of this interaction for the simplified case of uniform biaxial stretching. They use a nonlinear material model to show how under some operating conditions, natural rubber can outperform 3 M VHB acrylic, a material that offers favorable properties for many dielectric elastomer applications [3]. Since natural rubber is a very low-cost elastomer, the potential economic benefits are substantial.

In addition to the dielectric material, the overall performance of a dielectric generator is also based on the electrode material that coats the surfaces of the films. In general, it is desirable to make the electrode as compliant as possible. Because dielectric elastomers typically operate at high-voltage and low-current conditions, it is acceptable to use relatively high resistance materials for the electrodes. Electrode materials for dielectric elastomers typically include various carbon particles in polymer binders or patterned or corrugated metal coatings [17]. Most research on electrodes has been oriented toward actuation. For energy generation, the requirements are similar, except that the materials may have to undergo even larger strains.

Recently, silicone dielectric elastomer material already coated with a compliant electrode material (corrugated silver) was introduced to the market [18]. We also note that the 3 M VHB acrylic (uncoated dielectric elastomer) is also available in large quantities. The fact that such materials can be manufactured in large-scale roll-to-roll operations supports the feasibility of large-scale power generation.

16.3.1 Transducer Configurations

The basic operational element of Fig. 16.1 must be incorporated into a transducer or structure that allows the stretching of the film to be coupled with the forces that cause stretching. Kornbluh [19] has surveyed a variety of configurations for actuators. These same configurations can also be applied to generators. Figure 16.4 shows several important configurations, many of which have been used in the application examples in the following section.

The selection of the best configuration depends on many factors, including the type of driving force and mechanical transmission, operating strain, total amount of film needed, and the desired form factor. It is desirable for the boundary conditions to impose a uniform strain over the entire range of operation (stretch the film evenly) so that there are no concentrations of electric field or mechanical stress that would prematurely damage the film. The examples in Fig. 16.4 come close to this ideal, but it is difficult to avoid some stress or field concentrations at the edges.

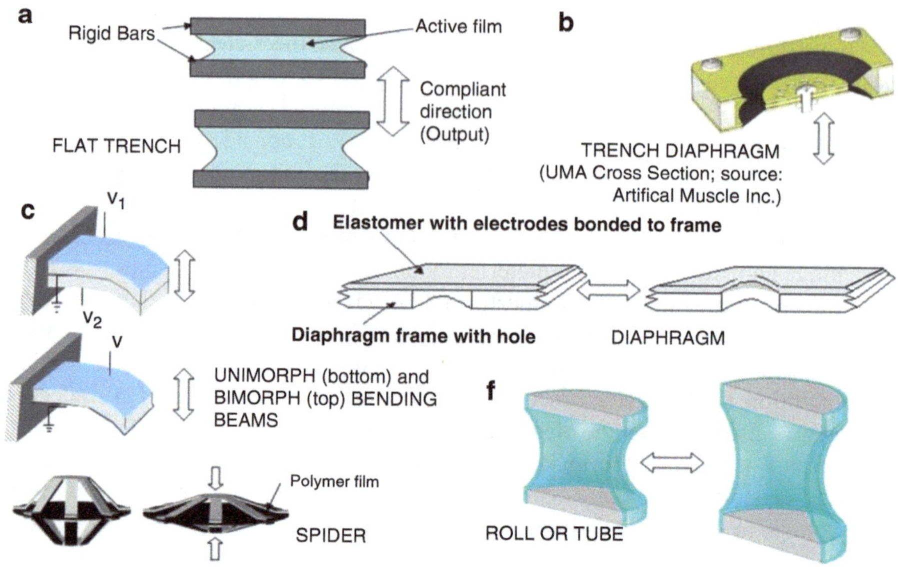

Fig. 16.4 Configurations for dielectric elastomer generators. For the *flat trench*, *bending beams*, and *roll* or *tube* configurations, the dielectric elastomer film is indicated with a *bluish color*. For the trench diaphragm, and spider configurations, the dielectric elastomer film is colored *black*. For the diaphragm configuration, the dielectric elastomer film is colored *grey* (*Source*: SRI International)

16.4 Unique Capabilities of Dielectric Elastomers for Energy Harvesting

This section illustrates the unique capabilities that dielectric elastomers can offer for energy harvesting. First, we quantitatively compare the technology to more common energy-harvesting technologies and discuss the potential advantages for different types of energy-harvesting applications. We then introduce specific examples of how dielectric elastomers can be applied to several of these application areas.

16.4.1 Comparison with Other Technologies

We have already touched on some of the unique properties of dielectric elastomers and the implications for energy harvesting. Table 16.1 quantifies some of these properties and compares them with common power generation technologies.

Other electronic (electric-field responsive) electroactive polymers besides dielectric elastomers, such as ferroelectric polymers (which often include a polyvinylidene fluoride component) and composites that include piezoelectric ceramics, have not shown the capacity for large energy densities (e.g., Liu [22], Jean-Mistral et al. [21]).

Table 16.1 Comparison of dielectric elastomer with other power generation technologies (adapted from [20, 21])

Technology	Typical stiffness	Maximum specific energy density (J/g)	Typical maximum efficiency	Comment
Dielectric elastomer	0.1–10 MPa	0.4 (0.05 for long lifetime operation)	>50%	Low-cost nontoxic materials, low stiffness, size scalable
Electromagnetic	NA	0.004	<20%	Low-energy density, constant-frequency electromagnetic generators can have much higher efficiency, but need variable frequency transmission for higher efficiency over a range of frequencies
Piezoelectric ceramic	50–100 GPa	0.01	>50%	Requires significant additional mass for support or motion amplification, expensive (and often toxic) materials

Source: SRI International

Jean-Mistral et al. also noted that wet (ionic) electroactive polymers have not shown high energy densities. These include conductive polymers and ionic polymer metal composites. Further, these materials are generally more expensive than dielectric elastomers and cannot yet be readily made into the large-area films needed for large-scale power production.

A great many potential energy generation applications can take advantage of the benefits of dielectric elastomer generators. Table 16.2 highlights the potential benefits of dielectric elastomers for several categories of energy sources. The following sections give examples of dielectric elastomer generators in the first three application categories of Table 16.2. Typical maximum operating voltages for these devices were up to 5 kV (although longer lifetime would reduce this value [14]).

16.4.2 Human Activity: Heel-Strike Generator

The proliferation of mobile electronics for the general public, military personnel, and emergency first responders has put demands on the life of batteries and has introduced the need to simplify the logistics of recharging systems. Harvesting the energy of human activity can help.

The current authors have developed a "heel-strike generator" that can be located in a normal shoe or boot [7]. The compression of the heel during normal walking was selected as the means of harvesting power from human activity because it does not add any physical burden to the wearer. Further, proper tuning of the amount of energy absorption at the heel could actually increase the comfort or walking efficiency of the wearer by absorbing and returning the optimal amount of energy per step. This device, shown in Fig. 16.5, produced an electrical output of 0.8 J per step, or about 1 W. The dielectric elastomer generator was a diaphragm type that used a fluid (or gel) coupling to transfer the compression of the heel into deflection of the diaphragm. The diaphragms in this device consisted of 20 stacked layers of dielectric elastomer films. While intended primarily for battery charging, the device also directly powered night vision goggles with the high-voltage output from the energy-harvesting circuit stimulating a photomultiplier tube. This device used prestrained VHB 4910 acrylic and performed with a maximum energy density of about 0.3 J/g [22].

This power level far exceeds outputs demonstrated by many other more complex, more costly, and heavier heel-strike generators based on direct deformation of piezoelectric elements [23], as well as that which would be possible with direct-drive electromagnetic devices (note the 100× difference in specific energy shown in Table 16.1). The power level therefore supports the claims of high efficiency and energy density possible with dielectric elastomers. By means of comparison, one can roughly estimate the available energy per step as weight times the maximum deflection of the heel. Based on this example, 2.4 J of energy is available from an 80 kg person with a maximum of 3 mm diaphragm deflection.

Table 16.2 Advantages and challenges of dielectric elastomers for various energy sources

Generator application	Competing technology	Dielectric elastomer potential advantage	Dielectric elastomer potential challenge
Human activity (e.g., heel strike or knee brace)	Electromagnetics, piezoelectrics	High energy density and low stiffness allows good load matching and eliminates much mechanical complexity, mass, and bulk	Electronics are more complex than electromagnetics
Environmental sources (waves, water flow, and wind)	Electromagnetics	Good matching to load; low-cost materials allow for large-scale and highly distributed energy harvesting	Long lifetimes for large film areas, electronics cost
Fuel (or other heat source) engine-driven generators	Electromagnetics	Higher energy density, lower cost, good low-speed performance, higher temperature performance	Electronics cost and weight (very small engines); lifetime
Parasitic energy harvesting (vibrations) for remote sensors or other device	Electromagnetics, piezoelectrics	Good load matching to some available energy sources enables simpler designs; lower cost	Electronics cost an issue for some applications; high efficiency at low duty cycles

Source: SRI International

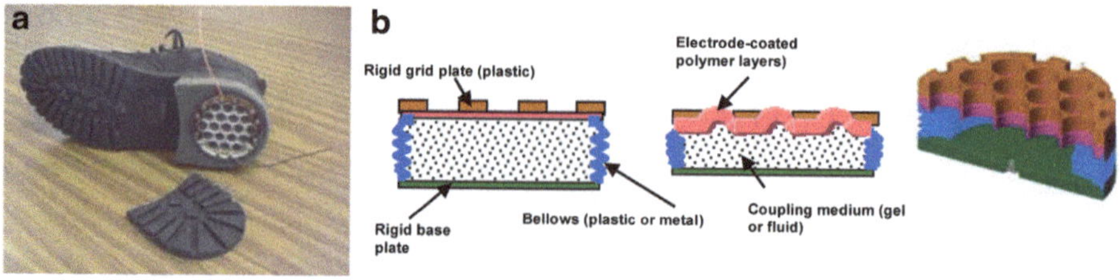

Fig. 16.5 Heel-strike generator based on a dielectric elastomer: photo of the device installed in boot (**a**), cross section of the device (**b**). (*Source*: SRI International)

16.4.3 Environmental Sources: Wave Energy Harvesting

New clean and renewable sources of electric power are critical as the world moves toward a more secure and sustainable energy future. Ocean wave power has the potential to produce clean, renewable energy in an environmentally sound manner that offers greater reliability than solar or wind, and lower visual and auditory impact than wind. Further, this energy source tends to be available near many centers of population and industry. The Electric Power Research Institute estimates that wave energy could potentially meet 10% of total worldwide electric demand [17]. Recently, the U.S. Department of Energy estimated that more than 25% of our nation's electrical power needs could, in principle, be met by harvesting ocean wave energy [24].

Ocean wave power is not yet used for electrical power generation to any significant degree. Widespread adoption of wave power harvesting is hampered by certain economic and logistical factors. For instance, the primary converter structure of conventional ocean wave power-harvesting systems must be overengineered to deal with high sea events (such as storms that cause high wave activity), and, as a result, these systems are very expensive. Similarly, efficient power take-off systems (the structure and transmission systems needed to convert the hydrodynamic energy into electrical power) are typically highly complex and expensive. Dielectric elastomers can potentially address these issues by enabling a simple, low-cost power take-off system.

The use of dielectric elastomers for harvesting the energy of ocean waves has been demonstrated by the authors. This work included two sea trials during which a complete energy-harvesting system was deployed at sea. The first system was based on a suspended proof mass that stretched the spring-like dielectric elastomer material as the buoy heaved on the waves. The roll type of configuration was used (see Fig. 16.4) [8]. The system was a proof-of-principle demonstration of how a buoy, such as a navigation buoy, might use ocean waves to power its onboard lighting or instrumentation and communications systems. The proof-mass approach is not practical for large-scale, grid-level power generation due to the large proof mass that would be needed. Therefore, we developed a proof-of-principle system

Fig. 16.6 Dielectric elastomer ocean wave power generator based on an articulated, multibody system buoy at sea trial site (**a**); concatenated rolls in a generator module (**b**). When a wave passes, the outriggers move relative to the buoy and stretch the rolls using a lever arm. The *green-edged black* material between the rings [visible in the *inset* photo (**b**)] is the electrode-coated dielectric elastomer material. Photos courtesy of SRI International

based on the direct conversion of hydrodynamic energy to mechanically stretch and contract the dielectric elastomer [19]. This system is shown in Fig. 16.6. For logistical convenience, the system used the same oceanographic buoy platform as the proof-mass system. An optimum system would likely not use such a platform.

The system was tested at sea in the Pacific Ocean near Santa Cruz, CA. The device produced an output of more than 25 J in laboratory testing. It used about 220 g of active dielectric elastomer material for a corresponding energy density of more than 0.1 J/g. At sea, the maximum voltage applied to the dielectric elastomer was deliberately limited to conservatively guard against inadvertent failure and the system only produced about half the energy density. The energy-harvesting circuit used in the sea trial was 78% efficient; that is, it harvested 78% of the expected energy for the particular energy-harvesting cycle used. This performance level suggests that dielectric elastomers may indeed be practical for large-scale power generation.

The ocean wave energy-harvesting buoys described previously were proof-of-principle systems whose structure and mechanics were not optimized for maximum efficiency or economic benefit. The low cost and simplicity of using dielectric elastomer materials for energy harvesting can enable fundamentally new system designs. Figure 16.7 shows a conceptual design of such a generator. The basic harvesting element is similar to that used in the single buoy device shown in Fig. 16.6, but here it is built into a highly modular system that can be easily assembled and transported (The individual modules can each be transported by truck) [19]. Its size can be tailored to the prevailing wave conditions (e.g., open ocean deep water waves versus waves that might hit an existing seawall or breakwater).

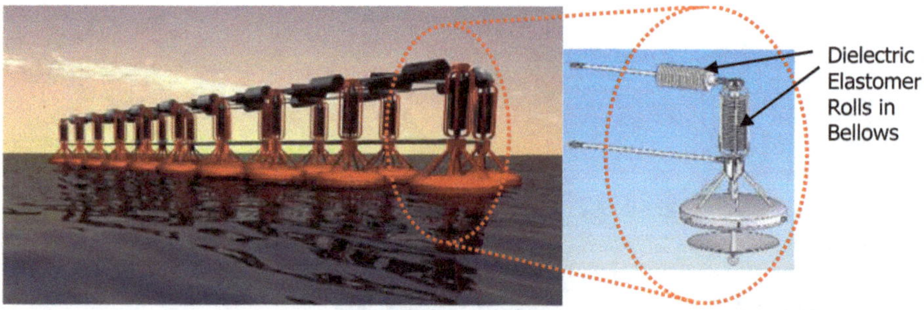

Fig. 16.7 Highly modular distributed ocean wave energy-harvesting system based on dielectric elastomer transducers. System is comprised of individual floating buoys (shown in *red*) connected by dielectric elastomer rolled transducers. Additional buoys can be added to make the system loner or wider as desired

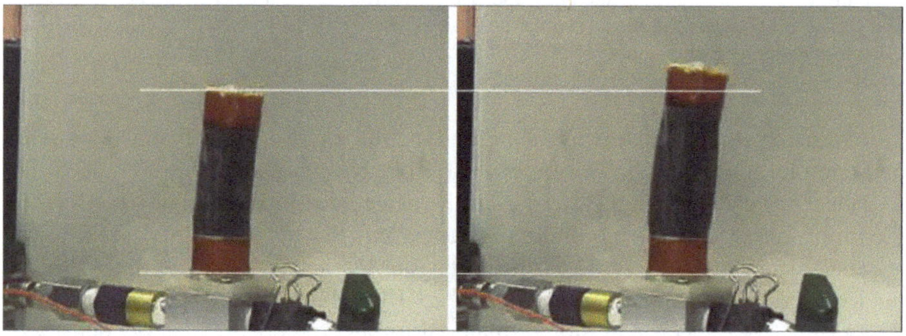

Fig. 16.8 Dielectric elastomer cylinder of a "polymer engine" undergoing 23% linear expansion during internal combustion. A rolled transducer made with an acrylic elastomer was used. Butane was the combustion fuel (*Source*: SRI International)

16.4.4 Fuel or Heat Sources: Polymer Engine Generator

The authors and colleagues have demonstrated a dielectric elastomer heat engine by making the cylinder itself out of dielectric elastomer [9]. In other words, expanding gases directly drive the expansion of the dielectric elastomer. In addition to minimizing mass and structure, this approach allows for greater efficiency of a small engine because of fewer losses from fuel leakage or friction of sliding seals, less wear, and potentially less heat loss for the same mass, since the polymer is a better thermal insulator. This work demonstrated that a polymer cylinder can indeed sustain the temperature of combustion and can provide 11% fuel-to-mechanical efficiency—a good value for a small (<20 W) engine. Figure 16.8 shows the expansion of a rolled dielectric elastomer actuator due to combustion of butane. A small, milliwatt-level amount of electrical power was generated with this proof-of-principle device.

In addition to rolls, diaphragms and tubes were also demonstrated as cylinders of a polymer engine. Such engines could use a variety of hydrocarbon or hydrogen fuels as an energy source and might also harvest solar energy or waste heat. This kind of simple engine design can enable unique energy-harvesting systems.

16.5 Summary and Discussion of Remaining Challenges

Dielectric elastomer generators are capable of good characteristics and performance, both in theory and in experimental and demonstration devices. Properties of these generators include high energy density and high efficiency. Devices such as the heel-strike generator and ocean wave power harvester have also demonstrated that simple, low-compliance devices can directly couple to the mechanical energy source. The ocean wave energy harvesters have shown that devices that incorporate large amounts of film can produce significant amounts of power. No other direct drive, smart material technology has produced as much energy per stroke as have dielectric elastomer systems. Despite this potential and progress, several challenges remain.

Lifetime and reliability issues may be the greatest challenge to adoption of dielectric elastomers. We will likely need to trade off performance versus lifetime and reliability. While these trade-offs have not been fully characterized, we note some exemplary lifetimes. Rolled transducers identical to the individual elements used in the ocean wave-harvesting buoy have survived for more than five million cycles with an energy density of 0.01 J/g. While this energy density is more than an order of magnitude less than that achievable with small amounts of film, or over short lifetimes [6], dielectric elastomer transducers can still outperform many competing technologies, considering overall system mass and complexity. Since dielectric elastomers are still a relatively young technology, many further improvements can be expected. Fault tolerance or self-healing capabilities will likely be needed for large-scale power outputs. Kornbluh et al. [14] discuss this data and lifetime issues in more detail.

Modeling is another area with remaining challenges. While progress is being made, fully modeled systems that include the full range of electromechanical coupling effects and environmental sensitivities do not yet exist. Further, the necessary software tools to model or solve for the material behavior, nonlinear electrical effects, and complex interactions with the environment are not available. Better modeling tools would not only allow for better design and material selection, but also help guide the development of new materials.

The design of energy-harvesting circuits is another area that has opportunities for further development. In many cases, tradeoffs will be necessary between circuit complexity (to get high efficiency) and simplicity or cost. Further, the optimal energy-harvesting cycle cannot be implemented unless the material is well characterized and modeled. Large-scale energy-harvesting systems might benefit from numerous simple energy-harvesting circuits as opposed to more centralized and sophisticated circuits. Again, integrated modeling can help address this issue.

In some cases, energy-harvesting circuitry could be too large and/or too expensive for a given application, negating many of the benefits of using dielectric elastomers. To date, there has been little market for transistors suited to the relatively high voltages used (typically several kilovolts) in electroactive polymer energy-harvesting circuits. As a result, few such transistors are available in the marketplace. As better high-voltage transistors become available and harvesting circuits are refined, the shortcomings of today's circuitry can be overcome.

How achievable is the promise of more economical and convenient power generation with a simple, low-cost rubbery material? Physically small applications will likely be first, because the technological and economic barriers are lower. To enable physically large applications, such as wave power harvesting to be practical, we will need advances in large transducer fabrication, operational lifetime, energy-harvesting circuitry, modeling, and system engineering.

Acknowledgements The authors wish to thank their colleagues at SRI International, whose efforts contributed to the work presented here. We would also like to thank the numerous clients and government funding agencies whose support over the past 20 years has enabled much of this work. In particular, Shuiji Yonemura and Mikio Waki of HYPER DRIVE Corp. have generously supported our development of the ocean wave power-harvesting systems.

References

1. Bar-Cohen Y (ed) (2004) Electroactive polymer (EAP) actuators as artificial muscles: reality, potential and challenges, vol 2. SPIE Press, Bellingham, WA
2. Pelrine R, Kornbluh R, Pei Q, Joseph J (2000) High-speed electrically actuated elastomers with over 100% strain. Science 287(5454):836–839
3. Kornbluh R, Pelrine R, Pei Q, Oh S, Joseph J (2000) Ultrahigh strain response of field-actuated elastomeric polymers. In: Proceedings of SPIE, smart structures and materials 2000: electroactive polymer actuators and devices (EAPAD), vol 3987, p 51
4. Brochu P, Pei Q (2010) Advances in dielectric elastomers for actuators and artificial muscles. Macromol Rapid Commun 31(1):10–36
5. Carpi F, DeRossi D, Kornbluh R, Pelrine R, Sommer-Larsen P (2008) Dielectric elastomers as electromechanical transducers. Fundamentals, materials, devices, models and applications of an emerging electroactive polymer technology. Elsevier, Amsterdam, The Netherlands
6. Pelrine R, Kornbluh R, Eckerle J, Jeuck P, Oh S, Pei Q, Stanford S (2001) Dielectric elastomers: generator mode fundamentals and applications. In: Proceedings of SPIE, smart structures and materials 2001: electroactive polymer actuators and devices (EAPAD), vol 4329, p 148
7. Ashley S (2003) Artificial muscles. Sci Am 289(4):52–59
8. Chiba S, Waki M, Kornbluh R, Pelrine R (2008) Innovative power generators for energy harvesting using electroactive polymer artificial muscles. In: Proceedings of SPIE, electroactive polymer actuators and devices (EAPAD), vol 6927, p 692715
9. Prahlad H, Kornbluh R, Pelrine R, Stanford S, Eckerle J, Oh S (2005) Polymer power: dielectric elastomers and their applications in distributed actuation and power generation. In: Proceedings of ISSS 2005 international conference on smart materials structures and systems, Bangalore, India, 28–30 Jul 2005, pp SA-100–SA-107
10. Graf C, Maas J, Schapeler D (2010) Energy harvesting cycles based on electro active polymers. In: Proceedings of SPIE, electroactive polymer actuators and devices (EAPAD), vol 7642, pp 764217–1

11. Graf C, Maas J, Schapeler D (2010) Optimized energy harvesting based on electro active polymers. 10th IEEE international conference on solid dielectrics, pp 752–756
12. Pelrine R, Kornbluh R, Joseph J (1998) Electrostriction of polymer dielectrics with compliant electrodes as a means of actuation. Sens Actuators A Phys 64(1):77–85
13. Kornbluh R, Pelrine R (2008) Chapter 4: fundamental configurations for dielectric elastomer actuators. In: Carpi F, DeRossi D, Kornbluh R, Pelrine R, Somer-Larsen P, Carpi F, DeRossi D, Kornbluh R, Pelrine R, Somer-Larsen P (eds) Dielectric elastomers as electromechanical transducers. Fundamentals, materials, devices, models and applications of an emerging electroactive polymer technology. Elsevier, Amsterdam, The Netherlands, pp 33–42
14. Kornbluh R, Wong-Foy A, Pelrine R, Prahlad H, McCoy B (2010) Long-lifetime all-polymer artificial muscle transducers. In: MRS proceedings, vol 1271, p 1271-JJ03-01. doi:10.1557/PROC-1271-JJ03-01
15. Kofod G, McCarthy DN, Stoyanov H, Kollosche M, Risse S, Ragusch H, Rychkov D, Dansachmuller M, Wache R (2010) Materials science on the nano-scale for improvements in actuation properties of dielectric elastomer actuators. In: Proceedings of SPIE, electroactive polymer actuators and devices (EAPAD), vol 7642, p 76420J. doi:10.1117/12.847281
16. Koh SJA, Keplinger C, Li T, Bauer S, Suo Z (2011) Dielectric elastomer generators: how much energy can be converted. IEEE ASME Trans Mechatron 16:33
17. Electric Power Research Institute (EPRI) Ocean tidal and wave energy, renewable energy technical assessment guide, (TAG-RE 1010489, 2005)
18. Benslimane M, Kiil H-E, Tryson MJ (2010) Electromechanical properties of novel large strain polypower film and laminate components for DEAP actuator and sensor applications. In: Proceedings of SPIE, electroactive polymer actuators and devices (EAPAD), vol 7642, p 764231
19. Kornbluh R, Pelrine R, Prahlad H, Wong-Foy A, McCoy B, Kim S, Eckerle J, and Low T (2011) From boots to buoys: promises and challenges of dielectric elastomer energy harvesting. In: Proceedings of SPIE, electroactive polymer actuators and devices (EAPAD), vol 7976, p 797605. doi:10.1117/12.882367
20. Carpi F, Kiil H-E, Kornbluh R, Sommer-Larsen P, Alici G (2010) Electroactive polymer actuators: from lab to market. In: Borgmann H (ed) Proceedings of actuators, pp 405–417
21. Kofod G, Somer-Larsen P (2008) Chapter 7: compliant electrodes: solutions, materials and technologies. In: Carpi F, DeRossi D, Kornbluh R, Pelrine R, Somer-Larsen P (eds) Dielectric elastomers as electromechanical transducers. Fundamentals, materials, devices, models and applications of an emerging electroactive polymer technology. Elsevier, Amsterdam, The Netherlands, pp 69–76
22. Kornbluh R (2008) Chapter 8: fundamental configurations for dielectric elastomer actuators. In: Carpi F, DeRossi D, Kornbluh R, Pelrine R, Sommer-Larsen P (eds) Dielectric elastomers as electromechanical transducers. Fundamentals, materials, devices, models and applications of an emerging electroactive polymer technology. Elsevier, Amsterdam, The Netherlands, pp 79–90
23. Pelrine R, Prahlad H (2008) Chapter 15: generator mode: devices and applications. In: Carpi F, DeRossi D, Kornbluh R, Pelrine R, Sommer-Larsen P (eds) Dielectric elastomers as electromechanical transducers. Fundamentals, materials, devices, models and applications of an emerging electroactive polymer technology. Elsevier, Amsterdam, The Netherlands, pp 146–155
24. EPRI. Mapping and assessment of the United States Ocean Wave Energy Resource, Palo Alto, CA, 1024637. http://www1.eere.energy.gov/water/pdfs/mappingandassessment.pdf
25. Jean-MistralC, Basrour S, Chaillout J-J (2010) Comparison of electroactive polymers for energy scavenging applications. Smart Mater Struct 19(8): 085012
26. Liu Y, Ren KL, Hofmann HF, Zhang Q (2005) Investigation of electrostrictive polymers for energy harvesting. IEEE Trans Ultrason Ferroelectr Freq Control 52(12):2411
27. Paradiso JA, Starner T (2005) Energy scavenging for mobile and wireless electronics. IEEE Pervasive Comput 4(1):18

Chapter 17
Materials and Devices for MEMS Piezoelectric Energy Harvesting

Miso Kim, Seung-Hyun Kim, and Seungbum Hong

Abstract Piezoelectric vibration energy harvesters (PVEHs) for microelectromechanical systems (MEMS) have received considerable attention as an enabling technology for self-powered wireless sensor networks. MEMS-PVEHs are particularly attractive because of the potential to deliver power required for sensor nodes and their ability to be integrated concurrently with the microfabrication of electronic circuits such as sensor nodes. This chapter consists of four subsections, starting with Sect. 17.1, where various piezoelectric materials commonly used for MEMS-scale PVEHs are reviewed. Typical device configurations of PVEH systems are introduced in Sect. 17.2, followed by analytical modeling of different configurations in Sect. 17.3 to link material characteristics to device performance: standard capacitor type electrodes for {3–1} mode of operation and interdigitated electrodes (IDTEs) for {3–3} mode of operation. In the last section, fabrication and characterization of MEMS-scale PVEHs in both of these modes are presented with model–experiment comparisons.

M. Kim (✉)
Department of Materials and Science, Massachusetts Institute of Technology, Cambridge, MA 02139, USA

Division of Industrial Metrology, Center for Safety Measurement, Korea Research Institute of Standards and Science, 267 Gajeong-Ro, Yuseong-Gu, Daejeon, 305-340 Republic of Korea
e-mail: misokim@kriss.re.kr

S.-H. Kim
School of Engineering, Brown University, Providence, RI 02912, USA
e-mail: seunghyun_kim@brown.edu

S. Hong
Materials Science Division and Center for Nanoscale Materials, Argonne National Laboratory, Argonne, IL 60439, USA
e-mail: hong@anl.gov

N. Elvin and A. Erturk (eds.), *Advances in Energy Harvesting Methods*,
DOI 10.1007/978-1-4614-5705-3_17,

17.1 Piezoelectric Materials for MEMS Energy Harvesting

Piezoelectric materials consist of ferroelectric materials, such as $Pb(Zr,Ti)O_3$, $BaTiO_3$, and $LiNbO_3$, and non-ferroelectric materials, such as AlN and ZnO [1, 2]. One of the defining traits of a piezoelectric material is that the molecular structure is oriented such that the material exhibits a local charge separation, known as an electric dipole. In general, the electric dipoles throughout a material are oriented randomly, but for ferroelectrics, the dipoles can be oriented such that when the material is heated slightly below the Curie temperature and/or a very strong field is applied, the electric dipoles align themselves to the electric field; this process is termed *poling*. Once the material is cooled, the dipoles maintain their orientation and the material is then said to be poled. After the poling process is completed, the material can exhibit a relatively high piezoelectric effect.

Energy conversion using piezoelectric materials is possible because mechanical strain in a piezoelectric material induces deformation of the electric dipoles, forming electrical charges that can be removed from the material and used to power various devices. Such coupled mechanical and electrical behavior of piezoelectric materials can be described using linear piezoelectric constitutive equations that contain relevant material property constants, one type of which expression is as follows:

$$T_i = c_{ij}^E \mathrm{S}_j - e_{ij} E_j \tag{17.1}$$

$$D_i = e_{ij} \mathrm{S}_j + \varepsilon_{ij}^S \mathrm{E}_j \tag{17.2}$$

S_j and T_i are the mechanical strain and stress, while E_j and D_i are the electric field and the electrical displacement. c_{ij}^{E} represents the elastic stiffness coefficient and ε_{ij} is the permittivity value. The superscripts S and E denote that the respective parameter is evaluated at constant strain and constant electric field, respectively. Piezoelectric coefficient, e_{ij}, is the measure of piezoelectric coupling of the given active materials. Piezoelectric materials typically exhibit anisotropic characteristics; thus, the properties of the material differ depending upon the direction of forces and orientation of the polarization and electrodes, defined by the subscripts in Eqs. (17.1) and (17.2). For a more complete description of these constants, the reader is referred to the IEEE standards [3]. Since piezoelectric materials acquire charge when directly strained, their performance, lifetime, and transduction efficiency are dependent upon their materials properties.

There is a wide range of piezoelectric materials available for different application environments [1, 2, 4–6]: single crystals, polycrystalline ceramics, polymers, composites, relaxor-type ferroelectrics, etc. Single crystal materials including quartz, lithium niobate ($LiNbO_3$), and lithium tantalate ($LiTaO_3$) are important functional materials in surface acoustic wave (SAW) devices and high-frequency filter applications [5]. However, little has been investigated on their use in energy harvesting. These materials have relatively high electromechanical coupling coefficients but small dielectric constants, generating less current than the ceramic material, PZT [7]. Also, these materials are only available in bulk single crystals.

Polycrystalline ceramics along with polymers are the most extensively explored piezoelectric energy harvesting materials [1, 2, 4]. Piezoelectric ceramic materials include ferroelectric materials with perovskite crystal structures such as barium titanate ($BaTiO_3$), lead titanate ($PbTiO_3$, PCT), lead zirconate titanate ($PbZr_xTi_{1-x}O_3$, PZT), and non-ferroelectric materials with wurtzite crystal structures such as ZnO and AlN. Among all these materials, lead zirconate titanate (PZT), a solid solution of ferroelectric $PbTiO_3$ and antiferroelectric $PbZrO_3$, is the most common type of piezoelectric used in energy harvesting applications due to its high piezoelectric coupling. The dielectric and piezoelectric constants of PZT depend strongly on materials composition and doping. In terms of composition, PZT films at the morphotropic phase boundary (MPB) with a Zr/Ti ratio of 52/48 have been shown to exhibit a maximum piezoelectric response and are typically used in MEMS device applications. Doping effects on PZT material properties of various dopant elements such as Nb, Ta, and Mn have been the focus of many research studies. Bulk poled PZT is used widely as a piezoelectric ceramic material in sensor, actuator, and transducer applications. PZT thin films are very competitive compared with bulk PZT for power generation due to their higher piezoelectric coefficients and suitability for MEMS integration [6, 8]. Although PZT thin films have served successfully in numerous devices due to advances of MEMS integration techniques, the growth of high-quality PZT thin films still needs more effort [1].

There has also been considerable interest in non-ferroelectric piezoelectric semiconductors such as ZnO and AlN harvesting materials. Their compatibility with conventional processing technologies for integrated circuit technology offer advantages to their development for MEMS-PVEHs. These materials are not ferroelectric and thus cannot be poled like perovskite materials. Although their piezoelectric coefficients are considerably lower than those of ferroelectric ceramics such as PZT, their semiconducting characteristics and potential application in biochemical sensors with improved sensitivity and selectivity (e.g., ZnO nanowire arrays) have brought a great deal of research attention to these materials [5, 9, 10].

There has been a growing interest in polymer piezoelectric materials [11] and poly(vinylidene fluoride) (PVDF) has been the most widely adopted flexible piezoelectric material or energy harvesting, one such example is the so-called "energy harvesting eel" [6]. Unlike brittle piezoelectric ceramic materials, polymer piezoelectric materials such as PVDF (PVF_2) or PVDF-TrFE offer considerable flexibility and thus can sustain large amounts of strain. In the development of polymer-based energy harvesters, durable and strong electrode layers are required to operate piezoelectric devices over long periods of time. To date, studies on electrode materials for PVDF have been done using not only inorganic materials such as platinum (Pt) and indium tin oxides (ITO) but also on organic coating such as poly(3,4-ethylenedioxy-thiophene)/poly(4-styrenesulfonate) (PEDOT/PSS).

Composite materials that typically consist of piezoelectric ceramic materials such as PZT rod and fibers embedded in a polymer matrix have also been used for energy harvesting [12]. Sodano et al. performed several studies to compare the harvesting ability and efficiency of macro-fiber composite (MFC) and quick-pack (QP) actuators with commercially available PZT ceramics [13–15].

Despite their extremely high electromechanical coupling coefficients and thus their potential for PVEHs with enhanced performance, little research has been undertaken on devices based on relaxor-type ferroelectrics [4, 5, 16]. Relaxor-type materials include $(Mg_{1/3}Nb_{2/3})O_3$ (PMN), $Pb(Zn_{1/3}Nb_{2/3})O_3$ (PZN), and binary forms of these systems coupled with $PbTiO_3$, PMN-PT, and PZN-PT, respectively. Last but not least, it should be noted that material properties of thin films differ from those of bulk materials, even for the same materials and thus, they should be evaluated differently [1, 2, 8]. Comprehensive reviews on the history of piezoelectric materials and their properties in bulk form and progress to date in piezoelectric thin film materials and devices are available in numerous books [17–21] and articles [1, 2, 5, 8].

17.2 Device Configurations for Piezoelectric Energy Harvesting Systems

A single PVEH device typically consists of piezoelectric layers, structural layers, electrode layers, and a proof mass. The most common geometric configurations are cantilever beams or plates, because they are geometrically compatible with MEMS fabrication processes and have proven to be easy to implement and effective for harvesting energy from ambient vibrations [6]. A cantilever is a compliant structure that can not only be designed to provide low resonant frequencies, particularly by the addition of a mass on the end of the beam/plate, but also they produce high strain, and thus more power generation, in comparison with other structural configurations [22]. There have been other efforts to enhance power performance of PVEH devices by modifying geometric configurations [4, 6]. Trapezoidal shapes by tapering or initially curved cantilevers were studied in order to improve conventional cantilever designs and to better suit other harvesting applications. Triangular or clamped circular plates, known as "cymbal" transducers, have also been of interest. S-, or T-shaped or modified membrane configurations have been explored to enhance their performance over a wide range of vibration frequencies [4, 22].

Depending on the number of piezoelectric element layers, the harvester structure can be categorized as unimorph, bimorph, or multilayer configurations [4]. A unimorph configuration comprises one piezoelectric layer sandwiched between two electrodes along with structural layers. Two piezoelectric layers are involved in bimorph configurations, and these two layers are connected electrically either in series or in parallel.

There are two practical modes of transduction according to the direction of electrical field and applied strain: {3–1} and {3–3} modes of operation [23]. Conventionally, the poling direction is always in the "3" direction, and thus this coincides with the direction of the induced electric field. In {3–1} mode, the voltage

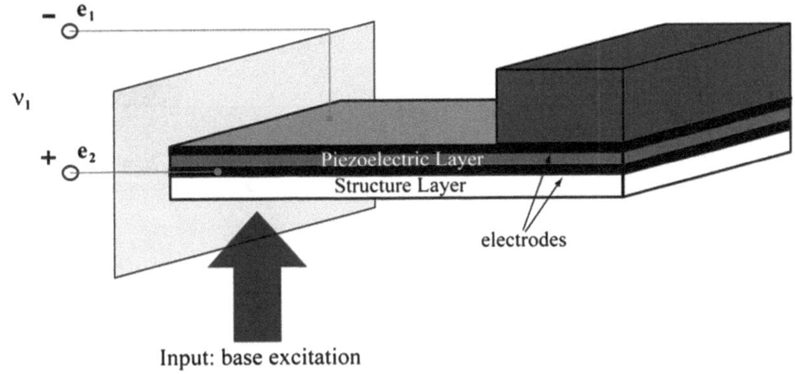

Fig. 17.1 Unimorph cantilevered piezoelectric energy harvester device in {3–1} mode of operation with standard electrode configuration

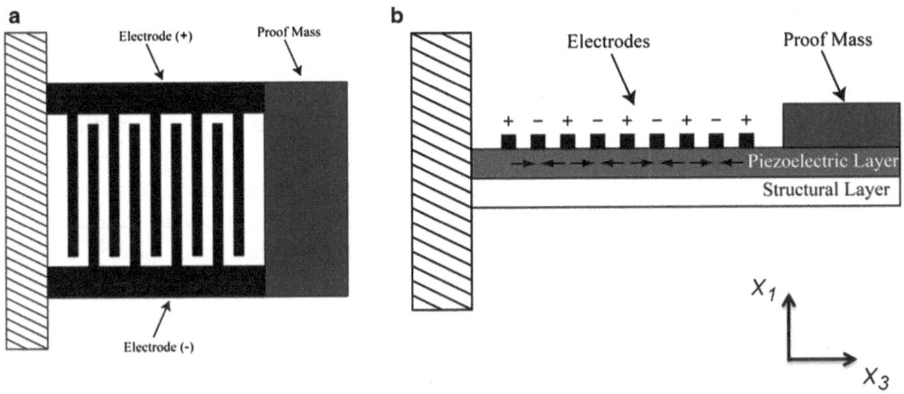

Fig. 17.2 Interdigitated electrode (IDTE) configuration in cantilevered piezoelectric energy harvesting {3–3} mode devices: (**a**) top view and (**b**) side view

(and therefore, electric field) acts in the "3" direction as the mechanical strain is applied in the "1" direction. In {3–3} mode, both strain and voltage occur in the same direction, i.e., the "3" direction. Choice of electrode configuration is dependent upon the modes of operations. Standard capacitor type electrodes are employed for the {3–1} mode of operation while interdigitated electrodes (IDTEs) are commonly used for {3–3} modes of operation, each of which is illustrated for cantilevered PVEH systems in Figs. 17.1 and 17.2, respectively [4, 13, 24, 25]. The {3–3} mode of operation is advantageous in that the generated voltage can be readily controlled. While the electrode spacing is determined by the thickness of the piezoelectric layer in the {3–1} mode, the electrode spacing determines the voltage produced in the {3–3} mode configuration and can therefore be varied in the design. In microsystems, there is a limitation in the thickness of piezoelectric layer that can be deposited

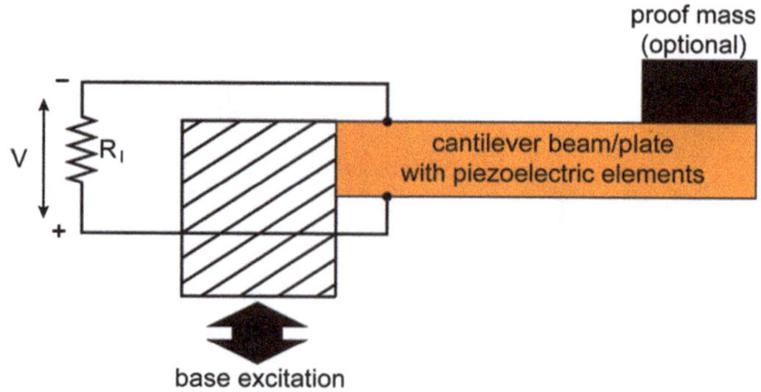

Fig. 17.3 Schematic of a cantilevered piezoelectric energy harvesting system with a simple resistive electrical load, R_l

due to the microfabrication processing, and thus, the voltage that can be generated from the {3–1} mode will be limited as well. However, once a series of PVEHs on a single die as a final system has been manufactured, it is possible to control the electrical output of the entire system by controlling the interconnections of individual devices [26]. Therefore, this implies that both modes of operations are potentially attractive for practical applications.

Once the individual device or arrays of energy harvester have been manufactured, they can be wired to the appropriate electrical circuit to extract the useful electrical properties such as voltage and power. In most cases, the input power source is provided by structural vibrations resulting from base excitation. A general architecture of a piezoelectric energy harvester is illustrated in Fig. 17.3.

17.3 Modeling of Various Electroded Piezoelectric Cantilevers for Energy Harvesting

While models for standard capacitor type electrodes—either approximated or detailed—are not hard to find in the literature, only a few modeling approaches have been attempted on IDTE configurations in PVEH devices. Jeon et al. demonstrated a MEMS-scale, {3–3} mode, piezoelectric micro-power generator with IDTEs in [21] where their calculation of output voltage and power is based only on a very simple approximation. Other prior modeling includes theoretical analysis by Mo et al. where they developed a model for unimorph piezoelectric benders with IDTEs and performed both numerical and parametric studies on energy, charge, and output voltage [27]. Their model encompasses only static considerations, and it does not consider electrode spacing. This presents a need to develop a dynamic model of PVEHs with IDTEs and refine the model to include electrode spacing. In this section, a model for a unimorph PVEH device in the {3–3} mode of operation with IDTEs is developed and briefly compared with the {3–1} mode using standard

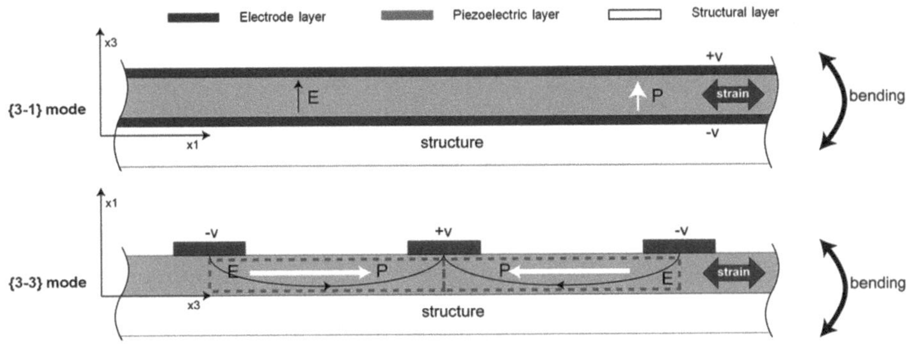

Fig. 17.4 {3–1} and {3–3} mode unimorph PVEH configurations: side-view elements of (*top*) a cantilevered PVEH in {3–1} mode of operation with standard electrodes, and (*bottom*) a cantilevered PVEH in {3–3} mode of operation with IDTEs

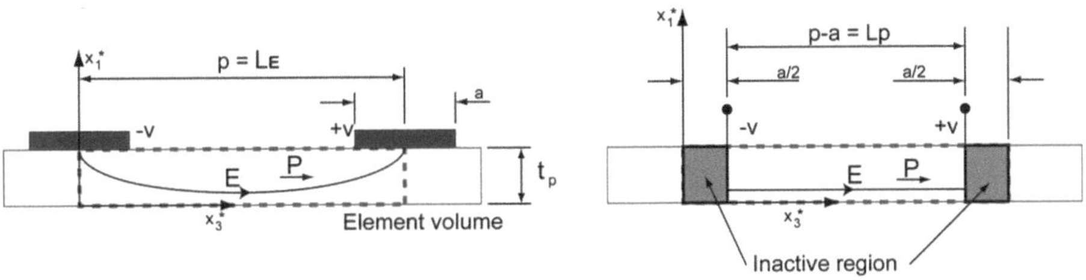

Fig. 17.5 Side view of a piezoelectric layer in a unimorph, cantilevered PVEH in {3–3} mode of operation: {3–3} mode of operation (*left*) and approximate model (*right*) of electric field (*E*) between interdigitated electrodes. *p* is the pitch of the electrodes and *a* is the width of the electrodes. x_1^* and x_3^* are the element local coordinates. *P* denotes polarization

electrodes. The comparisons are performed in terms of electrical parameters such as electrical potential, capacitance, and piezoelectric coupling. A model for a bimorph piezoelectric cantilever in {3–1} mode of operation was previously presented in [23]. All these models incorporate dynamic motion of a piezoelectric energy harvesting cantilever as well as parameters that define the electrode structures.

In {3–3} mode, the directions of the strain and the electric field are parallel to each other in the 3-direction. For simplicity of analysis in the {3–3} mode of operation, x_1 coincides with the beam thickness coordinate while x_3 corresponds to axial coordinate of the beam structure (see Fig. 17.4). The definition of this coordinate system is closely related to the poling directions, of which details are found elsewhere [26]. As demonstrated in Figs. 17.2 and 17.4, IDTEs are employed only on the top of a single layer of piezoelectric material in the {3–3} mode, thus eliminating the need for a bottom electrode. However, the {3–1} mode requires both top and bottom electrode layers. In the IDTE configuration, three geometric parameters are required: the thickness of the piezoelectric layer, t_p, the width of the electrodes, a, and the spacing between the centers of the electrodes (so called pitch), p. These are shown in Fig. 17.5. For the IDTE configuration, it is helpful to introduce several simplifying approximations. First of all, the piezoelectric element between the electrodes is assumed to be fully coupled in the {3–3} mode. Additionally, it is assumed that the region of the piezoelectric element under the electrode is electrically inactive. Although the electric field is not completely axial through the thickness of the piezoelectric element, nor is the region entirely inactive in practice, these effects are expected to compensate for each other to some extent. The geometry of this approximate model is illustrated in Fig. 17.5 (right). Then, as in the {3–1} mode, proper expressions for electrical potential that varies from +1 at the electrode (left side) to 0 to the other electrode (right side) gives a constant electric field between the electrodes.

Development of analytical modeling for IDTEs begins with consideration of the piezoelectric layer over a single element between one pair of electrode fingers, where element length L_{E} is defined as equal to pitch, p, i.e., $L_{\mathrm{E}} = p$, and the length of the piezoelectric layer not directly under the electrodes, L_{p}, is defined as $L_{\mathrm{p}} = p - a$, and t_{p} is the thickness of the piezoelectric layer. Then, the key governing equations over the single element, L_{E}, in base-excited PVEH structures operating in the {3–3} modes can be written as in Eqs. (17.3) and (17.4). The detailed derivation procedure is similar to that of {3–1} modes that can be found in [23, 28].

$$M\ddot{r} + C\dot{r} + Kr - \theta_{\mathrm{p}} v = -B_{\mathrm{f}} \ddot{w}_B \tag{17.3}$$

$$\theta_{\mathrm{p}} \dot{r} + C_{\mathrm{p}} \dot{v} + \frac{1}{R_{\mathrm{l}}} v = 0 \tag{17.4}$$

Each key parameter in Eqs. (17.3) and (17.4) above, including the mass (M), the stiffness (K), coupling (θ_{p}), and capacitive matrices (C_{p}), is as in Eqs. (17.5)–(17.9). The forcing function, B_{f}, accounts for inertial loading on the beam/plate structure due to the base excitations.

$$M = (m_{\mathrm{p}} + m_{\mathrm{s}}) \psi_r^2 L_{\mathrm{E}} \tag{17.5}$$

$$K = c_{\mathrm{s}} I_{\mathrm{s}} (\psi''_r)^2 L_{\mathrm{E}} + (c_{33}^{\mathrm{E}} *) I_{\mathrm{p}} (\psi''_r)^2 L_{\mathrm{E}} \tag{17.6}$$

$$\theta_{\mathrm{p}} = (e_{33} *) S_{\mathrm{p}} \psi''_r \tag{17.7}$$

$$C_{\mathrm{p}} = \frac{(\varepsilon_{33}^S *) A_{\mathrm{p}}}{L_{\mathrm{p}}} \tag{17.8}$$

$$B_{\mathrm{f}} = -(m_{\mathrm{s}} + m_{\mathrm{p}}) \psi_r L_{\mathrm{E}} \tag{17.9}$$

where I_s and I_{p} are moment of inertia for structural layer and piezoelectric layer, respectively. $A_{\mathrm{p}} = b t_{\mathrm{p}}$, and S_{p} is the moment of the area, A_{p}, about the neutral axis. ψ''_r is the mechanical mode shape at the center of the element (L_{E}), and v represents the voltage on the electrodes. Mass per length is denoted as m_p and m_s where subscripts p and s represent piezoelectric layer and structural layer, respectively. In Eq. (17.6), c_s represents the stiffness of structural layers in the axial beam direction, and $c_{33}^{\mathrm{E}}*$ denotes the effective piezoelectric material elastic stiffness for the beam or plate in the {3–3} mode of operation.

For modeling of the entire cantilever with length, $L = nL_\mathrm{E}$, summation or integration of each parameter (M, K, θ_p, C_p, and B_f) in Eqs. (17.5)–(17.9), over the entire length, L, is required. These integrated terms, in turn, can be substituted into the governing equations to predict device performances of the whole cantilever system such as mechanical displacement, voltage, and power. The total beam M, K, and B_f terms now include the piezoelectric mass, m_p, and stiffness, $c_{33}^{E}*$, which are readily combined with the basic structure terms, m_s and c_s. The total beam piezoelectric terms, θ_p and C_p, are summed up as:

$$\theta_\mathrm{p} = \sum_{i=1}^{n} (e_{33}*) S_\mathrm{p} \frac{\mathrm{d}^2\psi_r}{\mathrm{d}x_3^2} \approx \frac{(e_{33}*) S_\mathrm{p}}{L_\mathrm{E}} \frac{\mathrm{d}\psi}{\mathrm{d}x_3}(L) \tag{17.10}$$

$$C_\mathrm{p} = \sum_{i=1}^{n} \frac{(\varepsilon_{33}^{S}*) A_\mathrm{p}}{L_\mathrm{p}} = \frac{L}{L_\mathrm{E}} \frac{(\varepsilon_{33}^{S}*) A_\mathrm{p}}{L_\mathrm{p}} \tag{17.11}$$

where $n = L/L_\mathrm{E}$, and the approximate numerical integration relation below was introduced into Eq. (17.10), namely:

$$\int_0^L \frac{\mathrm{d}^2\psi}{\mathrm{d}x_3^2}\mathrm{d}x = \left[\frac{\mathrm{d}^2\psi(1)}{\mathrm{d}x_3^2} + \frac{\mathrm{d}^2\psi(2)}{\mathrm{d}x_3^2} + \ldots\right]\Delta x_3 = \frac{\mathrm{d}\psi(L)}{\mathrm{d}x_3} - \frac{\mathrm{d}\psi(0)}{\mathrm{d}x_3} \tag{17.12}$$

In the above, $\frac{\mathrm{d}\psi(0)}{\mathrm{d}x_3} = 0$ because of the cantilever boundary condition and also $\Delta x_3 = L_\mathrm{E}$. Note that the piezoelectric constant for longitudinal piezoelectric effect, $e_{33}*$, is used in Eq. (17.7) for modeling of {3–3} mode PVEH devices, whereas piezoelectric constant for transverse piezoelectric effect, $e_{31}*$, is suitable for modeling of {3–1} mode PVEH devices (see [23]). Equations (17.10) and (17.11) describe the coupling term (θ_p) and capacitance (C_p) over the entire cantilever that are obtained by summing Eqs. (17.7) and (17.8) over the cantilever length, L. Direct substitution of these expressions into the governing equations for PVEH devices enables modeling of a unimorph cantilevered PVEH device in the {3–3} mode of operation with IDTEs.

When modeling PVEH devices in the {3–1} mode of operation with the standard electrode configuration, the induced electric field is regarded as constant through the thickness of the piezoelectric layer [23, 28]. Since only one pair of electrodes exists in the standard electrode configuration (i.e., top and bottom), there is no need to use integration or summation of each element in {3–1} mode PVEHs. If the piezoelectric layer between the electrodes is regarded as a simple capacitor, the device in {3–1} mode consists of only one capacitor over the entire length (L) while the device in {3–3} mode can be thought of as a collection of multiple small capacitors. In comparison with {3–3} mode of operation, corresponding expressions to describe the coupling term (θ_p) and capacitance (C_p) of a unimorph, {3–1} mode, PVEH device with standard electrode configuration are as follows in Eqs. (17.13) and (17.14):

$$\theta_p = (e_{31}*)\frac{bz_N}{L}\frac{d\psi}{d\xi}(L) = \frac{(e_{31}*)S_p}{t_p}\frac{d\psi}{dx_1}(L) \tag{17.13}$$

$$C_p = \frac{(\varepsilon_{33}^S*)bL}{t_p} = \frac{(\varepsilon_{33}^S*)A_p}{t_p}\frac{L}{t_p} \tag{17.14}$$

where ξ is nondimensional length coordinate ($= x_1/L$), and z_N represents the distance between the centroid of area A_p and the neutral axis of the combined beam. These Eqs. (17.13) and (17.14) give coupling and capacitance for the entire beam, not just over one element, and these equations for the {3–1} mode are directly comparable to Eqs. (17.10) and (17.11) for the {3–3} mode, with no need to integrate when the beam is uniform in cross-section. Again, it should be noted that the piezoelectric constant, $e_{31}*$, is used in the {3–1} mode while $e_{33}*$ is used for the {3–3} mode. For application to a bimorph, {3–1} mode PVEH device, see the discussion about θ_p and C_p in [23].

17.4 Fabrication and Characterization of MEMS Piezoelectric Energy Harvesters

For applications in microsystems, several studies have focused on developing MEMS-PVEHs using established piezoelectric film processing [25, 29–31]. In macroscale, a number of devices have been successfully developed, tested, and even available commercially (e.g., the test device used in [23] to make a cantilever with a proof mass). While fabrication of MEMS-PVEH devices is an area of active research, not many microscopic prototype devices have yet been documented. In terms of materials, lead zirconium titanate, PZT, receives the most focus and its corresponding multilayer structure is typically deposited on a Si substrate. In general, SiO_2 and/or SiN_x are deposited first as a supporting layer to enhance the mechanical strength of the structure by compensating for the internal stress between the Si substrate and the other layers. ZrO_2 is often used as a diffusion barrier/buffer layer to prevent electrical charge diffusion from the piezoelectric layer (PZT). Pt and Ti comprise the electrode layers, and Ti has the role of improving adhesion between PZT and Pt.

For {3–1} mode harvesters, the bottom electrode Pt and interlayer Ti are deposited before the PZT layer, followed by a liftoff process for the top electrode. In comparison with {3–1} mode harvesters, only the top IDTEs consisting of Pt/Ti are deposited after the deposition of PZT layer in {3–3} mode harvesters.

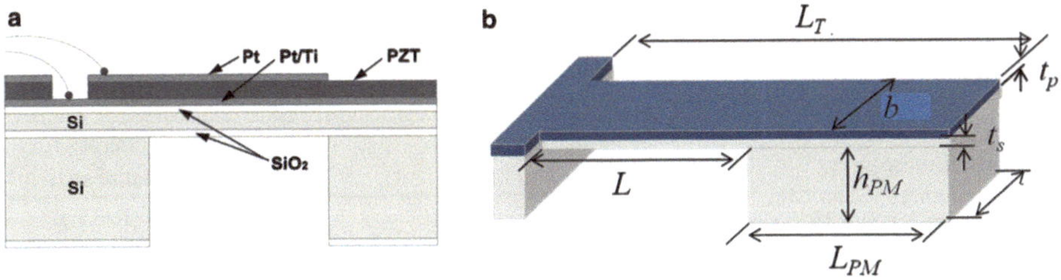

Fig. 17.6 A unimorph, MEMS-scale PVEH: (**a**) schematic of the side view of piezoelectric energy harvesting cantilever based on SOI wafer, (**b**) cantilever structure with indication of dimension parameters. Reprinted with permission from [25, 33]

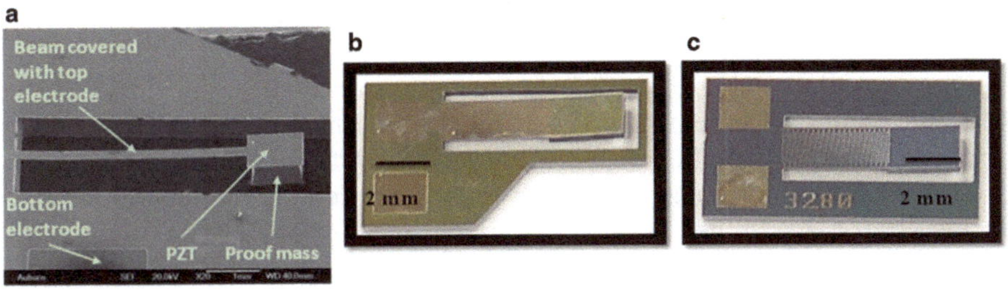

Fig. 17.7 Fabricated, unimorph, MEMS-scale PVEH: (**a**) SEM image and (**b**) optical image of a MEMS-scale PZT cantilever in {3–1} mode with a proof mass, (**c**) optical image of a MEMS-scale PZT cantilever with a proof mass in {3–3} mode with IDTEs. (**a**) Reprinted with permission from [33]. (**b, c**) Courtesy of Dr. Jung-Hyun Park

Here, MEMS-scale unimorph PZT-based energy harvester cantilevers, both in {3–1} mode and {3–3} mode using standard electrodes and IDTEs, are analyzed both experimentally and analytically. More details on the fabrication and evaluation of MEMS-scale cantilevered PVEH devices can be found in the literature [25, 32, 33]. One of the more recent studies includes a micromachined PZT cantilever based on SOI (silicon on insulator) structure with integrated Si proof mass for low-frequency vibration energy harvesting [33]. Compared to their previous PVEH devices based on a Si substrate, use of SOI allows more precise control of the device dimensions, especially thickness, resulting in much less discrepancy between calculated (designed) and measured resonant frequency. A schematic of a piezoelectric energy harvesting cantilever based on a SOI wafer is given in Fig. 17.6. The detailed fabrication process is beyond the scope of this chapter and can be found in [33]. Both {3–1} mode and {3–3} mode PZT-based devices are poled under the same condition at 200 KV/cm AC and at room temperature. Note that an under-hanging Si proof mass as shown in Figs. 17.6 and 17.7 is integrated through bulk etching of the Si wafer. Figure 17.7 contains images of fabricated prototype energy harvester devices in {3–1} and {3–3} modes of operation. These devices are designed to target low frequencies (60–200 Hz), and their geometric dimensions are provided in Tables 17.1 and 17.2.

Table 17.1 Dimension for a unimorph thin film PZT cantilever

Parameter	L_T	L	L_{PM}	b	b_{PM}	h_{PM}
Dimension (mm)	7	4	3	2	2	0.5

Table 17.2 Layers in microscale, unimorph, PZT-based energy harvester devices in {3–1} and {3–3} modes of operation

{3–1} mode of operation		{3–3} mode of operation	
Layer	Thickness (μm)	Layer	Thickness (μm)
Pt	0.12	Pt	0.2 (IDTE)
PZT	1.0	PZT	1.0
Pt/Ti	0.12/0.01	ZrO_2	0.12
SiO_2	0.5	SiO_2	0.5
Si	20.0	Si	20.0
SiO_2	0.5	SiO_2	0.5
Total	21.75	Total	22.12

The properties of piezoelectric materials are critical to the quality and the reliability of the devices, especially in microelectromechanical systems. While both mechanical and piezoelectric properties are well characterized in bulk piezoelectric ceramics, the material properties of thin films are unfortunately not well characterized. This is due to the unique geometry and small displacements of thin films that thus requires different measurement methods from bulk cases when considering material properties of piezoelectric thin films [1, 2, 34]. As these material constants are primary inputs in analytical modeling, the prediction capability of the model is highly dependent upon the accuracy of the material properties of the PZT thin films. The material properties of PZT thin films vary substantially depending on a number of factors such as composition, film orientation (texture), processing technique, thickness, etc. Therefore, evaluation of mechanical and dielectric constants as well as piezoelectric coefficients in the specific PZT thin films is essential in order to predict the performance of PVEH devices.

First, PZT thin film was deposited on a substrate in a stack structure that was the same as the final PVEH devices. Although the specific order, or material, of these layers can vary depending on design requirements, PZT thin films are typically fabricated on Pt (111)/Ti/SiO_2/Si substrates. Since we fabricated the final PVEH device on top of a SOI structure, the PZT thin films were thus characterized with the same structural substrate (i.e., the SOI wafer). Since film orientation (texture), composition, fabrication technique, and thickness all impact the piezoelectric coefficients of PZT thin films, it is necessary to define the specific conditions for the PZT thin film of interest. In this work, we used the following PZT thin films both for material evaluation and for the PVEH devices:

- Composition: $Zr_{0.52}/Ti_{0.48}$ ($Pb(Zr_{0.52}Ti_{0.48})O_3$), morphotropic phase boundary (MPB)
- Processing technique: sol–gel (multiple) coating method
- Thickness of PZT thin film: 1 μm

Appropriate composition selection is required according to the specific application. When compared with other film compositions (e.g., 30/70 or 70/30), MPB

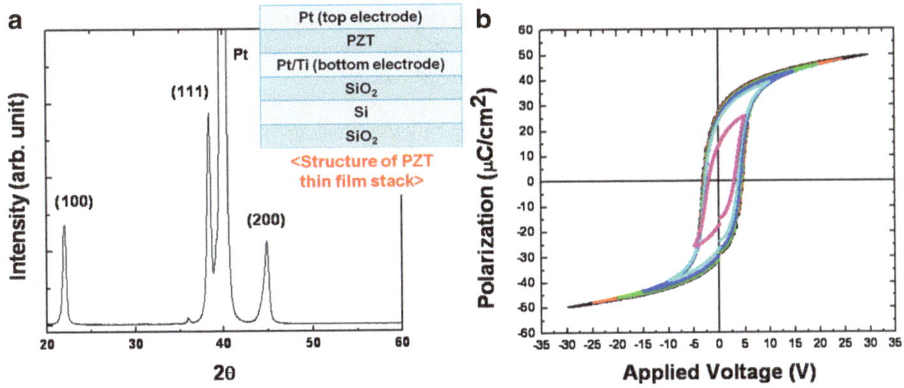

Fig. 17.8 Characterization of fabricated PZT thin films: (**a**) X-ray diffraction and (**b**) polarization voltage hysteresis loop for polycrystalline PZT thin films

(52/48) is known to exhibit high piezoelectric coefficient, low coercive voltage, and high dielectric constant, which are considered optimal for PVEH devices [35]. Crystalline texture (e.g., {111} preferred) and microstructure (e.g., columnar) are also factors that should be considered.

Second, multiple tests were undertaken to check the characteristics of fabricated PZT thin films. To determine the crystalline texture and phase of the PZT thin films, X-ray diffraction was used (Fig. 17.8a). Polarization as a function of applied voltage (P-V hysteresis loops) was obtained and used to see if the PZT thin films retain sufficient ferroelectric properties after processing (Fig. 17.8b). The measurement of the piezoelectric coefficients of PZT thin films is of major interest. Several evaluation techniques including the normal loading method, the impulse method, a wafer flexure technique, an interferometer method, and atomic force microscopy have been previously reported [1–3, 34–36]. (For bulk PZT, the Berlincourt method and the resonance method are widely utilized, which are not usually applicable for piezoelectric thin films.) In this work, the transverse mode piezoelectric coefficient of thin films, $e_{31,f}$, is determined by 4-point bending measurement (aixACCT aix 4PB, http://www.aixacct.com) while a pneumatic loading method is utilized to measure the longitudinal coefficient, $d_{33,f}$. The piezoelectric stress constant, $e_{31,f}$, was found to be -11.1 C/m^2 while d_{31f} was measured to be -104.3 pm/V by the 4-point bending method. Explicit knowledge of mechanical properties, particularly elastic moduli, is also required to implement the simulation of these PVEH devices. For example, published values for elastic moduli of PZT thin films are known to range from 37 to 400 GPa [36], which indicates that the disparities among piezoelectric coefficient values reported in the literature could result from calculation based on different elastic moduli. Liu et al. [37] presented the effects of the substrate and crystalline orientations on the mechanical properties of sol–gel processed $Pb(Zr_{0.52}Ti_{0.48})O_3$ thin films on a Si/SiO_2 substrate. Since the thin films used in [37] are very similar to the PZT thin films used in the present study, the same

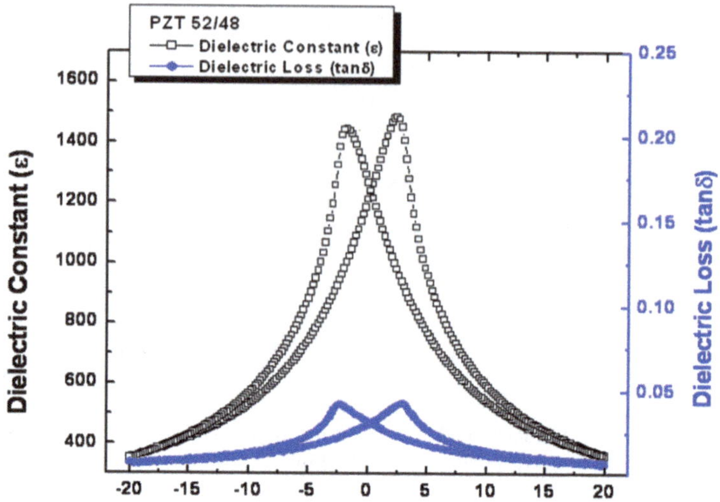

Fig. 17.9 Dielectric constant-voltage curves for the fabricated PZT thin films

elastic modulus of 140 GPa is adopted in the model simulations. Dielectric data was collected using a HP 4292A impedance analyzer and the resulting dielectric constant as a function of dc biased voltage is shown in Fig. 17.9, giving $1{,}550\varepsilon_0$ for the dielectric constant, ε_{33f}^{S}.

Models were implemented on the fabricated unimorph MEMS-PVEH devices both in {3–1} and {3–3} modes (the geometric dimensions are given in Tables 17.1 and 17.2). Material properties of each layer in both {3–1} and {3–3} mode devices are listed in Table 17.3. Key effective parameters such as mass, stiffness, capacitance, and system coupling that appear in the governing Eqs. (17.3) and (17.4) were computed, as listed in Table 17.4. Notice that the key parameters of MEMS-scale PVEHs are much smaller than those for macroscopic bimorph PVEHs [23], due to their significant difference in device scale. Comparison of the calculated results for {3–1} mode and {3–3} mode devices in Table 17.4 suggests that electrical and electromechanical terms differ depending on the type of electrode configurations, as expected. Those terms related to geometric dimensions and mechanical properties such as mass (M), stiffness (K), and forcing function (B_f) have similar values as these two devices are fabricated with similar dimensions (see Tables 17.1 and 17.2).

Model simulation permits prediction of both the resonant and the anti-resonant frequencies of MEMS-scale unimorph PVEH devices in both modes of operation. Note that throughout this chapter resonance refers to the short-circuited condition while anti-resonance refers to the open-circuit condition. For the unimorph, {3–1} mode device, 121.3 Hz is calculated as resonant frequency while 121.6 Hz is estimated as anti-resonant frequency. For the unimorph, {3–3} mode device with similar dimensions, 116.2 Hz and 116.8 Hz are obtained as resonant and anti-resonant frequencies, respectively. Several aspects are noteworthy here. First of all, the difference in resonant and anti-resonant frequencies between {3–1} and

Table 17.3 Materials properties of layers in the MEMS unimorph energy harvester device [26, 38]

Layer material	Density[a,b] (kg/m^3)	Modulus[a,b] (GPa)	Poisson's ratio[a,b], ν	Plate modulus[c] (GPa)
Pt	21440	170.0	0.39	200.5
Ti	4510	110.0	0.34	124.4
PZT (thin film)	7750	140.0	–	140.0
ZrO_2	6000	244.0	0.27	263.2
SiO_2	2300	69.0	0.15	70.6
Si	2329	129.5	0.28	140.5

[a]Measured
[b]Literature values
[c]Computed as $E/(1-\nu^2)$

Table 17.4 Key parameters for model implementation on MEMS unimorph energy harvester devices in {3–1} and {3–3} modes

	M (kg)	K (N/m)	θ_p (N/V)	C_p (F)	B_f	κ^2
{3–1} mode	8.35×10^{-5}	48.5	-1.98×10^{-4}	1.60×10^{-6}	2.44×10^{-5}	0.00995
{3–3} mode	8.32×10^{-5}	44.4	-3.07×10^{-6}	2.15×10^{-11}	2.44×10^{-5}	0.00509

{3–3} devices with similar dimensions arises because of the different constituent layers such as ZrO_2 in the {3–3} mode device and different electrode configurations. Particularly, in {3–3} mode devices with IDTE configuration, contributions of the electrode layers to the mass and stiffness of the entire system affect the resonant frequencies. As electrode layers don't cover the entire area of the cantilever, the ratio of the area covered by the fingered electrodes in IDTEs to the entire area is useful to estimate the partial contribution of these electrode layers to the mass and stiffness of the system, and thus resonant frequencies. The mass and stiffness of IDTE layer materials (Pt, here) are thus multiplied by the area ratio (i.e., 0.31, here) and then used to calculate the resonant frequency of the entire systems.

Another aspect to note is the low value of system coupling, κ^2. In contrast with system couplings for macroscopic PVEH devices shown in the previous work [23] with values around 0.1, the system coupling for the MEMS-scale PVEH devices considered here are in the range of 0.005–0.009 as shown in Table 17.4. The low value of system coupling, κ^2, here results in the small difference between the resonant frequencies and anti-resonance frequencies (e.g., 121.3 Hz and 121.6 Hz, respectively for the {3–1} mode device) of these MEMS-scale systems. This is due to the fact that anti-resonant frequencies of PVEH devices differ from their resonant frequencies depending on the magnitude of system coupling (κ^2). The low κ^2 here is likely due to lack of any systematic design for optimal PVEH performance—the design of these devices was driven mostly by the availability of fabrication techniques in the lab. When model-based design for optimal devices is implemented (see Sect. 17.3), it is possible to achieve MEMS-PVEH devices with much higher values of system coupling κ^2. For example, model-based optimization scheme can be used to calculate optimal device dimensions for operating power density (e.g., beam length, $L = 0.10$ mm and proof mass length, $L_{PM} = 0.68$ mm). These optimal dimensions give 0.06 for system coupling, κ^2, given the same materials properties and structures as those in Table 17.3. This highlights the considerable significance of model-derived optimal device design in order to realize PVEH devices with optimal materials and system parameters. Thus, future work will include further optimal device design studies to determine dimensions such as layer thickness, beam length, and proof mass length to enable fabrication of optimized PVEH devices.

"Resonant frequencies" were experimentally measured for these devices by finding the frequencies with the highest voltage outputs for two electrical resistances (11 kΩ and 4 MΩ). 128.3 Hz was obtained experimentally for the {3–1} mode device while 118.1 Hz was measured for the {3–3} device, which shows 5.8% and 1.6% difference from theoretically predicted values. However, the frequencies at which the peak voltages occur at certain electrical loading conditions do not necessarily correspond to either resonant or anti-resonant frequencies [28]. As part of a full characterization of energy harvester performance, measurement of natural frequencies of a system both at the short circuit condition (resonance) and the open-circuit condition (anti-resonance) is recommended using the appropriate equipment such as a laser vibrometer.

Electrical device responses, voltage and power, of the fabricated MEMS-scale device in {3–1} mode of operation were measured at the experimental "resonance" condition while keeping the base acceleration constant at $0.25g$ ($g = 9.8$ m/s^2). In Fig. 17.10, measured electrical performance is plotted against electrical resistances ranging from 0 to 100 kΩ (dots) along with the simulated results (lines) for voltage and power at various damping conditions. In contrast with the experiments on macroscopic system presented previously in [23, 28], it was not possible to estimate the mechanical damping ratios for these MEMS-scale systems due to the lack of measured data for mechanical performance such as a harvester tip displacement. Thus, here, several reasonable values were chosen for mechanical damping ratio or quality factor to best fit the experimental electrical results. Model-experiment comparison shows that trends of electrical behavior are well predicted regardless of quality factors. In terms of magnitudes, simulated voltage and power match well with the experimental results when the quality factor (Q) is close to 250. The predicted power results reveal that the optimal resistance values for maximal power vary depending on the magnitudes of quality factor, implying the significance of the operating environment, especially, damping conditions of the MEMS-scale system. According to the analytical modeling results, an output power of 3.1 μW can be generated when the harvester is driven at resonance, with $Q = 250$ and an electrical resistance of 5.1 kΩ. Experimentally, power of 3.0 μW was extracted at a load resistance of approximately 11 kΩ and an experimental resonant frequency of 128.3 Hz, as shown in Fig. 17.10b.

In Fig. 17.11, simulated results of electrical performance for the unimorph, {3–3} mode device are also graphically demonstrated at a damping condition of $Q = 250$. (Measured data were not available for the performance of the fabricated {3–3} mode devices with IDTEs.) It should be mentioned that fewer thin film material model

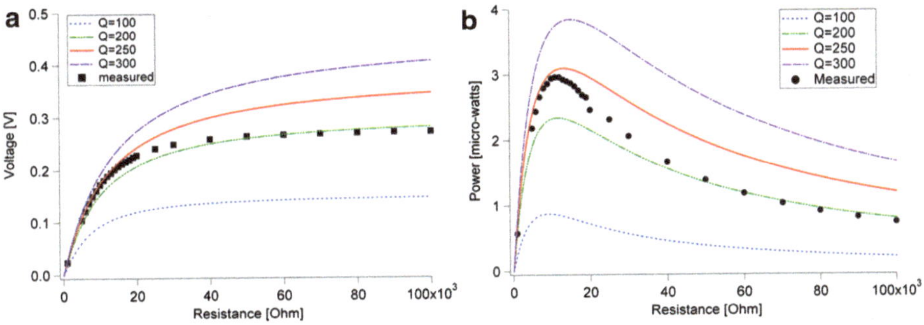

Fig. 17.10 Model-experiment comparisons for a MEMS-scale unimorph energy harvesting PZT cantilever with a proof mass in the {3–1} mode: (**a**) voltage versus electrical resistance and (**b**) power versus electrical resistance at resonances and various damping conditions. Q indicates the quality factor

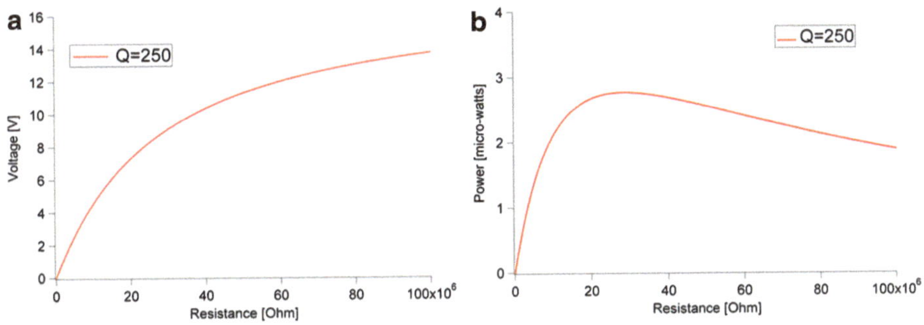

Fig. 17.11 MEMS-scale unimorph energy harvesting PZT cantilever with a proof mass in {3–3} mode with IDTEs: predicted (**a**) voltage versus electrical resistance and (**b**) power versus electrical resistance and $Q = 250$. No measured data available

parameters for PZT are available for {3–3} mode devices when compared with {3–1} mode devices. Therefore, considering that material properties in thin films are typically smaller than in the bulk material, bulk material constants for PZT-5A [23, 28] were used to obtain a rough estimate of the PZT thin film material properties of the {3–3} mode MEMS-PVEHs and then the calculated results of the device performance. Simulation results reveal that a maximum power of 3.5 μW can be generated from the {3–3} mode unimorph energy harvester device when operated either at 23.8 MΩ (i.e., closer to resonance) or 170 MΩ (i.e., closer to anti-resonance). Comparison of PVEHs of similar size but operating in different modes ({3–1} and {3–3}) is possible using the results shown in Figs. 17.10 and 17.11. While both mode devices are expected to produce similar maximum power of around 3.0–3.1 μW, much higher voltage (14 V) and required electrical resistance (20 MΩ) are observed for the {3–3} mode device when compared to the {3–1} mode device where maximum voltage is around 0.4 V and optimal electrical resistance is 13 kΩ.

The primary goal of this chapter is to make steps forward to realizing pervasive use of wireless, battery-less small electronics powered by MEMS-PVEHs. With this purpose in mind, an overview of piezoelectric materials and device configurations is provided for better understanding of PVEHs both at materials- and systems-levels. Furthermore, refined model for different electrode configurations in PVEHs, with special focus on IDTEs, is developed, implemented, and compared with the experimental test results of MEMS-PVEH device fabricated. The developed model for PVEH devices exhibits conservative predictive capability not only on macroscopic devices but also on MEMS-scale PVEHs. Future work will include (1) parametric study to investigate the effect of IDTE geometric dimensions on power performance of PVEH devices and (2) construction of model-derived piezoelectric materials design guidelines to aid in the design of optimal MEMS-PVEH systems.

Acknowledgements Part of this work (S. Hong) was carried out at Argonne National Laboratory (ANL), a US DOE Science Laboratory operated under contract no. DE-AC02-06CH11357 by UChicago Argonne, LLC.

References

1. Muralt P (2000) Ferroelectric thin films for micro-sensor and actuators: a review. J Micromech Microeng 10:136–146
2. Trolier-Mckinstry S, Muralt P (2004) Thin film piezoelectric for MEMS. J Electroceram 12:7–17
3. IEEE 1987 ANSI Standard 176–1987: IEEE Standard on Piezoelectricity
4. Cook-Chennault KA, Thambi N, Sastry AM (2008) Powering MEMS portable devices-a review of non-regenerative and regenerative power supply systems with special emphasis on piezoelectric energy harvesting systems. Smart Mater Struct 17:043001
5. Tadigadapa S, Mateti K (2004) Piezoelectric MEMS sensors: state-of-the-art and perspectives. Meas Sci Technol 20:092001
6. Anton SR, Sodano HA (2007) A review of power harvesting using piezoelectric materials (2003–2006). Smart Mater Struct 16:R1–R21
7. Funasaka T, Furuhata M, Hashimoto Y, Nakamura K (1998) Piezoelectric generator using a $LiNbO_3$ plate with an inverted domain. Ultrasonics symposium, Sendai, pp 959–962
8. Setter N et al (2006) Ferroelectric thin films: review of materials, properties, and applications. J Appl Phys 100:051606
9. Wang ZL, Song J (2006) Piezoelectric nanogenerators based on zinc oxide nanowire arrays. Science 312:242
10. Xu C, Wang S, Wang ZL (2009) Nanowire structured hybrid cell for concurrently scavenging solar and mechanical energies. J Am Chem Soc 131:5866–5872
11. Lovinger AJ (1983) Ferroelectric polymers. Science 220:1115–1121
12. Priya S (2007) Advanced in energy harvesting using low profile piezoelectric transducers. J Electroceram 19:165–182
13. Sodano HA, Inman DJ, Park G (2004) A review of power harvesting from vibration using piezoelectric materials. Shock Vib Dig 36(3):197–205
14. Sodano HA, Park G, Inman DJ (2004) An investigation into the performance of macro-fiber composites for sensing and structural vibration applications. Mech Syst Signal Process 18: 683–697

15. Sodano HA, Park G, Leo DJ, Inman DJ (2004) Model of piezoelectric power harvesting beam. In: ASME international mechanical engineering congress and exposition, Washington, DC, 15–21 November, vol 40, p 2
16. Kim H et al (2009) Piezoelectric energy harvesting, chapter 1. In: Priya S, Inman DJ (eds) Energy harvesting technologies. Springer Science & Business Media, LLC, New York, NY
17. Ikeda T (1996) Fundamental of piezoelectricity. Oxford University Press, New York
18. Schwartz RW, Ballato J, Haertling GH (2004) Piezoelectric and electro-optic ceramics. In: Buchanan RC (ed) Ceramics materials for electronics. Dekker, New York
19. Xu Y (1991) Ferroelectric materials and their applications. North-Holland, Amsterdam
20. Gady WG (1946) Piezoelectricity. McGraw-Hill, New York
21. Setter N (2005) Electroceramics-based MEMS: fabrication-technology, and applications. In: Tuller HL (ed) Electronics materials: science and technology. Springer, New York
22. Beeby SP, Tudor MJ, White NM (2006) Energy harvesting vibration sources for microsystems applications. Meas Sci Technol 13:R175–R195
23. Kim M, Hoegen M, Dugundji J, Wardle BL (2010) Modeling and experimental verification of proof mass effects on vibration energy harvester performance. Smart Mater Struct 19:045023
24. Jeon YB, Sood R, Jeong J-H, Kim S-G (2005) MEMS power generator with transverse mode thin film PZT. Sens Actuators A 122:16–22
25. Shen D, Park J-H, Ajitsaria J, Choe S-Y, Howard CW III, Kim D-J (2008) The design, fabrication and evaluation of a MEMS PZT cantilever with an integrated Si proof mass for vibration energy harvesting. J Micromech Microeng 18:055017
26. du Toit NE (2005) Modeling and design of a MEMS piezoelectric vibration energy harvester. Master's thesis, Massachusetts Institute of Technology
27. Mo C, Kim S, Clark WW (2009) Theoretical analysis of energy harvesting performance for unimorph piezoelectric benders with interdigitated electrodes. Smart Mater Struct 18:055017
28. duToit NE, Wardle BL (2007) Experimental verification of models for microfabricated piezoelectric vibration energy harvesters. AIAA J 45:1126–1137
29. Myers R, Vickers M, Kim H, Priya S (2007) Small scale windmill. Appl Phys Lett 90:3
30. Fang HB, Liu JQ, Xu ZY, Dong L, Wang L, Chen D, Cai BC, Liu Y (2006) Fabrication and performance of MEMS-based piezoelectric power generator for vibration energy harvesting. Microelectron J 37:1280–1284
31. Marzencki M, Charlot B, Basrour S, Colin M, Valbin L (2005) Design and fabrication of piezoelectric micro power generators for autonomous microsystems. DTIP' 05 symposium design, test, integration & packaging of MEMS/MOEMS, Montreux, Switzerland, pp 299–302
32. Ajitsaria J, Cho S-Y, Shen D, Kim DJ (2007) Modeling and analysis of a bimorph piezoelectric cantilever beam for voltage generation. Smart Mater Struct 16:447–454
33. Shena D, Park J-H, Noh JH, Choe S-Y, Kim S-H, Wikle HC III, Kim D-J (2009) Micromachined PZT cantilever based on SOI structure for low frequency vibration energy harvesting. Sens Actuators A 154:103–108
34. Muralt P, Polcawich RG, Trolier-McKinstry S (2009) Piezoelectric thin films for sensors, actuators, and energy harvesting. MRS Bull 34:658–664
35. Ledermann N, Muralt P, Baborowski J, Gentil S, Mukati K, Cantoni M, Seifert A, Setter N (2003) {100}-Textured, piezoelectric $Pb(Zr_x, Ti_{1-x})O_3$ thin films for MEMS: integration, deposition and properties. Sens Actuators A 105:162–170
36. Lefki K, Dormans M (1994) Measurement of piezoelectric coefficients of ferroelectric thin films. J Appl Phys 76(3):1
37. Liu D, Yoon SH, Zhou B, Prorok BC, Kim DJ (2009) Investigation of the crystalline orientations and substrates dependence on mechanical properties of PZT thin films by nanoindentation. Materials research society symposium proceeding, vol 1129
38. Mracek AM (2004) Towards an embeddable structural health monitoring sensor: design and optimization of MEMS piezoelectric vibration energy harvesters. Master's thesis, Massachusetts Institute of Technology

Chapter 18
Nonlinear Vibration Energy Harvesting with High-Permeability Magnetic Materials

Xing Xing and Nian X. Sun

Abstract In this chapter, we introduce the recent demonstrations of high energy density nonlinear vibration energy harvesting with high-permeability magnetic materials, which show great promise for compact and wideband vibration energy harvesting systems. Two generations of nonlinear vibration energy harvesting technology based on high-permeability magnetic material will be discussed in this chapter. The first generation energy harvester design consists of a high-permeability magnetic cantilever beam, in a solenoid, and a hard magnet pair that provides the biasing field. The mutual interaction between the vibrating highly permeable beam and the bias magnetic field of the magnets leads to maximized flux change and therefore a large induced voltage. This harvester has shown a maximum power output of 74 mW, a power density of 1.07 mW/cm^3 at 54 Hz under an applied acceleration of $0.57 \times g$ (with $g = 9.8$ m/s^2), and a bandwidth of 10 Hz (or 18.5% of the operating frequency). The second generation energy harvester design, which has two solenoids fixed on two sides of a spring supported hard magnet pair, has demonstrated a significant increase of the output power when compared with the first generation device. The improved design has an output voltage of 2.52 V, a power density of 20.84 mW/cm^3 at 42 Hz, and a half-peak working bandwidth of 6 Hz (or 14%). The coexistence of magnetostatic and elastic potential energies in both designs results in a nonlinear effect, which produces the wide working bandwidth. Details of the two generation energy harvester designs, their performance, and the origin of the nonlinear behavior are detailed in this chapter.

X. Xing • N.X. Sun (✉)
Department of Electrical and Computer Engineering, Northeastern University, 360 Huntington Avenue, 409 Dana, Boston, MA, 02115, USA
e-mail: n.sun@neu.edu; nian@ece.neu.edu

N. Elvin and A. Erturk (eds.), *Advances in Energy Harvesting Methods*,
DOI 10.1007/978-1-4614-5705-3_18,

18.1 Introduction

Mobile electronics and wireless sensors have been developing rapidly in several fields, including environmental, industrial and medical monitoring. Traditional power supply systems such as batteries and wired sources all have their limitations. Batteries need frequent recharging or replacement, which is costly and sometimes impossible, especially for wireless networks with thousands of physically embedded nodes [1]. Alternatively, the need for power supply cables can seriously limit the flexibility and cost for wired electronic devices. As a result, self-renewable power supplies are attracting more and more attention for powering mobile electronics and wireless sensors.

Energy harvesting, sometimes defined as "power scavenging" or "energy extraction," refers to methods of converting ambient energy to usable electrical energy. Mechanical energy associated with vibration has been one of the major energy sources for energy harvesting systems. Different harvesting mechanisms have been utilized for collecting ambient vibration energy, including electromagnetic [2], electrostatic [3], piezoelectric [4, 5], and magnetoelectric (ME) [6] mechanisms. In addition, different functional materials and structures have been used in harvesting vibration energy, including piezoelectric, magnetostrictive [7], and magnetoelectric composite beams. An electromagnetic micro-cantilever device with a volume of 0.15 cm^3 was reported to generate a maximum power 46 μW at 52 Hz and an acceleration of 0.59 m/s^2 [2]. Optimized electrostatic energy harvesting devices showed an output power of 1 mW under vibration amplitude of 90 μm at 50 Hz [8]. Piezoelectric cantilever harvesters were reported to achieve a maximum output power of 790 μW with a tip mass of 10 g at an acceleration of 9 m/s^2 and frequency of 72 Hz [9]. A magnetoelectric energy harvester with a beam consisting of both magnetostrictive and piezoelectric materials was demonstrated to generate an output voltage of 4 V under a vibration acceleration of $0.05 \times g$ at 20 Hz and an AC magnetic field amplitude of 2 Oe [10]. A cantilever with six adhesively bonded layers of magnetostrictive Metglas 2605SC ribbon in a solenoid was claimed to generate a maximum output power of 900 μW at $1 \times g$ [11].

Most conventional vibration energy harvesters are designed as linear resonant structures with narrow operating bandwidths, which severely limit their deployment in real world environments with wideband vibration frequencies. As a result, efforts have been made to explore nonlinear mechanisms that lead to vibration energy harvesters with wide working bandwidths. In this chapter, we describe two generations of nonlinear vibration energy harvesters employing high-permeability materials. In the first generation energy harvester [21], the strong magnetic coupling between the vibrating high-permeability cantilever and fixed magnets produce the nonlinear effect and hence the wide working bandwidth. Besides the wide working bandwidth, achieving high output power and power density will enable a much wider range of applications of vibration energy harvesting technologies. So far, piezoelectric-based vibration energy harvesters, which have received the greatest attention, have demonstrated much higher power densities than their magnetic-based counterparts, even though most piezoelectric-based energy harvesters show a narrow bandwidth or a limited operating frequency range of 2–5% of the center operating frequency. For example, a piezoelectric bare beam-based vibration energy harvester can generate a power of 6.63 mW/cm^3 (Volture Piezo Energy Harvester-V25W). Nevertheless, theoretically, the magnetostatic energy density $(1/2)\mu H^2$ in high-permeability magnetic materials is 10^5–10^6 times that of the electrostatic energy density $(1/2)\varepsilon E^2$ in piezoelectrics [12]. Vibration energy harvesters with hard or soft magnetic materials have been studied and tested [13, 14]. However, the full potential of achieving high power density in vibration energy harvesters with high-permeability magnetic materials has not yet been realized. The second generation energy harvester with more layers of high-permeability materials as the induction core than the first generation harvester was demonstrated to have a significantly increased output power density while maintaining a wide bandwidth [22].

18.2 First Generation: High-μ Cantilever Vibration Energy Harvester

18.2.1 Prototype and Testing Platform

A novel vibration energy harvesting model based on the strong magnetostatic coupling between a cantilever beam and a bias magnet pair was set up and experimentally verified. The inhomogeneous bias magnetic field enables the highly permeable beam to experience complete magnetic flux reversal twice in one vibration period, which leads to maximized magnetic flux change rate in the solenoid. At the same time, the magnetic potential energy makes the cantilever vibrate in a wider potential well than in the simple harmonic case, which allows a wide working bandwidth of the harvester. The schematic design of the vibration energy harvester is shown in Fig. 18.1. The key component of this energy harvester is a high-permeability (high-μ) single layer beam, with one end fixed and the other end vibrating inside the solenoid. Two identical rectangular hard magnets are placed in close proximity to the free end of the high-μ beam outside the solenoid to induce a nonhomogeneous bias field, which acts to induce magnetization reversal. The bias magnets are aligned in parallel with each other and parallel to the beam. As shown in Fig. 18.2, the magnets form a closed flux path, with the magnetizations antiparallel to each other. When the free end of the cantilever passes through the spatially inhomogeneous fringing magnetic field generated by the bias magnet pair, the magnetization of the magnetic cantilever would be reversed by the antiparallel bias magnetic field by 180°. The magnetization reversal in the magnetic cantilever leads to maximized flux change in the solenoid, resulting in an induced voltage which varies at the same frequency of the mechanical vibrating source. However, when the magnetization of the two bias magnets are in parallel, as shown in Fig. 18.3, the repulsive magnetic field keeps the beam magnetized in the same direction all the

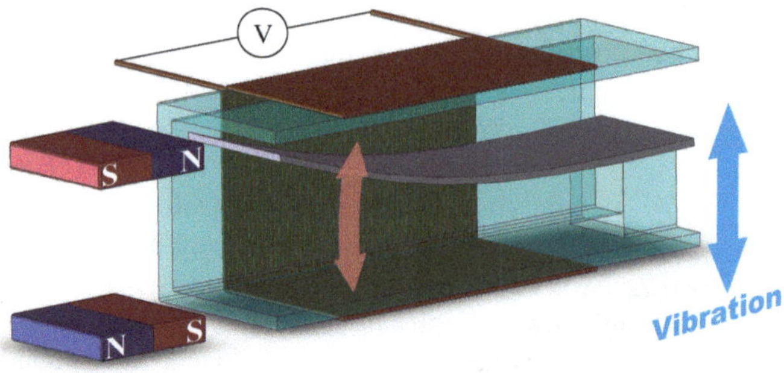

Fig. 18.1 The section view of the schematic design of the vibration energy harvesting device. Dimension of each part is: 4.4 cm × 3.2 cm × 4 cm for the solenoid, 1.25 cm × 2.2 cm × 1.5 cm for the magnet pair including the air gap, 1.3 cm × 1.5 cm × 2.5 cm for the mounting frame on one side and 0.5 cm × 1.5 cm × 0.6 cm on the other

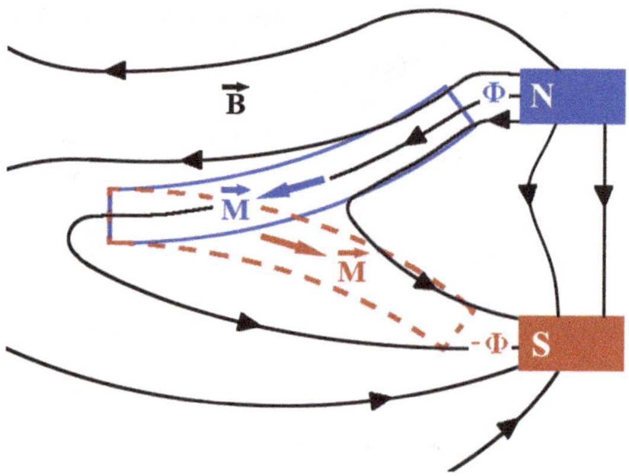

Fig. 18.2 Magnet pair with antiparallel magnetic moment provides closed magnetic field lines, ensuring maximum magnetic flux change, from Φ to $-\Phi$ during the vibration

time. As a result, the magnetic flux change will be smaller in this case compared to the antiparallel case in Fig. 18.2, leading to a lower output voltage and an induced AC voltage with a frequency that is double of that of the mechanical vibration frequency.

The harvester, including the MμShield® cantilever, the solenoid, and a SmCo magnet pair, is seated on a vibrating stage, as shown in Fig. 18.4. The stage is driven by an audio power amplifier connecting to a lock-in amplifier. The mechanical movement of the stage is monitored by an accelerometer. Voltage output of the harvester in time domain is monitored by a digital oscilloscope. Total volume of the energy harvester is 68.96 cm^3, including the solenoid with the enclosed beam, the magnet pair with the air gap, and the mounting structure. The coil resistance of the solenoid is 1 Ω, the inductive impedance is 2.86 Ω at 54 Hz, which is obtained by direct measurement. A lead tip mass of 0.5 g is attached on the free end of the cantilever to adjust the vibration amplitude and the intrinsic frequency.

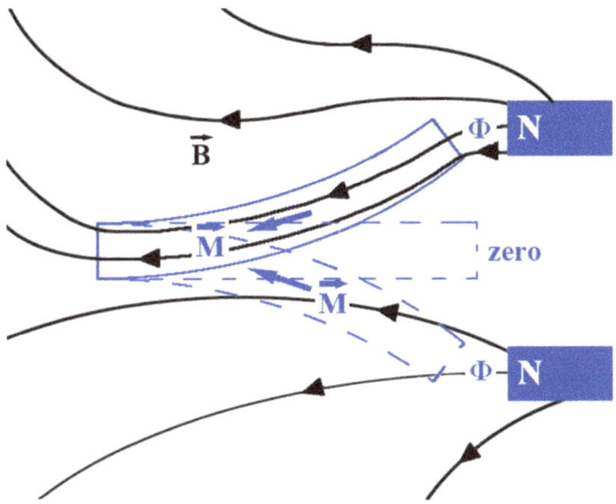

Fig. 18.3 Magnet pair with parallel magnetic moment provides repelling magnetic field lines, in which case the magnetic flux changes from Φ to 0 and back to Φ during the vibration

Fig. 18.4 Prototype of the wideband high-permeability vibration energy harvester

18.2.2 Theoretical Model

Since the magnetic field magnitude varies along the beam length, the total induced voltage across the coil can be calculated by integrating through the solenoid, as shown in Fig. 18.5. According to Faraday's law the open circuit voltage can be expressed by:

$$V_{\text{open}} = \frac{\mathrm{d}\phi(t)}{\mathrm{d}t} = \frac{\mathrm{d}\int \mu_0 \{H[x, y(x)] + M[x, y(x,t)]\} \cdot A \cdot dN}{\mathrm{d}t}$$
$$= \frac{\mathrm{d}\int \mu_0 M[x, y(x,t)] \cdot A \cdot dN}{\mathrm{d}t}, \tag{18.1}$$

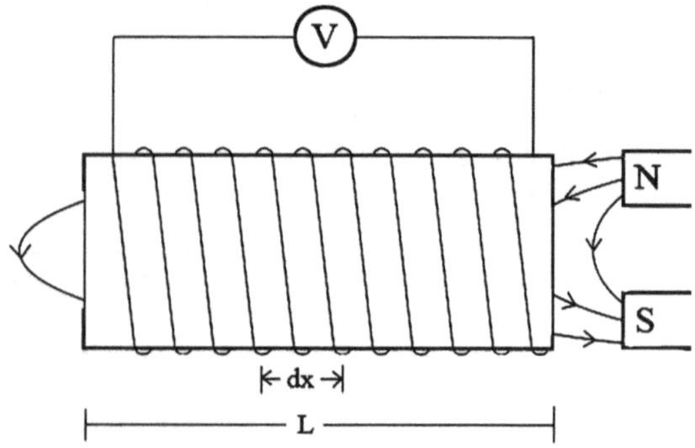

Fig. 18.5 The schematic of the calculation of induced voltage across the solenoid

where M is the magnetization in the beam, A is the cross-sectional area of the beam, and dN the number of loops in the infinitesimal unit length of the solenoid dx, shown in Fig. 18.5, and

$$dN = \frac{N_{\mathrm{L}}}{d_{\mathrm{w}}} dx \tag{18.2}$$

N_{L} is the number of coil loop layers in the solenoid and d_{w} is the copper wire diameter of the coil. The dimension of the beam in the experimental setup is 4.6 cm × 0.8 cm × 0.0254 cm, with a length: width: thickness ratio 181:31.5:1, which ensures that the length direction is the magnetic easy axis. The high-μ beam material has a maximum permeability of 300,000 and a saturation magnetization of 7,500 G. The magnetic hysteresis loop $M(H)$ along the length direction of the beam was measured, as shown in Fig. 18.6, in order to calculate the magnetization $M[x,y]$ at an arbitrary point on the beam in the nonuniform magnetic field. The vibration amplitude of an arbitrary point on the beam $y(x,t)$ is obtained by combining solutions of the thin-beam deflection equation $y(x)$ and the equation of motion for the cantilever $y(t)$. According to the thin-beam equation and for clamped-free boundary conditions, the static approximation of the cantilever deflection is $y(x) = \frac{3Lx^2 - x^3}{2L^3}$ [15], as shown in Fig. 18.7. By combining the time dependent term $a(t)$, the beam shape function at time t is

$$y(x,t) = a(t)\left(\frac{3Lx^2 - x^3}{2L^3}\right), \tag{18.3}$$

where L is the length of the beam and $a(t)$ is the amplitude at free end, which is determined by the following equation of motion

$$m_{\mathrm{eff}} \frac{\mathrm{d}^2 a(t)}{\mathrm{d}t^2} = -\frac{\mathrm{d}U}{\mathrm{d}a} - b\dot{a}(t) + F_{\mathrm{drive}}, \tag{18.4}$$

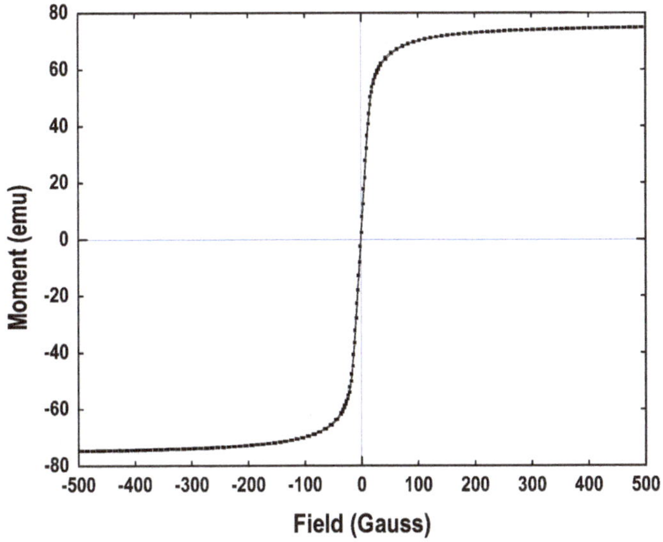

Fig. 18.6 Hysteresis loop of the MuShield beam with the dimension of 4.6 cm × 0.8 cm × 0.0254 cm

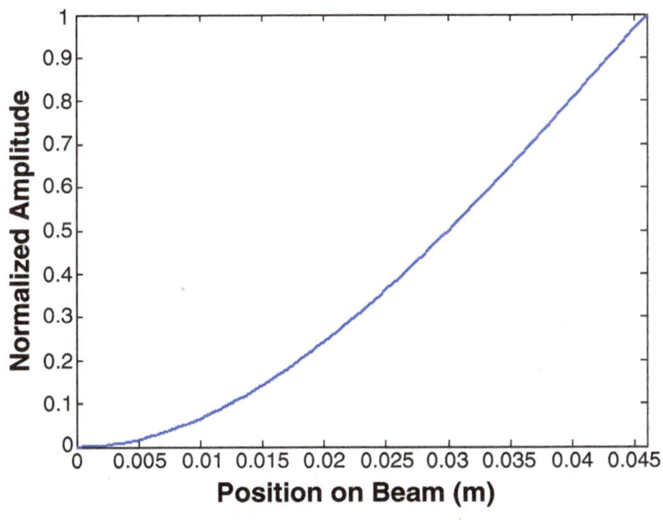

Fig. 18.7 Beam shape at its maximum deflection

where m_{eff} is the effective mass of the beam plus tip mass at the free end, which is 0.74 g. The first term on the right-hand side $-\mathrm{d}U/\mathrm{d}a$ is the force due to the total potential energy,

$$\frac{\mathrm{d}U}{\mathrm{d}a} = \frac{\mathrm{d}(U_{\mathrm{magnetic}} + U_{\mathrm{elastic}})}{\mathrm{d}a}, \tag{18.5}$$

where the elastic potential energy $U_{\text{elastic}} = (k/2)/a^2$, with k the elastic stiffness, which is experimentally determined to be 77 N/m, and U_{magnetic} the magnetic potential due to the bias magnet pair

$$U_{\text{magnetic}} = \int_0^L -\vec{B}[x, y(t)] \cdot \mathrm{d}\vec{m}, \quad (18.6)$$

in which $\vec{m}$ is the magnetic moment in the beam. The size of each of the identical SmCo hard magnet is 2.2 cm × 1.3 cm × 0.2 cm, providing a fringing magnetic field 500 Oe at the free end of cantilever and 10 Oe in the middle of the solenoid. The distribution of magnetic field is obtained with discrete spatial magnetic field measurements and fitted with simulation. The second term $-b\dot{a}$ in Eq. (18.4) is the mechanical damping term, with b the damping constant which is experimentally determined to be 0.0024 Ns/m. The third term F_{drive} stands for the vibration driving force applied on the fixed end, which is a sine wave input.

When a load resistance is connected across the coil, the vector voltage across each element has the relationship $\dot{V}_{\text{open}} = \dot{V}_{\text{load}} + \dot{V}_{\text{coilres}} + \dot{V}_{\text{coilind}}$. The solenoid is equivalent to a resistance in series with an inductance. The output power is optimized when the impedance is matched, $X_{\text{L}} = 2\pi f L = X_{\text{C}} = \frac{1}{2\pi f C}$ and $R_{\text{load}} = R_{\text{coil}}$, which could be done by inserting a capacitance in series with the rest of the circuit and adjusting the load resistance. In this way, the maximum output power is

$$P_{\max} = \frac{\left(\frac{V_{\text{open}}}{2}\right)^2}{R_{\text{load}}} = \frac{1}{4R_{\text{load}}}\left(A\mu_0\frac{N_{\text{L}}}{d_{\text{w}}}\right)^2\left(\int_0^L\left\{\frac{\mathrm{d}M\,[x, y(x,t)]}{\mathrm{d}t}\right\}dx\right)^2 \quad (18.7)$$

Equation (18.7) indicates that at a fixed frequency, the output power depends on the rate of change of magnetization in the beam.

18.2.3 Results and Analysis

Figure 18.8 shows calculated and measured results of the open circuit voltage in two cases. When the magnet pair is set to have antiparallel magnetization, the energy harvester shows a high open circuit voltage with a peak value of 544 mV at a vibration frequency of 54 Hz and an acceleration amplitude of $0.57 \times g$. As expected, for the case when the two bias magnets are arranged with parallel magnetizations, the output voltage has a significantly lowered peak value of 8 mV at double the driving frequency (i.e. at 108 Hz). It is interesting to note that the mechanical vibration source is a sine wave signal, while the output voltage is not, but with narrow peaks with a full width at half maximum of 1 ms. This is related to the nonuniform magnetic field spatial distribution, leading to the approximate square wave time varying magnetic flux, shown in Fig. 18.9. The free end amplitude dependent flux is also plotted. It is clear that the flux in the beam is reversed

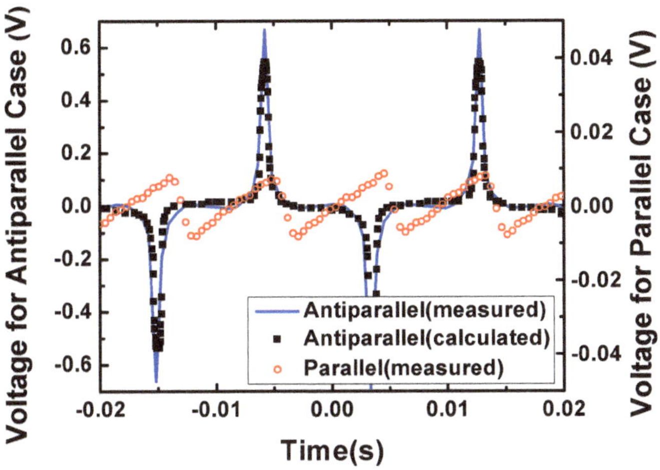

Fig. 18.8 Measured and calculated results of the open circuit voltage for the energy harvesting device at a mechanical vibration frequency of 54 Hz and an acceleration amplitude of $0.57 \times g$ ($g = 9.8$ m/s^2)

Fig. 18.9 Normalized magnetic flux as a function of time and free end amplitude, at vibration frequency of 54 Hz, acceleration amplitude: $0.57 \times g$

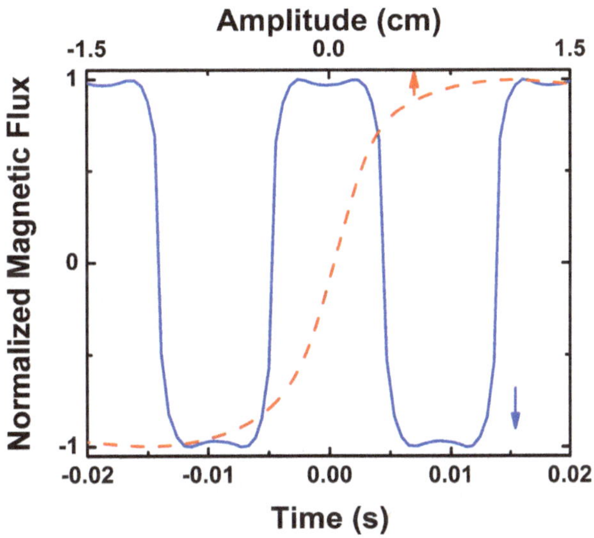

immediately while passing the nonstable equilibrium position in the middle of the hard magnet pair. The sharp drop and rise in flux is the reason for the large induced voltage.

Figure 18.10 shows the frequency response of the harvester. The maximum measured output power is 74 mW across a 1 Ω load and with a time average value 5 mW at an acceleration amplitude of $0.57 \times g$ corresponding to a maximum power density of 1.07 mW/cm^3 or 1.88 mW/(g cm^3).

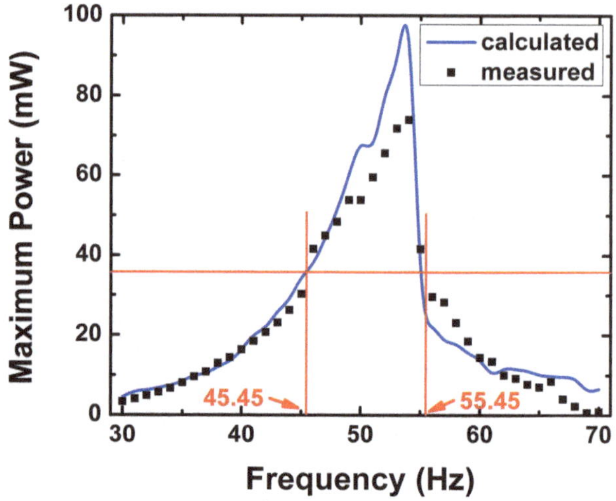

Fig. 18.10 Measured and calculated frequency response of the energy harvester

18.2.4 Nonlinear Effect

From Fig. 18.10, it can be seen that the working bandwidth is about 10 Hz, 18.5% of the central frequency, compared with 2.1 Hz (Ferro Solution VEH360) or 3.5% of the central frequency, for a typical linear oscillator harvester. The major reason for the large bandwidth is that the magnetic coupling is not linear to the displacement of the oscillator, so that the nonlinear effect provides the system with a wider working bandwidth [16]. As shown in Fig. 18.11, compared with the elastic potential energy, the magnetic potential energy curve has two potential wells distributed right next to each other, resulting in a wider total potential well. As long as the acceleration is larger than its threshold value, or when the cantilever is supplied with enough energy to get over the potential barrier, it can vibrate between the two shallow potential wells. Thus a relatively wide oscillation region at a particular driving frequency is obtained. However, if the oscillator is not supplied with enough energy or the damping is so strong that it is not able to overcome the potential barrier, the oscillation is limited in one well, instead of two. In this case, the behavior of the oscillator is more like that of a linear one, with a narrow working bandwidth.

Both the calculated and measured curves in Fig. 18.10 exhibit unsymmetrical peaks about the central frequency. This is because the performance of the oscillator at lower frequencies differs from that at higher frequencies. At lower frequencies, the oscillation behavior is dominated by the nonlinear effect. The output power decreases slowly as the frequency reduces due to the mismatch between the intrinsic and driving source frequency. Even at a frequency as low as 30 Hz, oscillation between the two potential wells was still observed. However, at a higher frequency range than 54 Hz, the performance is dominated by the linear effect and the

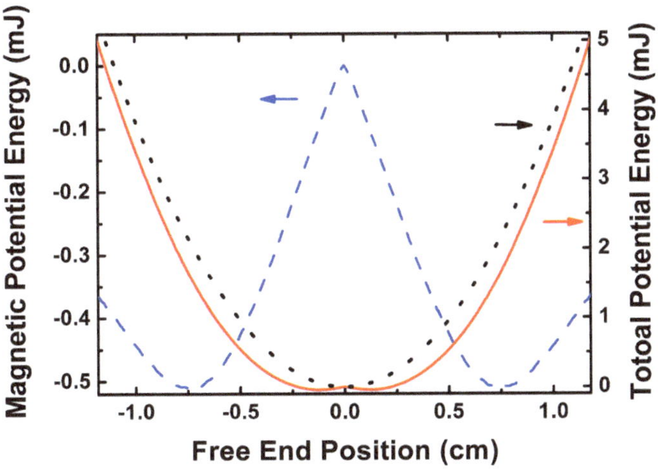

Fig. 18.11 Elastic potential energy (*dotted line*), magnetic potential energy (*dash line*), and total potential energy (*solid line*) of the oscillation system as functions of the free end displacement of the beam

oscillator was observed to be trapped in one well, leading to one sided narrow peaks in the voltage signal. This is because in the higher frequency range, a larger dynamic speed results in a larger damping force, so that the cantilever does not have enough energy to climb over the potential barrier. This single well oscillation dramatically decreases the harvesting efficiency, because the cantilever beam cannot reach flux reversal.

18.2.5 Discussion

Table 18.1 shows the figures of merit for vibrating energy harvesters with different types of working mechanisms and materials, including magnetoelectric, electrostatic, piezoelectric, magnetoelectric sensor based, magnetostrictive, and high-permeability material-based devices. Among all these different mechanisms, the wide bandwidth energy harvester based on the magnetic coupling between the high-μ material and the bias field of the hard magnets, generates a relatively high power density and a wide working bandwidth. The metallic high-permeability single-layer beam has advantages from the material point of view as well. First of all, it is mechanically more robust compared with most of the piezoelectric materials or glue bonded multilayer materials. Second, it does not have possible problems of degraded performance over the lifetime of the device.

Table 18.1 Comparison of several key figures of merit for different vibrating energy harvesting mechanisms

Mechanisms/products	f_{center} (Hz)	a (g)	P_{max} (mW)	R_{output} (Ω)	V_{total} (cm^3)	Power density (mW/cm^3)	HPBW[a] (Hz)
Magnetoelectric micro-cantilever [2]	52	0.06	0.046	4k	0.15	0.31	0.5
Electrostatic [8]	50	0.91	1.052	–	1.8	0.58	–
Magnetoelectric composite beam [10]	40	1	–	3M	–	0.4	–
Volture Piezo Energy Harvester-V25W	30	1.1	6.5	–	0.98	6.63	–
Magnetoelectric sensor-based system (Ferro Solution VEH360)	60	0.1	10.8	–	113	0.096	2.1
Magnetostrictive cantilever [11]	58.1	1	0.2	80	–	–	–
Perpetuum PMG37	21.9	1	92	–	130.67	0.7	12.9
KCF technologies VPH300	360	0.239	4.1	–	208.86	0.0196	5
High-μ cantilever	54	0.57	74	1	68.96	1.07	10

[a]Power density, in unit of mW/(g cm^3)

18.3 Second Generation High Output Power Vibration Energy Harvester with High-Permeability Material

18.3.1 Prototype and Testing Platform

The schematic design of the high output power energy harvester is shown in Fig. 18.12. Two identical solenoids with high-permeability/insulator multilayer cores were placed on two sides of a vibrating hard magnet pair with antiparallel magnetization. The key components of this energy harvester are the two identical solenoids with a high-permeability (high-μ) MuShield® core inside, which are placed at two sides of the magnet pair and fixed on a vibrating stage. The magnets have antiparallel magnetization and are supported by a regular circular cross-section spring with its bottom fixed on the surface. When the magnet pair moves up and down with respect to the vibrating stage, the magnetic field inside each solenoid periodically changes its direction. The magnetostatic coupling between the solenoids and the time varying inhomogeneous bias magnetic field results in a nonlinear oscillation and a complete magnetic flux reversal in the solenoids. The presence of highly permeable cores dramatically increases the magnitude of magnetic flux inside the coils. Thus, a large induced voltage is generated on both

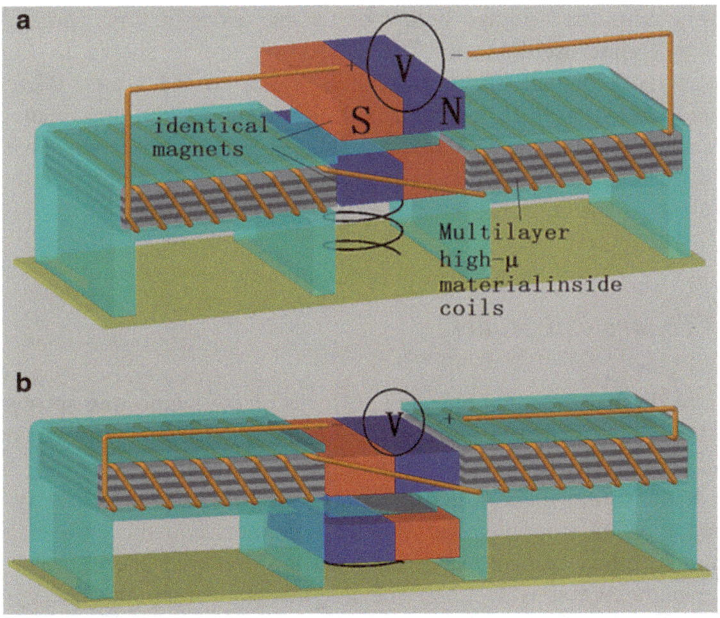

Fig. 18.12 The schematic design and working mechanism of the high power vibration energy harvester. (**a**) The magnet pair moves to the top. (**b**) The magnet pair moves to the bottom

Fig. 18.13 Structure of the vibration energy harvester. Dimension of each component is $2 \times 2.5 \times 1$ cm^3 for the solenoids and $1.25 \times 2.2 \times 1.5$ cm^3 for the magnetic pair which includes the air-gap

sides of the solenoid, and the output voltage can be doubled with serial connection of the two solenoids.

The entire device is powered by a vibrating stage, which is driven by an audio power amplifier. Voltage output of the harvester in time domain is monitored by a digital oscilloscope. The device picture is shown in Fig. 18.13. Each SmCo hard magnet has a dimension of 2.2 cm $\times$ 1.3 cm $\times$ 0.2 cm. Each solenoid core consists of 28-layers of high-permeability MuShield material, with a dimension of 2 cm $\times$ 2 cm $\times$ 0.002 in. Total volume of the energy harvester is 6.44 cm $\times$ 3.25 cm $\times$ 1.4 cm = 29.3 cm^3, which includes the solenoids, the magnet pair, and air-gap. The coil resistance of each solenoid is 1.3 Ω.

18.3.2 Theoretical Model

The mass of the hard magnet pair, the stiffness of the supporting spring, and the magnetostatic coupling between the solenoids and the hard magnet pair determine the resonant vibration frequency and the output voltage of the energy harvester. The equivalent spring-mass system becomes a nonlinear oscillation system due to the magnetostatic coupling between the solenoids and the hard magnet pair. This nonlinear effect can be explained from the potential energy point of view, as shown in the previous section. The magnetostatic potential energy has two identical minimum values due to the coupling between the magnets and solenoids. These minimums occur when the magnets move a short distance up or down from the equilibrium position in the middle of the hard magnet pair. As a result, the superposition of two different types of potential energies result in a nonlinear total potential, leading to a wider oscillation frequency range, as has been shown in Fig. 18.11.

As explained in the previous section, since the magnetic field magnitude varies along the solenoid axis, the open circuit voltage can be calculated by an integration through the solenoid.

$$V = 2\frac{\mathrm{d}\varphi(t)}{\mathrm{d}t} = 2\frac{\mathrm{d}\int\{H[x,y(t)] + 4\pi M[x,y(x,t)]\}\cdot A\cdot dN}{\mathrm{d}t} = 2\frac{\mathrm{d}\int 4\pi M[x,y(x,t)]\cdot A\cdot dN}{\mathrm{d}t}, \quad (18.8)$$

where A is the total cross-sectional area of the multilayer cores, dN is the number of loops in the infinitesimal unit length of the solenoid, and

$$dN = N_{\mathrm{L}}\cdot dx\,/d_{\mathrm{w}} \quad (18.9)$$

N_{L} is the number of loop layers and d_{w} is the copper wire diameter. Hence, the maximum output power, which happens when the load impedance equals the conjugate of the output impedance of the solenoid coil, is

$$P_{\max} = \frac{\left(\frac{V}{2}\right)^2}{R_{\mathrm{coil}}} = \frac{16S}{R_{\mathrm{coil}}}\left(A'\pi\frac{N_{\mathrm{L}}}{d_{\mathrm{w}}}\right)^2\left(\int_0^L\left\{\frac{\mathrm{d}M[x,y(x,t)]}{\mathrm{d}t}\right\}dx\right)^2, \quad (18.10)$$

where S is the number of layers in each core and A' is the cross-section area of one layer. Equation (18.10) indicates that the output power increases as the resonance frequency increases, if all other parameters are kept constant. Moreover, at a particular frequency, the output power depends on the total magnetic flux change in the solenoid in one oscillation period, which is directly related to the permeability of the magnetic cores. The solenoid with a soft magnetic MuShield® core, which has a high permeability, has a great potential for generating a high voltage output. Moreover, the multilayer structure of MuShield® material is expected to generate a much larger flux change than a single layer. The application of a single layer was discussed in the earlier section.

18.3.3 Results and Analysis

Figure 18.14 shows the measured open circuit voltage of the energy harvester with different springs and resonant frequencies. For spring #1, with resonant frequency of 27 Hz, the peak voltage is 1.18 V for an acceleration amplitude of $2\times g$; for spring #2, with resonant frequency of 33 Hz, the generated maximum voltage is 1.64 V for an acceleration of $3\times g$; spring #3, increased the peak voltage to

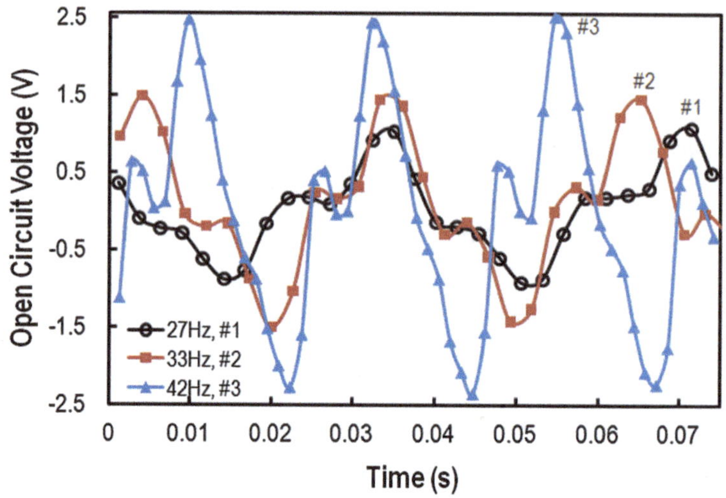

Fig. 18.14 Measured results of the open circuit voltage for the energy harvesting device with three different springs at respective resonant frequencies: spring #1 at 27 Hz; spring #2 at 33 Hz, and spring #3 at 42 Hz

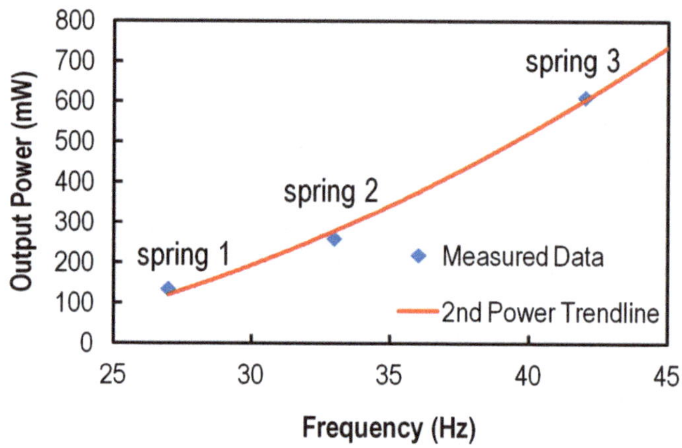

Fig. 18.15 Measured maximum output power of the harvester with three different springs and their associated resonant frequencies

2.52 V, with an intrinsic frequency of 42 Hz and an acceleration of $5 \times g$. Increasing acceleration values were applied to maintain the same source vibration displacement amplitude. The maximum output power on a 2.6 Ω load is 133.88 mW, 258.62 mW, and 610.62 mW, respectively, as shown in Fig. 18.15. Considering that the total practical volume of the device 29.3 cm^3, this device has good performance with a maximum power density of 20.84 mW/cm^3 at 42 Hz (using spring #3). The Q factor of the harvester at 42 Hz is 16, which was obtained from the decay curve

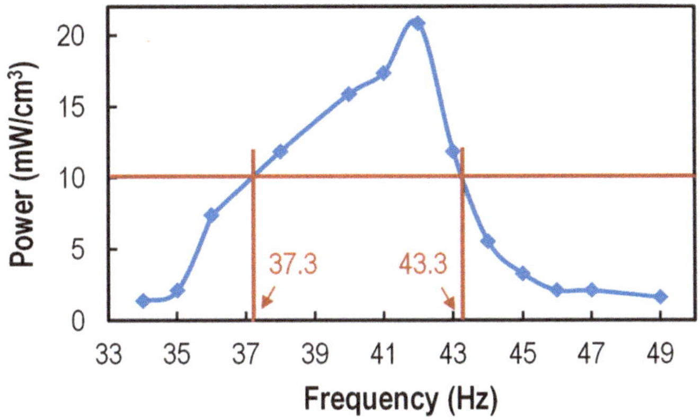

Fig. 18.16 Measured output power spectrum of the harvester with spring #3. Maximum output is 610.62 mW, obtained at 42 Hz, corresponding to a volume density of 20.84 mW/cm^3. This *curve* shows a half-peak working bandwidth of 6 Hz

of output voltage when turning off the source [2]. Almost all of the damping is generated by the mechanical collision between the spring supported magnets and the solenoid holder. That means a much lower input force or acceleration is needed and much higher Q factor could be reached by using a better manufacturing technique. A simple relation between frequency and power can be derived from Eq. (18.10), $P_{\max}{\sim}(\Delta M/\Delta T)^2{\sim}f^2$, if all other parameters kept constant, where ΔM is the flux change per period and ΔT is the period. In fact, the measured results agree with the parabolic curve fitting, as shown in Fig. 18.15. Clearly this vibration energy harvester design can accommodate different vibrating frequencies of the environment by changing the spring that is connected to the hard magnet pair. If the vibration amplitude of the testing stage is kept the same, the output power and power density are proportional to the second power of the vibration frequency. Hence, if this $P_{\max}{\sim}f^2$ relationship can be extrapolated to higher frequencies, much higher output power density can be achieved on condition that the ambient vibration amplitude is constant. Note that a large working bandwidth could still be obtained at high frequencies due to the nonlinear effect. These exciting data prove a promising future of the high-permeability material-based energy harvesting mechanisms.

In order to construct the frequency response curve, testing data were collected at different values of the source frequency. As indicated in Fig. 18.16, output power shows a gradual rise below 42 Hz and a rapid decline above this frequency. The major reason for the asymmetrical curve is the nonlinear oscillation with increasing mechanical damping as the frequency increases, as explained earlier. The half-power bandwidth of the device with spring #3 was measured to be 6 Hz, 15% of the central frequency, which is relatively large.

18.3.4 Discussion

Compared to the previous vibration energy harvester design based on a vibrating high-μ beam and a stationary bias hard magnet pair, this new generation device utilizes a vibrating hard magnet pair and a stationary solenoid pair with thick multilayer high-μ core materials. The multilayer high-permeability solenoids core leads to significantly increased flux change in the solenoid within one oscillation period without increasing the total volume of the device. In addition, the solenoid at both sides of the vibrating magnets makes full use of the spatially inhomogeneous bias magnetic fields on both sides of the magnets, leading to a doubled power output, and a dramatically enhanced power density by approximately 20 times over the previous energy harvester design with high-μ materials.

18.4 Summary

In this chapter, an energy harvesting platform based on high-permeability material was theoretically studied and tested. Two generations of devices were designed. The first generation consisted of a high-μ cantilever vibration energy harvester incorporated a vibrating high-permeability cantilever, a solenoid, and a bias magnet pair. Interaction between the high-permeability magnetic beam and the bias magnets leads to complete flux reversal of the high-permeability beam, which generates a maximum power of 74 mW and a high power density of 1.07 mW/cm^3 at an ambient vibration frequency of 54 Hz and at an acceleration of $0.57 \times g$. The inhomogeneous magnetic field leads to a nonlinear magnetic force on the high-permeability beam, resulting in a nonlinear oscillator with a wide working bandwidth of 10 Hz or 18.5% of the operating frequency. A second generation high-permeability material-based vibration energy harvester was demonstrated, achieving a power density larger than 20 mW/cm^3 when subjected to accelerations of $5 \times g$; this is over three times greater than the best power density data reported for vibration energy harvesters (Volture Piezo Energy Harvester-V25W). A wide working bandwidth of 14% (6 Hz half-power bandwidth at a 42 Hz central working frequency) is still maintained due to the nonlinear effect.

The vibration energy harvesters based on high-permeability magnetic material exhibit high output power density, high output power, as well as wide working bandwidth, which provides great opportunities for practical compact vibration energy harvesters.

Acknowledgements Financial supports from NSF awards 0824008, 0746810 and ONR awards N00014710761, N00014080526 are gratefully acknowledged.

References

1. Raghunathan V, SchurgersC V, Park S, Skrivastava MB (2002) Energy-aware wireless microsensor networks. IEEE Signal Process Mag 19:40–50
2. Beeby SP, Torah RN, Tudor MJ, Glynne-Jones P, O'Donnell T, Saha CR, Roy S (2007) A micro electromagnetic generator for vibration energy harvesting. J Micromech Microeng 17:1257
3. Roundy S, Wright PK, Pister K (2002) Micro-electrostatic vibration-to-electricity converters. In: Proceedings of IMECE2002-34309
4. Roundy S, Wright PK, Rabaey J (2003) A study of low level vibrations as a power source for wireless sensor nodes. J Comput Commun 26:1131
5. Shenck NS, Paradiso JA (2001) Energy scavenging with shoe-mounted piezoelectrics. IEEE Micro 21:30
6. Huang J, O'Handley RC, Bono D (2003) New high-sensitivity hybrid magnetostrictive/electroactive magnetic field sensors. Proc SPIE 5050:229
7. Zhao X, Lord DG (2006) Application of the Villari effect to electric power harvesting. J Appl Phys 99:08703
8. Despesse G, Jager T, Chaillout J, Leger J, Basrour S (2005) Design and fabrication of a new system for vibration energy harvesting. Ph.D. Research in Microelectronics and Electronics, vol 1, p 225
9. Gao RX, Cui Y (2005) Vibration-based energy extraction for sensor powering: design, analysis, and experimental evaluation. Proc SPIE 5765:794
10. Dong SX, Zhai J, Li JF, Viehland D, Priya S (2008) Multimodal system for harvesting magnetic and mechanical energy. Appl Phys Lett 93:103511
11. Wang L, Yuan FG (2008) Vibration energy harvesting by magnetostrictive material. Smart Mater Struct 17:045009
12. Judy JW (2001) Microelectromechanical systems (MEMS): fabrication, design and applications. Smart Mater Struct 10(6):1115–1134
13. Kulkarni S, Roy S, O'Donnell T, Beeby S, Tudor J (2006) Vibration based electromagnetic micropower generator on silicon. J Appl Phys 99:08P511
14. Kulkarni S, Koukharenko E, Torah R, Tudor J, Beeby S, O'Donnell T, Roy S (2008) Design, fabrication and test of integrated micro-scale vibration-based electromagnetic generator. Sens Actuators A 145–146:336–342
15. Beer FP, Johnston ER, DeWolf JT (2002) Mechanics of materials, 3rd edn. McGraw-Hill Book Company, New York
16. Cottone F, Vocca H, Gammaitoni L (2009) Nonlinear energy harvesting. Phys Rev Lett 102:080601
17. Ferro Solution VEH360. http://www.ferrosi.com/files/VEH360_datasheet.pdf
18. KCF technologies VPH300. http://www.kcftech.com/products/documents/WSKdatasheet.pdf
19. Perpetuum PMG37. http://perpetuum.isonlinehere.com/home.php?page_id=13
20. Volture Piezo Energy Harvester-V25W. http://www.mide.com/products/volture/v25w/v25w.php
21. Xing X (2009) Wideband vibration energy harvester with high permeability magnetic material. Appl Phys Lett 95:134103
22. Xing X (2011) High power density vibration energy harvester with high permeability magnetic material. J Appl Phys 109:07E514

Kolibri 360 GmbH
Ferdinand-Jühlke-Straße 7
99095 Erfurt, Deutschland
produktsicherheit@kolibri360.de